W9-AQJ-696

AROMATIC AMINO ACID HYDROXYLASES AND MENTAL DISEASE

AROMATIC AMINO ACID HYDROXYLASES AND MENTAL DISEASE

Edited by

M. B. H. Youdim

Technion–Israel Institute of Technology
Faculty of Medicine, Department of Pharmacology, Haifa, Israel

A Wiley–Interscience Publication

JOHN WILEY & SONS

Chichester · New York · Brisbane · Toronto

British Library Cataloguing in Publication Data:

Aromatic amino acid hydroxylases and mental disease
1. Hydroxylases 2. Amino acids
3. Aromatic compounds
I. Youdim, M. B. H.
612'.0151'8 QP603.HP 79–40642

ISBN 0 471 27606 5

Typeset by Latimer Trend & Company Ltd, Plymouth, in
Monophoto Times New Roman
and printed by The Pitman Press, Bath.

Contributors

K. Blau — *Bernhard Baron Memorial Research Laboratories, Department of Chemical Pathology, Queen Charlotte's Maternity Hospital, Goldhawk Road, London W6 OXG*

S. Bourgoin — *Groupe NB, INSERM U.114, Collège de France, 11 Place Marcelin Barthelot, 75231 Paris cedex 5, France*

B. L. Goodwin — *Bernhard Baron Memorial Research Laboratories, Instititute of Obstetrics and Gynaecology, Queen Charlotte's Maternity Hospital, Goldhawk Road, London W6 OXG*

M. Hamon — *Group NB, INSERM U.114, Collège de France, 11 Place Marcelin Barthelot, 75231 Paris cedex 5, France*

J. W. Maas — *School of Medicine, Department of Psychiatry, Yale University, 333 Cedar Street, New Haven, Connecticut 06510, USA*

H. M. van Praag — *Academisch Ziekenhuis Utrecht, University Clinic for Psychiatry, Nicollas Beetstraat 24, 3500 CG Utrecht, Postbus 16250, The Netherlands*

N. Weiner — *Department of Pharmacology, University of Colorado School of Medicine, Denver, Colorado 80262, USA*

M. B. H. Youdim — *Technion–Israel Institute of Technology, School of Medicine, 12 Haaliya Street, Bat-Galim, POB 9649, Haifa, Israel*

Contents

Preface

The three pterin-dependent aromatic amino acid hydroxylases, phenylalanine, tyrosine, and tryptophan, are very important in normal brain function and in the pathophysiology of various disease states. Although they show remarkable similarities in their kinetic properties these enzymes are relatively specific for the hydroxylation of phenylalanine in the liver and tyrosine and tryptophan within specific neurons in the brain. However, in the presence of naturally occurring pterin cofactor their specificity can be altered. The resulting hydroxylated amino cids can in turn be decarboxylated to yield biogenic monoamines and some like dopamine and serotonin (5-hydroxytryptamine) function as neutrotransmitters.

This book is intended to survey some of the important properties of aromatic amino acid hydroxylases and the impact they have had on understanding the function of monoamine neurotransmitters. Basic chemical and biochemical and regulatory aspects of the hydroxylases are dealt with at length in three chapters. Following each of these the impact of hydroxylases research in the understanding of disease processes is discussed by experts in the field of human pathology.

I would like to thank all the authors whose contributions have made this volume possible.

M. B. H. Youdim
1979

Aromatic Amino Acid Hydroxylases and Mental Disease
Edited by M. B. H. Youdim

CHAPTER 1

Introduction

M. B. H. Youdim

There is substantial evidence that biogenic monoamines, dopamine, noradrenaline, and 5-hydroxytryptamine function as transmitters or neuroregulators between nerve cells and play a central role in the normal as well as abnormal processes in the brain resulting in altered behaviour. These neuroregulators have been implicated in the pathogenesis of mental and neurological disorders. The central question is, how does altered amine biochemistry change behaviour? This is reflected in our understanding of the pathways of biogenic monoamine synthesis and degradation and the regulatory factors which may modify these processes *in vivo*. The rate at which the brain synthesises the neurotransmitters is controlled by the intraneuronal hydroxylation of the amino acids tyrosine, tryptophan and perhaps phenylalanine. However, the latter amino acid is almost exclusively hydroxylated in the liver. Tyrosine and tryptophan hydroxylases are the rate-limiting enzymes in the synthesis of catecholamines and 5-hydroxytryptamine respectively, while phenylalanine hydroxylase is rate-limiting in the hydroxylation of phenylalanine to tyrosine.

Although much information has been obtained on this subject, there remains an incomplete description of the physiological regulation of aromatic amino-acid hydroxylation. The complete characterization of hydroxylating enzymes in brain as well as in liver requires a lot of additional information, much of which may not be obtainable by the experimental approaches currently used.

Aromatic amino acid hydroxylating systems for phenylalanine, tryptophan and tyrosine remains one of the most intense and controversial areas of research. The three hydroxylating enzymes that are discussed in this book are called mono-oxygenases or 'mixed-function' oxidases. The latter name refers to the novel type of oxidative transformation, resulting in the different fates of the two atoms of oxygen in the oxygen molecule, one atom being incorporated into the compound that is undergoing oxidation, while the other is reduced to the level of water (see Kaufman, 1977).

Since the focus of attention is on the hydroxylation of tyrosine, tryptophan, and phenylalanine, within the monoaminergic neurons as well as in the liver, it seems clear that several questions formerly thought to have been resolved are still unanswered. For example, what factors regulate the specificity and the activities of the hydroxylases under physiological conditions which may have a bearing on the pathogenesis of mental diseases? How does the defect in the liver phenylalanine hydroxylase system lead to irreversible brain damage and mental retardation?

The three pterin-dependent hydroxylases show remarkable similarities in their kinetic properties. The behaviour of all three enzymes varies dramatically with the nature of the pterin cofactor used in *in vitro* studies. Kaufman (1974) has shown that regulatory properties are revealed only in the presence of the naturally occurring cofactor, tetrahydrobiopterin. Not only is the apparent affinity of the enzymes for this pterin much greater than it is for synthetic analogues, such as 6,7-dimethyltetrahydropterin or 6-methyltetrahydropterin, but the apparent affinity for the other substrates is also increased. Thus, the K_m of brain tryptophan hydroxylase, determined in the presence of 6,7-dimethyltetrahydropterin, has been reported to be in the region of 200–400 μM, a value that is greater than the reported concentration of tryptophan in the brain. However, in the presence of tetrahydrobiopterin, the K_m for tryptophan is 20–50 μM, a value close to the concentration of this amino acid in the brain. Furthermore, *in vitro* studies have indicated that in the presence of tetrahydrobiopterin not only is the K_m of enzymes for the other substrates decreased, but also a new property of the enzymes is revealed, that is, inhibition by high concentrations of the amino acid substrate. Furthermore, Kaufman (1974) has suggested that the substrate inhibition of hydroxylases found *in vitro* in the presence of tetrahydrobiopterin is relevant to the control of the enzyme activities *in vivo*. Another property of this group of enzymes that can be altered in the presence of naturally occurring cofactor is the substrate specificity. Thus in the presence of tetrahydrobiopterin phenylalanine and tryptophan have been reported to be excellent substrates for purified preparations of tyrosine and phenylalanine hydroxylases respectively and phenylalanine is also a substrate for the brain tryptophan hydroxylase (Kaufman 1974). One may ask whether these phenomena occur *in vivo*.

When dealing with such complex enzyme systems, results obtained using tissue homogenates or purified enzyme preparations demonstrate the differences in the rate of metabolism and the properties of enzymes studied in these preparations, when compared with those using isolated perfused organs where the cellular and catalytic organization are maintained (Youdim and Woods, 1975). During the study of rate controlling enzymes within metabolic pathways tissues are often subjected to manoeuvres such as slicing, homogenization or subcellular fractionation which destroy the catalytic environment. The experimental systems thus obtained are artificial in that many of the metabolic compartments and diffusion barriers which play a role in control of enzyme activity are destroyed, together with the intracellular arrangements or organelles many of which have specific metabolic functions. Thus, for phenylalanine hydroxylase, although the enzyme may be susceptible to substrate inhibition, in the isolated perfused liver, the 'inhibition' occurs at substrate concentrations much greater than those causing inhibition *in vitro*, and thus far above physiological substrate concentrations or those reported present in phenylketonuria (Youdim *et al.*, 1975; Woods and Youdim, 1977). The second feature of these findings is that isolated perfused liver, which can rapidly hydroxylate phenylalanine, is incapable of hydroxylating tryptophan or tyrosine to any significant extent (Woods *et al.*, 1979). These results are in contrast to what has been reported with purified enzyme preparations (Kaufman, 1974), where the reaction is fortified by added co-factor, tetrahydrobiopterin. Although final results are not available for the hydroxylation of phenylalanine in isolated perfused brain (Woods *et al.*, 1976) or adrenal gland, one can presume that the substrate specificities of tyrosine hydroxylase and tryptophan hydroxylase in these organs are retained. It cannot be claimed that the isolated perfused liver or brain is completely physiological, but results obtained using these preparations may have a closer relevance to the *in vivo* conditions.

Since the discovery of aromatic amino acid hydroxylating systems and the recognition of their rate-limiting role in a catecholamine and 5-hydroxytryptamine biosynthesis, considerable effort has been made in elucidating the possible regulatory mechanism of these crucial steps in neurotransmitter synthesis. The hydroxylases appear to be subject to many potential regulatory factors including levels of pterin co-factor, precursor amino acid, cyclic AMP, phospholipids, and cations, including calcium and membrane binding. Although the precise roles of calcium, phospholipid and precursor amino acid in regulating these enzymes is controversial, evidence has accumulated that protein phosphorylation by a cyclic AMP-dependent protein kinase is involved in the regulation of hydroxylase activity and this system is of physiological importance in modifying biogenic amine synthesis.

Although significant information has been accumulated on the properties of hydroxylases using purified or partially purified enzyme preparation, a complete description of their physiological regulation continues to elude us. An attempt has been made in this and the following chapters to put into perspective the possible *in vivo* significance of a host of regulatory factors which are known to affect the hydroxylating enzyme systems and thus the synthesis of biogenic monoamines.

REFERENCES

Kaufman, S. (1974). In *Aromatic Amino Acids in the Brain. Ciba Foundation Symposium 22* (G. E. W. Wolstenholme and D. W. Fitzsimons, Eds.), North Holland, Amsterdam, pp. 85–116.

Kaufman, S. (1977). In *Structure and Function of Monoamine Enzymes* (E. Usdin, N. Weiner and M. B. H. Youdim, Eds.), Dekker, New York, pp. 3–23.

Woods, H. F., Graham, C. W., Green, A. R., Youdim, M. B. H., Graham-Smith, D. G., and Hughes, J. T. (1976). *Neuroscience*, **1**, 313–323.

Woods, H. F. and Youdim, M. B. H. (1977). In *Structure and Function of Monoamine Enzymes* (E. Usdin, N. Weiner and M. B. H. Youdim, Eds.), Dekker, New York, pp. 263–278.

Woods, H. F., Youdim, M. B. H., Goodwin, B., and Rulhven, C. R. (1979). Submitted for publication.

Youdim, M. B. H. and Woods, H. F. (1975). *Biochem. Pharmacol.*, **24**, pp. 317–323.

Youdim, M. B. H., Mitchell, B., and Woods, H. F. (1975). *Biochem. Soc. Trans.*, **3**, 683–685.

Aromatic Amino Acid Hydroxylases and Mental Disease
Edited by M. B. H. Youdim

CHAPTER 2

Phenylalanine Hydroxylase

B. L. Goodwin

A HISTORICAL

Early in this century the first evidence for the existence of an enzyme that hydroxylates phenylalanine was obtained. Embden's group[83] perfused liver with solutions of phenylalanine and tyrosine, and they found that both compounds were converted to acetoacetic acid, which suggested that they may share a common catabolic pathway. Embden and Baldes later perfused dog liver with a solution of phenylalanine and found tyrosine in the perfusate.[82] This finding was not fully accepted at that time. Dakin,[69] for instance, tried to repeat this work and failed to find any tyrosine. The reason for this failure may well have been the rapid turnover of the tyrosine formed, and therefore most of it does not appear in the perfusate, but is further metabolized *in situ*. Some supporting evidence for the hydroxylation of phenylalanine was provided by Kotake *et al.*,[195] who found that when rabbits were fed phenylalanine *p*-hydroxyphenylpyruvic acid was excreted in the urine, a finding that was confirmed later by Shambaugh *et al.*[258] The next important piece of evidence is found in the biochemical study carried out by Medes just before 1930 on her classical case of tyrosinosis, where she showed that administered phenylalanine increased the urinary excretion of tyrosine metabolites, and there was circumstantial evidence for the excretion of tyrosine under these conditions.[213] During that era studies on amino acid requirements in animals demonstrated that phenylalanine is an essential amino acid,[295] whereas dietary tyrosine can be completely replaced by phenylalanine, and when phenylalanine and tyrosine are fed together tyrosine reduces the amount of phenylalanine required for normal growth in the absence of tyrosine, which strongly suggested that phenylalanine can be converted to tyrosine.

This body of circumstantial evidence was further strengthened by Fölling's discovery of phenylketonuria[91] in which tissue phenylalanine concentrations build up to levels far in excess of normal, because the normal enzyme for removing excess phenylalanine from the tissues is absent.

In 1940 Moss and Schoenheimer[226] published data that conclusively demonstrated the formation of tyrosine in rats. They fed labelled phenylalanine to these animals, and after killing them they isolated and hydrolysed proteins from various tissues. The tyrosine that they isolated was found to contain some of the label that was originally in the phenylalanine. Soon after this Bernheim and Bernheim demonstrated that when phenylalanine is incubated with liver slices a phenolic compound is produced whose properties correspond to those of tyrosine, although the identity of the compound was never rigorously proved by these workers.[22]

It was about 40 years after the original perfusion experiments before the enzyme was successfully studied in tissue extracts. This work, which was carried out by Udenfriend and Cooper,[282] was made possible by the use of a new, sensitive technique for the assay of tyrosine, which involves the formation of a yellow pigment by a reaction with α-nitroso-β-naphthol. Since that time the studies on phenylalanine hydroxylase have proliferated to the present state of the subject, first under the impetus of the relevence of this enzyme to phenylketonuria, and more recently as a result of the discovery of the so-called 'NIH shift', which has thrown considerable light on the mechanism of enzyme reactions that involve a hydroxylation step.

Various aspects of this subject have previously been reviewed;[103, 109, 161, 173, 179, 180] the present review examines the state of the subject in 1978.

B SPECIES AND TISSUE DISTRIBUTION

It can be assumed that phenylalanine hydroxylase (EC.1.14.16.1) is found in all mammalian species. All studies that have sought the enzyme in liver (at least, since adequate assay techniques became available) have been successful in detecting it, other than in phenylketonuric liver. It has also been found in other vertebrates, and in some plants and microorganisms. Mammalian species in which the enzyme has been demonstrated include man,[54, 97, 98, 100, 177, rat,[54, 97, 100, 169, 208, 282, 287] mouse,[100, 208, 287] guinea pig,[97, 100, 268, 282, 287] beef,[54, 163, 287] several species of monkey[57, 100, 287] rabbit,[282, 287] dog,[45, 282, 287] sheep, pig,[54, 287] ground squirrel, hamster, coati mundi, cat, galago,[287] and horse.[54] Among birds it has been demonstrated in domestic fowl,[267, 282, 287] pigeon,[20, 34, 287] goose, and ring dove.[287] The enzyme has been found in frog.[20] It has been detected in a wide range of fish including carp,[20] trout,[214, 287] bullhead, pumpkinseed, black crappie, blue gill, rock bass, and white bass.[287] It has been found in the caterpillar stage of *Celerio euphorbiae*, *Bombyx mori*, *Ephestia kühniella*, *Dixippus morosus*,[20] and *Calliphora erythrocephala*.[256] It has been detected in spinach leaf,[234] but not in peyote,[196] nor in the fruit of grapefruit.[146] An enzyme that catalyses this reaction is found in *Tetrahymena pyriformis*, a protozoan,[203] *Anabaena variabilis*, a blue-green alga,[139] *Comamonas*,[94] *Pseudomonas ATCC 11299a*,[129] *Pseudomonas atlantica*, *Claviceps purpurea*, *Volucrispora aurantiae*, *Beauvaria bassiana*, *Trichoderma viride* *Penicillium expansum*, *Fusarium oxysporum*, *Chaetomium cochliodes*, *Mucor genevensis*,[52] and *Geotrichum candidum*.[207] It has not been found in *Escherichia coli*, yeast, *Bacillus subtilis*, several *Streptomyces* species, *Rhodotorula*, *Aspergillus niger*, *Polyporus nidulans*, or *Stereum sanguinolentum*.[52, 282]

The enzyme is found in comparatively few tissues in mammalia. The main site of the enzyme is the liver.[10, 24, 54, 98, 169, 208, 282, 292] Early studies failed to demonstrate its presence in kidney,[282] but more recent studies have demonstrated that this organ is an important source of the enzyme.[10, 19, 24, 208, 282, 292]

The reliability of the claim that there is a renal phenylalanine hydroxylase has been placed in doubt by a report from Berry's group[229] which claims that these results are based on an artefact arising from the non-enzymic hydroxylation of phenylalanine in the presence of cofactor, at least in man and other primates. This may be incorrect, since Ayling *et al.*[7] find that autopsy specimens of human liver are usually inactive, whereas surgically removed specimens show about 25% of the activity found in liver. Similarly, Woolf[298] has found that most autopsy specimens of human liver are inactive, possibly because of a rapid post-morten inactivation of the enzyme. However, Ayling's group[7] has also examined post-morten liver for activity, and they noted that only one out of five livers showed any hydroxylase activity. This one was from a person who had died suddenly, but the others were from patients who had suffered a long illness. It has therefore been suggested that patients with a terminal illness lose the ability to hydroxylate phenylalanine. Despite these problems, Berry found the hydroxylase in rat liver and kidney.[229]

Human term placenta has been found to contain the hydroxylase, although the specific activity is very low when compared with that of liver.[201] The value is almost identical with that found in serum,[148] which suggests that the measured enzyme may have been entirely in blood.

The presence of the enzyme has been demonstrated in beef udder, but not in the mammary gland from various other species.[163] Rat intestinal mucosa has been shown to contain the enzyme,[292] although the intestinal system as a whole contains a negligible amount of the hydroxylase.[282] It has been claimed by some workers that it is present in the pancreas and spleen, although others are not in agreement with these findings.[19,198,282] One report claimed that in various areas of the brain an enzyme catalysed the hydroxylation of phenylalanine that did not appear to be tyrosine hydroxylase,[11] but most workers are now agreed that the brain does not contain a phenylalanine hydroxylase as such,[1,125,157,208,282] although brain tyrosine hydroxylase does catalyse this reaction to some extent.

Low but detectable levels of the enzyme have been found in plasma and serum, and it is claimed that the activity is high enough to permit a differential diagnosis of phenylketonuria to be made,[148] although it has been criticized on the grounds that the published results show a standard deviation for the assay method that is too high to allow valid comparisons to be made with the relatively small groups of subjects examined.[285]

Tryptophan hydroxylase isolated from rabbit brain is able to hydroxylate L-phenylalanine at a rate that is similar to the rate for L-tryptophan under the standard conditions for the assay, with tetrahydrobiopterin as the cofactor. With synthetic cofactors the rate of hydroxylation of phenylalanine is both lower than with the natural cofactor and lower than the rate with tryptophan as the substrate.[273]

The claim by Wapnir[291] that phenylalanine is rapidly hydroxylated in brain appears to have been based on an artifact.[1] Phenylalanine hydroxylase has not been detected in a wide range of organs, including heart, muscle,[208,280] lung,[280,282] brainstem, skin, thymus, and salivary glands.[280]

Although many organs do not contain any hydroxylase they still have the genetic potential to synthesize it. For instance, human foetal muscle does not show any phenylalanine hydroxylase activity. But if fibroblasts are removed from the foetal muscle and are grown in culture, phenylalanine hydroxylase activity is observed in the cultured cells, and this activity has been observed to persist through many subcultures. A day-to-day variation in the hydroxylase activity has been observed in the same culture which appears to follow an entirely random pattern; at least, no explanation has been offered for these variations.[272]

Cells from rat liver hepatoma show phenylalanine hydroxylase activity when they are grown in culture. The specific activity is dependent on the cell density in the culture; the activity is low in cultures with a low density, but the activity increases as the density of growth in the culture increases.[142]

The enzyme concentration found in blue-green algae is higher in cells that have been grown on a medium containing L-phenylalanine than in cells grown on an inorganic substrate. Its activity appears to be reduced during adaptation to dark conditions.[139]

The *Pseudomonas* enzyme is induced by growth on a medium that contains

0.1% of L-tyrosine or 0·2% of L-phenylalanine; L-asparagine is ineffective in inducing the enzyme.[128]

C ACTIVITY. ITS MEASUREMENT AND SOME RESULTS

1 Assay *in vitro*

A note of caution must be sounded about the assay of phenylalanine hydroxylase. Under conditions that are similar to those used during the enzyme assay a hydroxylation of phenylalanine that is not dependent on the hydroxylase, which is often non-specific in its course, has been observed. This reaction can lead to the formation of *o*- and *m*-tyrosine as well as tyrosine, and this can lead to spuriously high results.[66, 286, 301] If the rate of disappearance of the cofactor is used as a measure of the reaction rate it must be borne in mind that, depending on the conditions used, there may be considerable uncoupling between hydroxylation and the oxidation of cofactor when a synthetic cofactor is used with the enzyme.[84, 87, 266] The addition of ascorbate and Fe^{2+}, and tetrahydropteridines in particular, can cause non-enzymic hydroxylation of phenylalanine, and this is described in the section on non-enzymic hydroxylation (section F.1). When an assay is set up adequate controls must be carried out to obviate spurious results that arise from these causes.

The methods used for the assay of the enzyme depend either on the measurement of tyrosine formed during the reaction or on the rate of disappearance of the cofactor or other reactants. The measurement of tyrosine formed during the reaction is usually based on the formation of a yellow pigment which absorbs radiation at about 450 nm when tyrosine is incubated with α-nitroso-β-naphthol in the presence of nitrous and nitric acids.[282] This pigment can also be detected by its strong fluorescence in ultraviolet radiation, and this has been exploited in measuring low activities of the enzyme.[288] Folin's reagent has also been used in the measurement of the tyrosine formed in the reaction, but this is a very insensitive method.[240] Iodination of tyrosine has been used, either to form iodotyrosine, which can be measured as such by its optical density at a wavelength of 310 nm, or to release tritium from 3-tritiotyrosine, which is formed by the action of the enzyme on 4-tritiophenylalanine. All of these methods are based on the underlying assumption that the reaction rate for the formation of tyrosine is constant during the incubation period. This assumption is not necessarily correct. Many enzyme reactions show a lag period immediately after the reactants have been added. In some of the published assay methods the rate of reaction declines after some time, and the time period chosen has allowed for this observation. The magnitude of the error introduced by these factors is comparatively small. The method that measures the rate of oxidation of the cofactor[9, 10] overcomes this problem because the optical density of the reduced cofactor can be followed throughout the reaction, but the method suffers from the drawback that the reaction may not be stoichiometric, and therefore the rate of disappearance of cofactor may not strictly measure the rate of formation of tyrosine.

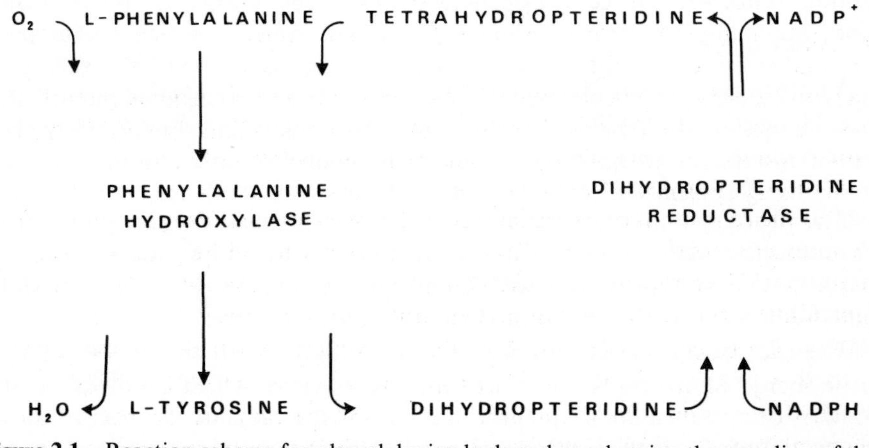

Figure 2.1 Reaction scheme for phenylalanine hydroxylase, showing the coupling of the hydroxylation to the hydrogen transport system via a pteridine cofactor and a nicotinamide cofactor

Before the discovery of the pteridine cofactors Kaufman[167] published an assay method in which dihydropteridine reductase, the so-called sheep liver enzyme, was added to the assay mixture in order to couple the hydroxylation reaction to the hydrogen transport system via NADPH, thus mimicking the situation found in the intact cell (Figure 2.1). In more recent procedures this coupling has been omitted and either the natural or a synthetic pteridine cofactor has been added to the reaction mixture to provide the cosubstrate.

Numerous papers have been published in which the tyrosine formed during the assay has been measured by the nitrosonaphthol method,[43, 123, 172, 208, 236, 296a] and some utilize the fluorescence of the reaction product.[3] This latter procedure has been successfully modified for use with an autoanalyser.[81]

A typical procedure for assaying the mammalian enzyme[43] is carried out at pH 7.7 in 0.1 M Tris buffer. The buffer contains 10 mM dithiothreitol, which regenerates the cofactor and protects the enzyme from inactivation by destroying peroxide that is formed by side reactions, 2 mM L-phenylalanine, and 0.2 mM cofactor (in this case 2-amino-4-hydroxy-6,7-dimethyltetrahydropteridine). Sufficient oxygen is dissolved in the buffer solution to make it unnecessary to pass oxygen through the solution. The reaction is stopped after 20 min. by the addition of trichloroacetic acid. The tyrosine in the supernatant is reacted with α-nitroso-β-naphthol. The standard procedure for this involves heating an ethanolic solution of α-nitroso-β-naphthol with the supernatant in dilute nitric acid containing a trace of nitrous acid at 55 °C for 30 min., and then extracting the excess reagent into ethylene dichloride. The optical density of the aqueous phase is measured at a wavelength of 450 nm. The rate of hydroxylation is constant for 30 min., but the reaction rate then decreases. Other thiols besides dithiothreitol will protect the enzyme against inactivation during the incubation, but many of these interfere with the colorimetry, which makes them unsuitable for use in the assay. Other protecting agents that have been used include ascorbate and catalase.[236] In other published procedures the pH used for incubating the enzyme has been 6.8, in a phosphate buffer.[236] Several minor modifications have been made that render the assay suitable for use with biopsy specimens.[16,17,46,125,177,304]

The procedure used for the assay of the *Pseudomonas* enzyme differs from that described above because this enzyme requires added Fe^{2+} as a cofactor. It is assayed[123] in Tris buffer at pH 7.3 in the presence of added Fe^{2+}, NADH, and 2-amino-4-hydroxy-6,7-dimethyltetrahydropteridine.

The radioassay method involves the formation of 3-tritiotyrosine from 4-tritiophenylalanine by the action of the hydroxylase, which catalyses the migration of tritium almost quantitatively from the *para* to the *meta* position (more than 95% of the label migrates when the rat liver enzyme is used, and about 85% with the *Pseudomonas* enzyme). Tritium is released from the labelled tyrosine as tritiated water by iodination with *N*-iodosuccinimide, which does not act on labelled phenylalanine to release this isotope, and the residual labelled amino acids are removed from the reaction mixture by passage through an ion-exchange column. The tritium in the effluent is then measured by its radioactivity using a scintillation counter.[125] A modification of this method has been used to assay tissue concentrations of the naturally occurring phenylalanine hydroxylase cofactor.[137]

The dynamic method for measuring hydroxylase activity has been applied to mammalian systems. The incubation system is essentially that described above for the nitrosonaphthol method with the mammalian enzyme, and the rate of conversion of 2-amino-4-hydroxy-6,7-dimethyl-5,6,7,8-tetrahydropteridine to its dihydropteridine analogue is determined by measuring the rate of change of optical density at 330 nm. It has been claimed that the oxidation of the pteridine cofactor is stoichiometric with the formation of tyrosine under the conditions used in this assay.[9,10]

2 Assay *in vivo*

Kaufman's group has recently developed two assays for the hydroxylase in rats, one of which is strictly an *in vivo* assay, and the other uses tissue slices, where the cellular architecture is still intact. The assay that uses tissue slices[220] uses ^{14}C-phenylalanine as the substrate (with added unlabelled phenylalanine). Liver slices are incubated in 2 ml of 0.1M pH 6.8 potassium phosphate buffer with 1 μCi of substrate, initially at 15 °C for 30 min. to allow equilibration of the substrate. At this temperature little tyrosine is formed because the temperature coefficient for hydroxylation is high. The incubation is then continued at 25 °C to assay the hydroxylase. The reaction is then stopped by the addition of 0.3 ml of 3 M perchloric acid, the tissue is homogenized, and after centrifugation an aliquot of the supernatant is chromatographed on a TLC cellulose plate, and the radioactivity of the tyrosine spot is measured. A correction is made for a small amount of hydroxylation observed with boiled tissue.

The hydroxylation was found to be linear with time. The kinetics were typical of a situation in which the Michaelis–Menten equation is obeyed, with inhibition at high substrate concentration. An apparent substrate inhibition was observed at a phenylalanine concentration of 2 mM and higher, and inhibition was observed at 0.4 mM concentration with 2-amino-4-hydroxy-6-methyl-5,6,7,8-tetrahydropteridine as the cofactor.

In the absence of added cofactor the rate of hydroxylation was very much lower than maximum, which suggests that *in vivo* the rate-limiting factor is the concentration of cofactor.

p-Chlorophenylalanine administered to rats prior to killing brought about a reduction in the hydroxylase activity that was similar to that observed with liver extracts obtained from similarly treated rats. When methotrexate was added to the slices the hydroxylation was inhibited, apparently because this compound interferes with the regeneration of the cofactor.

When 6-methyltetrahydropterin was injected into rats prior to killing an enhanced hydroxylase activity was rapidly established, which demonstrates that *in vivo* a pteridine cofactor can be taken up by the liver. It is possible that this observation may be of value in the treatment of phenylketonuria. In this disease there is a very low hydroxylase activity, but the administration of a suitable pteridine cofactor in sufficient quantity might increase the activity of the hydroxylase sufficiently to permit an adequate control of the tissue phenylalanine concentration without resort to the use of a low-phenylalanine diet.

The relative V_{max} for several pteridine cofactors was the same for the hydroxylase in tissue slices as for the isolated enzyme.

There was no detectable activity towards *m*-tyrosine in liver slices, which is also observed with the unactivated enzyme.[88] This result suggests that the enzyme *in vivo* is largely or entirely in an unactivated form, and that the hydroxylation of *m*-tyrosine has no *in vivo* significance.

The *in vivo* assay involves the administration of labelled phenylalanine to rats. A solution of either L-phenylalanine-D_5 or ring-labelled ^{3}H-phenylalanine is administered as its sodium salt by intraperitoneal injection in normal saline. The dose is about 0.5 g (kg body weight)$^{-1}$, which is considerably higher than the dose used in phenylalanine tolerance tests. At various times the animals are decapitated and the blood is collected into EDTA to prevent clotting. The water is isolated and either its deuterium content or its radioactivity is measured.[221]

It would appear that an essential step has been omitted from the description of the rationale of the method. The results obtained with the labelled substrate indicate that less than 2% of the tritium in the substrate is converted to water during the hydroxylation reaction; this is far too small to allow for an adequate assay. It would appear that the complete release of tritium (or deuterium) is brought about by the further degradation of tyrosine to CO_2 and water. This can only occur if there is a very rapid equilibration (within a few minutes) with the biosynthesized tyrosine to yield a steady state in which the rate of hydroxylation is equal to the rate at which the labelled tyrosine is further metabolized to CO_2 and water. This is because only one of the five deuterium atoms, and a negligible proportion of the tritium, is released during the hydroxylation. A second deuterium atom should be released during the formation of homogentisic acid, and the remainder are released after the ring fission of homogentisic acid, during the further oxidation of the compound. It takes a significant time from the injection of labelled phenylalanine until its labelled metabolites have reached a steady state, so that four more atoms of deuterium are being released for each molecule of phenylalanine that undergoes hydroxylation. From the results shown for this assay it would appear that this equilibrium is attained very rapidly in the rat.

The assay also requires that the rate of hydroxylation should be constant over the period of the experiment, and this will only happen if the substrate concentration is well above the K_m for the duration of the experiment, i.e. the substrate concentration must be about 3–5 mM.

Kaufman has suggested that methods of this sort may be of value as a substitute for the classical phenylalanine tolerance tests in man. Admittedly, this may be of value in distinguishing between phenylketonuria and hyperphenylalaninaemia, where metabolic pathways other than hydroxylation account for most of the phenylalanine metabolized, assuming, of course, that the abnormally high concentrations of tissue phenylalanine that would be found during this test did not cause any neurological damage to the patient. The present author sees only disadvantages, particularly expense, in using these techniques, compared with standard intravenous tolerance tests for distinguishing between normal subjects and heterozygotes for phenylketonuria. This technique would be difficult to carry out in man. It would require the intravenous infusion of 0.5 g of L-phenylalanine as its sodium salt per kilogram body weight. Besides any adverse effects that a dose of this size might have on the person being tested, the infusion itself is not without hazard. It has been found (L. I. Woolf, W. I. Cranston, and B. L. Goodwin, unpublished results) that considerable care must be taken to prevent a too rapid injection of this alkaline solution, otherwise there is a risk of local thrombosis. An infusion rate of 1.5 g min^{-1}, with a rapid blood flow to dilute the phenylalanine solution, was considered to be the maximum acceptable rate. Furthermore, there is the risk of a severe alkalosis when as much alkali as this is administered. If this technique is ever used in man it must be remembered that it will take much longer for the metabolic processes to reach equilibrium as compared with the rat, and longer time intervals should be used in the sampling of the blood as compared with rats. This author considers that it is far better to rely on the classical intravenous test which uses a much lower dose (15% of that used here, or less), since the information obtained from both is nearly identical.

The results obtained by the *in vivo* assay are very similar to those obtained by *in vitro* techniques, in particular the lower activity in female rats compared with male animals,[296a] and the *in vivo* inactivation of the enzyme by *p*-chlorophenylalanine or methotrexate.

3 Activity

The activity of the enzyme observed in various species is shown in Table 2.1. Some of the observed variations in activity found by different workers in the

Table 2.1 Measured activity of phenylalanine hydroxylase in various species

Species	Tissue, etc.	Activity (nmol (mg protein)$^{-1}$ min^{-1})	Activity (μmol (g tissue)$^{-1}$ min^{-1})	Reference
Rat	Liver		0.05	305
	Liver		0.94	54
	Liver	13		37
	Liver	2		97
	Liver (male)		0.98	208
	Liver (female)		0.72	208
	Liver		0.6	172
	Liver		0.31	227
	Liver	0.12		20
	Liver	6.7		287
	Liver		0.64	296a
	Kidney		0.17	208
	Kidney		0.10	227
Guinea pig	Liver	2		97
	Liver		0.03	305
	Liver	1.0–3.6		287
Mouse	Liver	1.5		287
	Liver		0.03	305
	Liver		0.66	208
	Kidney		0.36	208
Dog	Liver		0.03	305
	Liver		0.05	54
	Liver	0.5–3.2		287

[continued

TABLE 2.1 [*continued*

Species	Tissue, etc.	Activity (nmol (mg protein)$^{-1}$ min^{-1})	Activity (μmol (g tissue)$^{-1}$ min^{-1})	Reference
Man	Liver		0.43	54
	Plasma	0.000 06		148
	Liver		0.20	7
			0.23	7
	Kidney		0.047±0.011	7
Beef	Liver		0.07–0.13	54
	Liver	0.1–0.6		287
Pig	Liver		0.08	54
	Liver	0.2		287
Horse	Liver		0.04	54
Sheep	Liver		0.03	54
	Liver	0.2		287
Rabbit	Liver	0.8–1.0		287
Ground squirrel	Liver	0.8–1.2		287
Hamster	Liver	0.4–1.2		287
Coati mundi	Liver	3.3		287
Cat	Liver	1.0–1.6		287
Rhesus monkey	Liver	0.25		287
Squirrel monkey	Liver	2.2–3.0		287
Spider monkey	Liver	2.0		287
Pigeon	Liver	0.16		20
	Liver	0.05		54
	Liver	2.4–3.5		287
Ring dove	Liver	0.2		287
Goose	Liver	2.2		287
Domestic fowl	Liver	0.75–1.4		287
Frog	Liver	0.04		20
Trout	Liver	0.5–1.2		287
Bullhead	Liver	0.1		287
Pumpkin seed	Liver	0.6		287
Black crappie	Liver	1.1		287
Blue gill	Liver	0.6		287
Carp	Liver	0.08		20
Rock bass	Liver	0.6		287
White bass	Liver	3.8		287
Celerio euphorbiae	Caterpillar	0.09		20
	pupa	0		20
Bombyx mori	Caterpillar	0.03		20
	pupa	0		20
Ephestia kühniella	Caterpillar	0.6		20
Dixippus morosus	Young instar	0.8		20

same species may be due to differences in the assay conditions, e.g. temperature and pH. In some cases,[287] the text is somewhat unclear about the units used; the values given in Table 2.1 are, however, believed to be correct. The results of Belzecka *et al.*[20] are much lower than those of other workers; probably technical reasons underly the difference.

Waisman[287] has concluded from the results obtained by his group that there is no correlation between the hydroxylase activity and phylogenic classification in vertebrates.

Studies on the assay have demonstrated that the conditions of assay can cause considerable differences in the observed activity. For instance, during early studies, in which no pteridine cofactor was used, dialysis of the enzyme preparation prior to its assay was shown to cause a loss in activity of 17–81%, undoubtedly because of the loss of cofactor.[246] Similarly, the early studies gave low results because the concentration of cofactor was too low.

Liver enzyme does not show normal kinetics. After some purification its activity shows a sigmoid response to the concentration of phenylalanine.[85, 88] The point of inflection for rat liver enzyme is at about 0.25 mM phenylalanine concentration at 30 °C and about 0·1 mM at 37 °C. The kinetics are converted to normal (i.e. they obey the Michaelis–Menten equation) on the addition of lysolecithin or α-chymotrypsin. At pH 8 the specific activity decreases with increasing enzyme concentration due to a sharp increase in the apparent K_m for the natural cofactor. This is remedied by the addition of stimulating protein.[151] Thus, the assay must be carried out under standard conditions that do not introduce any of these effects as artifacts.

D PURIFICATION

Woolf's group has purified both human and rat enzymes to a high degree of purity, and Guroff's group has purified the *Pseudomonas* enzyme to some extent. Their methods illustrate the purification procedures used with this enzyme.

Human foetal enzyme was purified as follows.[296] All steps were carried out at 0–4 °C. A portion of foetal liver was homogenized with four volumes of 0.05 M potassium phosphate buffer at pH 7.0, containing 0.15 M potassium chloride, 1 mM dithiothreitol, 10 mM phenylalanine, and 5% glycerol to stabilize the enzyme. The homogenate was centrifuged at 35 000 **g**, and the supernatant was filtered through a plug of glass wool to remove suspended lipids. A saturated solution of ammonium sulphate that had been adjusted to pH 7.5 was added until a concentration of 26% saturation was obtained. The precipitate was centrifuged down and discarded. More ammonium sulphate solution was added to bring its concentration to 44% saturation and the precipitated protein was collected by centrifugation. The precipitate was dissolved in a 0.01 M Tris–HCl buffer at pH 7.2 to which 0.01 M potassium chloride, 0.1 mM dithiothreitol, 10 mM phenylalanine, and 5% glycerol had been added, using 0.4 ml of the buffer for each gram of liver used. This solution was desalted by passage through a column of Sephadex G-25 that had been previously equilibrated with the above Tris buffer. A column of DEAE cellulose (15 mm. × 200 mm. height for 20 g of liver) was equilibrated with the above buffer, the protein was adsorbed, and was eluted with a linear gradient of potassium chloride dissolved in the above buffer. It is not clear at what potassium chloride concentration the enzyme was eluted from this column, but the description of the method suggests that it was between 0.2 and 0.3 M. After precipitation of the enzyme by addition of an excess of ammonium sulphate and collection by centrifugation a solution of this precipitate was fractionated on a column of Sephadex G-200 (15 mm. × 900 mm. deep bed) that had been previously equilibrated with 0.1 M potassium phosphate buffer at pH 7.0 which contained 0.1 mM dithiothreitol, 10 mM phenylalanine, and 5% glycerol. A peak of activity was eluted from this column after the passage of 60 ml of buffer. The enzyme solution was then applied to a 10 mm. × 15 mm. long column of hydroxyapatite that had been equilibrated with the above buffer. Some protein was washed off the hydroxyapatite with this buffer, and the hydroxylase was recovered by elution with buffers whose composition was the same as that described above except that the potassium phosphate concentrations were 0.13 M and 0.19 M. The enzyme was eluted by using both these buffers, and it was essentially pure after this purification procedure.

Two procedures have been developed for the purification of rat liver enzyme.[110] One of these is almost identical with the above technique for human enzyme, and will therefore not be described further except to state that it also yields a preparation that is essentially pure. A second procedure was developed that yielded two active fractions with different molecular weights which represented the dimer and the tetramer. The procedure used was as follows. Rat liver (60 g) was homogenized in 300 ml. of 0.1 M pH 6.8 potassium phosphate buffer that contained 0.1 M potassium chloride and 5% glycerol. Cell debris was removed by centrifugation at 27 000 **g**. The supernatant was adjusted to pH 5 by the addition of cold 1 N acetic acid, and the precipitate was removed by centrifugation. The supernatant was adjusted to pH 7 by the addition of potassium hydroxide solution and excess ammonium sulphate was added to fractionate the enzyme. The protein that was precipitated between 30% and 50% saturation contained the hydroxylase activity and was collected by centrifugation. The precipitate was dissolved in 0.1 M potassium phosphate buffer at pH 6.8 that contained 5% glycerol, and this solution was desalted by passage through a column of Sephadex G-25. A portion containing 400 mg of protein was applied to a 20 mm × 200 mm deep column of DEAE cellulose that had previously been equilibrated with 0.02 M pH 7.2 potassium phosphate buffer that contained 5% glycerol and 10 mM phenylalanine. The column was eluted with a linear potassium chloride gradient in this buffer, and the enzyme was eluted at a potassium chloride concentration of 0.17–0.25 M. The protein was precipitated with excess ammonium sulphate and the precipitate was centrifuged down. A solution of this precipitate in 0.1 M pH 6.8 potassium phosphate buffer containing 5% glycerol and 10 mM phenylalanine was applied to a column of Sephadex G-200. Two peaks of activity were eluted. Both peaks were reckoned to be essentially pure phenylalanine hydroxylase.

The procedure used to partially purify *Pseudomonas* enzyme[130] started with harvesting the cultured cells, which were then washed and resuspended in pH 7.3 Tris buffer. A typical batch was prepared from 300 **g** of packed cells. The cells were then disrupted in a pressure cell. The homogenate was centrifuged at 12 000 **g** and the insoluble material was discarded. The protein that was precipitated at 50% ammonium sulphate saturation was collected by centrifugation and a solution of this was desalted by dialysis. Protein that was precipitated during dialysis was discarded. Phenylalanine was added to the supernatant to a concentration of 10 mM and then the solution was heated to 45–50 °C for 3 min. After cooling it was centrifuged to remove denatured protein, and the supernatant was fractionated with ammonium sulphate. Protein that was precipitated between 40% and 50% saturation was collected and dialysed against a 50 mM pH 6.0 acetate buffer and again protein that precipitated during this procedure was discarded. One tenth of a volume of 1 M sodium chloride was added to the supernatant and the solution was passed through a 25 mm × 160 mm deep column of DEAE Sephadex that had previously been equilibrated with 0.1 M sodium chloride in the acetate buffer. Some protein was washed off the column with more buffer, and the enzyme was eluted with a linear gradient of sodium chloride in this acetate buffer. The enzyme was eluted at a sodium chloride concentration of about 0.2 M. This procedure yielded about 30-fold purification, but the material was undoubtedly still very impure.

Kaufman[182] has developed a useful method for the preparation of rat liver enzyme in poor yield, but which is moderately pure. The fractionation steps used by him are (in order): fractionation of the liver extract with ethanol, precipitation with ammonium sulphate, adsorption onto a calcium phosphate gel followed by elution with a buffer of higher ionic strength, a second fractionation with ammonium sulphate, chromatography on DEAE cellulose, and finally gel filtration on Sephadex G-200. It is possible that this method would have yielded a material of higher specific activity in better yield if steps had been taken to protect the enzyme during purification similar to those used by Woolf's group.

Cotton[57] has developed a procedure for purifying the enzyme from *Macaca irus* liver. The steps in this procedure are the precipitation of acidic proteins with protamine sulphate, fractionation with ethanol, precipitation with ammonium sulphate, gel filtration on Sephadex G-200, chromatography on DEAE cellulose, and two adsorption steps on Brushite. This yielded a material that was purified by about 300-fold.

The details of the recovery and purification for some of these procedures are shown in Table 2.2.

Table 2.2 Enzyme purification. The results from several published methods

Step	Relative activity	Yield (%)
Man[296]		
Liver homogenate	1	100
Liver supernatant	2.7	98.5
Ammonium sulphate fraction	21	94.6
Sephadex G-25	21	86.8
DEAE-cellulose	81	52
Sephadex G-200	300	19.1
Hydroxyapatite	805	5.2
Rat[108]		
Liver homogenate	1	100
Liver supernatant	2.2	94
Acid precipitation	4.1	82
Ammonium sulphate fraction	22.5	50
DEAE-cellulose	64	40
Sephadex G-200		
Tetramer	36	15
Dimer	175	
Rat[182]		
Liver extract	1	100
Ethanol fraction	3.2	112
Ammonium sulphate	10.6	101
Calcium phosphate gel	24.5	51
Ammonium sulphate	28	31
DEAE-cellulose	86	11
Sephadex G-200	175	5
Pseudomonas[130]		
Cell extract	1	100
Ammonium sulphate fraction	3.2	124
Heat treatment	3.5	80
Ammonium sulphate	5.2	30
DEAE-Sephadex	29	12

Cotton[58] has prepared a chromatographic adsorbent by reacting CH-Sepharose 4B with the synthetic cofactor 2-amino-4-hydroxy-6,7-dimethyl-5,6,7,8-tetrahydropteridine. The underlying principle for this is that the immobilized cofactor should bind to phenylalanine hydroxylase at the cofactor binding site, but most enzymes that do not require a pteridine cofactor should not bind to it. Thus it is expected that when a crude solution of phenylalanine hydroxylase is passed through a bed of this adsorbent the hydroxylase should bind and other proteins should be readily washed out of the modified Sepharose.

This material has been used successfully in the purification of the hydroxylase from the liver of *Macaca irus*.[60]

E SPECIFICITY

The earliest reports on this enzyme indicated that it shows a high degree of specificity,[282] with no action at all on D-phenylalanine or several analogues of the substrate. L-Phenylalanine was the only substrate found in this study. The product of the reaction was L-tyrosine. More recently it has been demonstrated that many substituents in the *para* position, including tritium, migrate to the *meta* position, and this is described in the section on the NIH shift. Jepson's group has found that a small proportion of the phenylalanine undergoing hydroxylation (about 0.2%) is converted to *m*-tyrosine, but it is not clear whether this is due to an attack at the *meta* position rather than the *para* position, or whether it is due to the migration of the hydroxyl group in an intermediate instead of the hydrogen atom.[63, 64]

Early reports on the hydroxylase claimed that *m*-tryosine and L-tryptophan were not substrates.[172, 224] It was observed in later studies that the liver enzyme does in fact hydroxylate L-tryptophan, although at first it was considered to be due to a separate hydroxylase.[93, 94] It was then discovered that a preparation of rat liver phenylalanine hydroxylase slowly hydroxylates L-tryptophan.[248, 249] This has been amply confirmed in later studies[65, 266] and a comparison of the physical properties of the two hydroxylating systems suggests that they are identical.[14] In view of this ability to hydroxylate tryptophan it is surprising at first sight that the daily urinary excretion of 5-hydroxyindole-3-acetic acid is so low (3.0 ± 1.0 mg d^{-1})[116] since this compound is the major catabolite of 5-hydroxytryptophan. Similarly, in isolated perfused livers using a perfusion medium containing serum albumin very little tryptophan is hydroxylated.[296a, 303] This may be explained if it is assumed that the concentration of free tryptophan in plasma and in liver cells is similar. L-Tryptophan binds to serum albumin[211] and although it has been claimed that about 25% of the tryptophan in fasting plasma is unbound,[212] other workers[113] have found that only about 2% of L-tryptophan added at low concentration to serum is recovered on ultrafiltration. Thus, the concentration of the unbound compound in serum may be as low as 1 μM, which is about 0.01% of the K_m for tryptophan. Similarly, the V_{max} for tryptophan is about 3% of the value for L-phenylalanine.[249] *In vivo* in man about 50% of the phenylalanine in blood is hydroxylated in 90 min[113] with a reserve distributed through about 50 litres of tissue fluid. If the same relationships apply to L-tryptophan one would expect about 30% of the free L-tryptophan in blood to be hydroxylated in 24 h. with a similar volume of tissue fluid that contains a reserve of L-tryptophan. This is equivalent to a formation of as little as 15 μmol of 5-hydroxytryptophan in 24 h., which is entirely consistent with the amount of 5-hydroxyindole-3-acetic acid excreted.[116]

L-*m*-Tyrosine has now been shown to be a substrate for the liver hydroxylase [274] and it forms L-dihydroxyphenylalanine (L-DOPA) at about 10% of the rate of hydroxylation for phenylalanine. This hydroxylation, combined with the inefficient formation of *m*-tyrosine from phenylalanine by this enzyme, affords an inefficient alternative route for the biosynthesis of L-DOPA. It is doubtful whether this pathway has any physiological significance, since the hydroxylation does not take place where L-DOPA is required, and studies on the treatment of parkinsonism with L-DOPA have demonstrated that massive doses of this compound are required to generate proper nerve function when it is not synthesized in the nerves, where it is converted to catecholamines.

Halogenated analogues of phenylanine are hydroxylated by both the liver and *Pseudomonas* enzymes. These include 2-, 3-, and 4-fluorophenylalanine,[65, 79, 170, 172, 266, 296] 4-chlorophenylalanine,[67, 136] 4-bromophenylalanine,[123] and 4-iodophenylalanine.[67] For a large proportion of the substrate the *para* substituent migrates to the *meta* position when 4-chlorophenylalanine and 4-bromophenylalanine are hydroxylated, but not when 4-fluorophenylalanine or 4-iodophenylalanine are hydroxylated.[67, 123, 133, 136] The release of fluoride ions has been demonstrated with 4-fluorophenylalanine.[170] In the case of 4-chlorophenylalanine about 85% of the product is 3-chlorotyrosine, 10% is tyrosine, and less than 5% is oxidized to 4-chloro-3-hydroxyphenylalanine.[133, 136] 5-Fluorotryptophan is also a substrate.[62, 65] *p*-Methylphenylalanine is oxidized by the enzyme.[70, 127] The products of this reaction are *p*-hydroxymethylphenylalanine, *m*-methyltyrosine and smaller amounts of 3-hydroxy-4-methylphenylalanine when either *Pseudomonas* or rat liver enzyme is used. The reaction mixture used with the *Pseudomonas* enzyme contained added Fe^{2+}, but when liver enzyme was used without the addition of Fe^{2+} the amount of these products formed per unit of enzyme was appreciably less than when the *Pseudomonas* enzyme was used.[70] In view of the non-specific hydroxylation that was observed by Jepson's group with phenylalanine when a mixture of Fe^{2+} and 2-amino-4-hydroxy-6,7-dimethyl-5,6,7,8-tetrahydropteridine was incubated with this amino acid a question mark must hang over this unusual hydroxylation, since it could, at least in part, be non-enzymic.

The *Pseudomonas* enzyme is able to catalyse the hydroxylation of glycyl-L-phenylalanine.[129]

The following compounds have not undergone hydroxylation with this hydroxylase: D-phenylalanine, *o*-tyrosine, α-phenylalanine, *N*-acetyl-L-phenylalanine, *N*-chloroacetyl-L-phenylalanine, L-phenylalanine ethyl ester, aniline, acetanilide, phenylacetic acid, and *p*-methoxyphenylalanine.[172, 224, 250, 282]

Spinach enzyme shows a different specificity, since it hydroxylates *p*-fluoro-L-phenylalanine at about the same rate as L-phenylalanine, but L-tryptophan is not a substrate.[234]

The specificity for the cofactor is detailed in the section on the reaction mechanism.

F MECHANISM

The reaction mechanism is complex and the more studies that are carried out the greater the complexity that is revealed.

The basic reaction is

$$\text{L-Phenylalanine} + O_2 + \text{THP} \rightarrow \text{L-Tyrosine} + \text{DHP} + H_2O$$

where DHP is a dihydropteridine and THP is a tetrahydropteridine.

With one exception all the compounds that have been demonstrated to have significant cofactor activity are tetrahydropterins, i.e. they are derivatives of 2-amino-4-hydroxy-5,6,7,8-tetrahydropteridine.

1 Non-enzymic hydroxylation

The hydroxylation of phenylalanine can be mediated non-enzymically as well as enzymically. For instance, a general hydroxylating mechanism has been demonstrated by Udenfriend's group in which hydroxylation is observed when an aromatic compound is incubated with a solution of Fe^{2+}, ascorbate, and a chelating agent in air.[283] In the presence of a large excess of 2-amino-4-hydroxy-6,7-dimethyltetrahydropteridine and Fe^{2+}, phenylalanine is oxidized to DOPA, tyrosine, and several other compounds.[66] In the presence of Fe^{3+} a tetrahydropteridine causes the oxidation of phenylalanine to *o*-, *m*-, and *p*-tyrosine. The yield of tyrosine is increased by the addition of borohydride. This reaction has an optimum at pH 7,[286] which clearly means that this reaction could interfere with measurements of the enzyme activity. The reaction does not discriminate between the optical isomers of phenylalanine, and phenylalanine dipeptides are also substrates. It has been suggested that phenylalanine may be attacked by a peroxide of the tetrahydropteridine. Hydrogen peroxide is, however, unable to carry out this oxidation. These reactions make it imperative to carry out control experiments to eliminate the effects of these side reactions when an assay for phenylalanine hydroxylase is set up.[301]

It is possible that these non-enzymic reactions may throw some light on the enzymic reaction, but analogies between the two must be treated with caution. One area in which some light has been thrown on the enzymic reaction is the non-enzymic oxidation of aromatic compounds by peroxides with the migration of a substituent in the benzene ring during the hydroxylation.[160] Beyond this, non-enzymic hydroxylations do not yet appear to have shed much light on the mechanism of enzymic hydroxylation.

2 Mechanism relative to the substrate

Two postulated mechanisms appear to be incorrect. Kaubisch *et al.*[165] synthesized a compound that appeared to be phenylalanine-3,4-oxide, a possible intermediate in the hydroxylation reaction. If such an intermediate should be formed it might be expected to break down spontaneously to tyrosine. In fact, on treatment with acid there was no formation of tyrosine, and on this weak evidence it has been tentatively concluded that this epoxide is not an intermediate.

Soloway[263] postulated the formation of a peroxide bridge between C1 and C4 of the benzene ring in the hydroxylation of L-phenylalanine, which then breaks down to form L-tyrosine. This was shown to be incorrect by some results obtained by Guroff, Daly and coworkers.[126, 134, 136] They demonstrated a general hydroxylating mechanism when aromatic compounds are hydroxylated by microsomes. If the compound is substituted with, say, tritium at the site at which hydroxylation takes place, part of the tritium is retained by the molecule during hydroxylation, and is found *ortho* to the new hydroxyl group. When they repeated this work with phenylalanine labelled with deuterium or tritium *para* to the side chain much, although not quite all, of the label was retained during the reaction when either rat or *Pseudomonas* enzyme was used. The label was found *ortho* to the hydroxyl group, so it migrates during the reaction. The percentage retention is the same whether *p*- or *m*-tritiophenylalanine is used as the substrate. A similar degree of retention is obtained during the non-enzymic hydroxylation of phenylalanine, using either peroxide and Fe^{2+} or ascorbate and Fe^{2+} as catalysts, which suggests that the migration may occur during a non-enzymic step. The enzyme does have some influence, however, on this reaction step, since the degree of retention of the label does depend to some extent on the source of the enzyme, with a lower degree of retention when *Pseudomonas* enzyme is used, compared with liver enzyme.[125]

The mechanism postulated by Guroff, Daly and coworkers is outlined in Figure 2.2. In this the substrate is attacked by OH^+ at the *para* position, and a pair of π electrons becomes localized at the *para* position, which results in a net unit positive charge at the *meta* position. One of the *para* substituents, usually the hydrogen atom, then migrates to the *meta* position at the same time that the positive charge migrates to the *para* position. The hydroxyl group probably migrates in some molecules during the reaction, but this is not a favoured reaction; Jepson's group have found that only about 0.2% of the substrate is oxidized to *m*-tyrosine,[63, 64] and this is the upper limit for the migration of the hydroxyl group, since it is possible that some *m*-tyrosine is formed by an attack of OH^+ at the *meta* position.

The retention of tritium is about the same for both *m*- and *p*-tritiophenylalanine[136] (after allowing for the double labelling in *m*-tritiophenylalanine),

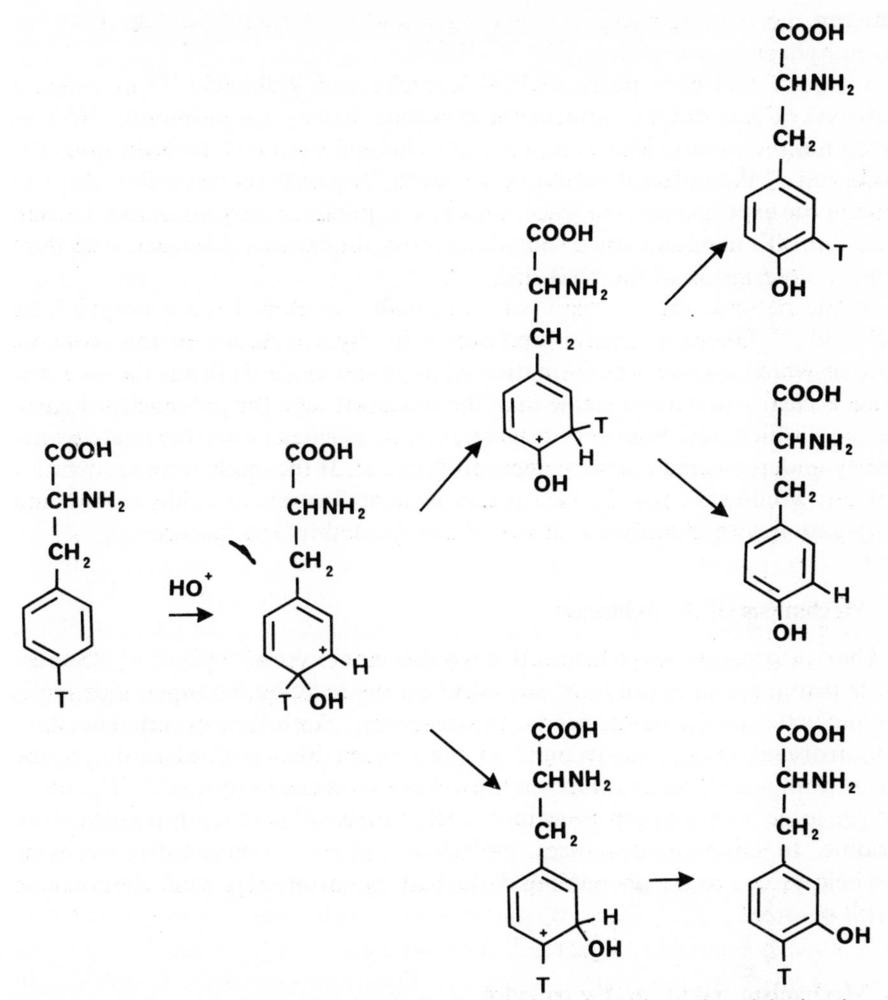

Figure 2.2 Postulated mechanism for the hydroxylation of *p*-tritio-L-phenylalanine, based on the NIH shift mechanism of Guroff *et al.*[136]

which indicates that the *meta* substituent is not displaced during the migration of the *para* substituent, but instead a transitory intermediate is formed in which there are two substituents on the *meta* position, one of which is eliminated preferentially, depending on the relative binding forces of the two substituents. This implies that the half-life of the intermediate is longer, and perhaps much longer, than the time taken for the migration, or the time for a bond-stretching vibration of the *meta* hydrogen. This mechanism is further supported by the relatively greater retention of deuterium compared with tritium when these species are used to label the *para* position of phenylalanine, since the difference between the binding energy for hydrogen and deuterium is smaller than between hydrogen and tritium.[125, 126, 135, 136]

A scheme has been postulated by Jaenicke and Wahlefeld,[155] in which a hydroxyl radical or OH^+ attacks the molecule at the *para* position to yield an oxyepinium cation (which is a positively charged epoxide). In both cases the oxidation of the cofactor would be involved. The authors claim that this can explain the experimental evidence. This is not, in fact, a very novel mechanism, since Guroff's team suggested something fairly similar as a side reaction in their original description of the NIH shift.[136]

At the present time the team who originally developed the concept of the NIH shift[34] favour a general mechanism for hydroxylation of the aromatic nucleus which involves the formation of an arene oxide. If this is the case, the oxide is clearly much less stable than the oxides of, say, the polynuclear hydrocarbons, which, although very reactive, can be isolated; thus, the oxide would clearly undergo rapid rearrangement to tyrosine. If the oxide were moderately stable it would take part in such reactions as hydrolysis to a dihydrodiol and conjugation with glutathione. It would also probably be a mutagen.

3 Mechanism of the oxidation

The oxidizing species in the reaction is molecular oxygen.[98, 126, 135, 136] Superoxide dismutase does not have any effect on the reaction, so superoxide ion is not the active oxidizing species for this reaction.[87] Labelling experiments have demonstrated that virtually none of the oxygen incorporated into tyrosine arises from water, but it is incorporated from molecular oxygen.[185] The other oxygen atom in the oxygen molecule usually oxidizes the cofactor to a dihydropteridine. In some circumstances, molecular oxygen is converted to peroxide (see below), due to the uncoupling of the hydroxylation step, while the cofactor is still oxidized.

4 Mechanism relative to the cofactor

This hydroxylase is a mixed-function oxidase, i.e. it not only oxidizes phenylalanine, but at the same time it oxidizes the cofactor from a reduced form to an oxidized form.

The existence of a cofactor other than the pyridine nucleotides was first suggested by an initial time lag in the reaction when NADP or NAD were used as the sole added cofactor[167] and the 'sheep enzyme' and pyridine nucleotide were later shown to be replaceable by a tetrahydropteridine.[169] Thus, although the early studies had shown that the hydroxylase is activated by NADPH as an apparent cofactor,[166] it is now known that this is simply a part of the hydrogen transport system.

The natural cofactor for the mammalian enzyme is 5,6,7,8-tetrahydrobiopterin[171, 174, 176] and for *Pseudomonas* enzyme it is reduced L-threoneopterin.[131] These are further described in the section on the natural cofactor (section G). Other (synthetic) pteridines are active,[8, 132, 169, 183, 266] including 2-amino-4-hydroxy-6,7-dimethyl-5,6,7,8-tetrahydropteridine, 2-amino-4-hydroxy-6-methyl-5,6,7,8-tetrahydropteridine, 2-amino-4-hydroxy-7-methyl-5,6,7,8-tetrahydropteridine, 2-amino-4-hydroxy-6-hydroxymethyl-5,6,7,8-tetrahydropteridine, 2-amino-4-hydroxy-6-hydroxymethyl phosphate-5,6,7,8-tetrahydropteridine, 2-amino-4-hydroxy-6-hydroxymethyl pyrophosphate-5,6,7,8-tetrahydropteridine, 2-amino-4-hydroxy-5,6,7,8-tetrahydropteridine, and 2,4-diamino-6,7-dimethyl-5,6,7,8-tetrahydropteridine. 2-Amino-4-hydroxy-6-carboxy-5,6,7,8-tetrahydropteridine appears to be active with the *Pseudomonas* enzyme, but not with the rat liver enzyme.[8, 132] It is noteworthy that all these compounds contain both a 2-amino and a 4-hydroxy group, with one exception in which the 4-hydroxy group is replaced by a 4-amino group. Inactive compounds include 4-amino-2-hydroxy-6-methyl-5,6,7,8-tetrahydropteridine, 4-hydroxy-6,7-dimethyl-5,6,7,8-tetrahydropteridine, 4-hydroxy-5,6,7,8-tetrahydropteridine, 2,4-dihydroxy-6,7-dimethyl-5,6,7,8-tetrahydropteridine, tetrahydroaminopterin, and 2-amino-4-hydroxy-6,7-diphenyl-5,6,7,8-tetrahydropteridine.[8, 183]

Tetrahydrofolate is a poor cofactor (about 1% of the activity for some of the above pteridines), but freshly prepared tetrahydrofolate contains a trace of 2-amino-4-hydroxy-6-methyl-5,6,7,8-tetrahydropteridine which is formed in a side reaction, and this accounts for most of the observed activity.[168, 183, 200]

Sepia pterin (2-amino-4-hydroxy-6-lactyl-7,8-dihydropteridine) can act as a cofactor in the reaction[171] but only after the side chain keto group has been reduced to yield dihydrobiopterin.[202]

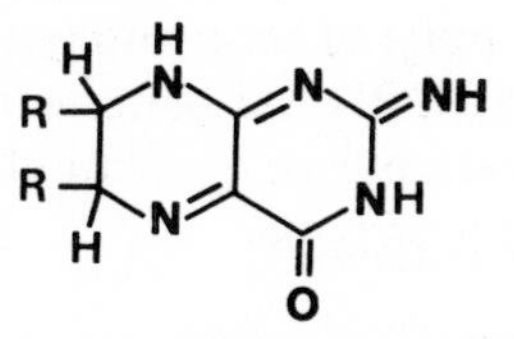

Figure 2.3 Possible quinonoid structure of dihydropteridines formed during the hydroxylation reaction

During the reaction the cofactor is oxidized to a dihydropteridine. The evidence indicates that the product has a quinonoid structure,[175] e.g. that shown in Figure 2.3, although there are other possible tautomeric forms. This is reduced back to a tetrahydropteridine by dihydropteridine reductase.[176] This enzyme is the so-called sheep liver enzyme that Kaufman used in his early studies on phenylalanine hydroxylase as an essential component of the system,[172] and its normal function *in vivo* is to couple the oxidation of phenylalanine to the hydrogen transport system via NADPH or NADH.[169] The existence of this enzyme was indicated by the early studies of Mitoma,[224] who separated the hydroxylase into two fractions, and Kaufman,[167] who extracted a second enzyme fraction, that was essential for the activity of the hydroxylase, from sheep liver. It would appear that the activation of the hydroxylase observed by Mitoma[224] with compounds such as alcohols and aldehydes, especially alcohols, was due primarily to regeneration of the nicotinamide cofactors by contaminating enzymes, e.g. alcohol dehydrogenase.

Dihydropteridine reductase is found in adrenals, as well as in liver.[231]

An assay has been devised for liver dihydropteridine reductase[237] isolated from rat, rabbit, sheep, beef, and cat. The principle of the assay is to keep the pteridine cofactor in the oxidized form by the action of hydrogen peroxide and peroxidase, and to follow the rate of oxidation of a nicotinamide cofactor by measuring its optical density. Solutions are prepared containing 10 mM hydrogen peroxide, peroxidase, and 0.01 mM 2-amino-4-hydroxy-6-methyldihydropteridine in 0.1 M Tris buffer at pH 6.8. The reductase is added before the pteridine, and NADH or NADPH is quickly added, and the mixture is incubated at 30 °C. The optical density is followed at 340 nm. A blank is carried through to correct for non-enzymic oxidation of the nicotinamide cofactor by the hydrogen peroxide.

Musacchio[230] has purified the beef adrenal medullary enzyme after the method of Nagatsu *et al.*[232] and studied some of its properties. It does not act on biopterin, since it appears to be mandatory that a pterin must be reduced to its dihydro analogue before it can act as a substrate. It is not inhibited by aminopterin or methotrexate at 0.1 mM. Its activity is rapidly destroyed at 80 °C.

There is some question about the use of nicotinamide cofactors in coupling this enzyme to the hydrogen transport system, which may just be a difference between the sources of the enzyme that results in a difference in properties. Musacchio[230] claims that NADPH is a better substrate than NADH for beef adrenal enzyme, whereas Nielsen *et al.*[237] state that NADH is the better substrate for the mammalian liver enzyme.

5 Stoichiometry

Basically, the reaction is stoichiometric, with the formation of 1 mol tyrosine from 1 mol each of L-phenylalanine, pteridine cofactor, and oxygen when the natural cofactor is used.[167, 266] With synthetic cofactors the situation can be quite different. In some situations, more cofactor is utilized than is required for hydroxylation, often by a factor of three. This effect has been observed with 2-amino-4-hydroxy-6-methyl-5,6,7,8-tetrahydropteridine and 2-amino-4-hydroxy-7-methyl-5,6,7,8-tetrahydropteridine.[266] The excess cofactor reduces oxygen to peroxide stoichiometrically.[87] Kaufman[266] has postulated that peroxide is the normal oxidant during the hydroxylation, at least at the enzyme level, but this remains unproved. The degree of uncoupling of hydroxylation that leads to the formation of peroxide is reduced by increasing either the concentration of the hydroxylase or buffer, or by reducing the temperature.[84] Kaufman suggested that this variation might be due to a reversible polymerization of the hydroxylase, with a smaller degree of uncoupling when the reaction is catalysed by the polymeric form. Kaufman[87,88] has demonstrated a different aspect of uncoupling in the absence of phenylalanine. When the hydroxylase is incubated with L- (but not with D-) tyrosine and either lysolecithin or α-chymotrypsin the hydroxylase oxidizes both 2-amino-4-hydroxy-6,7-dimethyltetrahydropteridine and tetrahydrobiopterin with the formation of 1 mol of hydrogen peroxide. The omission of tyrosine or lysolecithin or α-chymotrypsin prevents this reaction from taking place. It is just possible that this finding might have some clinical implications in neonatal tyrosinaemia in which the tissue concentration of tyrosine is raised by as much as 50-fold; it would, however, be necessary for some compound whose properties modify the action of the hydroxylase in a way similar to that of lysolecithin to be found in the liver. The author is not aware that any defect in phenylalanine hydroxylation has been described in neonatal tyrosinaemia. In tyrosinosis there is, in many cases, a defect in phenylalanine metabolism that results in an elevation of tissue phenylalanine concentrations, as well as an elevation of some other amino acids. It is not impossible that in tyrosinosis a compound like lysolecithin is present in the liver and kidneys, and as a result of the uncoupling of phenylalanine hydroxylation in the presence of high tyrosine concentrations abnormally high concentrations of peroxide are found in those organs, thereby causing the typical hepatic and renal lesions that are observed in the disease.

6 Mechanism relative to the enzyme

Comparatively little is known about this aspect of the subject. Jepson's group described results on the rat liver enzyme that appeared to indicate the presence of two enzymes, one of which acts on phenylalanine but not tryptophan and is not inhibited by 5-fluorotryptophan, and the other acting on phenylalanine and tryptophan and is inhibited by 5-fluorotryptophan.[62] These results have been reinterpreted to meet the requirement that just one enzyme species is present. This reinterpretation was made necessary by the observation that some indoleamino acids enhanced the hydroxylation of phenylalanine analogues, and phenylalanine analogues enhanced the hydroxylation of 5-fluorotryptophan. Two alternative interpretations have been placed on these results, one of which assumes the presence of two hydroxylating sites, one for benzenoids and the other for indoles, and each site is allosterically modified by binding at the other site in a way that lowers the activation energy for the hydroxylation.[65] The other model is a ligand exclusion model.[90] The results cannot be explained by a kinetic model in which the reactants can bind in any order to form the same ternary complex.[61]

7 Protein stimulator

A protein has been purified from rat liver that has an activating effect on the hydroxylation, yet does not itself catalyse the hydroxylation. It is also found in human liver, including liver from phenylketonurics.[152] The rat liver protein is able to activate human hydroxylase as well as rat hydroxylase.[98] It is likely that there are several different proteins that can bring about this activation, and it is not at present clear whether they act by the same mechanism. Bessman[25] finds that rat liver contains two activating fractions, one of which may be dihydropteridine reductase. They both alter the kinetics of the enzyme reaction, by transforming the sigmoid concentration–activity curve to what is presumably a hyperbolic curve that is consistent with the Michaelis–Menten equation. Without the stimulating protein, which can be purified from crude dihydropteridine reductase, the specific activity of the hydroxylase decreases as the concentration of the enzyme increases.[151, 152] Originally it was thought that the activating protein altered the reversible dimerization of the hydroxylase. It is now known that the decrease in enzyme activity is caused by a sharp increase in the apparent K_m for tetrahydrobiopterin without any change in V_{max}.[151] This effect is relieved by the addition of the stimulating protein. It has been suggested that this effect can be explained if an intermediate in the reaction sequence is released during the reaction, which can break down into the reaction products non-enzymically or (more quickly) recombine with the enzyme which also catalyses the formation of the reaction products, and the activating protein increases the rate of the non-enzymic step. The measured change in kinetics would then be an artefact arising from the addition of the different pathways for the breakdown of the intermediate. This intermediate could be either a biopterin–oxygen–phenylalanine complex or a precursor of the quinonoid dihydro form of biopterin. The change in kinetics is not in fact due to any change in molecular weight of the hydroxylase.

The molecular weight of the rat stimulating enzyme is 51 500, and it can be separated by electrophoresis into two active fractions with the same molecular weight. It can be broken down into four subunits with molecular weights of 12 500.[152]

The activator also alters the optimum pH of rat liver hydroxylase from a broad optimum at pH 5.8–6.8 to a slow increase from pH 5.8 to 7.4, with a small, sharp optimum at pH 7.6–8.2.[178]

Beef growth hormone, but not several other proteins, can also act as a stimulator for the hydroxylase.[178]

Comamonas enzyme is stimulated two-fold if it is incubated with bovine serum albumin during the preincubation with Fe^{2+}, but not if albumin is added after the preincubation.[138] This action is not stoichiometric. It has been suggested that metal ion binding is involved. The properties demonstrated during this procedure bear similarities to the protection of *p*-hydroxyphenylpyruvate oxidase during incubation with glutathione;[114] it is therefore possible that protection of the hydroxylase against oxidation may be involved.

8 Dehydroxylation of tyrosine

Although it is generally reckoned that the hydroxylation of phenylalanine is irreversible, and thermodynamic considerations certainly indicate that the equilibrium in the phenylalanine hydroxylase reaction is so far over to the side of hydroxylation as to make the reaction essentially irreversible, there has been a report of the reverse reaction[154] being observed. It has been observed that poultry suffer little ill-effect when they are fed a diet that is deficient in phenylalanine but contains an adequate amount of tyrosine. This phenomenon could be explained if tyrosine is converted to phenylalanine in poultry, and experiments were carried out to test this hypothesis. According to this work, when labelled tyrosine is injected into roosters that are then killed and protein is isolated from the birds and hydrolysed, although much of the label is found in tyrosine a small but significant amount of the activity is observed in the phenylalanine fraction, but not in any other amino acids. The results could not be explained on the grounds of either contamination with tyrosine or a rearrangement of the molecules that would allow the label to be transferred to phenylalanine by a mechanism other than the dehydroxylation of tyrosine. When the experiment was repeated with rats no label was detected in the phenylalanine isolated from protein.

These results suggest the existence of an enzyme other than phenylalanine hydroxylase, which, with the participation of an energy-rich source, e.g. ATP, removes the hydroxyl group. There are precedents for this type of reaction, particularly in some microorganisms, that can remove the *para* hydroxyl group from substituted catechols.[30,31,254,255,256] One group of workers has found that dopamine can be dehydroxylated in mammalian brain[32,33] to yield tyramine. Studies on poultry gut flora might demonstrate the site of tyrosine dehydroxylation.

G NATURAL COFACTORS

The identity of the natural cofactor (or more strictly, cosubstrate), depends on the species from which it is obtained. The mammalian cofactor (and presumably the cofactor in all vertebrate species) is 5,6,7,8-tetrahydrobiopterin, i.e. 2-amino-4-hydroxy-6-(1,2-dihydroxypropyl)-5,6,7,8-tetrahydropteridine. The cofactor isolated from *Pseudomonas* has properties that indicate that it is tetrahydro-L-threoneopterin, i.e. L-*threo*-2-amino-4-hydroxy-6-(1,2,3-trihydroxypropyl)-5,6,7,8-tetrahydropteridine, whose structure is closely allied to that of the mammalian cofactor. The structures of these compounds are shown in Figure 2.4.

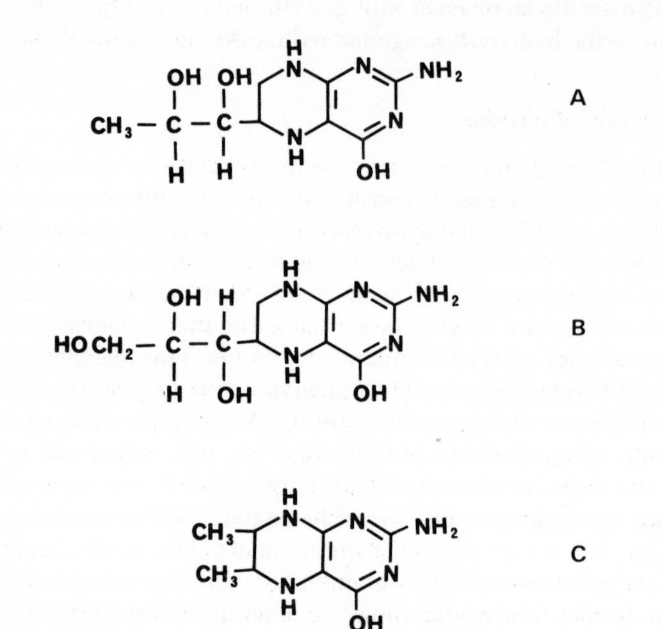

Figure 2.4 Some active cofactors. A, Natural mammalian cofactor, 5,6,7,8-tetrahydrobiopterin; B, natural *Pseudomonas* cofactor, 5,6,7,8-tetrahydro-L-threoneopterin; C, 2-amino-4-hydroxy-6,7-dimethyl-5,6,7,8-tetrahydropteridine

In the preparation of synthetic mammalian cofactor two diastereoisomers of *l-erythro*-tetrahydrobiopterin have been isolated. The structural difference occurs at position 6 of the nucleus. One of these is the natural cofactor. Both of them can be utilized by the hydroxylase with the same K_m, but V_{max} is four times greater for the natural cofactor. The natural cofactor shows substrate inhibition with the hydroxylase, whereas its isomer does not.[12]

A method has been developed by Guroff's group for the assay of the naturally occurring cofactor, using phenylalanine hydroxylase in excess as a reagent. The method is based on a modification of one of the methods used for the assay of phenylalanine hydroxylase,[124, 127] with the concentration of reagents adjusted so that the rate-limiting factor is the concentration of the cofactor. The *Pseudomonas* enzyme has been used in preference to the mammalian enzyme because of its better stability. A tissue extract is prepared that contains the cofactor, and mercaptoethanol is included as an antioxidant. This is centrifuged at 15 000 g. A portion of the supernatant (3 ml) is heated at 60 °C for 3 min. to inactivate enzymes, after displacement of dissolved oxygen by bubbling nitrogen through the solution, and the precipitated protein is centrifuged down. A portion of hydroxylase (about 250 μg in 20 μl) is incubated for 10 min. with 20 μl of 12.5 mM ferrous ammonium sulphate to activate the enzyme, and to this is added 20 μl of 1.5 M Tris buffer at pH 7.3, 1 μmol of L-phenylalanine in 10 μl, 2×10^5 counts min^{-1} of *p*-tritiophenylalanine in 10 μl, 10 μl of 0.1 M NADPH solution, 1 μmol of glucose in 10 μl, 1.2 mg of glucose dehydrogenase, 35 μg of dihydropteridine reductase, and the sample to be assayed to give a total volume of 0.25 ml. This is incubated at 30 °C for 60 min. in air, and the reaction is stopped by boiling. After cooling, acetate buffer at pH 5.5 and *N*-iodosuccinimide are added and the iodine displaces the tritium that migrated during the reaction from the tyrosine as labelled water. The remaining labelled amino acids are removed by passage through a column of ion-exchange resin and charcoal, and the radioactivity of the effluent is measured, and this represents the tyrosine formed on a mole-for-mole basis. It is necessary to carry blanks through the procedure to correct for unbound tritium in the labelled phenylalanine, and for non-enzymic hydroxylation.

This assay has demonstrated a cofactor concentration of 1–10 μg $(g\ tissue)^{-1}$ in rat and guinea-pig liver, kidney, and brain.[124]

A method has been developed for preparing the *Pseudomonas* cofactor biosynthetically from guanosine monophosphate, ATP, magnesium, and NADPH.[121]

In man the cofactor (and dihydropteridine reductase) are found in the liver of the foetus as early as 11 weeks gestation. Studies do not appear to have been undertaken to see how early it is present in the foetus.[157] However, it does not appear to be present in optimum concentration until just after birth, at least in rats.[42]

The cofactor is very susceptible to autoxidation, a problem that bedevilled early studies on phenylalanine hydroxylase before the problem was understood. It appears that autoxidation involves a superoxide ion, since the autoxidation is inhibited by the presence of superoxide dismutase.[87]

It appears that the structure of tetrahydrobiopterin is altered in some way by treatment with borohydride, since treatment with this reagent inactivates the cofactor when it is used in the hydroxylation of phenylalanine, but not when the substrate is tryptophan.[105]

H PHYSICAL PROPERTIES

1 Composition

The amino acid composition of purified rat liver phenylalanine hydroxylase has been determined on hydrolysates obtained from the enzyme.[89] This enzyme contains 1–2 mol iron per 10^5 **g** as the high-spin Fe^{3+} form. The signal that this ion produces vanishes on addition of phenylalanine or tetrahydropteridine, which indicates that these compounds interact with the metal ion during the reaction, even if oxidation and reduction of the iron is not an essential step in the hydroxylation reaction, a point that has yet to be resolved. Much of the iron (50–80%) can be removed by treating the enzyme with iron-chelating agents such as cysteine or *o*-phenanthroline, with a subsequent reduction in the activity of the enzyme, again demonstrating that the iron atom is essential for activity. This lost activity can be restored by the addition of low concentrations of Fe^{2+}. This ion cannot be replaced by Hg^{2+}, Ni^{2+}, Co^{2+}, Mn^{2+}, Cu^{2+}, Cr^{2+}, or Zn^{2+}. The holoenzyme contains five cysteine residues per subunit of molecular weight 50 000, whereas the apoenzyme contains only three. Presumably two cysteine residues are reversibly oxidized to a cystine residue on the removal of iron from the active centre of the molecule.

The human enzyme also contains iron (2 mol per mole). In addition it contains 1 mol copper, and studies with chelating agents to determine which inhibit the enzyme indicate, from the spectrum of ions that the effective agents can chelate, that both ions may be necessary for activity. Such conclusions must, however, be treated with some caution, since chelating agents are not necessarily specific for the ions that they will chelate. This enzyme also contains 1 mol of NADH, which is presumably involved in hydrogen transport during the reaction.[108] This enzyme contains five sulphydryl groups.

2 Molecular weight and isozymes

Numerous studies have been carried out on the molecular weight of phenylalanine hydroxylase, and these have brought a complex situation to light. Human liver enzyme has been purified to near homogeneity, and this has been isolated in two forms, whose elution profile from a column of molecular sieve indicates that they have a molecular weight of 108 000 and 250 000.[296, 298] Electrophoresis of the hydroxylase on polyacrylamide gel indicates that the enzyme has a molecular weight of 54 000, and this is considered to represent a monomeric form. The two forms obtained from the isolation procedure therefore behave like a dimer and a tetramer of the form observed on gel electrophoresis. The tetramer is only observed when lipid material is allowed to remain in contact with the enzyme during the purification procedure. If a solution of the crude enzyme is filtered through a glass wool plug to remove lipids early in the purification, the enzyme is obtained essentially in the dimeric form.

These results demonstrate the ease with which artefacts can be formed with this enzyme. Presumably either the dimer or the tetramer is an artefact, and there is at present no reason to believe that the tetramer is formed by the interaction of two molecules of the dimer rather than the interaction of the dimer with other polypeptide material. The extra polypeptide chains clearly do not foul the active centre of the hydroxylase molecule, since their removal does not alter the activity of the enzyme.

A similar picture is observed with the rat liver enzyme, which can be obtained as a monomer whose molecular weight has been determined to be 51 000–55 000, a dimer that has a molecular weight of 110 000, and a tetramer whose molecular weight has been determined by different workers to lie in the range 200 000–250 000.[14,108,182,276] The formation of the tetramer is similar to the analogous formation for the human enzyme. The enzyme is usually obtained mainly at the dimer, but if a filtration step with glass wool to remove lipids is omitted, or if dithiothreitol is not added during purification of the enzyme, part of the enzyme is obtained as the tetramer.[108] Conversion of the dimer to the tetramer is also brought about by incubation with L-phenylalanine at about 1.5 mM concentration for 10 min. at 25 °C. This conversion to the tetramer brings about activation of the enzyme. The activation follows an exponential course, defined by the equation

$$v_a = v_f + (v_0 - v_f)e^{-at},$$

where v_0 is the fully active rate before activation, v_f is the final equilibrium rate, t is the time, and a is a constant.[276] This activation is brought about only by L-phenylalanine.

These workers have postulated the existence of an activation site in the enzyme molecule that is distinct from the reaction site to explain the results, which presumably involves a molecule of phenylalanine forming a bridge between two molecules of the dimer.[276] These results contrast sharply with those of Woolf's group,[108] who found that the specific activity of the rat enzyme is markedly reduced when it is converted to the tetramer. The sedimentation constant for the non-activated (dimeric) form is 6.1 S, and for the activated form (tetramer) 8.14 S.[276] The formation of the activated tetramer may be an important control mechanism in the face of changing phenylalanine concentrations, it has been postulated.[276] The dimeric form of the enzyme dissociates into the monomer when its solution is diluted, or when the temperature is raised to 30° C.[182]

Rat liver and kidney enzymes are identical in molecular size, density, and shape as indicated by gel filtration and ultracentrifugation on a sucrose density gradient. The kidney enzyme can also be separated into two distinct forms on gel filtration which presumably also represent the dimer and tetramer.[10]

Cynomolgus monkey enzyme can be separated into two fractions in the final stages of its purification, both of which are required for activity. Both fractions have a molecular weight of 130 000, and one of them is yellow. It has been suggested that one of these may be similar to one of Bessman's stimulating factors.[57] Two active fractions have been obtained from this species by affinity chromatography.[60]

Pseudomonas enzyme has a molecular weight that lies in the range 60 000–70 000.[123,130]

There is a body of evidence for the existence of isozymes, some of which is indirect and is open to other possible explanations. For instance the kinetics of the rat liver enzyme relative to different substrates and inhibitors suggested to Jepson's group that there may be two phenylalanine hydroxylase isozymes in this organ, each with different specificities.[62] These results have since been reinterpreted as indicating instead that the kinetics of the hydroxylase are rather complex, and that there is no need to postulate the existence of more than one enzyme.[65] Similarly, Freedland[92] observed that the rat liver enzyme is activated by ascorbate and Fe^{3+} when L-tryptophan is used as the substrate, but not when L-phenylalanine is the substrate. Although these results might arise from the presence of more than one phenylalanine hydroxylase, Freedland, however, considered that these results could be explained without postulating the existence of more than one enzyme.[92]

Whatever the truth may be about the above explanations, more substantial evidence has been obtained for the existence of isozymes in rat liver. Bessman[25] found two, and then in a later study three, isozymes that he designated π, κ and υ, all with a molecular weight of about 200 000.[14] They all have the same optimum pH, but they differ in their Michaelis constant towards phenylalanine, their relative rates of heat inactivation, and the degree of inhibition by *p*-chloro-L-phenylalanine. These effects are described in detail in their respective sections in this review. The relative activities of these isozymes are in the ratio of 7:170:80 respectively. Rat liver enzyme that hydroxylates L-tryptophan shows the same pattern of properties for isozymes as phenylalanine hydroxylase, which has been cited as evidence for the hypothesis that the two activities reside on the same enzyme molecule. Adult liver, however, contains the above isozymes; the κ enzyme develops either late in pregnancy or at term.[13, 238, 239] Human foetal enzyme contains two isozymes on the same criteria as those described above.[13, 14]

Connellan and Danks[56] have described an insoluble form of human liver phenylalanine hydroxylase obtained at autopsy, as well as the soluble form. The insoluble form can be solubilized either by treatment with deoxycholate or by sonication, which suggested to these workers that the insoluble form may be linked to an insoluble cell fraction by means of a lipid linkage. After solubilization the insoluble form shows properties very similar to those of the soluble form. It is possible that the insoluble enzyme may be an artefact formed during the preparation of the tissue extract; there are examples of enzymes that are rendered insoluble in this way.

Tourian finds that there are three isozymes in rat liver, with isolectric points at pH 5.20, 5.30, and 5.60, whereas kidney contains one isozyme with an isoelectric point of pH 5.35.[278] The picture with isozymes has been further complicated by the work of Kaufman's team, which suggests that phosphorylation is involved in the enzyme activity. This work may, however, eventually lead to a clarification of the isozyme question.[2, 80, 159, 222]

This work has been carried out with purified rat liver enzyme that is considered to be pure. Assuming a molecular weight of 50 000 for the monomer, the enzyme contains 0.31 mol per mole of phosphate. Rat liver contains a phosphatase which, although inactive towards the classical substrates of phosphatases, will remove phosphate from the hydroxylase. This hydrolysis reduces the activity of the enzyme when the natural cofactor is used, but not when 6-methyltetrahydropterin is used. This reduction indicates that the group removed is in some way involved in binding of the cofactor to the enzyme molecule.

When the enzyme is exposed to the action of a phosphorylase, protein kinase, in the presence of ATP, cyclic AMP, and Mg^{2+}, a further 0.7 mol of phosphate is incorporated into the monomer, thus increasing its total content to 1 mol per mole. The activity of the enzyme is increased appreciably by the phosphorylation, but only when the natural cofactor is used. The product is different from the naturally occurring isozymes.

Two of the apparent isozymes have been treated in this way, and although both were phosphorylated, the degree of activation was different. Partial proteolytic digestion of the hydroxylase, a process which removes the phosphate, presumably by removal of the section of polypeptide chain that contains the phosphate, results in an enhanced activity.

A possible synthesis of these results is as follows, which covers the findings of Kaufman and Woolf, but possibly not those of Bessman:

The native enzyme exists *in vivo* as a dimer (or tetramer). The dimer contains two phospholipid chains, one attached to each monomeric unit. The phospholipid portion binds weakly to another polypeptide chain to form the tetramer, but there is not enough evidence to determine whether this occurs *in vivo*, or only during the isolation procedure. It is possible that lipid molecules not attached to the enzyme covalently may enhance this binding. The phospholipid is attached covalently via the phosphate to each hydroxylase monomer, and the phosphate bond is rather labile, so that it can be hydrolysed in part during the purification steps. This phospholipid unit is neutral, so it has only a minor effect on the overall properties of the enzyme. It is attached so that it interacts with the cofactor binding site in such a way that the phosphate unit interacts, probably by hydrogen bonding, with the side chain of the natural cofactor, whereas the side chain of the synthetic cofactor does not interact, either because the chain is too short, or because there is no group to make hydrogen bonding possible.

The phospholipid would also permit binding between the hydroxylase and any of the activating polypeptide chains, and phenylalanine might act to make another bridge, thus enhancing the activating effect.

The site to which the phospholipid is bound may be an amino group. If so, it would permit a molecule of lysolecithin to bind to it after removal of the natural phospholipid, as a salt, which, again, could cause activation of the enzyme. This possibility is not necessarily inconsistent with the claim that lysolecithin unmasks a sulphydryl group in the hydroxylase molecule.[88]

The fact that removal of the naturally occurring phosphate-containing moiety decreases the enzyme activity when the natural cofactor is used, and both phosphorylation and addition of lysolecithin increase the activity with this cofactor, and the markedly lower sensitivity of the enzyme to these constraints when the methyl-containing synthetic cofactor is used strongly suggests that all these phosphate-containing materials are acting at the same site.

Activation by partial proteolytic digestion could explain the presence of some of the other isozymes, and may even explain the presence of active isozymes in liver of patients with phenylketonuria. In fact, all of Bessman's isozymes could so arise from a parent molecule, and the activity of his various isozymes could reflect residual activity in fragments obtained by proteolysis of several allelic forms of almost inactive hydroxylase, in which case the overall pattern of isozymes rather than the individual isozymes observed in phenylketonuric patients might reflect the nature of the allele.

3 Michaelis constant

The measured values for this parameter are set out for various species in Table 2.3. Although there are considerable differences between the values obtained in different studies, several factors become apparent. In mammalia the lowest value of K_m for the cofactor is observed when the natural cofactor is used. There is a distinct interaction between the reactants to influence the values observed for the constants. For instance, the measured K_m for oxygen or phenylalanine is dependent on the cofactor used.[98] One study[86] indicates that if the results obtained with rat liver enzyme are extrapolated to saturation with the cofactor and substrate a true value of the K_m for oxygen is about 5% concentration in the gas phase. A linear relationship is observed between the measured K_m for oxygen and the degree of substrate inhibition by oxygen, that is independent of other factors in the assay, including those that cause the alterations in the K_m for oxygen. This suggests that the inhibition may be due to the binding of oxygen at the substrate binding site with a comparatively low affinity, and thereby inhibiting the enzyme.

Another study, in which variations in the above parameters were not examined, showed a K_m of 0.31 atm for oxygen with rat liver enzyme, using dimethyltetrahydropterin as the cofactor.[46]

The low apparent K_m for oxygen when tetrahydrobiopterin is used as the cofactor indicates that the tissue concentration of oxygen is not a limiting factor in the *in vivo* hydroxylation of phenylalanine.[86]

The isozymes that have been found in rat liver have different Michaelis constants. Isozymes π, κ, and υ have Michaelis constants of 0.87, 0.78, and 0.64 mM respectively for phenylalanine.[13, 14] Another group of workers has

Table 2.3 Michaelis constant of phenylalanine hydroxylase in different species

Species	Organ	Compound	K_m	Reference
Man	Liver	Tetrahydrobiopterin	3 μM	98
			57 μM	10
		2-Amino-4-hydroxy-6,7-dimethyl-5,6,7,8-tetrahydropteridine	40 μM	97
			5 μM	99
			50 μM	98
			57 μM	177
			40 μM	97
			157 μM	10
			70.2 ± 1.7 μM	296
		2-Amino-4-hydroxy-6-methyl-5,6,7,8-tetrahydropteridine	40 μM	98
			42–46 μM	256
		Oxygen[a]	0.95%[b]	98
			3.4%[c]	98
		L-Phenylalanine	40 μM[b]	98
			1.6 mM[c]	98
			0.9 mM[d]	179
			0.5 mM[c]	56
			1.0 mM[c]	177
			0.9 mM[c]	97
			0.61–0.64 mM[e]	256
			0.9–1.0mM[c]	244
			0.17 mM[c]	49
			0.62 mM[c]	10
			0.318 ± 0.016 mM[c]	296
			1.25 mM[c]	99
			0.9 mM[c]	97
		2-Amino-4-hydroxy-7-methyl-5,6,7,8-tetrahydropteridine	59–62 μM	266
	Kidney	L-Phenylalanine	0.84 mM[c]	10
		Tetrahydrobiopterin	57 μM	10
		2-Amino-4-hydroxy-6,7-dimethyl-5,6,7,8-tetrahydropteridine	144 μM	10
Macaca irus	Liver	L-Phenylalanine	0.57 mM[c]	57
		2-Amino-4-hydroxy-6,7-dimethyl-5,6,7,8-tetrahydropteridine	85 μM	57
Rat	Liver	Tetrahydrobiopterin	22 μM	7
			23 μM	10
		2,4-Diamino-6,7-dimethyl-5,6,7,8-tetrahydropteridine	127 μM	7
		2-Amino-4-hydroxy-6,7-dimethyl-5,6,7,8-tetrahydropteridine	64 ± 6 μM	39
			67 μM	38
			101 μM	10
			90 μM	9
			67 μM	46
			93 μM	7
			64–77 μM	13

[*continued*

TABLE 2.3 [*continued*

Species	Organ	Compound	K_m	Reference
		2-Amino-4-hydroxy-5,6,7,8-tetrahydropteridine	130 μM	7
		2-Amino-4-hydroxy-6-methyl-5,6,7,8-tetrahydropteridine	42–46 μM[f]	266
		L-Phenylalanine	0.83 mM[c]	10
			1.04 mM[c]	38
			0.76 mM[c]	9
			0.28–0.33 mM	156
			0.2 mM[g]	224
			1.12 mM[c]	46
			0.76±0.04 mM[c]	39
			0.64–0.87 mM[c]	13
			0.64–0.87 mM[c]	14
			1.01 mM	296a
		Oxygen[a]	20%[c]	86
			31±2%[c]	43, 46
		L-Tryptophan	6.8 mM[c]	249
			10–50 mM	94
	Kidney	L-Phenylalanine	0.95 mM[c]	10
		Tetrahydrobiopterin	21 μM	10
		2-Amino-4-hydroxy-6,7-dimethyl-5,6,7,8-tetrahydropteridine	107 μM	10
Pseudomonas		L-Phenylalanine	0.6 mM	129
			0.34–0.69 mM[c]	130
		2-Amino-4-hydroxy-6,7-dimethyl-5,6,7,8-tetrahydropteridine	44 μM	132
			0.1–0.3 mM	130
		Oxygen[a]	3.5%	130
		2-Amino-4-hydroxy-6-hydroxymethyl-5,6,7,8-tetrahydropteridine	78 μM	132
		2-Amino-4-hydroxy-6-hydroxymethyl phosphate-5,6,7,8-tetrahydropteridine	28 μM	132
		2-Amino-4-hydroxy-6-hydroxymethyl pyrophosphate-5,6,7,8-tetrahydropteridine	62 μM	132
		2-Amino-4-hydroxy-5,6,7,8-tetrahydropteridine	90	132
		2-Amino-4-hydroxy-6-carboxy-5,6,7,8-tetrahydropteridine	217 μM	132

[a] The concentration of oxygen in solution is that which is in equilibrium with the stated percentage concentration in the gas phase.
[b] With lysolecithin and tetrahydrobiopterin.
[c] With 2-amino-4-hydroxy-6,7-dimethyl-5,6,7,8-tetrahydropteridine.
[d] With 2-amino-4-hydroxy-6-methyl-5,6,7,8-tetrahydropteridine.
[e] With 2-amino-4-hydroxy-7-methyl-5,6,7,8-tetrahydropteridine.
[f] With 4-fluoro-L-tryptophan.
[g] With tetrahydrobiopterin.
Unless otherwise stated the amino acid substrate is L-phenylalanine.

studied the variations of K_m between the dimer and the tetramer obtained from rat liver.[108] They have obtained a value of 1.42 ± 0.02 mM and 1.22 ± 0.14 mM for the tetramer and the dimer respectively for L-phenylalanine as substrate. Another value, apparently obtained without isolating isozymes, is 1.12 mM.[46] Liver perfusion experiments indicate a K_m for phenylalanine in the rat of 1.5 mM, based on the rate of tyrosine formation.[296a, 303] A value of 28.8 ± 4.6μM and 55 ± 4 μM for 2-amino-4-hydroxy-6,7-dimethyl-5,6,7,8-tetrahydropteridine was obtained with these two polymeric forms respectively when L-phenylalanine was used as the substrate. With a mixture of isozymes Bublitz obtains the value of 67 ± 4 μM.[46] It appears that the Michaelis constant decreases slightly for both compounds when the concentration of the other reactant (other than oxygen, which has not been tested) is reduced. The decrease is possibly not statistically significant. The K_m for *p*-fluoro-L-phenylalanine has been found to be 1.33 mM and 2.5 mM for the tetramer and the dimer respectively, and when this substrate is used the K_m for 2-amino-4-hydroxy-6,7-dimethyl-5,6,7,8-tetrahydropteridine is 25 μM with both forms of the enzyme.

The protein stimulator has a distinct effect on the kinetics of the enzyme, including the measured values of the Michaelis constant. These effects are described in the section that deals with the protein activator (section F.7).

The addition of short-chain alcohols such as ethanol to rat liver phenylalanine hydroxylase at about 5% concentration in a buffer near the optimum pH decreases the K_m for phenylalanine and increases this parameter for 2-amino-4-hydroxy-6,7-dimethyl-5,6,7,8-tetrahydropteridine.[40]

4 V_{max}

Only a few studies of this parameter have been undertaken with pure or almost pure hydroxylase, but several studies have been carried out with the partially purified enzyme, mainly to examine the effect of a second parameter, and although the latter studies have not been able to give any information about the absolute values of V_{max}, some useful relative information has been obtained.

The absolute value has been measured by Woolf's team. For human foetal liver hydroxylase a value of 153 ± 0.6 nmol of tyrosine formed per minute per milligram of enzyme has been obtained at pH 7.0 and 25 °C, using 2-amino-4-hydroxy-6,7-dimethyl-5,6,7,8-tetrahydropteridine as the cofactor.[296] There are distinct differences between the dimer and the tetramer obtained in the purification of the rat liver enzyme. Although Tourian[276] finds that the tetramer is the more active form Woolf has found that the dimer is the more active form, with a V_{max} of 696 nmol of tyrosine formed per minute per milligram of protein at pH 6.8 and 25 °C using the same cofactor as above. The value obtained with the tetramer is only 160 nmol min^{-1}. When *p*-fluoro-L-phenylalanine is substituted for L-phenylalanine a lower V_{max} of 150 nmol mg^{-1} is obtained for the dimer and 29 nmole mg^{-1} for the tetramer. The ratio of the activity between the two enzyme forms is very similar for the two amino acids.[108]

Udenfriend's group determined that the value for V_{max} for L-tryptophan is only about 3% of the value obtained for L-phenylalanine[249] using rat liver enzyme, again with 2-amino-4-hydroxy-6,7-dimethyl-5,6,7,8-tetrahydropteridine as the cofactor.

Of several tetrahydropteridines tested with rat liver enzyme, tetrahydrobiopterin was the most active as determined from the values of the observed V_{max}, and in decreasing order come 2,4-diamino-6,7-dimethyl-5,6,7,8-tetrahydropteridine, 2-amino-4-hydroxy-6,7-dimethyl-5,6,7,8-tetrahydropteridine, and 2-amino-4-hydroxy-5,6,7,8-tetrahydropteridine, with about 15–30% of the activity observed with the natural cofactor.[8] A similar study using *Pseudomonas* enzyme has shown that with substituted 2-amino-4-hydroxy-5,6,7,8-tetrahydropteridines the most active as measured by the V_{max} of the enzyme is the 6,7-dimethyl analogue, with the 6-trihydroxypropyl analogue (tetrahydroneopterin) slightly less active. The 6-hydroxymethyl analogue shows about 70% of the activity observed with the dimethyl analogue. The side chain esters of the latter compound with phosphate or pyrophosphate have about 50% of the activity of the parent hydroxymethyl analogue. 2-Amino-4-hydroxy-5,6,7,8-tetrahydropteridine has about 25% of the activity of its dimethyl analogue, and the 6-carboxy analogue has about 15% of the activity observed with the dimethyl analogue.[132]

Experiments with α-tocopheryl phosphate demonstrated that this compound causes a decrease in the apparent V_{max} of the hydroxylase obtained from rat liver by about 10%, an effect that is considered to be statistically significant, although under the normal conditions of the assay an activation is observed because of a simultaneous decrease in the K_m for phenylalanine, and this more than counterbalances the decrease in V_{max}. In fact, this decrease in V_{max} appears to be an artefact arising from an increase in the K_m for the cofactor, which was not used in a sufficient excess in these experiments.[39] A similar set of correlations has been observed when short-chain alcohols such as ethanol are added to the reactants, but in this case the increase in V_{max} appears to be genuine.[40]

5 Optimum pH

This parameter has been measured by many workers in several species, and there is a consensus that the optimum pH for several mammalian species lies close to neutrality.[43, 46, 57, 282, 296, 302] In man the optimum is broad, the activity being maximal either at pH 7 or fractionally to the alkaline side of this pH.[49, 56, 296] The measured value varies slightly with the buffer used. The human foetal enzyme appears to have the same optimum pH as the adult enzyme.[49]

Cultures of human foetal fibroblasts contain a phenylalanine hydroxylase whose activity rises steadily over the pH range 6.3–8.3, which suggests that it may be different from the liver enzyme.[48]

The monkey (*Macaca irus*) liver enzyme has an optimum at pH 7.6.[57]

Rat liver enzyme also has an optimum which has been variously measured as pH 5.8–6.8,[178] 7.0,[43, 228, 282] 7.5,[19] and 7.5–8.0.[40] Addition of 7.5% ethanol to the enzyme is stated to lower the optimum pH of rat liver enzyme from 7.5–8.0 to 7.2.[40] This effect appears to be reversible, since ethanol is used in the enzyme preparation.

Addition of the activating protein to the hydroxylase results in a steady increase in activity over the pH range of 5.8–7.4, with a small sharp optimum superimposed on the activity curve at pH 7.6–8.2, whereas the same workers found that without the activating protein the optimum was broad, lying between pH 5.8 and 6.8.[178] The three tetrameric isozymes isolated by Bessman[14] all have the same optimum pH, that is reported to be 7.2. Rat liver and kidney enzymes have the same optimum pH, reported to be 7.0.[228]

Spinach enzyme differs markedly from the mammalian enzymes in that it has an optimum at pH 4.2,[234] but the *Pseudomonas* enzyme has an optimum at pH 6.8–7.3.[129]

6 Stability

Early work on phenylalanine hydroxylase indicated that the rat enzyme was highly unstable. A loss of more than 50% of the activity was observed when it was stored overnight, either frozen at −15 °C, or at 0 °C.[224] Since such a loss in activity is hardly consistent with a viable enzyme at physiological temperature (the rate of enzyme inactivation usually increases about 20-fold for every 10 °C temperature rise, and this would imply an extremely rapid loss of activity at physiological temperature), it is not surprising that the enzyme is not as unstable as these figures indicated. The loss in activity was in fact due to the destruction of cofactor, which had not at that time been identified.

More recent studies have shown that the rat liver enzyme is reasonably stable during storage, but the *Pseudomonas* enzyme is more stable than the rat liver enzyme, which makes it more suitable for use as a reagent, e.g. in the enzymic assay for the pteridine cofactor.[137]

At 50 °C, rat liver enzyme loses 50% of its activity in 4 min.,[99, 130, 228] and the rate of loss is increased by the addition of Fe^{2+}, but the rate is reduced by the addition of Fe^{2+} together with phenylalanine. Under the same conditions the *Pseudomonas* enzyme loses about 50% of its activity in 7 min. Its stability is decreased by Cd^{2+}, and is increased slightly by Hg^{2+}, but the stability is not altered significantly by Fe^{2+} or Cu^{2+}.[130] At 5 °C in the whole animal, guinea-pig enzyme loses 30–40% of its activity in 24 h., but homogenates containing human, mouse, rat, and guinea-pig enzymes lose relatively small but measurable amounts of activity in a week when stored frozen.[304] Rat kidney enzyme is as stable to heating as the liver enzyme, at least in the adult.[228]

The rat liver enzyme is protected against inactivation both by phenylalanine and some nucleosides at 55 °C.[45] The effective nucleosides include adenosine, cytidine, and guanosine. Most, but not all, of the effective compounds are inhibitors, which probably means that they are protecting the active centre of the enzyme. Dithiothreitol renders the enzyme more susceptible to heat inactivation at low concentrations of phenylalanine, but stabilizes it at higher concentrations of this compound. Mercaptoethanol and thioglycollate act in a similar manner, but α,α-dipyridyl, EDTA, and 8-hydroxyquinoline, which are also inhibitors, do not protect the enzyme from inactivation.

7 Temperature coefficient

This parameter for the rat liver enzyme has been found to lie between 4 and 5 for each 10 °C rise in temperature.[220] This is considerably higher than for most organic reactions, whose value usually lies close to 2. This value indicates that the activation energy for the hydroxylation is unusually high, and is in excess of 25 kcal. Murthy, however, obtained a value near 2 for this parameter.[227]

8 Structure relative to other enzymes

An antiserum to phenylalanine hydroxylase has been prepared by injection of the enzyme into rabbits[10] or sheep.[99, 100] The rabbit antiserum precipitates the rat hydroxylase but not the human hydroxylase.[10] The sheep antiserum that is prepared with the rat liver hydroxylase cross-reacts with phenylalanine hydroxylase from guinea pig, mouse, monkey, and human liver, and rat kidney, but it does not react with the *Pseudomonas* enzyme.[100] This indicates that the structure of the hydroxylase obtained from mammals must be similar, although not necessarily identical, whereas the structure of the *Pseudomonas* enzyme differs appreciably from these. This conclusion is predictable on evolutionary grounds. Although some evidence was accumulated to suggest a possible cross-reaction with tyrosine hydroxylase isolated from adrenal and brain,[100] an antiserum prepared in sheep to beef adrenal tyrosine hydroxylase does not cross-react with mammalian phenylalanine hydroxylase.[199] In addition, antiserum to phenylalanine hydroxylase does not cross-react with brain tryptophan hydroxylase.[100] Thus these three hydroxylases in mammalia do not appear to be minor modifications of a parent enzyme, but appear to be structurally distinct.

I ACTIVATION

When lysolecithin is added to a solution of the rat liver hydroxylase it activates the enzyme by a factor of 50-fold when the natural cofactor is used, but the degree of activation is negligible when the synthetic cofactor 2-amino-4-hydroxy-6,7-dimethyl-5,6,7,8-tetrahydropteridine is used.[85] A similar but not so marked activation is observed in the presence of the natural cofactor with the human hydroxylase according to some workers,[98] but others have found a slight inhibitory effect instead.[296] The kinetic changes that have been reported to be caused by this activation are described in section B above. It appears that during the activation lysolecithin unmasks a sulphydryl group, and by so doing it probably alters the conformation of the enzyme molecule.[88] In addition to this effect lysolecithin causes a tyrosine-dependent oxidation of the cofactor that is catalysed by the hydroxylase,[87] which could be mistaken for an authentic activation of the enzyme if the reaction rate were being measured by the rate of oxidation of the cofactor.

Several other lipids act in a similar way to lysolecithin, but not so effectively.[85] These include phosphatidylserine, sphingomyelin, and lysophosphatidylserine, but phosphatidylethanolamine and lysophosphatidylethanolamine are inactive.

α-Chymotrypsin has an effect that is very similar to the effect of lysolecithin, a process that appears to be mediated by proteolysis.[88] It is possible that proteolysis of the hydroxylase may be an important mechanism for controlling the *in vivo* activity of the enzyme. If this is in fact the case an active fragment of the enzyme might be expected to be present in the liver, with a molecular weight equivalent to that formed by the action of chymotrypsin on the purified enzyme. A fraction of this type does not appear to have been isolated from liver. The lysed enzyme shows catalytic activities that are similar to those of the lysolecithin-treated hydroxylase.[87]

The *Pseudomonas* enzyme is activated by several cations at low concentration. These include Hg^{2+}, Cd^{2+}, Cu^{2+}, and Fe^{2+}, but not Al^{3+}, Ba^{2+}, Co^{2+}, Ca^{2+}, Fe^{3+}, Mg^{2+}, Sn^{2+}, Sn^{4+}, Zn^{2+}, Pb^{2+}, Mn^{2+}, Ni^{2+}, Cr^{2+}, or Mo^{6+},[182, 198] in contrast with inhibition by several of these ions at a higher concentration, presumably by denaturation.[129] The pattern of this activation closely resembles the removal of an inhibitory chelating agent by the addition of cations that bind to the chelating agent,[115] but there is no evidence to suggest that this is in fact the mechanism involved. Of these ions it is considered that Fe^{2+} is essential for the enzyme to function.

The effect of Fe^{2+} is not straightforward. At low Fe concentrations, addition of *p*-hydroxyphenylpyruvate or homogentisate stimulate the enzyme, whereas at high Fe concentrations these acids are inhibitory.[197] The activation by Fe^{2+} is inhibited competitively by the amino acids tyrosine, phenylalanine, and tryptophan.

Low concentrations of ascorbate in the presence of iron activate the rat liver enzymes.[92] It has been claimed that under these conditions the rate of non-enzymic hydroxylation of phenylalanine is negligible. Other workers find that stimulation of the hydroxylase occurs in the presence of 0.4 mM ascorbate with or without added iron, but that ascorbate does not activate at a concentration of 4 mM.[105] It has been claimed that the effect of ascorbate is to stimulate the recycling of the cofactor,[265] but this does not preclude the possibility that it may in part protect a component of the system from oxidation by peroxide formed by oxidation of the cofactor. Ascorbate can be oxidized with the formation of peroxide;[106, 107] this may explain why the higher concentration of ascorbate (described above) fails to stimulate the hydroxylase activity, with stimulation of the hydroxylase by ascorbate balanced out by inactivation by peroxide. Similarly, dithiothreitol has been found by some workers to stimulate the enzyme when 2-amino-4-hydroxy-6,7-dimethyl-5,6,7,8-tetrahydropteridine is used as the cofactor, but not when the natural cofactor is used.[157] This is again due to regeneration of the cofactor, and it may also protect the enzyme against inactivation by destroying the peroxide formed as a byproduct when the synthetic cofactor is used.[43] Some workers, but not all, claim that dithiothreitol inhibits the reaction at high concentration.[9, 43] This compound is often used in assays of the enzyme in order to regenerate the cofactor.[43, 81]

Catalase has also been found to activate the hydroxylase.[119, 157] There are two views about this observation, the obvious one being the destruction of peroxide.[43] It has also been suggested that activation by catalase or peroxidase involves a different mechanism that resembles the action of the protein stimulator, which is described in section F.[108]

When rats are exposed to benzene vapour the activity of phenylalanine hydroxylase is increased by 35%. Probably this effect is due to an increase in the synthesis of the enzyme.[110]

Irradiation of chicks with X-rays (600–800 rad) 2 h. before death does not bring about any alteration in the activity of phenylalanine hydroxylase.[153]

The administration of γ-aminobutyric acid to dogs brings about an increase in the activity of phenylalanine hydroxylase.[47]

Cortisol or cortisone can activate the hydroxylase, according to some workers.[92] A study on the factors involved in this effect indicate that the adrenal is responsible for an important control mechanism.[194] In adrenalectomized animals the hydroxylase is markedly activated after an injection of cortisol, whereas in this study its activity decreased sharply after a cortisol injection in normal animals. It has been postulated that there are unidentified inhibitory and stimulating factors in the adrenal gland.

The administration of tetraiodoglucagon to rats for several days prior to a loading dose of phenylalanine led to a higher concentration of tyrosine and a reduced concentration of phenylalanine in the tissues of these animals as compared with a group of control rats. These results suggest that this modified hormone increased the activity of phenylalanine hydroxylase,[102] although it is possible that the results might be explained by a change in the permeability of the cell membranes. It is more likely that an increased dihydropteridine reductase activity is responsible. Brand and Harper[35, 36] have found that glucagon increases the rate of phenylalanine hydroxylation, and an increase in dihydropteridine reductase activity appeared to be responsible for this effect.

Tyrosine has no significant effect on the hydroxylase activity in normal rats,[194] whereas it causes an increase in activity when it is administered to adrenalectomized animals.

J INHIBITION

There have been two main purposes underlying the studies on inhibition of phenylalanine hydroxylase. The first has been to discover compounds that can be administered to animals which inhibit phenylalanine hydroxylase *in vivo*, and when they are administered in conjunction with a diet that is supplemented with high concentrations of phenylalanine cause a considerable rise in the tissue concentration of phenylalanine, thereby yielding a biochemical model of phenylketonuria. The second has been to obtain information about control and reaction mechanisms.

Some studies have been singularly unsuccessful in identifying inhibitors. The inactive compounds that have been studied *in vivo* include histidine and ephedrine.[270] Many phenylalanine analogues also fail to bring about inhibition,[72] but some phenylalanines substituted with halogens do cause some inhibition at a relatively high concentration.[68, 184] Cortisone has been found to be ineffective in causing inhibition by some workers, nor will it activate the enzyme,[270] in contrast to the result described in the preceding section. *In vitro* studies on a series of alkylating agents whose structure is based on the phenylalanine molecule, such as 4-bromoacetylphenylalanine, found that these compounds were ineffective in causing inhibition of the hydroxylase.[73, 75]

Several chelating agents, such as α,α-dipyridyl and 1,10-phenanthroline inhibit the rat liver enzyme at low concentration. Since these compounds chelate with iron in particular this inhibition was to be expected if the claim is correct that the hydroxylase is an iron-containing enzyme. Woolf's group has found that the rat liver enzyme is inhibited by both iron- and copper-chelating reagents,[108] although this does not necessarily prove that the enzyme contains both these ions, since chelating agents are not absolutely specific in their action. Hare *et al.*[145] have examined the effect of azadopamine analogues on the rat liver hydroxylase. They observed an inhibition that was uncompetitive relative to the substrate, and non-competitive relative to the cofactor. These workers have concluded that the inhibition was due, at least in part, to chelation with Fe^{2+} in the enzyme.

α-Methyltyrosine and α-methylphenylalanine are in fact very potent inhibitors of the hydroxylase.[118, 227] In the case of α-methylphenylalanine it would therefore appear that the α-methyl group causes the molecule to present itself at the active centre with the wrong configuration for hydroxylation, but in that configuration it will still bind firmly at the active centre. Since the methyl ester of α-methyltyrosine is a relatively poor inhibitor it would appear that the carboxyl group of the inhibitor is essential for the binding of the inhibitor, and the same probably applies to the substrate.

A range of cations has been found to inhibit the hydroxylase. These include Cu^{2+} for the mammalian enzyme,[172, 293, 302] and Ca^{2+}, Al^{3+}, Cd^{2+}, Zn^{2+}, Sn^{2+}, and Co^{2+} for the *Pseudomonas* enzyme, both before and after activation by incubation with Fe^{2+}.[129]

Several thiols have been reported to be inhibitors of the mammalian hydroxylase, including dimercaprol (BAL)[105, 172] and dithiothreitol,[43] although the latter claim has been disputed.[9] Inhibition is not caused by lipoic acid, lipoamide, dihydrolipoic acid, or 2-mercaptoethanol.[105] All the above inhibitions were observed with relatively high concentrations of the inhibitor.

Spinach leaf enzyme is inhibited strongly by 2-mercaptoethanol and BAL, but less strongly by cysteine.[234]

Many catechols are inhibitors of the mammalian enzyme. Aesculetin is one of the most potent inhibitors, and many of its analogues, particularly those substituted in the 4-position are as potent as, and sometimes more potent inhibitors than, the parent compound.[74] Some of these compounds have been found to be active inhibitors *in vivo*. Many other catechols inhibit to some extent, including 2,3-dihydroxynaphthalene, 1,2-dihydroxyanthraquinone, L-DOPA,[72] α-substituted dopacetamides, 3,4-dihydroxyphenylacetic acid, caffeic acid, α-methyl-DOPA, dopamine, and noradrenaline.[145, 219, 251] Noradrenaline inhibits competitively with the pteridine cofactor and non-competitively with phenylalanine. It also inhibits by a time-dependent process that is prevented by reducing agents, chelating agents, and aromatic amines. This inhibition is not prevented by catalase, Fe^{2+}, or mercaptoethanol, but it is partly reversed by ascorbate. This process requires the presence of oxygen, and may involve, for instance, the formation of a quinone from noradrenaline, which then tans the hydroxylase, thereby rendering it inactive.[44]

Colloidal melanin is an inhibitor.[233]

Some folic acid antagonists have been found to inhibit the hydroxylating system; these include aminopterin and amethopterin,[172, 183, 252] and are effective at a concentration of about 10^{-4} M. It appears that this inhibition occurs at the NAD stage in the hydrogen transport system.

A number of other compounds has been shown to be inhibitory. These include hydroxylamine,[302] dodecyl sulphate,[39, 296] and α-tocopherol,[296] although some workers[39] claim that, on the contrary, α-tocopheryl esters stimulate the hydroxylase. Azide, cyanide, arsenite, β-2-thienylalanine, cyclohexyl-*dl*-alanine, pyruvate,[282] *p*-chloromercuribenzoate,[105] and a number of phenylalanine metabolites and analogues, including homogentisic acid, hydrocinnamic acid, lactic acid, phenyllactic acid, phenylpyruvic acid, and *p*-hydroxyphenylpyruvic acid, are also inhibitory, but usually at a moderately high concentration. Addition of the cofactor reverses the inhibition caused by several of these compounds, which suggests that they react with the cofactor binding site. These inhibitions occur at a concentration that is highly unphysiological, and therefore these inhibitions probably have no clinical or biochemical significance *in vivo*.[206]

Persulphate is an inhibitor, but it acts by oxidizing the cofactor.[293]

Iodoacetate does not inactivate the hydroxylase, so it is thought unlikely that there is a sulphydryl group at the active centre, and therefore the inhibition brought about by heavy metals does not involve binding to the active centre by a sulphydryl group.[293]

Purified human foetal enzyme does not exhibit any substrate inhibition, although when the enzyme is not quite pure a small degree of inhibition is observed.[296] The spinach enzyme is, however, inhibited by the substrate at concentrations above 1 mM.[234] Isolated perfused liver inhibitor constant for phenylalanine is 10 mM,[296a] which is much higher than the *in vitro* studies.

As one might expect, the different substrates mutually compete to produce an inhibitory effect.[62,192] For instance, several analogues of phenylalanine, e.g. *p*-chlorophenylalanine, inhibit the hydroxylation of tryptophan. The addition of 2-amino-4-hydroxy-6,7-dimethyl-5,6,7,8-tetrahydropteridine causes a reduction in the degree of inhibition caused by the above phenylalanine analogues, both *in vitro* and with liver extracts from animals previously treated with these analogues. These results suggest that this inhibition is not entirely due to direct competition between different substrates.

Oxygen causes inhibition of the hydroxylase, possibly by means of competition with one of the other reactants.[86] Another process occurs in the presence of the synthetic cofactor, which appears to involve the oxidation of the cofactor. This inactivation is prevented by catalase, and the addition of dithiothreitol to a preparation of liver also protects the enzyme.[158] The mechanism by which this inactivation occurs is to be distinguished form the competitive inhibition that is observed with some synthetic pteridines.[8]

Rainbow trout fed with a diet that contains dieldrin show a decrease in the activity of liver phenylalanine hydroxylase. The loss in activity is not linear with the dose, but rather it appears to follow a logarithmic relationship. The mechanism involved in this process has not been identified, but it seems likely that a decrease in the synthesis rate for the enzyme is involved.[214]

The mammalian enzyme is inhibited by an unidentified compound that is found in the nucleo-mitochondrial fraction isolated from cells.[208]

K THE PHYSIOLOGY AND PATHOLOGY OF PHENYLALANINE HYDROXYLASE

1 Genetic lack of phenylalanine hydroxylase

L-Phenylalanine is an essential amino acid, and besides its function as a precursor (via tyrosine) of the adrenal and related hormones, and the thyroid hormones, it is an essential structural component of virtually every protein. It is present in proteins to an extent of about 5% of the amino acid residues, with only comparatively small variations from that figure. Thus if an adult human has a protein intake of about 50 g d^{-1}, this will yield about 3 g of L-phenylalanine after the protein has been digested. This is, of course, greatly in excess of any requirement for protein synthesis and the biosynthesis of hormones. From the point of view of the economy of an animal, it is necessary that this excess should be utilized efficiently for other purposes, for instance, in the production of energy. Furthermore, excessive amounts of phenylalanine that may be retained as such in the tissues are, directly or indirectly, toxic to the central nervous system. This became obvious when Fölling examined two siblings who were mentally retarded, and who were found to excrete large amounts of phenylpyruvic acid in their urine.[91] The structure of this compound is so similar to that of phenylalanine that it was immediately obvious that it arose from this amino acid. It is now known that it is formed from phenylalanine by a transamination reaction because of the high tissue concentration of phenylalanine in the tissues. Subsequent studies confirmed the association between the biochemical picture and the mental retardation. It is now clear that the mental retardation is caused by the high tissue concentration of phenylalanine. One of the mechanisms by which this occurs is the incomplete formation of myelin in the brain during development, a process which appears to be largely irreversible. This subject is fully described by K. Blau, Chapter 3 of this volume, which considers phenylketonuria and allied disorders. Clearly, in this clinical condition phenylalanine hydroxylase is not able to rid the body of the excess phenylalanine that is normally ingested, which is the primary function of this enzyme. In principle, this deficiency in the enzyme system might be caused by a failure of the enzyme to function because it is inactive, either because it is not synthesized or because it is present in an inactive form, or because the substrate is physically separated from the enzyme, for instance, by a permeability barrier. The enzyme could be inactive, either because it is essentially inactive because of a structural defect, analogous to the defect in haemoglobin S, which, because of the alteration of one amino acid residue in the protein chain, has properties that are sufficiently different from those of normal haemoglobin that it causes sickle cell anaemia, or because of the absence of another essential component in the hydroxylase system, such as the cofactor or dihydropteridine reductase.

Studies on liver from phenylketonurics showed at an early date that there was a defect in the enzyme system.[225, 281, 289] Addition of phenylalanine hydroxylase that had been isolated from rat liver to phenylketonuric liver extracts demonstrated that the cofactor and hydrogen transport coupling systems were viable, including the cofactor and dihydropteridine reductase. Although the addition of the hydroxylase to phenylketonuric liver extracts demonstrated that the cofactor system was fully active, there was no detectable hydroxylation in the absence of added hydroxylase. In more recent years further studies failed to detect any significant hydroxylase activity in phenylketonuric liver[164] until an assay which possessed a high degree of sensitivity was used by Kaufman's group.[101] They found that in one biopsy specimen of phenylketonuric liver the activity was 0.27% of that observed for a normal liver.

Bartholomé *et al.*,[17] using a normal assay procedure, were able to confirm the absence of the hydroxylase in 13 cases of phenylketonuria. In one further case, there appeared to be an activity similar to that in hyperphenylalaninaemia; this could have been due to an error in differential diagnosis.

A study on plasma,[148] in which very low but detectable amounts of the hydroxylase were measured in a series of normal control plasmas, failed to detect any significant amount of the enzyme in plasma from phenylketonurics.

Kaufman[181] has examined some of the properties of phenylalanine hydroxylase from a phenylketonuric with some residual hydroxylase activity. He found that V_{max} for the enzyme was about 0.3% of that for normal enzyme, and its K_m for phenylalanine was within the range for normal liver enzyme. Lysolecithin was much less effective in stimulating the phenylketonuric enzyme than normal enzyme; the difference was a factor of about four.

The effect of antiserum is not the same for both enzymes; the normal enzyme was inhibited by 63% under conditions that inhibited the phenylketonuric enzyme by only 18%. This confirms the above impression that there are structural differences in the phenylketonuric enzyme that are sufficiently marked to alter the effects of antiserum.[181]

Cotton and Danks[59] have isolated an inactive protein from phenylketonuric liver by a procedure that normally yields active enzyme. There is some doubt about these results, since there was, apparently, no attempt to confirm by independent means that the enzyme examined was, in fact, the enzyme sought.

Hoffbauer and Schrempf[147] have cultured fibroblasts from normal subjects and phenylketonuric patients. The control fibroblasts showed a measurable trace of activity, 45 pmol h^{-1} (mg protein)$^{-1}$, whereas four phenylketonurics showed a mean activity of only 8 pmol h^{-1} (mg protein)$^{-1}$. The variability in the assay was such that this figure did not differ significantly from zero. More recent work has not been able to confirm the presence of the hydroxylase in fibroblast cultures.[50, 53] This does not mean that the original work was incorrect; tissue culture is often subject to subtle changes in conditions of culture. It is probable that a factor of this sort is operating here. Therefore, at present it is not possible to diagnose phenylketonuria *in utero* by culturing fibroblasts from amniotic fluid.

The results of Grimm *et al.*[119, 120] require some comment. They found that the hydroxylase activity in phenylketonuric liver was only slightly lower than in normal liver in the absence of any added pteridine cofactor, but the addition of cofactor (2-amino-4-hydroxy-6,7-dimethyl-5,6,7,8-tetrahydropteridine) brought about a marked increase in the activity detected in normal liver, which is in agreement with the results observed by other workers, but in phenylketonuric liver extracts the cofactor caused a decrease in the rate of hydroxylation. It must be considered likely that the low residual activity measured in the phenylketonuric liver was an artifact that was caused by the conditions of the assay, e.g. a non-enzymic hydroxylation, whereas these workers consider that their results with the phenylketonuric liver were valid.

Bessman's group has examined several phenylketonuric livers for hydroxylase isozymes.[238, 239] At the last count they had examined 10 phenylketonuric livers. Three of them show no active isozymes, four have isozyme π, one has κ, and two have υ, whereas in a group of normal subjects seven possessed all three isozymes, three were missing isozyme κ, one lacked π, and one lacked both π and υ.

Kaufman's group has prepared an antiserum to rat liver phenylalanine hydroxylase, and they have tested it against extracts of phenylketonuric liver.[99] Its potency was sufficient to produce a precipitin line with phenylalanine hydroxylase at 10% of the normal liver concentration of the hydroxylase. No precipitin line was observed, which rules out the possibility that in phenylketonuria the liver contains an inactive, mutant hydroxylase whose antigenic properties are identical with those of the normal enzyme.

Later, Bartholomé and Ertel[15] used an antiserum prepared by injecting monkey liver enzyme into rabbits. They found that at least two precipitin lines formed with an extract from phenylketonuric liver. These results indicate that, although there appears to be a protein in phenylketonuric liver that corresponds to phenylalanine hydroxylase, it is probably altered sufficiently to change its reactivity towards a specific antiserum.

Allied to phenylketonuria there is another clinical disorder that at one time was mistaken for phenylketonuria, and which is now known as hyperphenylalaninaemia. It is distinguished from phenylketonuria by an elevation in tissue phenylalanine concentration that is not as marked as in phenylketonuria, and the clinical symptoms are therefore not so pronounced. In some cases, indeed, there are no symptoms, besides the elevated tissue concentration of phenylalanine, and the resultant urinary excretion of phenylalanine metabolites that are formed mainly by transamination. Studies on liver from patients with this condition show a hydroxylase activity that lies in the range of 0–80% of the activity in normal liver.[17, 119, 120, 164, 185] Some workers think that they fall into two distinct groups; the first group shows activity of less than about 9%

of the normal activity, with a median value of about 2.5–3%; the second group is smaller, with considerably higher, widely scattered activity. Since the measured activities observed in these liver samples are often far too high to be consistent with the comparatively high tissue phenylalanine concentration that is observed in this condition, some other factor must be present in many of these patients that modifies the kinetics of hydroxylation.

Kaufman[99] determined that the K_m for the hyperphenylalaninaemia phenylalanine hydroxylase was 0.62 mM for phenylalanine, using 2-amino-4-hydroxy-6,7-dimethyl-5,6,7,8-tetrahydropteridine as the cofactor, compared with 1.25 mM for the normal hydroxylase. The K_m of both enzymes was 5 μM for this cofactor. The hyperphenylalaninaemia hydroxylase was stimulated to a lesser extent by lysolecithin than was the normal hydroxylase, and it was more sensitive to heat inactivation than the normal human enzyme. It did not give a precipitin reaction with an antiserum prepared in sheep to rat liver phenylalanine hydroxylase, but instead it exhibited a 50% inhibition of its activity, which suggest that this hydroxylase is somewhat different from the normal hydroxylase, which might occur if there were a mutation that altered one amino acid residue in the hydroxylase molecule.

Studies on some heterozygotes for this condition showed anomalous kinetics for phenylalanine hydroxylation[299] with an apparent very low K_m for phenylalanine (probably below 0.25 mM). It is possible to interpret these results as indicating that the hydroxylase in hyperphenylalaninaemia undergoes a marked substrate inhibition.[300] The kinetics observed in the heterozygotes indicated that one parent was a normal phenylketonuria heterozygote, but the other parent carried one wild-type gene for phenylalanine hydroxylase and one gene for the hyperphenylalaninaemic hydroxylase. This hypothesis would explain why the tissue concentration of phenylalanine can rise so high in this condition. However, the concentration of phenylalanine that is used in the assay of the hydroxylase is often much higher than that found in hyperphenylalaninaemia;[296a] therefore some other explanation may have to be sought to explain why the assayed hydroxylase activity in this condition is often so high, such as a permeability barrier that restricts the access of phenylalanine to the hydroxylase.

Kaufman's group have examined liver biopsy specimens from six parents of hyperphenylalaninaemics.[185] They find that their activity lies in the 7–30% range of normal, and they have produced complex hypotheses to explain their results involving either regulator gene mutations or complex protein–protein interactions. However, they used 2 mM phenylalanine in their assay, which is 20 times higher than the maximum concentration we used in our phenylalanine tolerance tests. In our tests, at 0.1–0.2 mM blood phenylalanine the heterozygotes and normal subjects showed a similar absolute reaction rate, but the normals demonstrated first-order kinetics, whereas some of the heterozygotes showed a very close appoximation to zero-order kinetics. Thus, Kaufman's results are consistent with ours; when an extrapolation is made on our results to his assay concentration (assuming that zero-order kinetics are obeyed for the heterozygotes at higher phenylalanine concentrations, which may not be true), assuming a value for K_m of rather less than 1 mM for the control subjects, an activity of about 10% normal would be expected for the heterozygotes with 2 mM phenylalanine.

The population distributions of hyperphenylalaninaemia and phenylketonuria in Austria are different, which suggests that they are different entities.[271]

In principle, it is possible to treat phenylketonuria by administering phenylalanine hydroxylase in a suitable form, for instance, bound to a larger matrix. A matrix of this type has been prepared by coupling the rat liver enzyme to CH Sepharose-4B,[294] or by trapping on polyacrylamide.[293] This causes only minor alterations to the properties of the enzyme, but it does not improve its stability; in fact it renders the enzyme rather less stable. Therefore, the practical aspects of long-term enzyme transplants (even if it is possible to assemble the whole hydroxylating system) still exhibit considerable problems.

2 Effect of different clinical conditions

Clayton's group has examined the effect of various liver diseases on the activity of phenylanine hydroxylase in children,[210] using a biopsy specimen. In a number of these conditions the hydroxylase activity was reduced at least in some of the cases examined. These include biliary atresia, glycogen storage disease, obstructive jaundice, thalassaemia with splenomegaly, septicaemia with liver abscess, giant cell hepatitis, recurrent hepatitis, and lymphocytosis with splenomegaly. No reduction was observed in cases of histocytotic infiltration of the liver and the lymph nodes, abdominal cyst, hepatosplenomegaly, neonatal hepatitis, and various surgical treatments. With the exception of biliary atresia and a very few of the others only one case of each was examined for a particular disease condition, and in those where more than one case was examined, and in which there was a general decrease in the hydroxylase activity, some individuals showed a normal hydroxylase activity. Because of this biological variation these results cannot be used with any degree of certainty in predicting what might happen in any individual case.

In kwashiorkor the tissue concentration of phenylalanine has been found to be raised, which suggests that there may be an impairment in the phenylalanine hydroxylase activity. This anomaly is rectified by treatment of the condition with a diet that contains an adequate supply of protein. Similarly, when patients with this condition are given a loading dose of phenylalanine their urine contains an abnormally high concentration of phenylalanine metabolites.[71, 260] This is strong circumstantial evidence that there is an impairment in the activity of phenylalanine hydroxylase in this condition.

At least two cases have been described in which an elevated phenylalanine concentration was caused by an absence of dihydropteridine reductase. They showed a developmental regression with seizures. One died in the second year.[181, 223]

A clinical condition has been observed that is similar to phenylketonuria which does not respond to treatment. It is believed that there is a deficiency in the reductase system in this patient.[262] Another atypical case appears to have an intact hydroxylase system, but with a very high blood phenylalanine concentration.[18] The child also showed a deficit in the 5-hydroxyindole pathway. Clinically, the child, who had been treated with a low-phenylalanine diet, improved with a dietary supplement of DOPA and 5-hydroxytryptophan. One may speculate that this child had a defect in transport of phenylalanine, at least in the liver and possibly in the brain.

Tumours can often cause an abnormally low activity of phenylalanine hydroxylase, particularly when there is some liver involvement. Clayton's group found that the activity was slightly reduced in one case of neuroblastoma that showed some liver involvement.[210] Livers from rats bearing rhodamine sarcoma, hepatoma, or nodular hyperplasia showed a marked reduction in the activity of phenylalanine hydroxylase. Animals that had received a transplant of a hepatoma did not, however, show such a decrease in the activity of the liver enzyme. When normal liver was perfused with a solution that was supplemented with tyrosine and tryptophan the hydroxylase activity was observed to decrease. When liver from animals bearing these tumours were similarly perfused the hydroxylase activity did not show such a decrease. The activity measured in the hepatoma tumour is lower than that measured in liver.[261, 269]

Leukaemic rats do not show any phenylalanine hydroxylase activity.[5]

In rat a pronounced folic acid deficiency leads to a decrease (less than 50%) in the activity of phenylalanine hydroxylase cofactor. In man, phenylalanine tolerance tests indicate that amethopterin, a folic acid antagonist, interferes with phenylalanine metabolism,[112] which may be due to interference with the cofactor. Mammalian phenylalanine hydroxylase might be a flavoprotein. Therefore, rats were kept on a riboflavin-free diet to determine the effect on the hydroxylase. It was found that the activity of the liver hydroxylase was decreased by about 50%,[259] which suggests that under these conditions some of the enzyme might be synthesized as an inactive form which lacks a flavin moiety. Further experiments must be carried out to clarify this theoretically important point.

3 Development

For a long time it was thought that phenylalanine hydroxylase develops late in gestation, because studies carried out on human foetal liver showed little or no activity.[188] Later, it became apparent that the natural cofactor is unstable, and these studies were carried out without the addition of any pteridine cofactor.

Most studies on foetal liver extracts fortified with added pteridine cofactor have shown a different picture, which varies from species to species.

With the advent of wide-scale legal abortion (at least within Great Britain) human foetal liver has become readily available for study. Examination of this material for phenylalanine hydroxylase activity has shown the presence of the enzyme in foetal liver as early as 8 weeks gestation,[49, 77, 157, 177, 215, 244, 245, 253] At this age the activity appears to be quite low. In the age range of about 12–24 weeks gestation the activity (per unit weight of liver) is fairly constant, and is at a level that is similar to, but slightly lower than, the adult level, without any significant trend in the activity with age. At birth it appears that the activity may have fallen to about 50% of the adult level, although this result may be an artefact that is caused by delays in assaying the liver enzyme.[244] Bessman finds that the foetal activity is very variable and bears very little relationship to age, except as the foetus approaches birth, when the activity falls to zero. These results underline the problems that Woolf described in obtaining autopsy liver specimens that show any hydroxylase activity.[298]

It would appear that in very rare cases development of the hydroxylase may be delayed. For instance, one child has been described[264] who showed a markedly elevated blood phenylalanine concentration at the age of 1 month, and was therefore treated with a low-phenylalanine diet. Over the next few months the phenylalanine requirement increased to normal, and at an age of 6 months it was found necessary to place the child on a normal diet in order to keep the tissue phenylalanine concentration from falling too low. This was because the phenylalanine requirement had returned to normal. There were no indications that the blood phenylalanine concentration then rose to an abnormal level. Presumably, in the interim the hydroxylase had become active, and the simplest explanation is that synthesis of the hydroxylase had commenced during that period.

It appears that the human foetal enzyme is not to be found in kidney or the intestines. Thus, the renal enzyme appears to have a different development pattern from the hepatic enzyme, and this may mean that there is a different controller gene for the liver and kidney enzymes.[296]

The development pattern for the guinea-pig hydroxylase is similar to that observed in man, in that the hydroxylase can be detected during the later stages of pregnancy. In the final two weeks of pregnancy the activity of the liver enzyme is about 30% of that found in the adult. The activity reaches the adult level at birth.[24, 97]

Phenylalanine hydroxylase activity is present in *Macaca mulatta* liver after 100 days gestation, which suggests that the pattern of development may be similar to that in man. It is also found in the liver from mouse foetus.[24]

The development pattern for the rat liver hydroxylase is quite different from that of the guinea pig or man. The hydroxylase is not present early in pregnancy, but it develops shortly before or at birth. Numerous studies have been carried out on the pattern of development, and, although there are marked discrepancies between different studies in the details of the results, the overall pattern is plain. The discrepancies appear to be partly technical. In some early studies[189, 246] no pteridine cofactor was added, and in these studies little or no activity was observed until several days after birth, probably because there was not enough cofactor in the liver extracts to activate the hydroxylase. In more recent studies, discrepancies are found that are difficult to explain on the basis of experimental error. Differences in diet, strain of animal, or some other biological variation may have been responsible for the anomalies.

No activity is found in rat liver until shortly before birth. In one study[95] some was detected 5 days before birth, but this result was an exception. Most studies have failed to detect any significant hydroxylase activity until about 20 days gestation,[97, 188, 208, 228, 253] and there is some debate about the day on which it is first detected, possibly because of differences in the sensitivity of the assays used. There is a rapid increase in hydroxylase activity at birth. The time of the increase has been shown to be just before and during delivery.[13, 97] The activity may be as high as the adult level at birth, or, according to some workers, as much as 60% higher than the adult activity.[97, 291] Some workers, however, find that the activity in the newborn is appreciably lower than the adult level.[13, 156, 208] According to some publications the hydroxylase activity continues to increase in the first few days after birth.[13, 156, 188] Others have reported a decrease in activity.[157] One group of workers found that the hydroxylase activity decreased until the animals were between 10 and 15 days old, and after that it increased.[191] A peak of activity is observed at about the time of weaning, or, according to one group of workers, at an age of 10 days.[188] At weaning the activity is near the adult level,[95, 157, 188] so the increase from birth to weaning is in fact comparatively small. The course of development beyond weaning appears to be open to controversy. One report states that the activity doubles between weaning and an age of 8 weeks.[78] Another study claims that at an age of 45 days the activity is at the adult level.[191] One publication states that the activity rises steadily to a maximum at an age of 50–60 days, which is followed by a decline, the rate of decrease in activity being more rapid with female rats than with male animals. The activity finally stabilizes with a lower activity in females than in male animals.[95] A further report claims that the activity decreases steadily during the first 10 weeks after weaning, and after that it remains steady.[96]

Clearly, from the preceding results the actual pattern of development of the hydroxylase after birth in the rat is very poorly documented. Until this study is repeated far more rigorously with an adequate number of animals at each age and using litter-mates over the full range of ages to diminish the effects of biological variation, the most that can be said about the development of the rat hydroxylase is that after birth there are no radical alterations in its activity.

The rat kidney hydroxylase has been detected in the foetus 1 day later than the liver enzyme.[228] Thus, the control mechanism for this enzyme may be different from that found in man, and it may be the same as that found in liver.

The foetal rat enzyme is similar to, if not identical with, the adult enzyme. It does, however, appear to be slightly more sensitive to heat inactivation than the adult enzyme.[228]

The activity of the rat hydroxylase is influenced *in vivo* by cortisol at some stages of development, depending on the age of the animal. An injection (e.g. 100 mg (kg body weight)$^{-1}$) causes a 50% increase in the activity of the hydroxylase up to an age of about 10 days. At an age of 15 days cortisol has been found to increase the hydroxylase activity according to some workers, but others have found that it causes a decrease in activity. It appears to have no effect on the activity of the liver enzyme in adult rats,[191, 209, 210] and the same appears to be true for the kidney enzyme.

In domestic poultry the enzyme is first found during incubation of the eggs.[97, 153, 267] In the Vantress-Leghorn variety the hydroxylase can be detected after 12 days incubation[267] and reaches a peak of activity after incubation for 18 days. Three days after hatching the activity is about 10% lower than the pre-hatching peak. This decrease is considered to be greater than experimental error. The activity then remains constant throughout the rest of the first week after hatching.

In one insect (*Calliphora erythrocephala*) the hydroxylase is highly active during the third instar (larval form), and it then decreases during pupation. The activity again increases at the time at which the imago emerges. There are, however, no distinct maxima in its activity.[257]

Studies with cultures of foetal rat liver cells have failed to demonstrate the development of any phenylalanine hydroxylase activity in cultures grown from cells taken from the liver of 19–20-day-old foetuses. In cultures of cells from the liver of 21–22-day-old foetuses the activity that had developed prior to culturing decayed with a half-life of about 6 h., which is consistent with loss of activity due to inactivation, without any replacement of the hydroxylase. This suggests that phenylalanine hydroxylase may develop in foetal liver as a consequence of an extrahepatic signal, which may possibly be hormonal in nature.[277] Studies with cultures and with newborn rats have suggested that cyclic AMP, hydrocortisone, and corticosterone may be involved in the induction of the hydroxylase.[142, 208, 245] Glucagon[37] does not appear to be involved in the neonatal induction of the rat liver enzyme.

4 Circadian rhythm

When rats are kept in the light from 6 a.m. to 6 p.m. a circadian rhythm is observed in the activity of phenylalanine hydroxylase, which shows a maximum activity at about 10 p.m. An increase in the intake of phenylalanine increases the sharpness of the peak activity.[51] A similar rhythm has been observed in male guinea pigs, but this rhythm is sex dependent; it is not found in female guinea pigs.[268] When the guinea pigs are kept in light from 6 a.m. to 6 p.m. the peak of activity is observed at about 4 a.m. and the minimal activity at about 4 p.m. This pattern is not reversed by reversing the lighting schedule, but when the lighting pattern has been reversed for a period of 2 weeks it does cause some disturbances in the absolute level of the hydroxylase activity. As a result of these fluctuations a similar diurnal rhythm has been measured in guinea pigs of phenylalanine and tyrosine concentrations. This rhythm is out of phase with the rhythm of the hydroxylase activity, and its pattern is somewhat complex. This suggests that although the activity of the liver enzyme is an important factor in controlling this rhythm, there are probably other controlling factors, such as the time of eating.

5 The effect of diet

The influence of variations in the diet on the activity of phenylalanine hydroxylase has been extensively studied in rats, as one of the major purposes for altering the diet has been to obtain animals that can be used as a biochemical model of phenylketonuria. This is usually carried out by feeding the animals with a diet that has been supplemented with phenylalanine, far in excess of the normal dietary intake of phenylalanine, with or without the addition of a phenylalanine hydroxylase inhibitor.

The addition of phenylalanine to the diet has been consistently found to cause a decrease in the activity of hepatic phenylalanine hydroxylase when it is administered for a period of several weeks. Deprivation of an adequate supply of this amino acid also has an effect on the activity of the hydroxylase. In one study two groups of animals were fed a diet that contained 9% and 17% of protein (measured as the proportion of calories supplied)[111] for a period of 2 months. At the end of this period the group that received the lower protein intake showed a lower hepatic phenylalanine hydroxylase activity than that found in the other group.

The supplementation of the diet with phenylalanine reduces the activity of phenylalanine hydroxylase, the extent of the effect depending on the amount of phenylalanine added to the diet. In one study there was little or no reduction in the activity when a 1 or 2% phenylalanine supplement was added to the diet.[297] With a 3% supplement there was about a 30% reduction in activity, and the decrease in activity was about 50% when a 5% supplement was used. Several other studies[21, 28, 92, 96, 190, 243, 247, 290] have confirmed these findings, and have shown that when the diet is supplemented with 5–7% of phenylalanine the decrease in hydroxylase activity usually lies between 50% and 75%. The use of these diets impairs the growth of the rats,[297] but this appears to be reversed by adding vitamin A to the diet.[243] The use of this diet impairs the hydroxylase activity, it would appear, for as long as the supplement is used, however long that may be.[96]

Administration of a high-phenylalanine diet that is deficient in vitamins A and E also brings about a decrease in the activity of phenylalanine hydroxylase. Supplementation of this diet with vitamin A appears to have little effect on the activity of the hepatic hydroxylase (expressed as either activity per gram of tissue or per gram of protein); there is an increase in activity with a supplement of vitamin E that appears to be statistically significant, but a supplement of vitamins A and E together has little effect on the hydroxylase activity. These results contrast with those obtained in the absence of added phenylalanine, in that a supplement of vitamin E depresses the activity of the hydroxylase, and vitamin A or a mixture of vitamins A and E increase the activity, and in the latter case the increase is highly significant.[243]

In pregnant rats the hepatic hydroxylase activity was reduced by the addition of 5–7% of phenylalanine to the diet.[29]

The reduction in hydroxylase activity that is brought about by the high phenylalanine intake, combined with the increased amount of phenylalanine that requires to be metabolized, causes the tissue phenylalanine concentration to rise appreciably above the normal range, and this has enabled biochemical studies to be carried out on an experimental model of phenylketonuria which resembles the human disease in some ways. The administration of stanozolol to rats that are receiving a supplement of phenylalanine causes the hydroxylase activity to return to normal,[187] possibly by a mechanism that involves induction.[186] The injection of L-*erythro*-neopterin also prevents an experimental phenylketonuria in rats that were receiving a high intake of phenylalanine, again apparently by increasing the activity of the hydroxylase.[55]

When L-tryptophan or a mixture of L-tryptophan and L-tyrosine at about 3% concentration is added to the diet an increase in phenylalanine hydroxylase activity is observed that parallels tryptophan hydroxylase activity. The extent of the measured increase is about 30%.[92] Tyrosine on its own has little or no effect according to some workers,[92, 190] but others[96] claim that it increases the hydroxylase activity, whereas some have found that it reduces the activity.[292] The further addition of a supplement of phenylalanine reverses the effect of tryptophan.[96] Those workers who find that tyrosine increases the activity of phenylalanine hydroxylase[96] find that the effect of tyrosine and tryptophan are not additive. One study on tumour-bearing animals indicated that an injection of tyrosine and tryptophan together (but not alone) caused a radical reduction in the activity of the hydroxylase.[270]

Some studies have been carried out on pregnant rats, as models to determine the effect of management and mismanagement of maternal phenylketonuria during pregnancy. When pregnant rats were deprived of phenylalanine during the final week of pregnancy the activity of the hepatic hydroxylase in the newborn remained consistently lower than in animals in control litters.[27] When pregnant rats were fed Lofenalac, which is a commercially prepared diet that is restricted in phenylalanine, plus a phenylalanine supplement during and after pregnancy, the hydroxylase activity in the newborn rat livers depended on the amount of phenylalanine added. The highest activity at weaning was observed when a 0.2% phenylalanine supplement was fed, and the activity was lower in animals whose mothers were receiving a higher supplement. After weaning the activity fell slowly to a constant level in all groups that were fed Lofenalac, irrespective of the amount of phenylalanine supplement, whereas in a control group it rose after weaning.[78] These workers reckon that the mothers were suffering from a general dietary deficiency, as well as an abnormal phenylalanine intake, which presumably led to an abnormal composition of their mother's milk.

Although the hepatic enzyme is largely inactivated by feeding large amounts of phenylalanine there is a small but apparently significant increase in the activity of the rat renal and intestinal mucosal hydroxylases on this regime,[292] which again suggests that these enzymes are subject to a control mechanism that is different from the one that controls the activity of the hepatic enzyme.

6 Effect of stress

When rats are stressed by keeping them at an oxygen tension equivalent to an altitude of 6100 m the hydroxylase activity increases by about 60% in 24 h. or more.[235] If, however, the animals are treated with cycloheximide at the time at which the pressure is reduced, the increase in hydroxylase activity is prevented. The mechanism by which cycloheximide acts is not known. The effect of pressure reduction may not be just a response to a stressful environment. It may be due to the reduction in oxygen tension causing an induction of the hydroxylase.

7 Phenylalanine tolerance tests; an indirect assay of phenylalanine hydroxylase

It may be necessary to determine the activity of phenylalanine hydroxylase *in vivo*, and, because of the ethical and practical difficulties involved in obtaining biopsy specimens from human liver, the usual technique for assessing the hydroxylase activity is to carry out a tolerance test in which a test dose of L-phenylalanine is administered. The principle that underlies this test is that when phenylalanine is administered, the tissue concentration of this amino acid is raised. Its concentration can be monitored by measuring its blood concentration at intervals, and the rate at which the concentration returns to normal is a measure of the rate at which phenylalanine is being metabolized. Other metabolic pathways are available to phenylalanine, such as transamination, decarboxylation, and excretion in the urine, but these are quantitatively unimportant compared with hydroxylation when the hydroxylase is intact, and therefore to a good first approximation these tests measure the hydroxylase activity. In general, phenylalanine tolerance tests have been confined to man, and have been used primarily in studies on the relationship between the genetics of phenylalanine hydroxylase and the kinetics of phenylalanine hydroxylation, i.e. how fast phenylalanine is metabolized in normal human subjects, heterozygotes for phenylketonuria and phenylketonuric subjects, and in allied clinical conditions. Similar tests have also been used in studies on such diseases as kwashiorkor.

In early studies[4, 23, 112, 149, 150, 162] the phenylalanine load was administered orally. This was essentially unsatisfactory since intestinal absorption occurs over a period of time that is comparable with the rate constant for the metabolism of phenylalanine, and the rate of absorption also varies from person to person. The pattern of results obtained with these early tests showed a rise in the blood phenylalanine concentration that reached a peak in normal subjects and in heterozygotes about 1–2 h. after the dose was administered. There was a tendency for the peak to be later and higher in the heterozygotes, but individual variations were large in comparison with the difference between the two groups. At the same time an increase in the tyrosine concentration was observed, that was in general lower, and with a later peak concentration in the heterozygote group.

To obviate these problems an intravenous technique has been devised[113, 120] (also L. I. Woolf, B. L. Goodwin, and D. N. Wade, unpublished results), and by the use of this technique values have been obtained for the rate constants involved in the metabolism of phenylalanine in man. In general, it was found that normal subjects showed about twice the phenylalanine hydroxylase activity found in heterozygotes for phenylketonuria when the rate of disappearance of phenylalanine under standard concentration conditions was used as the measure of activity, and the rate of metabolism (not hydroxylation) was very slow in phenylketonurics. In addition, there was a difference between the sexes. For instance, in male subjects the spread of results (measured as the time taken for the phenylalanine concentration to fall half-way to the value measured prior to the dose) was considerably greater than for females among the normal subjects, with the result that it was easier to discriminate between normal subjects and heterozygotes in the female group than in the male group. There were other differences between normal subjects and heterozygotes, some of which made it difficult to be certain that the apparent phenylalanine hydroxylase activity was the true activity. For instance, the volume through which phenylalanine is distributed was found to be very close to the volume of total body water in the case of heterozygotes, but it was about 20% greater than the volume of the total body water in the case of normal subjects. Again, there were differences between the sexes. These results imply that phenylalanine was being concentrated in some organs in normal subjects, but not in the heterozygotes. If the main site of concentration should be the liver (there is no reason to believe that it is) then it is clear that the kinetics of phenylalanine metabolism would be upset by the high concentration in liver cells in normal subjects, but not in heterozygotes. However, the observed kinetics for phenylalanine metabolism fit so closely to the pattern expected for normal subjects and heterozygotes on theoretical grounds that it seems reasonable to believe that the observed kinetics in fact reflect the total phenylalanine hydroxylase activity, and therefore that the site at which phenylalanine is being concentrated is not the liver.

There was a difference between heterozygotes and normal subjects in the rate of release of tyrosine into the blood after a dose of phenylalanine. A mathematical model indicated that the results were consistent with the same rate of absorption for tyrosine into liver in both genetic groups (where it is further metabolized), but there is a considerably smaller rate of release of tyrosine formed from phenylalanine into the bloodstream in the heterozygote group as compared with the normal subjects. In some heterozygotes the rate of release was almost undetectable. At one time phenylketonurics were diagnosed on the basis of a negligible increase in the blood tyrosine concentration after a phenylalanine load. On this basis, some heterozygotes would have been classified as phenylketonurics, despite a normal blood phenylalanine concentration. The amount of tyrosine released into the blood from the liver in all the groups examined is considerably less than the amount expected if the exchange of tyrosine between blood and liver were rapid and complete.

Despite the greater phenylalanine hydroxylase activity in the normal subjects the fasting blood phenylalanine concentration was only slightly lower in the normal subjects than in the heterozygotes. Presumably this reflected maintenance of this parameter by tissue protein hydrolysis, which released phenylalanine into the blood.

The use of a radioactive tracer instead of a large phenylalanine load has been considered as an alternative to the standard phenylalanine tolerance test. Not surprisingly, this has been found to give results that are quite different from a standard test[241] since a tracer dose is affected by other parameters, particularly exchange reactions between the phenylalanine pool and tissue proteins. Since it is not possible to sort out these different reactions, it is not possible to obtain information about the rate of conversion of phenylalanine to tyrosine by this approach. The labelled tyrosine formed cannot be used as a guide, since it is itself rapidly metabolized.

8 Inactivation by *p*-chlorophenylalanine

This reagent has been found useful in developing a biochemical model for phenylketonuria in rats, since when it is administered to animals phenylalanine hydroxylase is largely inactivated. Although the hepatic enzyme is inactivated, the renal and intestinal mucosal enzymes are not appreciably affected,[292] or at least, not to the same extent.[76] Typically, after a single dose of 360 mg of this compound per kilogram body weight the hydroxylase activity decreased to 10% of the control activity after 1 day; at 2 days it was 11% of the control activity; by 3 days it had risen to 21%; and by 6 days it had returned almost to normal.[104,122,227,275] When the activity at 3 days after the dose was used as a measure of the effectiveness of the compound, 30 mg kg^{-1} had little effect, 100 mg kg^{-1} caused 20–40% reduction in hydroxylase activity, 300 mg kg^{-1} caused 60–75% reduction, and 1 g kg^{-1} brought about 80% reduction in activity.[242] Other workers claim that maximal inactivation (>85%) is obtained with a dose of more than about 200 mg kg^{-1} in suckling rats.[76]

The degree of inactivation is different for different isozymes.[14] Both the D- and L- isomers of the compound are effective in causing inactivation.[275] The time course for inactivation is similar to that observed after the administration of ethionine or puromycin, which inhibit protein synthesis, and this suggests that protein synthesis is involved in this inactivation of phenylalanine hydroxylase.[122] Administration of cycloheximide reduces the degree of inactivation.[104] If treated rats are then injected with phenylalanine the activity of the hydroxylase is increased by $2\frac{1}{2}$-fold in 9 h., both in adult and suckling rats, whereas control rats receiving a dose of phenylalanine do not show this effect.[117] There is an increase in activity in treated rats that are then injected with cortisol, and the effects of cortisol and phenylalanine are additive, so they appear to act by different mechanisms. This is further confirmed by the action of actinomycin, which deletes the effect of cortisol, but has no effect on the activation that is brought about by phenylalanine.

Similarly, rats receiving a dietary supplement of *p*-chlorophenylalanine for 30 days show a larger decrease in hydroxylase activity than those receiving an additional supplement of phenylalanine.[227] The kidney enzyme shows a similar pattern, although at a lower level of activity.

Protein that has been isolated from rats after the administration of *p*-chlorophenylalanine yields this compound on hydrolysis,[104] and disc gel electrophoresis of the hydroxylase shows an abnormal pattern. These results all suggest that *p*-chlorophenylalanine is incorporated into the enzyme molecule to produce an altered enzyme with a low or negligible hydroxylase activity. This is probably not the only effect. Some workers are of the opinion that the locus at which *p*-chlorophenylalanine acts is either the synthesis or the degradation of phenylalanine hydroxylase.[117] It does appear certain, however, from the slow rate of recovery in activity when *p*-chlorophenylalanine administration is stopped that new hydroxylase must be synthesized to replace whatever is missing during treatment with this compound.

3-Amino-1,2,4-triazole, an irreversible inhibitor of catalase, does not affect the activity of the hydroxylase *in vivo*, but when it is administered with *p*-chlorophenylalanine it potentiates the effect of this amino acid.[38] Since halogenated phenylalanines interact with phenylalanine hydroxylase to yield hydrogen peroxide,[266] and peroxide inhibits the enzyme,[158] this result suggests that the mechanism by which *p*-chlorophenylalanine functions is to produce hydrogen peroxide at or near the active centre of the hydroxylase, and thus inactivate it.

This mechanism might also explain some of the abnormalities in the structure of the hydroxylase obtained from animals treated with *p*-chlorophenylalanine.

The inhibitory effect of aesculetin *in vivo* is additive to the effect of *p*-chlorophenylalanine, so they appear to function by different mechanisms.[284]

In contrast, the effect of *p*-chlorophenylalanine *in vitro* is quite small.[6, 68, 198] In the standard assay using 2-amino-4-hydroxy-6,7-dimethyl-5,6,7,8-tetrahydropteridine as the cofactor the value for K_i is 1.5 mM, but if the natural cofactor is used *p*-chlorophenylalanine binds far more firmly.[6] Although this may account for some of the difference between the *in vivo* and the *in vitro* results it would not appear to account for the whole difference.

m-Fluorophenylalanine has been compared with *p*-chlorophenylalanine.[193] It is a potent lethal agent and convulsant in rats. Its action is antagonized by phenylalanine hydroxylase inhibitors such as *p*-chlorophenylalanine, *p*-fluorophenylalanine and α,β,β-trimethyl-DOPA. For this reason phenylalanine hydroxylase has been suggested as the site of toxicity, and the toxic agent may be *m*-fluorotyrosine. These workers did, however, find discrepances between the activity of the hydroxylase and the clinical effects of this compound in quantitative terms.

Another method has recently been developed to produce a model for phenylketonuria in rats. Suckling rats are given a high phenylalanine diet that contains α-methylphenylalanine. When this compound is administered at a level of 12 μmol (10 g body weight)$^{-1}$ the hydroxylase is inhibited by 75%.[118]

9 Tissue culture studies

In the last few years numerous studies have been carried out with cultured cells. Cultures of rat hepatoma show phenylalanine hydroxylase activity in some clones, but not in others. Although the number of clones examined is relatively small it would appear that most clones do not contain phenylalanine hydroxylase, which could be an explanation for the relatively low hydroxylase activity in hepatoma tumour. Stimulation of this enzyme is brought about by hydrocortisone, cortisone, dexamethasone, and N^6, $O^{2'}$-dibutyryl-3′,5′-cyclic AMP, an effect that is prevented by cycloheximide, which indicates that *de novo* synthesis of the hydroxylase is involved in the stimulation.[142, 218]

Clones that contain hydroxylase activity can be isolated by incubation of cells with a modified Ham's F-12 medium that is free from tyrosine. Since tyrosine is an essential amino acid in the absence of phenylalanine hydroxylase, only those cells that contain this enzyme are able to grow. By this approach one group of workers[141] has been able to demonstrate the presence of the hydroxylase in the cell lines H4-II-E-C3 and MH_1C_1, but none in the lines HTC or BRL. Other workers have demonstrated activity in Reuber H4 and Cloudman S91 cells,[41, 217] and in mouse mastocytoma cells.[53]

The measured amount of hydroxylase activity depends greatly on the condition of the culture, in particular the cell density in a monolayer culture. When Reuber hepatoma cells are subcultured there is a very rapid drop in activity, and this reduced level is maintained until the cells reach confluency. After this the specific activity increases to a high value.[204, 205] The activity approaches an exponential function of the cell density. Immunological tests with an antigen to phenylalanine hydroxylase show that the amount of protein per cell that reacts with the antigen is constant, irrespective of the cell density. This suggests that there may be an inactive form of the enzyme in the non-confluent cells.[205]

If serum or glucocorticoids are withdrawn from a culture of H4-II-E-C3 cells that has a high activity of hydroxylase, the activity decays with a half-life of 22 h., and the activity levels off after about 60 h. This effect is reversible: depending on the amount of activator used, the activity is restored to its high value over a period of about 24 h. It appears that only a part of the activation that is brought about by the addition of serum to the culture is caused by glucocorticoids. The remainder of the activation appears to be due to a high molecular weight material that possesses a moderate degree of stability to heat. Several other hormones were tested for stimulatory properties, but none of them was able to substitute for this unidentified material. This stimulation only takes place when there is ongoing protein synthesis. Thus, stimulation is blocked by the addition of 3-(2-(3,5-dimethyl-2-oxocyclohexyl)-2-hydroxyethyl)glutarimide.[144]

Tourian[279] has found that calf serum or insulin can induce the hydroxylase. The effect of these two agents is additive, so it is probable that they act by different mechanisms. The induction is biphasic. The first phase lasts about 3 h. and is inhibited by cycloheximide, but not by actinomycin. The second phase (12–24 h.) is inhibited by both these compounds.[279]

The physical properties of the hydroxylase that is obtained from cultures of hepatoma H4-II-E-C3 cells are very similar to those of normal rat liver enzyme.[140] There is no difference between them in K_m for the substrate or cofactor, but there are minor immunological differences. Tourian has found one isozyme in this culture with an isoelectric point of pH 5.20, which is the same as that for one of the isozymes he obtains from rat liver.[278] In cultures of Reuber H4 cells, pretreatment with cortisol increases the number of isozymes from one to three, and these appear to be very similar to the isozymes that are found in adult rat liver.[218] When *p*-chlorophenylalanine is added to the culture medium two of the three isozymes are eliminated, including the one that was originally present in the untreated hepatoma. The activity of the third isozyme is reduced.[217]

When 5 mM *p*-chlorophenylalanine is added to the culture in which Reuber H4 cells are growing, the activity of phenylalanine hydroxylase is reduced by 90% in 3 days.[216] There is no evidence for a direct inhibition by this compound, or the formation of an inhibitory reaction product. The activity is restored after removal of the *p*-chlorophenylalanine only when new protein synthesis has been allowed to take place. In the presence of this compound the stimulation brought about by cortisol still takes place, but the stimulation that is caused by the increasing cell density is prevented.

L SUMMARY

Phenylalanine hydroxylase has been shown to be an essential enzyme in the control of the concentration of L-phenylalanine in mammals. Without it the tissue phenylalanine concentration is controlled only by subsidiary processes, which permit dietary phenylalanine to accumulate to toxic levels, as can be seen in man, in patients who suffer from phenylketonuria and hyperphenylalaninaemia. The basic properties of the enzyme have been studied in detail both in man and in animals, especially in rats, with the purpose of determining the biochemical defect in phenylketonuria. This has been shown to reside in the hydroxylase itself, rather than in the enzymes that couple the hydroxylase to the hydrogen transport system. At present it is not known whether the defect is due to a mutation on the phenylalanine hydroxylase gene, which causes the biosynthesis of an altered, inactive phenylalanine hydroxylase, or whether it is due to the presence of a vanishingly small amount of the normal hydroxylase because of a mutation on a controller gene.

In the distant future, phenylketonuria may be successfully treated by genetic engineering, by the transfer of genetic material for the synthesis of phenylalanine hydroxylase to those who suffer from this condition. In the meantime, it may be possible to administer some compound that activates the hydroxylase *in vivo*, just as lysolecithin can activate it *in vitro* under some conditions.

The substrate specificity of the hydroxylase has been shown to be narrow. It acts on L-phenylalanine and a few other L-amino acids, including *m*-tyrosine and tryptophan. It is not known whether the latter hydroxylations have any physiological significance, and work needs to be undertaken to determine how important these reactions are quantitatively *in vivo*, especially in man. The cofactor specificity is equally narrow; with few exceptions the essential structure for activity is 2-amino-4-hydroxy-5,6,7,8-tetrahydropteridine.

The mechanism of the reaction has been thoroughly studied at a substrate level, and it has been shown that the substituent is not entirely eliminated from the position on the molecule at which hydroxylation occurs, but, depending on the nature of the substituent, some of it migrates *ortho* to the hydroxylation site. At the enzyme level, the mechanism is complex. The kinetics do not always obey the Michaelis–Menten equation. The enzyme can be activated by a liver protein and by some phospholipids, such as lysolecithin.

The enzyme contains phosphate, probably as a phospholipid. Removal of this group, which appears to occur in part during purification, results in a loss of activity. Phosphorylation of the dephosphorylated enzyme brings about an increase in activity, but some of the other properties of the enzyme are changed by phosphorylation. There are polymeric forms of the enzyme whose interrelationship appears to depend on lipids. This field requires much further study.

In vivo, phenylalanine hydroxylase is largely inactivated by administration of *p*-chlorophenylalanine, and some evidence suggests that this compound is incorporated into the enzyme molecule. Further study is needed on this subject, particularly to define what structural alteration, if any, does take place.

Considerable work has been carried out on the development of phenylalanine hydroxylase *in vivo* at and around the time of birth. It appears that the time course of development depends on the species. In some species the enzyme appears just before birth, and in others in early pregnancy. Much of this work, however, has left the field in confusion because of contradictory results. It is to be hoped that future studies may be carried out to clarify the position, and that the experiments may be designed so as to reduce the scope of random fluctuations. There is also a need to know how the hydroxylase develops in a wider range of species.

Considerable effort has been devoted to the isolation and culture of cell lines, particularly those obtained from rat hepatoma, that possess the capacity to hydroxylate phenylalanine. This work has not only complemented some of the studies on the enzyme obtained from the whole animal, but has yielded an insight into some of the control mechanisms that govern the activity of the hydroxylase. At this time, it is not possible to culture fibroblasts with any certainty of successful production of the hydroxylase. Therefore, prenatal diagnosis of phenylketonuria by this means is at present not possible.

The technology of purification of the hydroxylase has reached an exciting stage, where preliminary results have shown that it can be isolated by affinity chromatography. It will be of great practical importance if this technique can be applied to the isolation of the enzyme on a large scale. Other important isolation techniques such as electrophoresis and electrofocusing do not appear to have been used; it will be of considerable interest to workers in this field to know what results can be obtained by these approaches.

M REFERENCES

1. Abita, J.-P., Dorche, C., and Kaufman, S. (1974). *Pediat. Res.*, **8**, 714–717.
2. Abita, J.-P., Milstien, S., Chang, N., and Kaufman, S. (1976). *J. Biol. Chem.*, **251**, 5310–5314.
3. Ambrose, J. A., Sullivan, P., Ingerson, A., and Brown, R. L. (1969). *Clin. Chem.*, **15**, 611–620.
4. Anderson, J. A., Gravem, H., Ertel, R., and Fisch, R. (1962). *J. Pediat.*, **61**, 603–609.
5. Auerbach, V. H., and Waisman, H. A. (1958). *Cancer Res.*, **18**, 536–542.
6. Ayling, J. E., and Helfand, G. D. (1974). *Biochem. Biophys. Res. Commun.*, **61**, 360–366.
7. Ayling, J. E., Helfand, G. D., and Pirson, W. D. (1975). *Enzyme*, **20**, 6–19.
8. Ayling, J. E., Boehm, G. R., Textor, S. C., and Pirson, R. A. (1973). *Biochemistry*, **12**, 2045–2051.
9. Ayling, J., Pirson, R., Pirson, W., and Boehm, G. (1973). *Anal. Biochem.*, **51**, 80–90.

10. Ayling, J. E., Pirson, W. D., Al-Janabi, J. M., and Helfand, G. D. (1974). *Biochemistry*, **13**, 78–85.
11. Bagchi, S. P., and Zarycki, E. P. (1972). *Biochem. Pharmacol.*, **21**, 584–589.
12. Bailey, S. W., and Ayling, J. E. (1978). *J. Biol. Chem.*, **253**, 1598–1605.
13. Barranger, J. A. (1976). *Biochem. Med.*, **15**, 55–86.
14. Barranger, J. A., Geiger, P. J., Huzino, A., and Bessman, S. P. (1972). *Science*, **175**, 903–905.
15. Bartholomé, K., and Ertel, E. (1976). *Lancet ii*, 862–863.
16. Bartholomé, K., and Lutz, P. (1976). *Monatsschr. Kinderheilk.*, **124**, 421–422.
17. Bartholomé, K., Lutz, P., and Bickel, H. (1975). *Pediat. Res.*, **9**, 899–903.
18. Bartholomé, K., Byrd, D. J., Kaufman, S., and Milstien, S. (1977). *Pediatrics*, **59**, 757–761.
19. Basu, M., and Guha, S. R. (1972). *Indian J. Biochem. Biophys.*, **9**, 168–170.
20. Belzecka, K., Laskowska, T., and Mochnacka, I. (1966). *Acta Biochim. Polon.*, **11**, 191–196.
21. Belzecka, K., Jakubiec, A., and Puzynska, L. (1967). *Acta Biochim. Polon.*, **14**, 209–220.
22. Bernheim, M. L. C., and Bernheim, F. (1944). *J. Biol. Chem.*, **152**, 481.
23. Berry, H. K., Sutherland, B., and Guest, G. M. (1957). *Amer. J. Human Genet.*, **9**, 310–316.
24. Berry, H. K., Cripps, R., Nicholls, K., McCandless, D., and Harper, C. (1972). *Biochim. Biophys. Acta*, **261**, 315–320.
25. Bessman, S. P., and Huzino, A. (1969). *Fed. Proc. Fed. Amer. Socs Exptl Biol.*, **28**, 408.
26. Bessman, S. P., Wapnir, R. A., and Towell, M. E. (1977). *Biochem. Med.*, **17**, 1–7.
27. Bessman, S. P., Wapnir, R. A., Pankratz, H. S., and Plantholt, B. A. (1969). *Biol. Neonat.*, **14**, 107–116.
28. Boggs, D. E., and Waisman, H. A. (1964). *Arch. Biochem. Biophys.*, **106**, 307–311.
29. Boggs, D. E., and Waisman, H. A. (1964). *Proc. Soc. Exptl Biol. Med.*, **115**, 407–410.
30. Booth, A. N., and Williams, R. T. (1963). *Biochem. J.*, **88**, 66P–67P.
31. Booth, A. N., Robbins, D. J., Jones, F. T., Emerson, O. H., and Masri, M. S. (1965). *Proc. Soc. Exptl Biol. Med.*, **120**, 546–548.
32. Boulton, A. A., and Quan, L. (1970). *Canad. J. Biochem.*, **48**, 1287–1291.
33. Boulton, A. A., and Wu, P. H. (1972). *Canad. J. Biochem.*, **50**, 261–267.
34. Boyd, D. R., Campbell, R. M., Craig, H. C., Watson, C. G., Daly, J. W., and Jerina, D. M. (1976). *J. Chem. Soc., Perkin Trans. I*, **1976**, 2438–2443.
35. Brand, L. M., and Harper, A. E. (1974). *Biochem. J.*, **142**, 231–245.
36. Brand, L. M., and Harper, A. E. (1974). *Fed. Proc. Fed. Amer. Socs Exptl Biol.*, **33**, 652.
37. Brand, L. M., and Harper, A. E. (1974). *Proc. Soc. Exptl Biol. Med.*, **147**, 211–215.
38. Brase, D. A., and Loh, H. H. (1976). *Proc. West. Pharmacol. Soc.*, **19**, 172–176.
39. Brase, D. A., and Westfall, T. C. (1972). *Biochem. Biophys. Res. Commun.*, **48**, 1185–1191.
40. Brase, D. A., and Westfall, T. C. (1973). *Biochim. Biophys. Acta*, **309**, 271–279.
41. Breakefield, X. O., Castiglione, C. M., Halaban, R., Pawelek, J., and Shiman, R. (1978). *J. Cell. Physiol.*, **94**, 307–314.
42. Brenneman, A. R., and Kaufman, S. (1965). *J. Biol. Chem.*, **240**, 3617–3622.
43. Bublitz, C. (1969). *Biochim. Biophys. Acta*, **191**, 249–256.
44. Bublitz, C. (1971). *Biochem. Pharmacol.*, **20**, 2543–2553.
45. Bublitz, C. (1971). *Biochim. Biophys. Acta*, **235**, 311–321.
46. Bublitz, C. (1977). *Biochem. Med.*, **17**, 13–19.
47. Bunyatyan, G. Kh., Kazaryan, B. A., Safaryan, E. Kh., and Gevorkyan, G. A. (1971). *Vop. Med. Khim.*, **17**, 73–76.

48. Cartwright, E. C., and Danks, D. M. (1972). *Biochim. Biophys. Acta*, **264**, 205–209.
49. Cartwright, E. C., Connellan, J. M., and Danks, D. M. (1973). *Austral. J. Exptl Biol. Med. Sci.*, **51**, 559–563.
50. Cassano, V., Krooth, R. S., and Worthy, T. E. (1978). *Amer. J. Hum. Genet.*, **30**, 90.
51. Castells, S., and Shirali, S. (1971). *Life Sci.*, **10** Pt 2, 233–239.
52. Chandra, P., and Vining, L. C. (1968). *Canad. J. Microbiol.*, **14**, 573–578.
53. Choo, K. H., Cotton, R. G. H., and Danks, D. M. (1976). *Exptl Cell. Res.*, **101**, 370–382.
54. Christensen, P. J. (1962). *Scand. J. Clin. Lab. Invest.*, **14**, 623–628.
55. Collombel, C., Cotte, J., Coquet, B., and Perrot, L. (1969). *Ann. Biol. Clin. (Paris)*, **27**, 659–670.
56. Connellan, J. M., and Danks, D. M. (1973). *Biochim. Biophys. Acta*, **293**, 48–55.
57. Cotton, R. G. H. (1971). *Biochim. Biophys. Acta*, **235**, 61–72.
58. Cotton, R. G. H. (1974). *FEBS Letters*, **44**, 290–292.
59. Cotton, R. G. H., and Danks, D. M. (1976). *Nature*, **260**, 63–64.
60. Cotton, R. G., and Grattan, P. J. (1975). *Eur. J. Biochem.*, **60**, 427–430.
61. Coulson, W. F., and Hughes, C. M. (1971). *Biochem. J.*, **123**, 22P.
62. Coulson, W. F., Wardle, E., and Jepson, J. B. (1967). *Biochem. J.*, **103**, 15P–16P.
63. Coulson, W. F., Henson, G., and Jepson, J. B. (1968). *Biochem. J.*, **107**, 17P–18P.
64. Coulson, W. F., Henson, G., and Jepson, J. B. (1968). *Biochim. Biophys. Acta*, **156**, 135–139.
65. Coulson, W. F., Wardle, E., and Jepson, J. B. (1968). *Biochim. Biophys. Acta*, **167**, 99–109.
66. Coulson, W. F., Powers, M. J., and Jepson, J. B. (1970). *Biochim. Biophys. Acta*, **222**, 606–610.
67. Counsell, R. E., Chan, P. S., and Weinhold, P. A. (1970). *Biochim. Biophys. Acta*, **215**, 187–188.
68. Counsell, R. E., Desai, P., Smith, T. D., Chan, P. S., Weinhold, P. A., Rethy, V. B., and Burke, D. (1970). *J. Med. Chem.*, **13**, 1040–1042.
69. Dakin, H. D. (1922). *Oxidations and Reductions in the Animal Body*, 2nd ed., Longman Green, London, p. 86.
70. Daly, J., and Guroff, G. (1968). *Arch. Biochem. Biophys.*, **125**, 136–141.
71. Dean, R. F. A., and Whitehead, R. G. (1963). *Lancet i*, 188–191.
72. DeGraw, J. I., Cory, M., Skinner, W. A., Theisen, M. C., and Mitoma, C. (1967). *J. Med. Chem.*, **10**, 64–66.
73. DeGraw, J. I., Cory, M., Skinner, W. A., Theisen, M. C., and Mitoma, C. (1968). *J. Med. Chem.*, **11**, 225–227.
74. DeGraw, J. I., Cory, M., Skinner, W. A., Theisen, M. C., and Mitoma, C. (1968). *J. Med. Chem.*, **11**, 375–376.
75. DeGraw, J. I., Brown, V. H., Skinner, W. A., LeValley, S., and Mitoma, C. (1972). *J. Med. Chem.*, **15**, 781–783.
76. DelValle, J. A., and Greengard, O. (1976). *Biochem. J.*, **154**, 613–618.
77. DelValle, J. A., and Greengard, O. (1977). *Pediat. Res.*, **11**, 2–5.
78. Dierks Ventling, C., Wapnir, R. A., and Braude, M. C. (1968). *Proc. Soc. Exptl Biol. Med.*, **127**, 121–127.
79. Dolan, G., and Godin, C. (1966). *Biochemistry*, **5**, 922–925.
80. Donlon, J., and Kaufman, S. (1977). *Biochem. Biophys. Res. Commun.*, **78**, 1011–1017.
81. Edwards, D. J., and Blau, K. (1971). *Biochem. Med.*, **5**, 457–463.
82. Embden, G., and Baldes, K. (1913). *Biochem. Zeit*, **55**, 301–322.
83. Embden, G., Salomon, H., and Schmidt, Fr. (1906). *Beitr. Chem. Physiol. Path.*, **8**, 129–155.
84. Fisher, D. B., and Kaufman, S. (1970). *Biochem. Biophys. Res. Commun.*, **38**, 663–669.

85. Fisher, D. B., and Kaufman, S. (1972). *J. Biol. Chem.*, **247**, 2250–2252.
86. Fisher, D. B., and Kaufman, S. (1972). *J. Neurochem.*, **19**, 1359–1365.
87. Fisher, D. B., and Kaufman, S. (1973). *J. Biol. Chem.*, **248**, 4300–4304.
88. Fisher, D. B., and Kaufman, S. (1973). *J. Biol. Chem.*, **248**, 4345–4353.
89. Fisher, D. B., Kirkwood, R., and Kaufman, S. (1972). *J. Biol. Chem.*, **247**, 5161–5167.
90. Fisher, H. F., Gates, R. E., and Cross, D. G. (1970). *Nature*, **228**, 247–249.
91. Fölling, A. (1934). *Hoppe-Seyler's Zeit. Physiol. Chem.*, **227**, 169–176.
92. Freedland, R. A. (1963). *Biochim. Biophys. Acta*, **73**, 71–75.
93. Freedland, R. A., Wadzinski, I. M., and Waisman, H. A. (1961). *Biochem. Biophys. Res. Commun.*, **5**, 94–98.
94. Freedland, R. A., Wadzinski, I. M., and Waisman, H. A. (1961). *Biochem. Biophys. Res. Commun.*, **6**, 227–231.
95. Freedland, R. A., Krakowski, M. C., and Waisman, H. A. (1962). *Amer. J. Physiol.*, **202**, 145–148.
96. Freedland, R. A., Krakowski, M. C., and Waisman, H. A. (1964). *Amer. J. Physiol.*, **206**, 341–344.
97. Friedman, P. A., and Kaufman, S. (1972). *Arch. Biochem. Biophys.*, **146**, 321–326.
98. Friedman, P. A., and Kaufman, S. (1973). *Biochim. Biophys. Acta*, **293**, 56–61.
99. Friedman, P. A., Kaufman, S., and Kang, E. S. (1972). *Nature*, **240**, 157–159.
100. Friedman, P. A., Lloyd, T., and Kaufman, S. (1972). *Mol. Pharmacol.*, **8**, 501–510.
101. Friedman, P. A., Fisher, D. B., Kang, E. S., and Kaufman, S. (1973). *Proc. Nat. Acad. Sci. USA*, **70**, 552–556.
102. Fuller, R. W., and Baker, J. C. (1974). *Biochem. Biophys. Res. Commun.*, **58**, 945–950.
103. Gál, E. M. (1972). *Adv. Biochem. Psychopharmacol.*, **6**, 149–163.
104. Gál, E. M., and Millard, S. A. (1971). *Biochim. Biophys. Acta*, **227**, 32–41.
105. Gál, E. M., Chatterjee, S. K., and Marshall, F. D. Jr. (1964). *Biochim. Biophys. Acta*, **85**, 495–498.
106. Gilbert, D. L., Gerschman, R., Cohen, J., and Sherwood, W. (1957). *J. Amer. Chem. Soc.*, **79**, 5677–5680.
107. Gilbert, D. L., Gerschman, R., Ruhm, K. B., and Price, W. E. (1958). *J. Gen. Physiol.*, **41**, 989–1003.
108. Gillam, S. S., Woo, S. L. C., and Woolf, L. I. (1974). *Biochem. J.*, **139**, 731–739.
109. Goldstein, M. (1966). *Recent Results Cancer Res.*, **2**, 66–70.
110. Gontea, I. and Dumitrache, S. (1972). *Igiena*, **21**, 589–598.
111. Gontea, I., Dumitrache, S., and Rujinski, A. (1973). *Fiziol. Norm. Patol.*, **19**, 9–16.
112. Goodfriend, T. L., and Kaufman, S. (1961). *J. Clin. Invest.*, **40**, 1743–1750.
113. Goodwin, B. L. (1964). D.Phil. thesis, Oxford.
114. Goodwin, B. L. (1972). *Anal. Biochem.*, **47**, 302–305.
115. Goodwin, B. L., and Werner, E. G. (1973). *Experientia*, **29**, 523–525.
116. Goodwin, B. L., Ruthven, C. R. J., Weg, M. W., and Sandler, M. (1975). *Clin. Chim. Acta*, **62**, 439–442.
117. Greengard, O., and DelValle, J. A. (1976). *Biochem. J.*, **154**, 619–624.
118. Greengard, O., Yoss, M. S., and DelValle, J. A. (1976). *Science*, **192**, 1007–1008.
119. Grimm, U., Knapp, A., Tisher, W., and Schlenzka, K. (1972). *Acta Biol. Med. Germ.*, **28**, 549–552.
120. Grimm, U., Knapp, A., Schlenzka, K., and Reddeman, H. (1975). *Clin. Chim. Acta*, **58**, 17–21.
121. Guroff, G. (1964). *Biochim. Biophys. Acta*, **90**, 623–624.
122. Guroff, G. (1969). *Arch. Biochem. Biophys.*, **134**, 610–611.
123. Guroff, G. (1970). *Methods Enzymol.*, **17A**, 597–603.
124. Guroff, G. (1971). *Methods Enzymol.*, **18B**, 600–605.
125. Guroff, G., and Abramowitz, A. (1967). *Anal. Biochem.*, **19**, 548–555.

126. Guroff, G., and Daly, J. (1967). *Arch. Biochem. Biophys.*, **122**, 212–217.
127. Guroff, G., and Daly, J. (1968). *Fed. Proc. Fed. Amer. Socs Exptl Biol.*, **27**, 587.
128. Guroff, G., and Ito, T. (1963). *Biochim. Biophys. Acta*, **77**, 159–161.
129. Guroff, G., and Ito, T. (1965). *J. Biol. Chem.*, **240**, 1175–1184.
130. Guroff, G., and Rhoads, C. A. (1967). *J. Biol. Chem.*, **242**, 3641–3645.
131. Guroff, G., and Rhoads, C. A. (1969). *J. Biol. Chem.*, **244**, 142–146.
132. Guroff, G., and Strenkoski, C. A. (1966). *J. Biol. Chem.*, **241**, 2220–2227.
133. Guroff, G., Kondo, K., and Daly, J. (1966). *Biochem. Biophys. Res. Commun.*, **25**, 622–628.
134. Guroff, G., Levitt, M., Daly, J., and Udenfriend, S. (1966). *Biochem. Biophys. Res. Commun.*, **25**, 253–259.
135. Guroff, G., Reifsnyder, C. A., and Daly, J. J. (1966). *Biochem. Biophys. Res. Commun.*, **24**, 720–724.
136. Guroff, G., Daly, J. W., Jerina, D. M., Renson, J., Witkop, B., and Udenfriend, S. (1967). *Science*, **157**, 1524–1530.
137. Guroff, G., Rhoads, C. A., and Abramowitz, A. (1967). *Anal. Biochem.*, **21**, 273–278.
138. Guroff, G., Karadbil, M., and Dayman, J. (1970). *Arch. Biochem. Biophys.*, **141**, 342–345.
139. Gusev, M. V., Nikitina, K. A., and Ushakova, N. A. (1972). *Mikrobiologiya*, **41**, 959–963.
140. Haggerty, D. F., Popják, G., and Young, P. L. (1976). *J. Biol. Chem.*, **251**, 6901–6908.
141. Haggerty, D. F., Young, P. L., and Buese, J. V. (1975). *Dev. Biol.*, **44**, 158–168.
142. Haggerty, D. F., Young, P. L., and Buese, J. V. (1974). *Dev. Biol.*, **40**, 16–23.
143. Haggerty, D. F., Young, P. L., Popják, G., and Carnes, W. H. (1973). *J. Biol. Chem.*, **248**, 223–232.
144. Haggerty, D. F., Young, P. L., Buese, J. V., and Popják, G. (1975). *J. Biol. Chem.*, **250**, 8428–8437.
145. Hare, L. E., Lu, M. C., Sullivan, C. B., Sullivan P. T., and Counsell, R. E. (1974). *J. Med. Chem.*, **17**, 1–5.
146. Hasegawa, S., and Maier, V. P. (1972). *Phytochemistry*, **11**, 1365–1370.
147. Hoffbauer, R. W., and Schrempf, G. (1976). *Lancet ii*, 194.
148. Hoffbauer, R. W., Schrempf, G., and Mönch, E. (1976). *Lancet ii*, 1031.
149. Hsia, D. Y.-Y., Driscoll, K., Troll, W., and Knox, W. E. (1956). *Nature*, **178**, 1239–1240.
150. Hsia, D. Y.-Y., Paine, R. S., and Driscoll, K. W. (1957). *J. Ment. Defic. Res.*, **1**, 53.
151. Huang, C. Y., and Kaufman, S. (1973). *J. Biol. Chem.*, **248**, 4242–4251.
152. Huang, C. Y., Max, E. E., and Kaufman, S. (1973). *J. Biol. Chem.*, **248**, 4235–4241.
153. Hwang, U. K. (1968). *Experientia*, **24**, 683–684.
154. Ishibashi, T. (1972). *Agr. Biol. Chem.* **36**, 596–603.
155. Jaenicke, H., and Wahlefeld, A. W. (1968). In *Biochemical Aspects of Antimetabolite Drug Hydroxylation, 5th FEBS Meeting*, pp. 249–260.
156. Jakubiec-Puka, A., and Mochnacka, I. (1969). *Acta Biochim. Polon.*, **16**, 321–331.
157. Jakubovič, A. (1971). *Biochim. Biophys. Acta*, **237**, 469–475.
158. Jakubovič, A., Woolf, L. I., and Chan-Henry, E. (1971). *Biochem. J.*, **125**, 563–568.
159. Jedlicki, E., Kaufman, S., and Milstien, S. (1977). *J. Biol. Chem.*, **252**, 7711–7714.
160. Jerina, D., Daly, J., Landis, W., Witkop, B., and Udenfriend, S. (1967) *J. Amer. Chem. Soc.*, **89**, 3347–3349.
161. Jerina, D. M., Daly, J. W., and Witkop, B. (1971). *Med. Res. B*, **5**, 413–476.
162. Jervis, G. A. (1960). *Clin. Chim. Acta*, **5**, 471–476.
163. Jorgensen, G. N., and Larson, B. L. (1968). *Biochim. Biophys. Acta*, **165**, 121–126.
164. Justice, P., O'Flynn, M. E., and Hsia, D. Y.-Y. (1967). *Lancet i*, 928–929.
165. Kaubisch, N., Daly, J. W., and Jerina, D. M. (1972). *Biochemistry*, **11**, 3080–3088.

166. Kaufman, S. (1957). *Fed. Proc. Fed. Amer. Socs Exptl Biol.*, **16**, 203.
167. Kaufman, S. (1957). *J. Biol. Chem.*, **226**, 511–524.
168. Kaufman, S. (1958). *Biochim. Biophys. Acta*, **27**, 428–429.
169. Kaufman, S. (1959). *J. Biol. Chem.*, **234**, 2677–2682.
170. Kaufman, S. (1961). *Biochim. Biophys. Acta*, **51**, 619–621.
171. Kaufman, S. (1962). *J. Biol. Chem.*, **237**, PC2712–PC2713.
172. Kaufman, S. (1962). *Methods Enzymol.*, **5**, 809–816.
173. Kaufman, S. (1963). *Enzymes*, **8**, 373–383.
174. Kaufman, S. (1963). *Proc. Nat. Acad. Sci. USA*, **50**, 1085–1093.
175. Kaufman, S. (1964). *J. Biol. Chem.*, **239**, 332–338.
176. Kaufman, S. (1967). *J. Biol. Chem.*, **242**, 3934–3943.
177. Kaufman, S. (1969). *Arch. Biochem. Biophys.*, **134**, 249–252.
178. Kaufman, S. (1970). *J. Biol. Chem.*, **245**, 4751–4759.
179. Kaufman, S. (1970). *Methods Enzymol.*, **17A**, 603–609.
180. Kaufman, S. (1971). *Adv. Enzymol.*, **35**, 245–319.
181. Kaufman, S. (1976). *Biochem. Med.*, **15**, 42–54.
182. Kaufman, S., and Fisher, D. B. (1970). *J. Biol. Chem.*, **245**, 4745–4750.
183. Kaufman, S., and Levenberg, B. (1959). *J. Biol. Chem.*, **234**, 2683–2688.
184. Kaufman, S., Max, E. E., and Kang, E. S. (1975). *Pediat. Res.*, **9**, 632–634.
185. Kaufman, S., Bridgers, W. F., Eisenberg, F., and Friedman, S. (1962). *Biochem. Biophys. Res. Commun.*, **9**, 497–502.
186. Kawada, S., Imada, H., Taniguchi, Y., and Ikegami, Y. (1965). *Ann. Rept Nat. Inst. Nutr. (Tokyo)*, **1965**, 13–16.
187. Kawada, S., Imada, H., Taniguchi, Y., and Ikegami, Y. (1965). *Ann. Rept Nat. Inst. Nutr. (Tokyo)*, **1965**, 17–19.
188. Kenney, F. T., and Kretchmer, N. (1959). *J. Clin. Invest.*, **38**, 2189–2196.
189. Kenney, F. T., Reem, G. H., and Kretchmer, N. (1958). *Science*, **127**, 86.
190. Kerr, G. R., and Waisman, H. A. (1967). *J. Nutr.*, **92**, 10–18.
191. Khodorovskaya, E. D., and Akopyan, Zh. I. (1968). *Vop. Med. Khim.*, **14**, 628–631.
192. Koe, B. K. (1967). *Med. Pharmacol. Exp.*, **17**, 129–138.
193. Koe, B. K., and Weissman, A. (1967). *J. Pharmacol. Exptl Ther.*, **157**, 565–573.
194. Koller, C., Klinger, R., and Anderson, G. (1974). *Enzyme*, **17**, 155–159.
195. Kotake, Y., Masai, Y., and Mori, Y. (1922). *Hoppe-Seyler's Zeit. Physiol. Chem.*, **122**, 195–200.
196. Leete, E. (1966). *J. Amer. Chem. Soc.*, **88**, 4218–4221.
197. Letendre, C. H., Dickens, G., and Guroff, G. (1975). *J. Biol. Chem.*, **250**, 6672–6678.
198. Lipton, M. A., Gordon, R., Guroff, G., and Udenfriend, S. (1967). *Science*, **156**, 248–250.
199. Lloyd, T., and Kaufman, S. (1973). *Mol. Pharmacol.*, **9**, 438–444.
200. Lloyd, T., Mori, T., and Kaufman, S. (1971). *Biochemistry*, **10**, 2330–2336.
201. Matalon, R., Justice, P., and Deanching, M. N. (1977). *Lancet i*, 853–854.
202. Matsubara, M., Katoh, S., Akino, M., and Kaufman, S. (1966). *Biochim. Biophys. Acta*, **122**, 202–212.
203. Mavrides, C., Whitlow, K. J., and D'Iorio, A. (1973). *J. Protozool.*, **20**, 342–344.
204. McClure, D., Miller, M. R., and Shiman, R. (1975). *Exptl Cell. Res.*, **90**, 31–39.
205. McClure, D., Miller, M. R., and Shiman, R. (1976). *Exptl Cell. Res.*, **98**, 223–236.
206. McCormick, D. B., Young, S. K., and Woods, M. N. (1965). *Proc. Soc. Exptl Biol. Med.*, **118**, 131–133.
207. McEvoy, J. J. (1974). *Antonie van Leeuwenhoek*, **40**, 409–416.
208. McGee, M. M., Greengard, O., and Knox, W. E. (1972). *Biochem. J.*, **127**, 669–674.
209. McGee, M. M., Greengard, O., and Knox, W. E. (1972). *Biochem. J.*, **127**, 675–680.
210. McLean, A., Marwick, M. J., and Clayton, B. E. (1973). *J. Clin. Pathol.*, **26**, 678–683.

211. McMenamy, R. H., and Oncley, J. L. (1958). *J. Biol. Chem.*, **233**, 1436–1447.
212. McMenamy, R. H., Lund, C. C., Van Marke, J., and Oncley, J. L. (1961). *Arch. Biochem. Biophys.*, **93**, 135–139.
213. Medes, G. (1932). *Biochem. J.*, **26**, 917–940.
214. Mehrle, P. M., DeClue, M. E., and Bloomfield, R. A. (1972). *Nature*, **238**, 462–463.
215. Millard, S. A., and Gál, E. M. (1972). *J. Neurochem.*, **19**, 2461–2464.
216. Miller, M. R., McClure, D., and Shiman, R. (1975). *J. Biol. Chem.*, **250**, 1132–1140.
217. Miller, M. R., McClure, D., and Shiman, R. (1976). *J. Biol. Chem.*, **251**, 3677–3684.
218. Miller, M. R., and Shiman, R. (1976). *Biochem. Biophys. Res. Commun.*, **68**, 740–745.
219. Miller, M. R., and Shiman, R. (1976). *J. Biol. Chem.*, **251**, 3671–3676.
220. Milstien, S., and Kaufman, S. (1975). *J. Biol. Chem.*, **250**, 4777–4781.
221. Milstien, S., and Kaufman, S. (1975). *J. Biol. Chem.*, **250**, 4782–4785.
222. Milstien, S., Abita, J.-P., Chang, N., and Kaufman, S. (1976). *Proc. Nat. Acad. Sci. USA*, **73**, 1591–1593.
223. Milstien, S., Holtzman, N. A., O'Flynn, M. E., Thomas, G. H., Butler, I. J., and Kaufman, S. (1976). *J. Pediat.*, **89**, 763–766.
224. Mitoma, C. (1956). *Arch. Biochem. Biophys.*, **60**, 476–484.
225. Mitoma, C., Auld, R. M., and Udenfriend, S. (1957). *Proc. Soc. Exptl Biol. Med.*, **94**, 634–635.
226. Moss, A. R., and Schoenheimer, R. (1940). *J. Biol. Chem.*, **135**, 415–429.
227. Murthy, L. I. (1975). *Life Sci.*, **17**, 1777–1783.
228. Murthy, L. I., and Berry, H. K. (1974). *Arch. Biochem. Biophys.*, **163**, 225–230.
229. Murthy, L. I., and Berry, H. K. (1975). *Biochem. Med.*, **12**, 392–397.
230. Musacchio, J. M. (1969). *Biochim. Biophys. Acta*, **191**, 485–487.
231. Musacchio, J. M., D'Angelo, G. L., and McQueen, C. A. (1971). *Proc. Nat. Acad. Sci. USA*, **68**, 2087–2091.
232. Nagatsu, T., Levitt, M., and Udenfriend, S. (1964). *J. Biol. Chem.*, **239**, 2910–2917.
233. Nagatsu, T., Numata, Y., Kato, T., Sugiyama, K., and Akino, M. (1978). *Biochim. Biophys. Acta*, **523**, 47–52.
234. Nair, P. M., and Vining, L. C. (1965). *Phytochemistry*, **4**, 401–411.
235. Namboodiri, M. A., and Ramasarma, T. (1975). *Biochem. J.*, **150**, 263–268.
236. Nielsen, K. H. (1969). *Eur. J. Biochem.*, **7**, 360–369.
237. Nielsen, K. H., Simonsen, V., and Lind, K. E. (1969). *Eur. J. Biochem.*, **9**, 497–502.
238. Parker, C. E., Barranger, J., Newhouse, R., and Bessman, S. (1977). *Ann. Clin. Biochem.*, **14**, 122–123.
239. Parker, C. E., Barranger, J., Newhouse, R., and Bessman, S. P. (1977). *Biochem. Med.*, **17**, 8–12.
240. Pokrovskii, A. A., Usacheva, N. T., Milova, G. N., Ermolaev, M. V., and Ermolov, A. S. (1970). *Byull. Eksp. Biol. Med.*, **69**, 122–124.
241. Pollitt, R. J. (1978). *Clin. Chim. Acta*, **83**, 270–272.
242. Prichard, J. W., and Guroff, G. (1971). *J. Neurochem.*, **18**, 153–160.
243. Promkasetrin, N., and Harrill, I. (1970). *Nutr. Rep. Int.*, **1**, 281–289.
244. Räihä, N. C. R. (1973). *Pediat. Res.*, **7**, 1–4.
245. Räihä, N. C. R., and Schwartz, A. (1973). *Enzyme*, **15**, 330–339.
246. Reem, G. H., and Kretchmer, N. (1957). *Proc. Soc. Exptl Biol. Med.*, **96**, 458–460.
247. Rendina, G., Ryan, M. F., DeLong, J., Tuttle, J. M., and Giles, C. E. (1967). *J. Ment. Defic. Res.*, **11**, 153–168.
248. Renson, J., Goodwin, F., Weissbach, H., and Udenfriend, S. (1961). *Biochem. Biophys. Res. Commun.*, **6**, 20–23.
249. Renson, J., Weissbach, H., and Udenfriend, S. (1962). *J. Biol. Chem.*, **237**, 2261–2264.
250. Renson, J., Weissbach, H., and Udenfriend, S. (1965). *Mol. Pharmacol.*, **1**, 145–148.
251. Ross, S. B., and Haljasmaa, Ö. (1964). *Life Sci.*, **3**, 579–587.

252. Ross, S. B., and Haljasmaa, Ö. (1966). *Acta Pharmacol. Toxicol.*, **24**, 55–72.
253. Ryan, W. L., and Orr, W. (1966). *Arch. Biochem. Biophys.*, **113**, 684–686.
254. Sandler, M., Karoum, F., Ruthven, C. R. J., and Calne, D. B. (1969). *Science*, **166**, 1417–1418.
255. Sandler, M., Goodwin, B. L., Ruthven, C. R. J., and Calne, D. B. (1971). *Nature*, **229**, 414–416.
256. Scheline, R. R., Williams, R. T., and Wit, J. G. (1960). *Nature*, **188**, 849–850.
257. Schloerer, J., Sekeris, C. E., and Karlson, P. (1970). *Hoppe-Seyler's Zeit. Physiol. Chem.*, **351**, 1035–1040.
258. Shambaugh, N. F., Lewis, H. B., and Tourtellotte, D. (1931). *J. Biol. Chem.*, **92**, 499–511.
259. Shimomura, K., Fukushima, T., and Danno, T. (1972). *J. Biochem.*, **71**, 547–550.
260. Siqueira, E., Lalitha, K., Hariharan, K., and Krishnaswamy, P. R. (1967). *Indian Pediat.*, **4**, 1–10.
261. Smirnova, E. V. (1961). *Voprosy Onkol.*, **7**, 17–23.
262. Smith, I., Clayton, B. E., and Wolff, O. H. (1975). *Lancet i*, 1108–1111.
263. Soloway, A. H. (1966). *J. Theoret. Biol.*, **13**, 100–105.
264. Stephenson, J. B. P., and McBean, M. S. (1967). *Brit. Med. J. iii*, 579–581.
265. Stone, K. J., and Townsley, B. H. (1973). *Biochem. J.*, **131**, 611–613.
266. Storm, C. B., and Kaufman, S. (1968). *Biochem. Biophys. Res. Commun.*, **32**, 788–793.
267. Strittmatter, C. F., and Oakley, G. (1966). *Proc. Soc. Exptl Biol. Med.*, **123**, 427–432.
268. Sullivan, D. M., Carver, M. J., and Copenhaver, J. H. (1972). *Life Sci.*, **11** Pt 2, 533–539.
269. Terawaki, A., Sato, M., Yamanouchi, M., Fukuyama, T., Ito, N., and Nakajima, S. (1967). *Gann*, **58**, 185–191.
270. Terawaki, A., Yasui, O., Yamanouchi, M., Fukuyama, T., and Nakajima, O. (1967). *Gann*, **58**, 177–183.
271. Thalhammer, O. and Scheiber, V. (1972). *Neuropaediatrie*, **3**, 358–361.
272. Thomas, P. E. and Hutton, J. J. (1971). *J. Nat. Cancer Inst.*, **47**, 1025–1031.
273. Tong, J. H., and Kaufman, S. (1975). *J. Biol. Chem.*, **250**, 4152–4158.
274. Tong, J. H., D'Iorio, A., and Benoiton, N. L. (1971). *Biochem. Biophys. Res. Commun.*, **43**, 819–826.
275. Tong, J. H., D'Iorio, A., and Benoiton, N. L. (1972). *Canad. J. Biochem.*, **50**, 151–153.
276. Tourian, A. (1971). *Biochim. Biophys. Acta*, **242**, 345–354.
277. Tourian, A. (1973). *Biochim. Biophys. Acta*, **309**, 44–49.
278. Tourian, A. (1976). *Biochem. Biophys. Res. Commun.*, **68**, 51–55.
279. Tourian, A. (1976). *J. Cell. Physiol.*, **87**, 15–24.
280. Tourian, A., Goddard, J., and Puck, T. T. (1969). *J. Cell. Physiol.*, **73**, 159–170.
281. Udenfriend, S., and Bessman, S. P. (1953). *J. Biol. Chem.*, **203**, 961–966.
282. Udenfriend, S., and Cooper, J. R. (1952). *J. Biol. Chem.*, **194**, 503–511.
283. Udenfriend, S., Clarke, C. T., Axelrod, J., and Brodie, B. B. (1954). *J. Biol. Chem.*, **208**, 731–739.
284. Valdivieso, F., Gimenez, C., and Mayor, F. (1975). *Biochem. Med.*, **12**, 72–78.
285. van der Heiden, C. (1977). *Lancet i*, 422.
286. Viscontini, M., and Mattern, G. (1970). *Helv. Chim. Acta*, **53**, 372–376.
287. Voss, J. C., and Waisman, H. A. (1966). *Comp. Biochem. Physiol.*, **17**, 49–58.
288. Waalkes, T. P., and Udenfriend, S. (1967). *J. Lab. Clin. Med.*, **50**, 733–736.
289. Wallace, H. W., Moldave, K., and Meister, A. (1957). *Proc. Soc. Exptl Biol. Med.*, **94**, 632–633.
290. Wang, H. L., and Waisman, H. A. (1961). *Proc. Soc. Exptl Biol. Med.*, **108**, 332–335.
291. Wapnir, R. A., Hawkins, R. L. and Stevenson, J. H. (1971). *Biol. Neonate*, **18**, 85–93.

292. Wapnir, R. A., Hawkins, R. L. and Lifshitz, F. (1972). *Amer. J. Physiol.*, **223**, 788–793.
293. Weiss, B., Hui, M., and Lajtha, A. (1977). *Biochem. Med.*, **18**, 330–343.
294. Weiss, B., Hui, M., and Lajtha, A. (1977). *Res. Commun. Chem. Pathol. Pharmacol.*, **18**, 709–721.
295. Womack, M., and Rose, W. C. (1934). *J. Biol. Chem.*, **107**, 449–458.
296. Woo, S. L. C., Gillam, S. S., and Woolf, L. I. (1974). *Biochem. J.*, **139**, 741–749.
296a. Woods, H. F., and Youdim, M. B. H. (1977). In *Structure and Function of Monoamine Enzymes* (Eds., E. Usdin, N. Weiner, and M. B. H. Youdim), Dekker, New York, pp. 263–278.
297. Woods, M. N., and McCormick, D. B. (1964). *Proc. Soc. Exptl Biol. Med.*, **116**, 427–430.
298. Woolf, L. I. (1976). *Biochem. Med.*, **16**, 284–291.
299. Woolf, L. I., Cranston, W. I., and Goodwin, B. L. (1967). *Nature*, **213**, 882–885.
300. Woolf, L. I., Goodwin, B. L., Cranston, W. I., Wade, D. N., Woolf, F., Hudson, F. P., and McBean, M. S. (1968). *Lancet i*, 114–117.
301. Woolf, L. I., Jakubovič, A., and Chan-Henry, E. (1971). *Biochem. J.*, **125**, 569–574.
302. Yamanouchi, M. (1968). *Nara Igaku Zasshi*, **19**, 489–499.
303. Youdim, M. B. H., Mitchell, B., and Woods, H. F. (1975). *Biochem. Soc. Trans.*, **3**, 683–684.
304. Zannoni, V. G. (1976). *Biochem. Med.*, **16**, 251–253.
305. Zannoni, V. G., Weber, W. W., Van Valen, P., Rubin, A., Bernstein, R., and LaDu, B. N. (1966). *Genetics*, **54**, 1391–1399.

Aromatic Amino Acid Hydroxylases and Mental Disease
Edited by M. B. H. Youdim

CHAPTER 3

Phenylalanine Hydroxylase Deficiency: Biochemical, Physiological, and Clinical Aspects of Phenylketonuria and Related Phenylalaninemias

Karl Blau

A INTRODUCTION

Like so much else in life, the importance of the phenylalanine hydroxylase system in human metabolism was not realized until it became apparent how serious its absence could be. The story of the discovery of 'oligophrenia phenylpyruvica' by the Norwegian physician Asbjørn Følling in 1934 is well known,[143] and towards the end of his life he gave a charming and modest account of it, which showed his full awareness of all the major aspects to be discussed in this chapter.[146] He was an able biochemist: not many physicians would lightly undertake, almost unaided, the biochemical characterization of an inborn error of metabolism, and Følling did it without modern means of separation and identification of unknown compounds. 'Følling's syndrome' is now known as classical phenylketonuria, and we owe our knowledge of the basic defect as a deficiency of hepatic phenylalanine-4-hydroxylase (EC 1.14.16.1) to Jervis[218] and Mitoma, Auld, and Udenfriend.[306] Although phenylketonuria is a rare disorder, it has been extensively studied: aspects of its natural history have illuminated human metabolism and genetics, clinical medicine, neurology, and enzymology; interest in the disorder is undiminished, in spite of fears to the contrary.[38] Despite this sustained interest there are still unanswered questions. The most important question is: what is the mechanism that causes irreversible brain damage and mental retardation? We have a considerable body of facts, and their consideration may give tentative answers to this and to other unanswered questions concerning phenylketonuria and its variants, and we shall try to see how far we have already come in answering these questions. For convenience Bickel's term 'phenylalaninemia' will generally be used to describe conditions in man or in experimental animals where blood phenylalanine concentrations are increased above the normal range. 'Phenylketonuria' is strictly speaking the appearance of ketones, or more precisely of phenylpyruvic acid, in the urine, a specific biochemical effect that is transient and may be modulated by the manipulation of blood phenylalanine levels. Classical phenylketonuric patients who have been identified and treated early enough may never exhibit 'phenylketonuria' unless treatment is discontinued, and this may be a very important benefit of dietary treatment.

B GENETICALLY DETERMINED DEFECTS OF THE HEPATIC PHENYLALANINE HYDROXYLATING SYSTEM

While the early studies on patients with phenylketonuria identified the deficient enzyme system,[143, 218, 306] later experience, particularly after the introduction of mass-screening programmes, showed up the genetic heterogeneity of the enzyme deficiencies, which are probably due to different mutations in the DNA bases of the gene coding for the enzyme. Nowadays it is preferable to consider these mutant forms, whether relatively rare abnormal forms leading to clinical symptoms or more common normal variants, in molecular terms, rather than in genetic terms as 'allelism at specific gene loci'.

The constituents of the liver phenylalanine hydroxylating system, and how they act to convert phenylalanine and its cosubstrate oxygen into tyrosine, are described elsewhere in this book. Two enzymes are involved: phenylalanine hydroxylase (EC 1.14.16.1) itself, whose deficiency in the phenylalaninemias has been known for some time, and dihydropteridine reductase (EC 1.6.99.7), whose deficiency in some patients with hyperphenylalaninemia was discovered more recently.

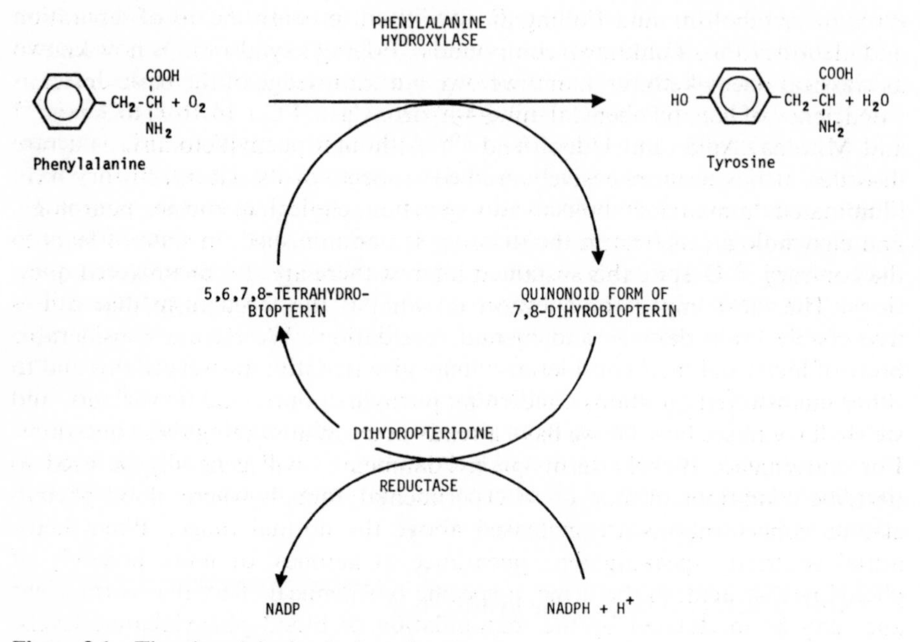

Figure 3.1 The phenylalanine hydroxylase system

1 Classical phenylketonuria and the defect in phenylalanine hydroxylase

The work of Jervis[218] and of Mitoma *et al.*[306] showed that the molecular defect in classical phenylketonuria was the virtual absence of liver phenylalanine hydroxylase activity. Kaufman's group found no material cross-reacting immunologically with a specific sheep antiserum to rat liver phenylalanine hydroxylase known also to cross-react with normal human phenylalanine hydroxylase.[152] This ruled out the possibility that an inactive protein was synthesized to any significant extent. In a later report[154] describing studies on material from the liver of a phenylketonuric patient, and the use of natural biopterin as cofactor, 0.27% of normal activity was found. This minimal residual activity was, however, enough for careful characterization of the enzyme. The authors excluded significant contributions to the production of tyrosine from phenylalanine by other pterin-dependent hydroxylases. Thus, tyrosine hydroxylase was inhibited with 3-iodotyrosine, and shown not to exceed 14% of total activity, and tryptophan hydroxylase was inhibited with 6-fluorotryptophan, and found not to exceed 9% of total activity. The behaviour of the residual enzyme towards inhibition with 4-chlorophenylalanine (PCPA), and towards activation with lysolecithin, as well as its immunological similarity to rat liver phenylalanine hydroxylase, was consistent with it consisting of at least 85% phenylalanine hydroxylase. Friedman *et al.*[154] compared the mutant enzyme with the normal one, and found three significant differences: it was not subject to substrate inhibition above 0.1 mM phenylalanine (the normal enzyme is inhibited above this level); it was not stimulated so well by lysolecithin (only 100% as compared with 300–400% for the normal enzyme); and it was not inhibited by antiserum as much as the normal enzyme (18% as compared with 63%, using the same antibody concentration). However, they were unable to define other differences: thus the mutant enzyme had the normal K_m for phenylalanine, it had the expected iron content, and the phenylketonuric liver extract had the normal concentrations of phenylalanine hydroxylase stimulator protein, tertrahydrobiopterin, and dihydropteridine reductase. The authors concluded that the finding of small but significant residual phenylalanine hydroxylase activity in the liver of a patient with phenylketonuria should eventually tell them 'whether the low activity is due to a small quantity of normal hydroxylase arising from a regulation mutation or due to a normal or low amount of a structurally altered hydroxylase'. Their work already suggests that the latter is the more likely alternative, but other mutations with different effects on the enzyme may yet be discovered.

2 Other defects in phenylalanine hydroxylase

There is a spectrum of clinical conditions, to be described in greater detail later,[43, 44, 74, 180, 207, 359, 423] that is characterized by phenylalaninemia of varying degree. Where enzyme studies have been carried out, some residual phenylalanine hydroxylase activity has been found.[228, 230] Such hyperphenylalaninemias are thought to be collectively less common than the classical mutation, and their diversity makes it hard to know how much one can generalize from a limited number of atypical cases. Friedman *et al.*[152] compared extracts from the liver of a patient with hyperphenylalaninemia with extracts from normal liver. The results are similar to those described for the extract from phenylketonuric liver, except that the residual enzyme activity in hyperphenylalaninemia was 5–7% of normal. Although double immunodiffusion against the sheep antiserum to the rat liver enzyme showed no precipitin line with the extract from the hyperphenylalaninemia liver, cross-reactivity was demonstrable: after incubation with antiserum about half of the residual activity was inhibited. This is evidence for the presence of less than 10% of the normal amounts of cross-reacting material. The mutant enzyme also had a lower apparent K_m for phenylalanine (0.62 mM as compared with 1.25 mM for the normal enzyme), and differed in the degree of stimulation by lysolecithin (less than 100%, as distinct from 300–400% for the normal enzyme). The loss of activity on heat inactivation for the mutant enzyme was also greater than normal (58 and 68% after 2 and 4 min., as distinct from 27 and 59% for the normal enzyme). There seems therefore to be a structural gene mutation that leads to an altered enzyme, presumably with the molecular defect in a part of the enzyme molecule less crucial for the retention of catalytic activity.[464] It seems that in both classical phenylketonuria, and in phenylalaninemia variants, less than the normal amount of immunologically determinant groupings are present, and this might arise from a variety of molecular defects. The rate of synthesis of enzyme protein may be diminished, or its rate of degradation may be increased, or both. Antigenically significant portions of the molecule may not be in the right conformation for cross-reaction with antiserum. However, the evidence suggests that a grossly reduced amount of structurally altered, but still somewhat catalytically active, enzyme protein is present in the livers of patients with classical phenylketonuria and with hyperphenylalaninemia.

3 Hyperphenylalaninemia and dihydropteridine reductase deficiency

Several patients have been described with phenylalaninemia due to inability to convert phenylalanine to tyrosine in spite of normal phenylalanine hydroxylase activity.[24, 404, 406, 407] (For a report on these 'malignant hyperphenylalaninemias' see Danks *et al.*[106]) Since blood phenylalanine could be normalized by dietary means without improving the progressive neurological disturbances, which were much more acute than anything seen in classical phenylketonuria, it was thought that perhaps in these patients there might be a deficiency of dihydropteridine reductase, and this was confirmed.[63, 69, 179, 238, 303, 363] Such a deficiency leads to a lack of tetrahydrobiopterin, cofactor not only of liver phenylalanine hydroxylase, but also of tyrosine hydroxylase (EC 1.14.16.2), which catalyses the rate-limiting step in the biosynthesis of catecholamines (Figure 3.2A);[259] tryptophan hydroxylase (EC 1.14.16.4), which catalyses the rate-limiting step in the biosynthesis of 4-hydroxytryptamine (Figure 3.2B);[172] and, according to Nishikimi,[323] of tryptophan pyrrolase (EC 1.13.11.11), which catalyses the ring-cleavage of tryptophan and of indoleamines (Figure 3.3). The activity of dihydropteridine reductase modulates the rate of formation of 3,4-dihydroxyphenylalanine (DOPA) via the hydroxylation of tyrosine (Figure 3.2A).[315] Deficiency of dihydropteridine reductase and thus of tetrahydrobiopterin must affect the availability of important neurotransmitters and hormones,

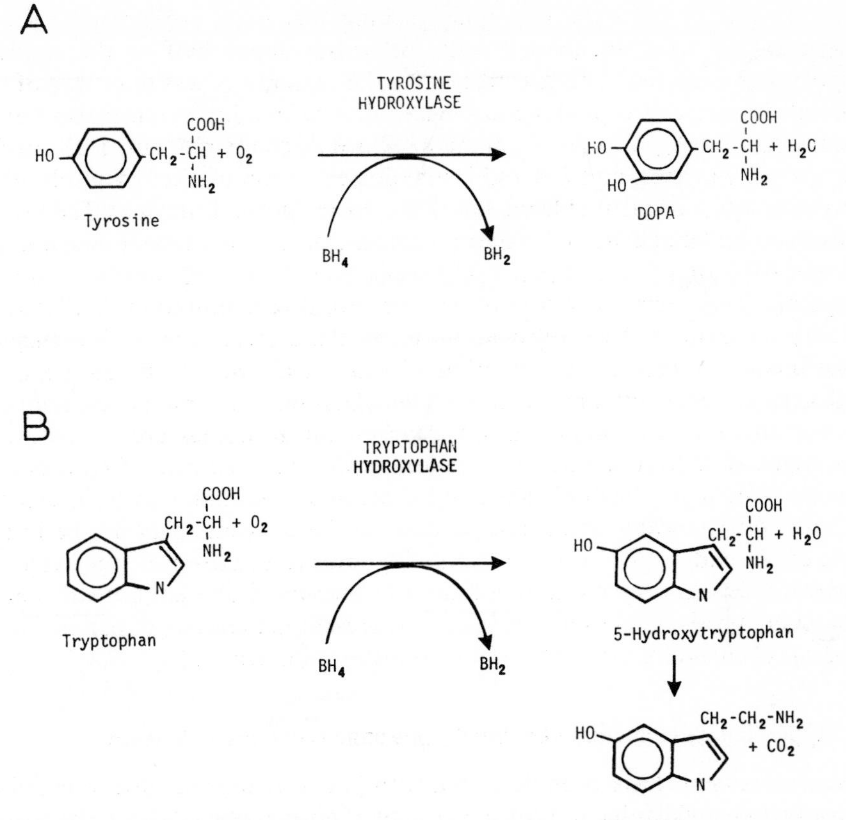

Figure 3.2 Hydroxylations using tetrahydrobiopterin as cofactor. A, Hydroxylation of tyrosine; B, hydroxylation of tryptophan

and also nicotinamide, since tetrahydrobiopterin is cofactor of anthranilate-3-hydroxylase (EC 1.14.16.3) on the pathway of nicotinamide biosynthesis. Defective synthesis of neurotransmitter amines has been demonstrated in patients with dihydropteridine reductase deficiency.[70]

4 Hyperphenylalaninemia and defects in the biosynthesis of biopterin

As Kaufman[234] pointed out, a functional deficit in the availability of biopterin may also result from defects in its biosynthesis, and a patient with such a defect has been discovered.[231, 235, 304] These phenylalaninemias that result from a lack of reduced pteridine cofactor are much rarer than phenylketonuria or other phenylalaninemias caused by deficiencies of phenylalanine hydroxylase activity, but it will be some time before their incidence becomes established.

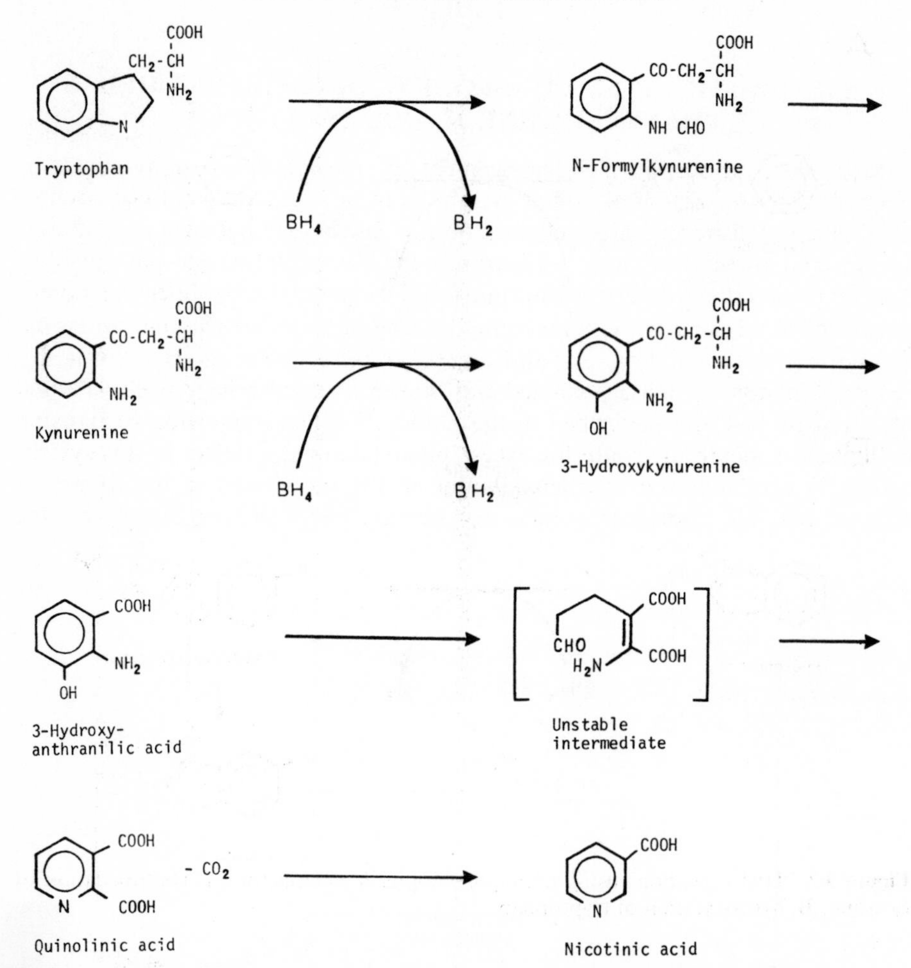

Figure 3.3 Tetrahydrobiopterin as cofactor in the catabolism of tryptophan and in the biosynthesis of nicotinic acid

5 Hyperphenylalaninemia and inability to hydroxylate aromatic amino acids

An enigmatic variant has been described by Bartholomé *et al.*, who discovered a patient with normal phenylalanine hydroxylase and dihydropteridine reductase activities and with normal concentrations of reduced pteridine cofactor in the liver, but who was nevertheless unable to hydroxylate phenylalanine, tyrosine, or tryptophan.[26] It was suggested that this might be due to inhibition of the aromatic amino acid hydroxylases by a substance accumulating as a result of some other metabolic lesion.

C BIOCHEMICAL CONSEQUENCES OF DIMINISHED PHENYLALANINE HYDROXYLATION

Most descriptions of the derangement of phenylalanine metabolism in phenylketonuria present a unified metabolic map of the pathways of phenylalanine catabolism, but such schemes do not distinguish between normal and abnormal pathways. Figure 3.4 outlines the relatively simple pathways of normal phenylalanine metabolism: most of the surplus dietary phenylalanine is converted to tyrosine, with minor routes leading to phenylethylamine, and some being used for protein synthesis and some excreted in the urine. Hydroxylation of phenylalanine is mainly carried out in the liver, although phenylalanine hydroxylase has been described in the kidney.[18] Since conversion to tyrosine is the major metabolic route for excess phenylalanine, a defect in this system results in accumulation of phenylalanine in the blood and in the tissues of affected patients, and leads to increased urinary phenylalanine excretion. Not

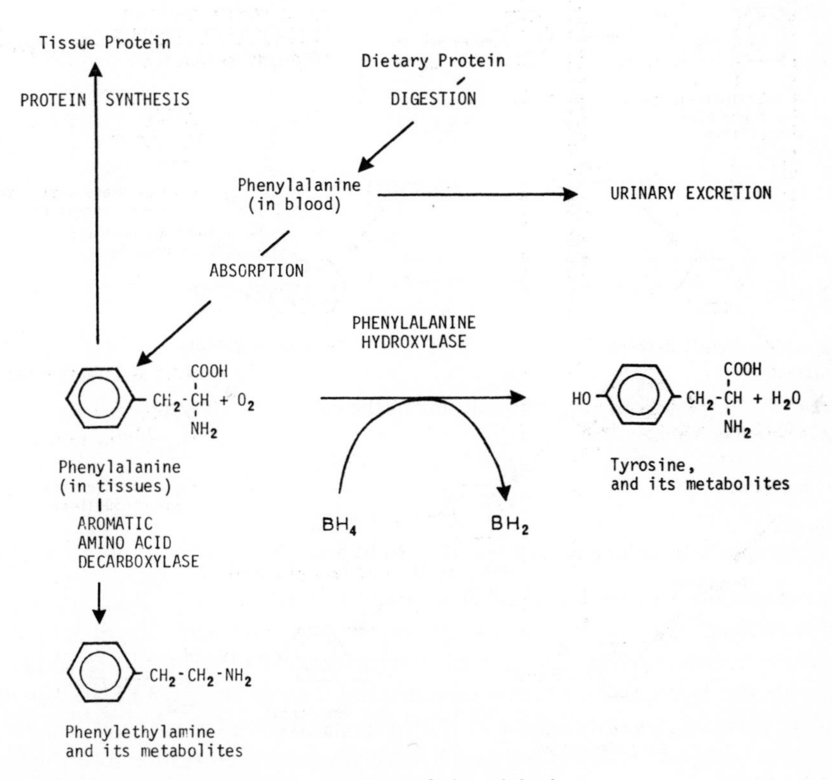

Figure 3.4 The normal human metabolism of phenylalanine

only is more phenylalanine channelled through the normal metabolic pathways of decarboxylation,[216, 325] but transamination, acetylation, and other pathways of phenylalanine metabolism, which are not usually significant, become apparent. It is unlikely that these are all normal pathways used excessively at high phenylalanine levels: phenylalanine at abnormally high concentrations is more probably successfully competing as substrate for enzymes that do not use it to any significant extent within its normal concentration range. Thus a whole new area of abnormal phenylalanine metabolism opens up (Figure 3.5), which has proved of great interest, but whose precise physiological and clinical significance is still being explored.

We must therefore distinguish between metabolites whose normal production increases as blood and tissue phenylalanine concentrations increase and those compounds that are not normally detectable and which only appear when blood and tissue phenylalanine concentrations pass a critical value. This distinction must be kept in mind, because the analytical methods used to

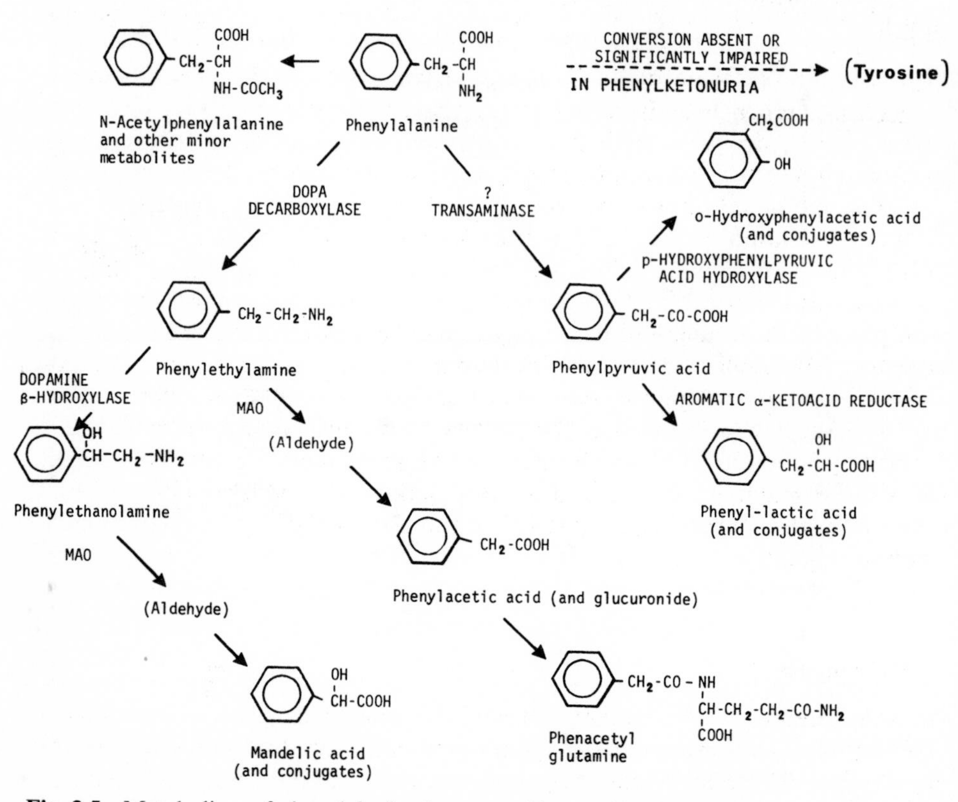

Fig. 3.5 Metabolism of phenylalanine in states of hyperphenylalaninemia

determine the metabolites of phenylalanine tend to conceal it. If we analyse the phenylalanine metabolites excreted in the urine of a normal person, we will detect very little except phenylalanine itself, and traces of aromatic acids[77, 360] and of phenylethylamine.[55, 140] However, in patients with phenylketonuria we will see a progressive increase of all of these as blood phenylalanine concentrations rise, but at a critical point the aromatic acid profile will now also include increasing amounts of *o*-hydroxyphenylacetic acid, β-phenyllactic acid, and phenylpyruvic acid; and indole-3-acetic acid will eventually also appear. Yet there are actually three different mechanisms for the production of aromatic acids and their appearance in the urine: normal metabolic pathways, already described, leading to the excretion of phenylacetic and mandelic acids; transamination of phenylalanine to phenylpyruvic acid and the *o*-hydroxyphenylacetic and phenyllactic acids derived from it; and indole-3-acetic acid arising from the action of the intestinal microflora on tryptophan, whose absorption from the gastrointestinal tract is impaired by competition with increasing amounts of phenylalanine. To complicate the picture, some of these acids may reach the urine by more than one pathway. Figure 3.5 includes normal phenylalanine metabolites which are not usually detectable because they represent an insignificant proportion of the total: these, as well as the abnormal metabolites, become more prominent as the phenylalanine concentration rises. Phenylalanine and its metabolites have been found to exert widespread inhibitory effects throughout the organism, leading to extensive quantitative distortions, quite apart from any specific toxic effects.

1 Decarboxylation of phenylalanine

Decarboxylation of phenylalanine to β-phenylethylamine is catalysed by a widely distributed enzyme or enzymes which have been variously called phenylalanine decarboxylase, aromatic α-amino acid decarboxylase or DOPA decarboxylase (EC 4.1.1.26).[84, 277] Phenylethylamine is widely distributed as a normal tissue constituent,[56, 95, 318, 373] and may have specific physiological functions, both in the nervous system, acting as a stimulatory neuromodulator in some as yet ill-defined fashion,[57, 122, 211, 312] and also peripherally, acting as a 'tonic principle'.[213] Although much of the evidence has come from animal studies, several authors have proposed that phenylethylamine is involved in states of alerting and wakefulness, and that its deficiency may lead to clinically significant effects.[140, 374, 375] Some of the phenylethylamine is converted to phenylethanolamine via hydroxylation catalysed by dopamine β-hydroxylase (EC 1.14.17.1).[122] Monoamine oxidase-catalysed oxidation of these amines leads to production of phenylacetic and mandelic acids[46] as shown in Figure 3.5. The further conjugation of phenylacetic acid in man leads to N^{α}-phenacetylglutamine.[214, 215] Metabolic schemes usually show phenylacetic acid coming from phenylpyruvic acid, but this probably only occurs to a very minor extent, if at all. We found that pargyline treatment of rats with phenylketonuria-like characteristics[121] abolished at least 90% of the phenylacetic acid excretion. Thompson *et al.*[428] have shown that phenylacetic acid can come from phenylpyruvic acid in procedures used for the isolation of aromatic acids, and this may also account for some of the mandelic acid. Urinary phenylacetic acid excretion is therefore an index of phenylethylamine turnover, while mandelic acid excretion reflects phenylethanolamine turnover, and the output of these acids and of phenacetylglutamine is not directly linked with that of phenylpyruvic acid and the other acids derived from it except through their common precursor, phenylalanine. This explains why the output of the aromatic acids derived from decarboxylation of phenylalanine may vary independently of those derived by its transamination. The decarboxylation products come from a normal pathway, and are normal urinary constituents, even though their normal output is small; the transamination products, on the other hand, are more an index of deranged phenylalanine metabolism.

2 The transamination pathways

The phenylketone found in the urine of affected patients is phenylpyruvic acid,[143] whose production is catalysed by an enzyme whose identity and location have not been established unequivocally. Several aminotransferases use phenylpyruvic acid as cosubstrate with an amino acid such as tryptophan or glutamate to yield phenylalanine. If the concentration of phenylalanine were to approach its K_m for such an enzyme, the reaction would go into reverse to yield phenylpyruvic acid. There is no direct evidence for this in phenylketonuria. It has often been proposed that there is a specific phenylalanine transaminase, with a lower K_m for phenylpyruvic acid than for phenylalanine, which goes into reverse at high phenylalanine concentrations.[17] It has also often been suggested that tyrosine transaminase (EC 2.6.1.5) could use phenylalanine in these circumstances, but transamination of phenylalanine occurs even in the absence of this enzyme.[134, 135, 157] Brand and Harper[62] showed that when the enzyme was induced with glucagon to nearly 12-fold normal activity, rats were able to convert a load of phenylalanine to phenylpyruvic acid less well than normal, not better. These contradictions are best explained by the hypothesis of Fellman *et al.*[135] that phenylpyruvic acid production in patients with phenylketonuria is catalysed by a mitochondrial aspartate:2-ketoglutarate aminotransferase (EC 2.6.1.1). The major conversion is envisaged as taking place in heart, muscle, and brain, because these tissues comprise a major proportion of the body mass. They lack *p*-hydroxyphenylpyruvic acid oxidase ('PHPPO', EC 1.13.11.27) which would further metabolize phenylpyruvic acid to *o*-hydroxyphenylacetic acid (see Figure 3.5). Fellman's hypothesis is supported by the finding that this mitochondrial enzyme from pig heart can indeed transaminate phenylalanine,[399] and similar findings for rat brain.[30, 326, 392] The K_m for phenylalanine is relatively high,[30, 392] explaining why phenylpyruvic acid excretion only occurs above a phenylalanine concentration threshold.[46, 47, 100] When normal subjects are given phenylpyruvic acid, the equilibrium was found to be greatly in favour of phenylalanine.[17] However, in phenylketonuria, with increasing blood and tissue phenylalanine concentrations, the K_m for the enzyme is approached, transamination begins, and phenylpyruvic acid spills over into the urine. This phenylpyruvate threshold has been variously quoted, and may be subject to individual, and also to age- and possibly genetically-determined variations, but in phenylketonuric infants it usually lies at a blood phenylalanine concentration of 0.5 mM (8 mg %).[46, 47, 362]

As Fellman *et al.*[135] point out, while brain, heart, and muscle lack PHPPO and can only produce phenylpyruvic acid, liver and kidney contain PHPPO and produce *o*-hydroxyphenylacetic acid via the 'NIH shift' reaction in which the entering hydroxyl displaces the side chain to the adjacent position in the ring[182] with simultaneous oxidative decarboxylation. Phenylpyruvic acid is reduced to phenyllactic acid by aromatic α-ketoacid reductase,[455, 476] which is widely distributed (for these pathways, see Figure 3.5). Of these two systems competing for phenylpyruvate, PHPPO has a high affinity and a limited cap-

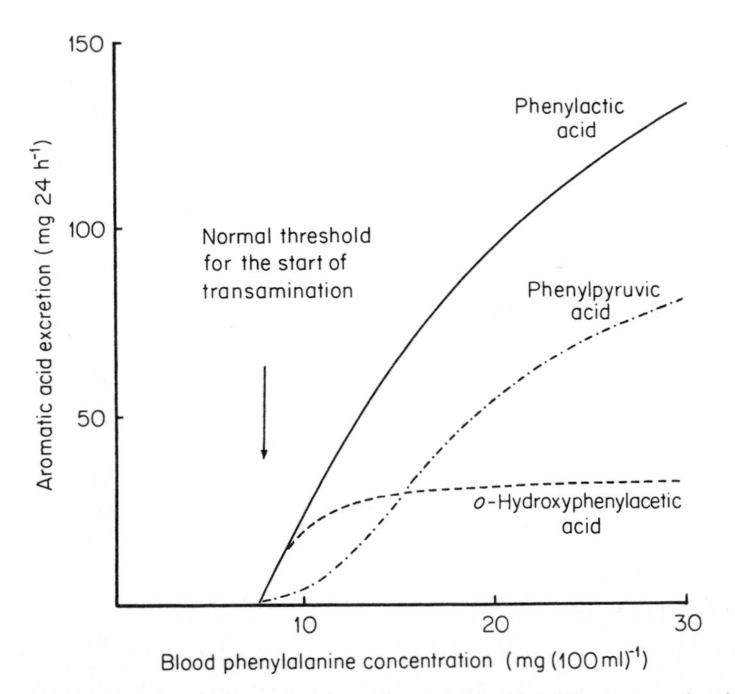

Fig. 3.6 Variations in the urinary excretion of aromatic acids of the transamination pathway with blood phenylalanine concentration in patients with phenylalaninemia

acity, while the reductase has a lower affinity, but a higher capacity (see Figure 3.6). The relative amounts of the three aromatic acids excreted in the urine are therefore governed by the properties and activities of three enzymes (see Table 3.1), which may well be subject to individual variability.

We have found (K. Blau, G. T. Cardwell, and G. K. Summer, in preparation) that *o*-hydroxyphenylacetic and phenyllactic acids are excreted in the form of conjugates with glucuronic and sulphuric acids in the urine of patients with phenylketonuria who are excreting much larger quantities of these acids in the unconjugated form, but these conjugates represent only a minor proportion of the total excretion of these acids.

Table 3.1 Transamination of phenylalanine with production of aromatic acids, according to Fellman *et al.*[135]

Tissue	Aspartate Amino-transferase	PHPPO	Aromatic α-ketoacid reductase	Products
Liver	+ + +	+ + +	+ +	*o*-Hydroxy-phenylacetate + phenyllactate
Kidney	+ +	+	+ +	*o*-Hydroxy-phenylacetate + phenyllactate
Muscle	+	−	+ +	Phenylpyruvate + phenyllactate
Heart	+ + +	−	+ + +	Phenylpyruvate + phenyllactate
Brain	+ +	−	+	Phenylpyruvate + phenyllactate

3 Minor metabolites

Goldstein[169] reported the excretion of *N*-acetylphenylalanine in patients with phenylketonuria. Van Sumere *et al.* found traces of a blue-fluorescent metabolite of phenylalanine in normal urine, which they identified as 7-hydroxycoumarin.[439] Goldstein found α-ureidocinnamic acid in phenylketonuric urine, and gave some evidence arguing against the obvious interpretation that this might be an artifact of storage.[170] Boscott and Bickel[54] found 5-benzalhydantoin in phenylketonuric urine, but suspected that it might be artifactual. 3′,4′-Deoxynorlaudanosoline carboxylic acid, a condensation product of phenylpyruvic acid with dopamine, was found in the urine of children with phenylketonuria, and of rats with experimentally induced hyperphenylalaninemia.[254] This substance appeared to accumulate in the cerebellum and cortex of the experimental animals, and was also taken up by the brain from the peripheral circulation. Phenylpyruvic acid might also condense in an analogous fashion with other biogenic amines, although such condensations would be less favoured energetically than that with dopamine. Loo isolated a condensation product of phenylethylamine and pyridoxal from phenylketonuric urine.[274]

In addition to the metabolites derived from phenylalanine, phenylketonuric patients have been found to have higher than normal plasma levels of biopterin derivatives.[255, 369] However, in patients with hyperphenylalaninemia due to dihydropteridine reductase deficiency, plasma levels and urinary excretion of biopterin derivatives are markedly lower than normal.[257]

4 Interference with actively-mediated membrane transport of amino acids

A great deal of attention has been focused on the mechanisms by which phenylalanine and other amino acids derived from the diet reach their ultimate destination in the body, because the transport of amino acids across membranes is an active metabolic process, which is inhibited by abnormally high levels of phenylalanine. Much of our knowledge of amino acid transport mechanisms comes from studies of reabsorption in renal proximal tubules,[198] but similar systems, under genetic control like other proteins, are widespread,[195, 319] and active transport systems carrying phenylalanine across membranes also carry other large monoamino monocarboxylic acids, such as the branched-chain amino acids and the aromatic amino acids tyrosine and tryptophan.[76] It was shown in experimental animals that when transport systems are overloaded with phenylalanine, it monopolizes them to the partial exclusion of the other amino acids they carry, leading to an amino acid imbalance in crucial areas such as the central nervous system.[5, 76, 264, 268, 269, 448]

The point of entry of phenylalanine from dietary protein is absorption from the gastrointestinal tract into the bloodstream, and both animal experiments and clinical investigations indicate that phenylalanine accumulation diminishes the absorption of the other amino acids that share the same carrier system.[264, 268, 269, 448] The most significant result of this competitive transport inhibition by phenylalanine is probably on the body's tryptophan economy: when competition by increased intracellular phenylalanine concentrations diminishes tryptophan uptake from the intestinal lumen into the bloodstream,[448] an increased amount of tryptophan remains with the intestinal contents, and the resulting increased production by the intestinal microflora of tryptophan metabolites leads to a 'Hartnup-like' syndrome.[39, 217] There is increased production of those tryptophan metabolites with an unaltered indole ring (see Figure 3.7), and a decrease in the ring-cleaved products of the kynurenine pathway,[10, 39, 217, 281, 440] possibly because of inhibition of tryptophan pyrrolase by *o*-hydroxyphenylacetic acid.[10] This distorted tryptophan metabolism is reflected in the urinary excretion pattern of its metabolites.[14, 39]

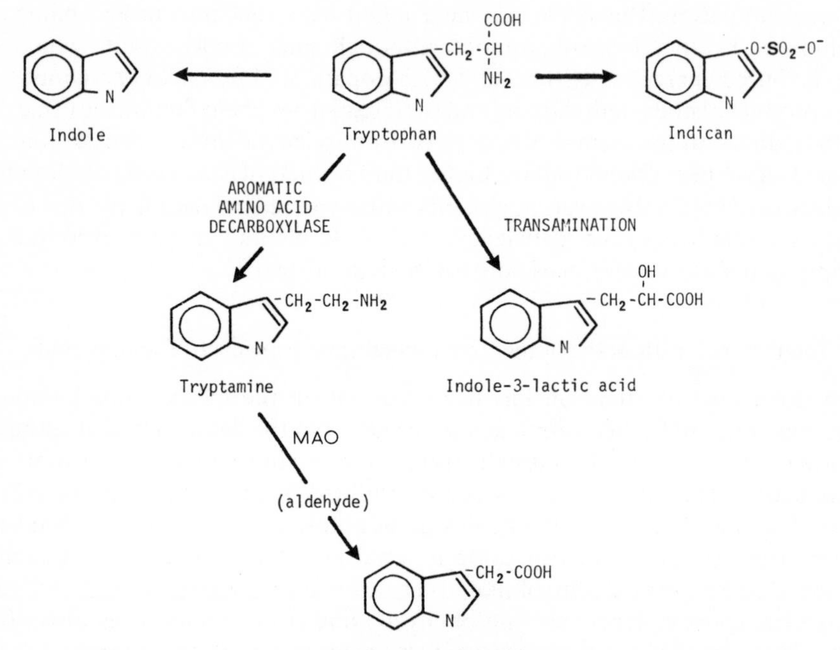

Fig. 3.7 The metabolites of tryptophan which are excreted in the urine in increased amounts in phenylketonuria. The urinary excretion of metabolites of the serotonin (Figure 3.2B) and kynurenine (Figure 3.3) pathways is diminished

Blood amino acid concentrations are generally only slightly decreased in phenylketonuria (except of course that of phenylalanine), in spite of the competition of phenylalanine for transport sites.[64, 88, 124, 269, 292, 435] The most reliable value for the decrease in blood tryptophan concentration comes from McKean and Peterson.[292] However, the uptake of amino acids from the bloodstream into the central nervous system is diminished because of competition from phenylalanine,[12, 19, 22, 223, 278, 292, 435] leading to a significant fall in brain tryptophan concentration.[12, 22] Tryptophan is the amino acid whose deficiency might first lead to a crucial amino acid imbalance and a resultant interference with the synthesis of brain proteins in the correct proportions.[12, 400, 471] Aoki and Siegel showed that a phenylalanine injection in 7-day-old rats caused disaggregation of brain polyribosomes and diminished protein synthesis in proportion to the tryptophan depletion in brain, though not in liver.[12, 400, 471] However, a more recent report on the tryptophan content of cerebrospinal fluid in human phenylketonuria patients found raised tryptophan in patients not under dietary control, and this fell to normal with dietary treatment.[356] This discrepancy may result from subtleties in the hydrodynamics of cerebrospinal fluid.[110, 160] It has been suggested that disaggregation of polyribosomes is the result, and not the cause, of inhibition of protein synthesis.[289] The protein whose synthesis is crucially affected in phenylketonuria, and also in model animal studies, is the basic protein of myelin,[5] the proteolipid 'insulation' of nerve fibres.

In 1957 Pare, Sandler, and Stacey reported a defect of 5-hydroxyindole metabolism in phenylketonuria,[332] confirmed several times since[294, 333, 433] in humans, and also in animal experiments.[474, 475] This has been ascribed to inhibition of serotonin biosynthesis at the amino acid decarboxylation step,[273] or more plausibly to inhibition of the hydroxylation of tryptophan and of its transport.[474, 475] Neuhoff's group found both 5-hydroxytryptophan and serotonin concentrations in the cerebrospinal fluid of patients with phenylketonuria to be significantly increased compared with normal control values,[28, 322] and suggested that extracellular 5-hydroxylated indole derivatives were elevated because of diminished penetration into and increased efflux out of neurons, in other words that the cerebrospinal fluid concentrations do not reflect those inside the nerve cells. Phenylalanine accumulation also affects the renal tubular reabsorption of amino acids.[64, 88, 124, 267] Lines and Waisman found distinctly increased urinary output of specific neutral amino acids, due to inhibition of renal tubular absorption by high intracellular phenylalanine concentrations.[185, 264] We found (K. Blau, J. Graham, and W. Grant, unpublished observations) that when a phenylketonuric patient, controlled by dietary treatment, was challenged with a standard phenylalanine load (0.1 g (kg body weight)$^{-1}$) there was a transient but distinct amino aciduria accompanying the increased phenylalanine output, consistent with the phenylalanine competitively inhibiting renal tubular reabsorption of amino acids, but this effect may be missed unless urinary amino acids are determined at the right time after the load. Lines and Waisman found an even more pronounced amino aciduria in some patients with phenylalaninemia, and speculated whether such patients were phenylketonuric, but protected to some extent against phenylalanine accumulation by their better ability to get rid of it in the urine. Güttler and Rosleff were unable to confirm this in a separate group of children with hyperphenylalaninemia.[185] These effects of phenylalanine on active transport systems in the intestine, kidney, or brain, and indeed in cells generally, must be considered bearing in mind the developmental stage of the organism, for they will have a maximal effect at certain crucial stages of development. This has been extensively documented,[12, 278, 327, 435] and accords with the general concept of critical periods of development.[108]

D THE PATHOLOGY OF PHENYLKETONURIA

A study of the clinical aspects of liver phenylalanine hydroxylase deficiency reveals a range of potentially pathological effects of the toxicity of phenylalanine and of its metabolites, and the data come from clinical observation, post-mortem studies on material from affected patients, and model animal experiments. We can rely on the clinical material and on the post-mortem studies, but the results of the animal experiments must be interpreted more cautiously, because the validity of the animal models depends on factors that will be discussed when we come to the design of such models of phenylketonuria.

1 The clinical symptoms of phenylketonuria and of related disorders

The worst effect of uncontrolled hyperphenylalaninemia in man is mental retardation, and few infants with classical phenylketonuria, and chronically elevated blood levels of phenylalanine from birth, escaped it. The damage was done in the first months of life, most of it probably in the first year, and it is suggestive that this is the period of brain myelination,[108] although of course other systems are developing simultaneously. The degree of retardation was variable, presumably depending on the level of blood phenylalanine, on the timing, and on the susceptibility of the individual patient, but untreated patients were generally severely retarded.

Physical signs often include fair hair and blue eyes in whites, or dilution of normal pigmentation in other races.[27, 93, 126, 174, 357, 396] Dry skin,[27] eczema, and other non-specific skin lesions have often been reported. There may be microcephaly[288, 331] and other pathological signs such as abnormal EEG,[142, 219, 331, 377, 450] epileptiform seizures, and disturbed sleep patterns.[141] Other neurological signs may include stooped posture, muscular hypertonicity, and rocking. There are usually behavioural and psychological disturbances, and untreated patients may be hyperactive, irritable, subject to temper tantrums and erratic aggressive behaviour, and they may also be clumsy both in speech and in motor skills.[186, 340] The neurological and behavioural deficits in phenylketonuria have been ably reviewed.[94] Changes in bone age have been described in some patients.[131]

These signs and symptoms refer to the classical form of phenylketonuria, with very high circulating phenylalanine concentrations and massive phenylpyruvic acid excretion. Conditions of atypical phenylketonuria, mild persistent hyperphenylalaninemia, and other phenylalaninemias in which the blood phenylalanine concentrations are lower than in classical phenylketonuria, are considerably less damaging to the affected child, and where blood phenylalanine concentration does not exceed 1.27 mM (20 mg %) the symptoms may be negligible, and treatment unnecessary. However, we know much less about the various conditions of phenylalaninemia than about the classical variant, because they are rarer and have not been studied as long or as widely. Each patient is unique and needs to be managed individually: a blood phenylalanine concentration below 1.2 mM (<20 mg %) is not a guarantee of immunity from brain damage or mental impairment.[454]

The clinical symptoms of dihydropteridine reductase deficiency are acute and more severe than those of phenylalanine hydroxylase deficiency, and are those of a progressive neurological illness with febrile episodes and seizures, uncontrolled movements, and myoclonus.[24, 106, 404, 406, 407] Greasy skin and hypersalivation have also been described.[25] These effects are ascribable to deficits in brain neurotransmitter amines, because they are reversible on treatment with the amine precursors DOPA and 5-hydroxytryptophan, as will be described in section G. Most of these patients have received treatment, but in spite of normalization of blood phenylalanine levels, the progressive motor impairment may not be arrested. The clinical picture for the patient with biopterin deficiency[235] was similar to that of the patients with dihydropteridine reductase deficiency.

2 The visible effects of phenylketonuria in the central nervous system

A search for physical changes in post-mortem material from patients with phenylketonuria in mental institutions has shown significant abnormalities only in the central nervous system. Crome[98] has given a valuable summary of the findings, and points out some of the difficulties: it is not possible in a post-mortem brain to do a detailed survey of the whole central nervous system, and, in any case, some of the subtler changes are likely to escape notice. Nevertheless, some very significant changes have been repeatedly described: the most important is a moderate (*c*. 10%) reduction in total weight compared to normal in phenylketonuric patients who have died after a lifetime without dietary treatment. Of the microscopic changes, attention has been drawn to myelin, because of the hypothesis that myelination is impaired in phenylketonuria, and the histopathology certainly supports the hypothesis: the total ratio of white to grey matter is decreased,[96] fibrous gliosis is frequently present,[288, 353] and this may indicate loss of axis cylinders and of myelin sheaths.[97, 99] There is also loss of pigment in those regions of the brain that are usually pigmented.[132] Oteruelo has described inclusion bodies of a characteristic lamellar type in oligodendroglial cells in the brains of two patients with phenylketonuria.[330]

3 Maternal phenylketonuria

Since the clinical picture of phenylketonuria is variable, some women with uncontrolled phenylketonuria have been sufficiently normal to get married and to raise families. Also, now that treated phenylketonurics are coming to maturity, they too will marry and have children. It has been repeatedly found that children of untreated phenylketonuric mothers are born with brain damage, although such children are heterozygous, and would not be expected to be clinically affected. The damage is ascribed to the high phenylalanine concentrations to which the foetus was exposed *in utero*, for phenylalanine is elevated in the circulation and tissues of the mother[11, 66, 137, 147, 150, 202, 282, 283, 284, 299, 347, 350, 385, 417, 431, 446, 459, 473] and also in amniotic fluid[125, 427] and the mother's milk.[436] The symptoms in the affected children, who have suffered irreversible damage before birth, do not differ significantly from those of homozygous phenylketonurics, and microcephaly is a frequent finding. Dietary treatment after birth is ineffective, because these children are metabolically normal and do not accumulate phenylalanine. However, provided the problem is recognized in time, dietary control of the mother during pregnancy may prevent intrauterine damage to the foetus,[16, 68, 129] even if a phenylalanine-depleted diet is not started until the pregnancy is under way.[129] There is no danger of mental retardation to the offspring of pregnant heterozygotes for phenylketonuria, because their blood phenylalanine concentrations never get high enough to harm the foetus.[37]

E PHENYLALANINE ACCUMULATION AND HYPOTHESES ABOUT THE PATHOGENESIS OF THE MENTAL DEFICIENCY

In the initial stages of uncontrolled classical phenylketonuria the blood phenylalanine levels may rise to 20–30 times normal, i.e. up to 3–3.5 mM (50–60 mg %). Phenylalanine is an essential amino acid, but one can have too much of a good thing, and the human organism shows biochemical and clinical evidence of overloading long before such extreme levels are reached. It is difficult to assess the toxic effects of phenylalanine and its metabolites because of the variety of the investigations that have produced the evidence. Ideally we ought to restrict ourselves to the affected patient, and to the critical period of development when susceptibility to these toxic effects is greatest.[115] Many of the clinical studies have, however, been done on institutionalized patients who are not being treated with phenylalanine-depleted diets and who are past the crucial developmental stage; on post-mortem material; on animals at various stages of development; on isolated organs; on cells in culture; on tissue homogenates; or on partially or highly purified enzymes. For these reasons it is not easy to obtain a unified view of the effects of phenylalanine in man, and it is therefore difficult to interpret all the results and to relate them to the clinical picture. There are three main hypotheses for the pathogenic mechanism producing the brain damage and mental retardation in phenylketonuria.

The *defective myelination hypothesis* is based on the evidence of post-mortem material from phenylketonuric patients, and on animal studies, for a reduction in myelin and reduced myelination in phenylalaninemic states. It postulates that interference with the biosynthesis of myelin components (both the basic protein and the lipid components) results in a decreased rate of myelin assembly, a delay in the myelination of axons, and possibly even demyelination.

The *monoamine hypotheses* attempt to explain brain damage by a decrease in the concentration of brain neurotransmitter amines. Alternatively, increased turnover of phenylethylamine is supposed to be accompanied by various toxic effects.

Finally, the *glutamine hypothesis* proposes that the obligatory metabolic conjugation of phenylacetic acid with glutamine leads to a chronic depletion of glutamine, which may play a part in damaging the developing human brain.

1 The defective myelination hypothesis

Myelin is a complex proteolipid which consists of a basic protein combined with phospholipid and cholesterol. During myelination layers of myelin are wrapped coaxially around nerve fibres to produce the myelin sheath, which eventually becomes a tightly-wound insulating covering several layers thick. Myelin is metabolically relatively inert[410] and makes up a significant proportion of the brain mass, predominantly in the white matter, and is thought to contribute to the proper functioning of axons, and in particular to the velocity of propagation of nerve impulses. Therefore the synthesis and assembly of all the components of myelin, the correct developmental timing of myelination, and the maintenance of proper myelin sheathing around axons must be essential for the proper performance of the brain. It must follow that interference with any of these processes could have serious consequences. We have seen that the brains of phenylketonurics examined *post mortem* showed deficient myelination,[96, 97, 288, 353] and the hypothesis, first formulated by Alvord *et al.*,[7] therefore correlates this deficiency with the possibility that accumulation of phenylalanine and its metabolites interferes with the normal biosynthesis, assembly, deposition, and maintenance of myelin, particularly because the time-course of myelination parallels the time-course of the brain damage in uncontrolled human phenylketonuria.[109] This hypothesis has been very fruitful, although much of the work done to test it has had to be done on animals, animal tissues, and purified enzymes from animals.

The previously described effects of high phenylalanine concentrations on the active membrane transport of amino acids were shown to lead to what is evidently a crucial amino acid imbalance in the brain, and we know from other evidence that this leads to interference with protein synthesis via a variety of mechanisms.[189] If we visualize the synthesis of a given protein to require an optimal 'mix' of amino acids, then the suboptimal concentration of only one of these is apparently enough to interfere with the biosynthesis of that protein.[40] Disaggregation of polyribosomes for lack of tryptophan has already been mentioned.[12, 401] There seems to be, additionally, interference by phenylalanine with the correct charging of transfer RNA, by what may be a mass-action effect.[13] Hughes and Johnson[209] have shown that the inhibition of neural protein synthesis may be due to a reduction of the initiator transfer RNA, $tRNA_f^{Met}$. These various effects of hyperphenylalaninemia may explain the interference with the synthesis of the basic protein of myelin (the arguments apply equally to the synthesis of brain proteins generally), but additionally a wide range of effects of phenylalanine on lipogenesis have been found, which may also contribute to deficiencies in myelin biosynthesis and myelination.

To the overall defects of myelination, both in phenylketonuric brains and in phenylalaninemic rat brain,[20, 85, 97, 98, 288, 298, 353, 354] we must add effects on specific myelin constituents. Shah, Peterson, and McKean have shown a reduction in the synthesis of cholesterol,[393] but, since the myelin had a grossly normal constitution, concluded that the reduced availability of cholesterol may have been responsible only for a decrease in the rate of myelin biosynthesis.[394] This is confirmed by the finding of reduced cholesterol and phospholipids in the cerebrospinal fluid of phenylketonurics,[34] broadly correlating with the motor disturbances of the patients.[339, 340] Sulphated galactocerebroside production was also reduced.[412] Although the gross lipid composition of the diminished myelin in patients with phenylketonuria appears to be the same as that of normal myelin, Cumings, Grundt, and Yanagihara[102] found slight changes in the fatty acid composition of myelin in phenylketonuria, as well as an overall decrease of cerebroside, mirroring the decreased amounts of myelin. Thus there was a relative decrease of unsaturated fatty acids, mainly 24=1, and the effect of phenylpyruvic and phenyllactic acids on desaturation of stearyl-CoA and on the production of unsaturated fatty acids found by Shah and Johnson[395] confirms these results. In fact, the effect of phenylpyruvic acid on lipid biosynthesis seems to be most marked at the point of synthesis of free fatty acids.[167, 252] Other enzymes involved in lipogenesis have been found to be inhibited by phenylpyruvate, including pyruvate carboxylase (EC 6.4.1.1) in rat and human brain,[335, 336] pyruvate dehydrogenase complex (EC 1.2.4.1),[58, 251, 336] and also citrate synthase (EC 4.1.3.7), acetyl-CoA carboxylase (EC 6.4.1.2),[252] and NADP–malate dehydrogenase (EC 1.1.1.40).[337] Chase and O'Brien also found a decrease in the rate of brain sulphatide formation via inhibition of the formation of adenosine-3′-phosphate-5′-phosphosulphate.[81]

There are many steps in the biosynthesis of something as complex as myelin, but phenylalanine or phenylpyruvate have been shown to inhibit so many of the enzymes involved that it might at first seem that the defective myelin hypothesis has been amply proved. However, the Barbatos[20] do caution that the *in vitro* experiments with animal tissues and added inhibitors always seem to need far higher phenylalanine or phenylpyruvate concentrations to demonstrate inhibitory effects than ever occur in patients with the most fulminant uncontrolled phenylketonuria. Others have made the same observations, and this suggests that *in vivo* the levels of phenylalanine and of phenylpyruvate, though abnormally high, fail to reach the point at which inhibition really occurs. Against this it can be argued that phenylalanine or phenylpyruvate concentrations at key locations, such as within the Schwann cells that synthesize myelin, may be far greater than we would deduce when we average out our analytical results by determining phenylalanine in whole brain or even in brain regions.

The clinical and histological findings, the time-course of myelination and the period in which the damage caused in phenylketonuria occurs, the fact that this damage is irreversible, and that the myelin deficiency is the only permanent structural defect so far found in phenylketonuria, all taken together make a strong case for the association between myelin deficiency and the brain damage and mental retardation of affected patients. However, Geison and Siegel have concluded that myelination is significantly inhibited only when other parameters of brain growth and development are affected.[162]

2 Hypotheses concerning monoamines

The monoamine neurotransmitters adrenaline, noradrenaline, dopamine, and serotonin are involved in the transmission of nerve impulses in both the periphery and in the central nervous system, and we can list three main hypotheses involving them: deficiency of catecholamines; deficiency of serotonin; and increased turnover of phenylethylamine. To some extent these effects are interrelated.

Weil-Malherbe and Bone showed that patients with phenylketonuria had low plasma adrenaline,[456] and Fellman[132] showed that the enzyme which catalyses the first step in catecholamine biosynthesis. DOPA decarboxylase (EC 4.1.1.28), was inhibited by phenylpyruvate, phenyllactate, and phenylacetate (see Figure 3.8). Nadler and Hsia showed that all three catecholamines were significantly decreased in plasma and urine of patients with phenylketonuria.[316] Not only catecholamines, but vanilmandelic acid, the main product of catecholamine catabolism, is depressed in phenylketonuria.[78] The rate-limiting step in catecholamine biosynthesis, hydroxylation of tyrosine to DOPA catalysed by tyrosine hydroxylase (EC 1.14.16.2, see Figure 3.8), is inhibited by high concentrations of phenylalanine *in vitro*.[210] More recently Curtius *et al.*, in elegant isotopic studies using gas chromatography–mass spectrometry, showed that after feeding deuterated tyrosine the metabolites homovanillic acid, vanilmandelic acid, and 3-methoxy-4-hydroxyphenyl glycol, the main catecholamine metabolites found in the urine, were depressed in phenylketo-

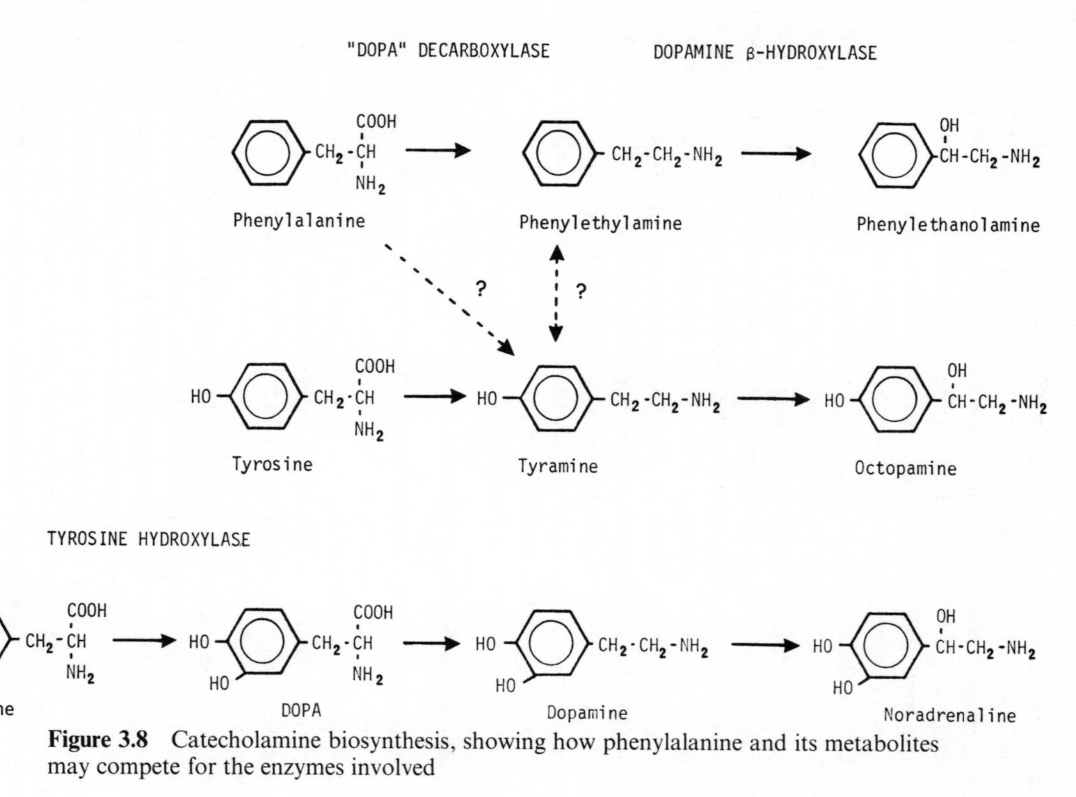

Figure 3.8 Catecholamine biosynthesis, showing how phenylalanine and its metabolites may compete for the enzymes involved

nuric patients with high circulating phenylalanine concentrations.[102] When the phenylalanine was normalized by diet, the output of the catecholamine metabolites went up, and this was attributed to a lifting of tyrosine hydroxylase inhibition, because administration of DOPA, and so bypassing the supposed block, also led to an increased output of catecholamine metabolites. The norlaudanosoline carboxylic acid produced by the condensation of dopamine with phenylpyruvic acid[254] proved to be an inhibitor of dopamine β-hydroxylase (EC 1.14.17.1), and could therefore contribute to a functional deficiency of catecholamines, and by analogy of other biogenic amines in affected patients.

Deficiency of serotonin in phenylketonuria[332] can be explained by the inhibition of tryptophan-5-hydroxylase, which catalyses the rate-limiting step in serotonin biosynthesis, at high phenylalanine concentrations.[173] The next step, decarboxylation of 5-hydroxytryptophan to serotonin, was also found to be inhibited, this time by phenylpyruvate, phenyllactate, and phenylacetate[107] (see Figure 3.2B). Decreases in brain serotonin and catecholamines caused by phenylalaninemia were described by McKean.[294] Although the decrease in brain serotonin has been held responsible for the loss in learning ability in children with phenylketonuria,[468, 474] animals who were given the specific tryptophan hydroxylase inhibitor *p*-chloroamphetamine, which produced a significant fall in the brains of immature rats, did not show any startling loss of learning ability.[384] There were, however, indications that serotonin depletion at a crucial developmental stage did have effects on behaviour. These effects were, in the long-term, reversible, and this is inconsistent with the serotonin depletion hypothesis. A further objection is that sleep is correlated with serotonin levels, and lowering serotonin should therefore influence sleep and sleep EEG patterns. In fact, however, these were almost identical in untreated phenylketonurics before and after 4–6 weeks of dietary treatment.[387] A shortcoming of monoamine-depletion theories is that the reversal of the metabolic disturbance by diet leads to a reversal of the monoamine depletions without permanent after-effects, so it would appear that minor effects on monoamines cannot be responsible for the permanent retardation of phenylketonuria. The evidence from the disease leading to a functional biopterin deficiency is difficult to interpret: a virtual lack of serotonin and the catecholamines since the intrauterine stage clearly has very severe effects on the central nervous system, but it is hard to know at what degree of postnatal monoamine depletion one would begin to see neurological disturbances. It must be remembered that a child with phenylketonuria is born perfectly normal, because the excess phenylalanine is dealt with by the mother while the child is *in utero*. The relatively minor changes in brain monoamine concentrations that occur after birth, and which we have seen are reversible, probably have more of an effect on behaviour and affective states rather than producing mental retardation, although learning and thought may be impaired while the imbalance of brain monoamines is present. On subsequent normalization the deficits in learning may not be made up, and would thus prove permanent.

Similar objections apply to the hypothesis that increased phenylethylamine turnover in phenylketonuria[122, 216, 325, 341] may have neurotoxic effects. This amine, with its amphetamine-like structure and pharmacological properties, is known to deplete stored brain monoamines.[224, 273] Theories underlying its excitatory properties have been mentioned earlier, and its rapid turnover may also tend to monopolize machinery normally used for the synthesis of catecholamines (see Figure 3.8), leading to the synthesis of increased amounts of phenylethanolamine and octopamine,[122, 470] amines that could exert 'false neurotransmitter' effects. There is evidence that this increased turnover of phenylethylamine leads to the formation of a Schiff base with pyridoxal,[274] which has neurotoxic properties,[272, 274] and may also deplete vitamin B_6. Kurtz *et al.*[248] showed that vitamin B_6 deficiency in suckling rats leads to a 30–50% decrease in cerebral lipids. If this were applicable to the clinical situation, it might be an additional biochemical mechanism for interference with lipid synthesis and myelination in the central nervous system.

Inhibition of glutamic acid decarboxylase (EC 4.1.1.15) by phenylalanine and phenylpyruvic acid has also been reported.[188, 320] This would lead to a deficiency of γ-aminobutyric acid, an inhibitory neurohumour, which might contribute to the excitability and short attention span observed in affected patients additionally to the excitatory effects of an excess of phenylethylamine.

3 The glutamine hypothesis

This hypothesis was originally proposed by Meister *et al.*[297] in relation to the increased output of phenacetylglutamine (PAG) in phenylketonuria (see Figure 3.5), and the biosynthesis of PAG was worked out by Moldave and Meister:[308, 309, 310]

$$\text{Phenylacetate} + \text{ATP} \rightarrow \text{Phenacetyl-AMP} + \text{PP}_i$$

$$\text{Phenacetyl-AMP} + \text{CoA} \rightarrow \text{Phenacetyl-CoA} + \text{AMP}$$

$$\text{Phenacetyl-CoA} + \text{Glutamine} \rightarrow \text{PAG} + \text{CoA}$$

The pathway appears to be adaptive. This hypothesis was revived in its present form by Perry *et al.*[346] They found that the plasma glutamine concentration of an untreated mentally defective patient with phenylketonuria was significantly lower than that of his equally untreated but clinically normal phenylketonuric brother, and on extending the study low plasma glutamine turned out to be a general feature of severely affected phenylketonuric patients. Since plasma glutamine was the only biochemical parameter that was found to be different between these two brothers, and since the plasma glutamine in the affected brother was difficult to normalize by oral glutamine administration, it was proposed that chronic glutamine deprivation in the brain, especially during the critical developmental period, might contribute to neurological damage.[346] These suggestions have not been supported by the results of others. However, glutamine is not readily determined accurately and tends to decompose on storage[345] and there has been confirmation of decreased levels of glutamine in the plasma of affected patients.[191] Glutamine is important, because it is involved in the protection of the brain against the accumulation of the neurotoxic ammonia:

$$\alpha\text{-Ketoglutarate} + NH_4^+ + NADH \xrightarrow[\text{dehydrogenase}]{\text{Glutamate}} \text{Glutamate} + NAD^+$$

$$\text{Glutamate} + NH_3 + ATP \xrightarrow[\text{synthetase}]{\text{Glutamine}} \text{Glutamine} + ADP + P_i$$

Both these reactions use up ammonia and diminished availability of glutamine may slow up both of these reactions. The patient with uncontrolled phenylketonuria is already transaminating abnormally large amounts of phenylalanine to phenylpyruvic acid, and, if the resulting increased amounts of ammonia cannot be removed as effectively as normal, hyperammonemia may occur, which we know from other clinical states is highly toxic to the central nervous system.

Most of the patients with hyperphenylalaninemia studied by Perry *et al.*[346] were adults, and, since the brain damage in phenylketonuria happens mainly during the first postnatal months, the reasons for any given patient being affected or escaping mental deficiency must lie in infancy, and may relate to dietary and other influences at that time, leading to eventual more permanent consequences. Our understanding of the factors that govern susceptibility, resistance, or tolerance of phenylalanine and its metabolites is very limited. The glutamine hypothesis is distinctive in postulating a permanent metabolic rather than structural defect. This might be a genetically determined pre-existing difference between the two brothers studied by Perry *et al.*, segregating independently, by which the normal brother has a rare mutant gene which in some way confers protection on him. The affected brother, like most other untreated phenylketonurics, falls victim to the disease mechanisms. Such a polymorphism would be unlikely to be noticed in the population at large, nor would it be easy to identify or study, which will make it difficult to confirm or disprove the glutamine hypothesis, but it does offer a basis for further investigations.

4 Other systems affected by accumulation of phenylalanine and its metabolites

No matter how harmless a compound may normally be, if its concentration in blood and tissues rises to abnormal levels, toxic effects will eventually appear. Phenylalanine, and phenylpyruvic acid, are, however, relatively toxic in any case, and have been found to interfere with a wide range of enzymes and other systems, as already described. Other places where their toxic effects have been described and put forward as possible contributions to the pathogenic mechanisms in phenylketonuria are set out here: Patel and Arinze have given a good summary.[337] Pyruvate kinase (EC 2.7.1.40) has been found to be affected, both *in vitro* and *in vivo*,[73, 165, 166, 301, 372, 389, 441, 451, 452, 453] and the effect appears to be specific to phenylalanine, acting as an allosteric inhibitor.[372] This inhibition leads to important changes in glycolytic metabolites in brain and liver,[165, 301] notably increases in dihydroxyacetone phosphate, fructose diphosphate, phosphoenolpyruvate, and lactate. Glycolysis has also been shown to be affected via inhibition of hexokinase (EC 2.7.1.1),[42, 165, 166, 452, 453] and Kamaryt and Mrskos[226] explained the hypoglycaemic episodes sometimes seen in fasting phenylketonurics by showing that glutamate:pyruvate transaminase (SGPT, EC 2.6.1.2) in patient liver biopsies was inhibited. Oxidative phosphorylation too can be affected,[159] and brain metabolic activity generally was found to be depressed in model rat experiments, mainly through stimulation of brain ATPase by phenylalanine.

Various enzyme systems outside the brain have been found to be inhibited by phenylalanine and its metabolites. Phenylpyruvate has been shown to inhibit gluconeogenesis in kidney[246] and liver.[58] The dilution of pigmentation of untreated phenylketonurics[93, 126, 174, 357, 396] is due to a fall in melanin biosynthesis caused by inhibition of tyrosinase (EC 1.14.18.1).[104, 307] Phenylalanine has also been found to be a specific inhibitor of the intestinal form of alkaline phosphatase (EC 3.1.3.1).[141]

Significant amounts of phenyllactate and phenylpyruvate were found in the brains of rats in model experiments[121, 168, 276] and also in patients with phenylketonuria.[339] Some toxic effects of phenyllactic acid, notably inhibition of decarboxylases,[107, 132] have been described, and effects on lipid metabolism[393, 395] and on respiration have also been found.[112] Since significant brain phenyllactate concentrations may occur in states of phenylalaninemia, further investigation of the toxic effects of this acid may be worth while. Lähdesmaki and Oja showed that phenyllactate, as well as phenylpyruvate (and homogentisate), inhibited protein synthesis by affecting the formation of aminoacyl-tRNAs.[250] Silberberg[350] did not find phenyllactic acid toxic in his cerebellum cultures, but phenylacetic acid was toxic, and this acid occurs in the central nervous system, both normally, and in increased amounts in rats with induced phenylketonuria-like characteristics[121, 276] as the end-product of phenylethylamine metabolism. The increased phenylethylamine turnover in phenylketonuria and related conditions may result in significant increases in brain phenylacetic acid, and local concentrations at the site of MAO-catalysed oxidation of phenylethylamine may be particularly high. This may be relevant in connection with the glutamine hypothesis, since phenylacetic acid may act to deplete brain glutamine via obligatory metabolic conjugation to form phenacetyl glutamine.

5 Tyrosine deficiency

If the body cannot make tyrosine from phenylalanine, then dietary tyrosine becomes an essential amino acid, and since it is precursor to so many important substances (thyroid hormones, catecholamine neurotransmitters, melanin pigments, etc.), a relative shortage of tyrosine may contribute to the pathology of phenylketonuria. We have already discussed the imbalance in amino acid composition caused by an excess of phenylalanine, and in particular its interference with the membrane transport of aromatic amino acids leading to a relative decrease in brain tryptophan and tyrosine.[40, 263] Although no direct evidence has appeared to suggest that a lack of tyrosine contributes to the symptomatology of phenylalaninemia, Snyderman *et al.*[411] showed that supplementation with tyrosine deepened the hair colour of phenylketonurics, but had no other clinical benefits. Concern has been expressed that patients on a phenylalanine-depleted diet for phenylketonuria may not be getting enough tyrosine,[65, 243] but there is at present no real evidence of this.

6 Discussion

We ought now to be able to draw some conclusions about the cause of the mental deficiency in phenylketonuria, and about the biochemical mechanism that leads to the brain damage, but in fact we have no definite proof for any such mechanisms. We know some of the toxic agents: inhibition by phenylalanine and its metabolites is widespread, though it may be oversimplifying to suppose that there is a unique mechanism of pathogenesis. Inhibitory effects may influence the organism in several ways, some more severe, others more subtle. The dietary restriction to bring phenylalanine concentrations down to more normal levels eliminates the toxic metabolites and does away with all these influences, and this is the beauty of the treatment. The toxic effects are chiefly felt in the central nervous system, in the white rather than the grey matter, and possibly most crucially in the cerebellum.[3, 161, 345] The cerebellum is sensitive to metabolic accumulations in other inborn errors of metabolism, for instance in the mucopolysaccharidoses.[119] Silberberg used cerebellar cultures to test the toxicities of several substances known to accumulate in uncontrolled phenylketonuria.[402] It is suggestive that in a description of patients with congenital lactic acidosis due to deficiency of the pyruvate decarboxylase portion of the pyruvate dehydrogenase complex (a system also inhibited by phenylpyruvate), symptoms included microcephaly, cerebellar ataxia, motor disturbances, and mental retardation.[58, 130]

Deficient myelination is the most probable explanation for the brain damage and mental retardation, because it coincides in time with the period when we know the damage is done, and because it is characteristic of the disease. On the other hand, subsequent demyelination may not be too relevant, because patients with demyelinating diseases do not generally become retarded, but again, the timing may be important.

The question has been asked: 'Which is more toxic, phenylalanine or phenylpyruvic acid? Or some other metabolite?' An answer to this might provide a more effective approach to treatment, by specifically getting rid of the toxic agent. Treatment with a phenylalanine-depleted diet works well, but has drawbacks and imposes stresses that may distort the family environment, so an alternative treatment, or a way that would enable patients to manage with some simple protein limitation, would be attractive. A consideration of the transamination of phenylalanine to phenylpyruvic acid is relevant to this point. Transamination normally occurs when the blood phenylalanine increases to over 0.5 mM (8 mg %), so that patients with uncontrolled phenylketonuria, with phenylalanine at over 2 mM (33 mg %), also produce a lot of phenylpyruvate. Perry *et al.*,[346] in connection with their glutamine hypothesis, described several adult patients with normal intelligence who had hyperphenylalaninemia, and who, according to genetic studies on their families, and from other data, are regarded as being homozygous for the classical mutation or perhaps may be compound heterozygotes for the classical mutation and a hyperphenylalaninemia gene. We investigated their aromatic acid excretion (T. L. Perry, K. Blau *et. al.*, unpublished) and were surprised to find in all of them a threshold of transamination of over 1 mM (over 16 mg %), and relatively much lower phenylpyruvic acid output. In contrast, Wadman *et al.*[444] described patients who produced phenylpyruvic acid without phenylalaninemia or apparent deficiency of phenylalanine hydroxylase: their threshold of transamination lay below 0.1 mM (less than 1.5 mg %) and they too had normal intelligence. Although we cannot build much on a few atypical cases, it could be that what is damaging is not phenylalanine on its own, nor phenylpyruvic acid and the transamination metabolites on their own, but that it may be the combination of phenylalanine with the transamination products acting together that is so toxic to the organism. This might be tested by prevention of phenylpyruvate formation through inhibition of the transaminase by such drugs as aminooxyacetic acid or isoniazid.

While we cannot definitely identify the precise biochemical mechanisms by which the developing brain suffers damage in uncontrolled phenylketonuria, we do have a lot of evidence of ways in which it may occur. More than one of these may be operative simultaneously, or they may be happening at different stages of development.

F SCREENING, INCIDENCE, AND GENETICS

1 Mass screening for phenylketonuria

When Bickel, Gerrard, and Hickmans first described successful treatment of a patient with phenylketonuria, using a phenylalanine-depleted casein hydrolysate,[41] the whole picture of phenylketonuria changed, because there was now something that could be done for affected patients. Experience with the treatment made it clear that brain damage already present could not be undone by treatment, and that the patients therefore had to be treated early in life, before damage could occur.[229] The only way to find the patients before they became affected was to test all infants for the biochemical features of phenylketonuria in time to start treatment of those found to have the condition. This led to the setting up of mass-screening programmes to test all infants soon after birth, and required decisions to provide finance, resources, and manpower indefinitely. The rationale and factors underlying such screening programmes have been widely discussed, particularly the need for satisfactory follow-up, confirmatory, and treatment facilities to be available before the screening programme is started,[1] but some critics felt that the phenylketonuria programmes were not widely enough discussed beforehand, and were set up too hastily.[90] The fact that 37 of the states in the USA legislated for phenylketonuria programmes[83] was resented as an intrusion of the law into medical matters.[38] On the other hand, those pressing for the programmes were daily seeing children deteriorating—needlessly as they saw it. The programmes, for whatever reason, were 'sold' on the financial benefit/cost ratio, which is clear-cut,[83, 391, 413, 438] let alone the prevention of wasted lives and of anguish to the families. Pilot studies[163] using Følling's ferric chloride test[143] for phenylpyruvic acid in urine were found to be unreliable,[415] and we now know that this was because transamination of phenylalanine may not occur to any significant extent in affected patients for some time after birth, through delay in maturation of the enzyme system responsible.[15] There were technical problems, sensitivity was not high enough, the programmes were administratively awkward, and their coverage disappointingly low. For these reasons tests were developed that would determine blood phenylalanine. Although the prospect of dealing with large numbers of blood samples may have been daunting, Guthrie's development of sample collection in the form of blood spotted on filter paper and allowed to dry, made his microbiological inhibition assay a mass-screening method which is now the most widely used. Other methods for the determination of blood phenylalanine included paper chromatography using small heparinized glass capillary tubes for blood collection,[33, 123, 390] which had the advantage that other aminoacidopathies were screened for at no extra cost. The fluorimetric method of McCaman and Robins[291] was automated[193] and found to have some advantages,[200] but used Guthrie's blood spot samples. Combined with an automated blood tyrosine determination[48] it has many applications additional to screening.[49] Thin-layer chromatography[4, 35, 101, 245, 249, 351] and thin-layer electrophoresis[127, 376] have also been applied to screening programmes, with improved speed and resolution, although like paper chromatography these procedures are not quantitative. Screening programmes for phenylketonuria (and in many places, for other inborn errors of metabolism) using these procedures are now widespread in developed countries (for example in North America;[149, 175, 192, 199, 236, 262, 279, 445] in Europe;[21, 72, 280, 296, 422] and in Australia.[349])

Since the establishment of these programmes there has been improvement in their coverage, which in most places now exceeds 90% of the newborn population.[190] With experience came improvements in the delay between detection and treatment. Benefits of screening included the discovery of other types of phenylalaninemia, bringing, however, dilemmas of management and diagnosis, and there is now a realization of the genetic heterogeneity of phenylalaninemias, and of inborn errors of metabolism generally. Few would now advocate stopping mass-screening for phenylketonuria, but with the experience gained, other mass-screening programmes, such as for hypothyroidism, are being approached more cautiously. There are three main reservations about screening programmes: that it is often not easy to assess whether the screening test will identify the right patients; that the test may uncover many basically normal patients who do not need treatment, but will have to be individually evaluated, and may get treated unnecessarily; and that there are more urgent needs. As an example of the problems exercising all concerned we may mention the present position with histidinemia: only half the patients may need treatment, but we cannot at present readily determine which half.[244, 352, 424]

2 The incidence of phenylketonuria and related disorders in various populations

The screening programmes have given an indication of the incidence of phenylketonuria and related disorders in various populations. There are significant differences, with the highest incidence reported from Ireland, 1:5000 live births,[72, 75] Scotland, and Iceland,[432] most probably due to contributions from people of Celtic origins.[381] Incidence varies in other European countries: 1:6275 in some East German populations,[286] 1:9000 in Denmark,[280] 1:12 000 in England and Wales,[296] 1:33 000 in Sweden,[477] and less than 1:100 000 in Finland.[442] In the North American populations the average incidence is in the region of 1:15 000 to 1:20 000.[149, 175, 236, 413] The proportion in the Negro population is much lower.[126, 174] Figures from Israel indicate a lower incidence in Ashkenazi as distinct from Sephardic populations.[87] Some cases have been reported from Japan,[396, 419] where the incidence is about 1: 60 000, and from India,[27, 357] but without indication of incidence. In Australia the incidence is 1:11 000.[349]

It has been suggested[38] that not all phenylketonurics become mentally deficient, and that they may be living undetected in the population at large, but in a large survey of the adult population in Massachusetts it was found that the few uncontrolled adult phenylketonurics detected were in fact mentally retarded.[260] Others have reported 'phenylketonurics' of normal or borderline normal intelligence in the general population, but they have usually been persons with phenylalaninemias rather than classical phenylketonurics.[171, 205] The incidence of the various phenylalaninemias cannot be reliably assessed, since there is no universally accepted classification, but taken together they are probably about half as common as the classical form.

Since screening programmes provide figures for the incidence of the phenylketonuria gene in different populations, this information may be used as a genetic marker to draw conclusions about the relationships of different populations. Such anthropological studies led Saugstad[382] to conclude that the Viking raids of the tenth and eleventh centuries brought western influences and people to the shores of Norway, bringing with them not only the goods found in their graves, but also the gene for phenylketonuria. Analysis of the birthplaces of phenylketonurics and of their grandparents showed a good correlation with coastal areas known to have been settled in Viking times, and with the locations of burial sites that have yielded bronze objects of western origins.

A surprising finding of the screening programmes was that there appeared to be an unequal sex ratio in the number of patients detected: about 2:1 in favour of males,[116, 208] whereas an autosomally inherited disorder ought to have a 1:1 sex ratio. This caused considerable discussion.[240, 261, 328] However, the expected sex ratio was found in ascertained patients[71, 190] and it seems that the unequal proportions were an artefact of screening caused by the arbitrary time at which blood is taken for screening. Male phenylketonurics tend to have a steeper initial rise in blood phenylalanine, and if samples are taken too early, will be preferentially detected, while some females may be missed.[201] These differences may be due to hormonal effects on residual phenylalanine hydroxylase or on some of the other enzymes whose activities govern the blood phenylalanine concentration.[177, 295] It is important not to collect samples for screening until both female and male phenylketonurics can be clearly distinguished from normals, which appears to be some time after the fourth day of life.

The best comment on screening is an encouraging paper by MacCready[285] documenting the decline in admissions of patients with phenylketonuria to institutions for the mentally retarded since the screening programmes were started: one hopes there will be many more such reports, a vindication of the effort that has gone into the detection and treatment of patients with phenylketonuria.

3 Association of phenylketonuria with other rare disorders

Phenylketonuria is bad enough, but reports appear from time to time describing the association, in a single patient, of phenylketonuria with some other rare inherited disorder. The probabilities, though extremely low, are finite, and given random mating and the known incidence of the separate disorders, can easily be calculated. It is surprising that an infant who has two inborn errors of metabolism can survive, but it does happen if neither of the diseases is severe enough to be life-threatening.[321] Analogous associations incompatible with survival probably occur. Some of these associations that have been reported are listed in Table 3.2, and Blascovics has quoted the association of phenylketonuria with other rare conditions.[5]

Table 3.2 Association of phenylketonuria with other rare disorders, and estimated probabilities of such associations

Disorder	Estimated probability of association	References
Down's syndrome (mongolism)	3×10^{-6}	51, 136, 164
Hereditary enamel hypoplasia	10^{-9}	321
Hypophosphatasia	5×10^{-10}	45
Klinefelter's syndrome	5×10^{-8}	29
Leucodystrophy	Unknown	96
Juvenile amaurotic idiocy (Spielmeyer–Vogt–Batten syndrome)	Unknown	364
Cystinuria	6×10^{-8}	305
Rubinstein–Taybi syndrome	Unknown	386
Duchenne muscular dystrophy	3×10^{-8}	368
Maternal von Gierke's disease (glycogen storage disease, type 1)	10^{-6}	128

4 The genetics of phenylketonuria and of related disorders, and their classification

Family studies of many kindreds indicate that phenylketonuria and the phenylalaninemias are transmitted via the autosomal recessive mode of inheritance.[238] A familial association has been reported between phenylketonuria and milder forms of phenylalaninemia.[92, 205] The severity of the symptoms in affected infants ranges over a continuum, from the severe defect of the classical variant with virtually no hepatic phenylalanine activity to mild persistent phenylalaninemia with residual enzyme activity of several per cent of normal, but there does not seem to be a clear-cut relationship between the amount of residual activity and the intelligence of the affected child. Such studies are now practically impossible, since all infants who need treatment should now receive it, and liver biopsies may be difficult to justify in normal patients. However, the enzyme is expressed in kidney[18, 314] and also in human term placenta[289] and there are unconfirmed reports that it has been detected in human skin fibroblasts.[196] In this connection it is interesting that there is a polymorphism of normal human phenylalanine hydroxylase,[23] with three isozymes charmingly designated π, κ, and υ (pi, kappa, and upsilon). The activities of these isozymes were determined in patients with phenylketonuria and other phenylalaninemias, who showed deficiencies of one, two, or all three isozymes.[354] There has been insufficient pedigree analysis to establish how these isozymes are transmitted, nor is there any immediately obvious correlation between isozyme pattern and levels of enzyme activity. Genetically determined polymorphisms of this kind may underlie variations in the degree of phenylalanine accumulation and of the resulting toxic effects. Similar polymorphisms may exist in other enzymes of phenylalanine metabolism and could contribute to individual variability in resistance or susceptibility to these toxic effects, as suggested already in connection with the different transamination thresholds.

There is a real need for rationalization of the evidence, both clinical and biochemical, obtained from patients, and we also need some established basis for trying to establish their dietary needs. Paediatricians have tried to classify patients into systematic schemes for this purpose, and some generally accepted principles are emerging. Everyone puts classical phenylketonuria as the first type, agreeing that these children need treatment. And everyone concludes with the transient phenylalaninemia and tyrosinemia due to immaturity of enzyme maturation which usually resolves spontaneously and needs no treatment. In between are phenylalaninemias giving difficulties both in classification and in management, because they are so rare that each patient almost seems to fit into a separate category. A single mutation at each of the loci coding for one of the three isozymes of phenylalanine hydroxylase, together with independent segregation and random mating, can result in a wide variety of distinct genotypes, and so these difficulties of classification very possibly rest on real phenotypic differences.

An early attempt at classification was based on the blood phenylalanine levels of the untreated infants, with some consideration of the extent of excretion of phenylpyruvaic acid and other aromatic acids.[74] The authors distinguished classical phenylketonuria from atypical phenylketonurias that levelled off below 1.8 mM (less than 30 mg %) and possibly needed treatment, and mild phenylalaninemias with levels lower than 1.2 mM (less than 20 mg %) needing no treatment. They included phenylalanine transaminase deficiency with normal hydroxylase activity but little or no aromatic acid in the urine as also needing no treatment except a close watch on protein intake, and the delayed enzyme maturation phenylalaninemia secondary to transient tyrosinemia. Essentially similar classifications have been proposed by Hsia[207] and by Charpentier *et al.*[82] Classifications on the basis of a standard phenylalanine tolerance test with a load of administered phenylalanine (0.1 g (kg body weight)$^{-1}$) had led to similar schemes.[180, 359] Attempts have been made to relate the activity of the phenylalanine hydroxylase system to the degree of clinical impairment of the patients, in spite of experimental difficulties, but have given disappointing results.[178] A scheme using prolonged challenges with normal diet has been proposed by Blaskovics, who plotted the patients' responses over the 72–96 h. test period.[43, 44] These responses (blood phenylalanine plotted against time) were classified into seven distinct types, and the classical variant was subdivided into two disease entities (types 1 and 2) depending on the severity of the clinical outcome in untreated patients. Then come variant types 3, 4, and 5, characterized by different time-courses of blood phenylalanine encountered during development to maturity, followed by type 6, the transamination defect, and finishing with phenylalaninemia type 7, secondary to neonatal tyrosinemia of delayed enzyme maturation. Actually there is also a type 8, tyrosinosis, with secondary phenylalaninemia, although tyrosinosis and tyrosinemia syndromes must be dealt with as diseases of distinct metabolic schemes. The most recent of the classifications seeks to include all the variant forms, including those of dihydropteridine reductase deficiency and other states of dihydropteridine deficiency[430] (see also Danks *et al.*[106]). Unanimity about these classifications is lacking: the real need is to have objective biochemical as well as clinical criteria for assigning any given patient to a specific category with its appropriate treatment and management. This is easier said than done because the symptoms tend to shade into one another, and because patients at different stages of development may move from one category to another.

Blood phenylalanine is a better, and certainly a more convenient, criterion for diagnosis than liver phenylalanine hydroxylase activity, but we may have to consider other factors, including transamination, to decide how best each patient ought to be managed. Transamination will perhaps repay more detailed study: not in the simplistic way used hitherto of measuring the aromatic acid at any given moment, but in greater detail and in a wider context. What does increased transamination involve? Where does the ammonia go? What cofactors are involved? What is the energetic cost? What are the consequences of increased glutamate or aspartate production? In phenylketonuria it appears that some important metabolic pumps may be running backwards, and we must begin to look at what they could be geared to, because already there are tantalizing hints of fascinating answers to these questions.

5 Detection of carriers of the phenylketonuria gene

Følling himself first tried to detect carriers of the gene for phenylketonuria in the families of his patients[144, 145] by giving his subjects a load of phenylalanine and looking for phenylpyruvic acid in the urine with the ferric chloride test. This did not work, because unfortunately D,L-phenylalanine was given, and half of it was deaminated to phenylpyruvic acid by D-amino acid oxidase (EC 1.4.3.3) anyway. Since then, however, the basic design of the phenylalanine tolerance test has been almost universally used to detect heterozygotes, but there have been many different ways of analysing the responses, and in evaluating the genetic status of the subjects. Figure 3.9 is a diagrammatic representation of the three types of phenylalanine tolerance curve obtained from phenylketonurics, normals, and heterozygotes. The ranges of curve obtained for heterozygotes overlap with that from normals, and a proportion of subjects cannot be reliably scored. Ways of improving the discrimination either chemic-

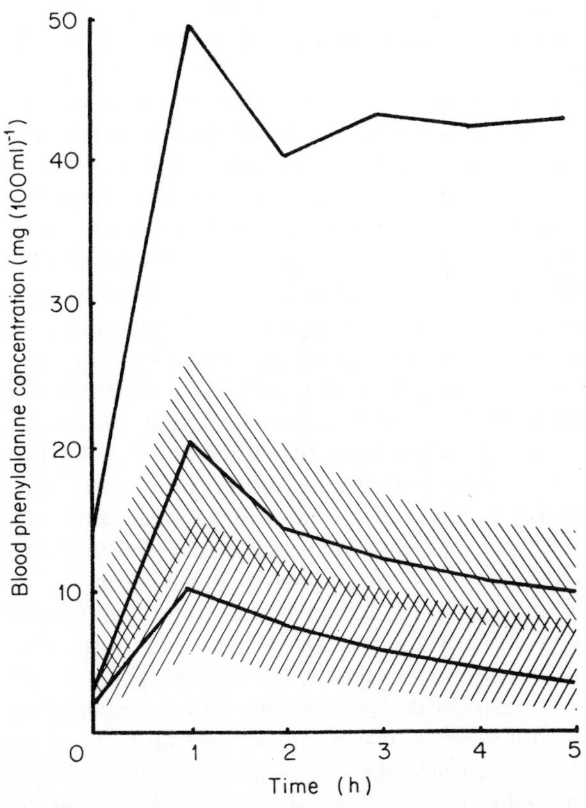

Figure 3.9 Phenylalanine tolerance curves of homozygotes (top) and heterozygotes (middle) for phenylketonuria, and of normals (bottom), showing the zone of uncertainty between heterozygotes and normals

ally or mathematically have been explored. Since phenylalanine is the starting material for the critical conversion to tyrosine, to get a measure of the *in vivo* hydroxylating capacity of the organism one should also consider the time-course of blood tyrosine concentration.[9, 220] This approach was used to develop a discriminant score mathematically from the numerical values of blood phenylalanine and tyrosine concentrations at various times after the administration of the phenylalanine load, and this method of scoring reduced the error to about 5%.[9] Perry *et al.* tried to improve the test by intravenous infusion of the load,[343] and also found that the phenylalanine-to-tyrosine ratio, either at intervals after the load[343] or fasting[344] was a more useful parameter. The ratio $[\text{Phenylalanine}]^2/[\text{Tyrosine}]$ was used by Griffin *et al.*[176] These procedures depend on accuracy and precision in the amino acid analyses, and ideally ion-exchange chromatography must be used. Mathematical processing of the data by discriminant analysis[358] gave a classification error of less than 1%, and Kääriainen and Karlsson[199] used logarithmic variables and computerized discriminant analysis to achieve a clear-cut unambiguous separation. Since there are very few phenylketonurics in Finland, they had to use the data of Rampini *et al.*[358]

Extraneous factors may affect phenylalanine tolerance curves. Chloramphenicol inhibits intestinal absorption of phenylalanine and therefore tends to lower the tolerance curves.[458] Oral contraceptives lower the tyrosine response,[67] probably because of induction of tyrosine aminotransferase, which catalyses the rate-limiting step of tyrosine catabolism.[365] The same mechanism may also be responsible for the lowered tyrosine response in pregnancy, where glucocorticoid steroids are increased, and where the phenylalanine response is more variable than normal, both factors impairing the accuracy of tests for heterozygosity in pregnancy.[472] Obesity is a factor which has also been considered in relation to the excessive phenylalanine load (0.1 g $(\text{kg body weight})^{-1}$, which includes inactive depot fat).[67] Parameters other than the amino acid response have been measured after a phenylalanine tolerance test: Levine *et al.* used the urinary excretion of phenylethylamine and tyramine, but found it to be unreliable, presumably because of the additional uncertainty of a further two metabolic steps.[258] The excretion of aromatic acids, particularly of *o*-hydroxyphenylacetic acid, after a phenylalanine load has been followed by several authors.[32, 100] Excretion of this acid depends on the extent and duration of the elevation of blood phenylalanine above the transamination threshold.[100] This is negligible in normal subjects, while heterozygotes for phenylketonuria excreted small but measurable amounts. We reinvestigated the earlier reports, and found a sex difference between fathers and mothers of phenylketonuric children, but were able to eliminate this by pretreatment with prednisolone, and were then able to achieve good discrimination.[50] Similar results even without the steroid pretreatment have been reported.[242, 328] This test may also help to distinguish carriers for variant forms of phenylalaninemia.[242]

The real value of tests for heterozygosity is in family studies, where it is now practicable to identify the genetic status of the relatives of patients with considerable confidence. Although theoretically it would be preferable to screen for heterozygotes rather than homozygotes, so that marriages at risk could be identified before the first pregnancy, and advised of the potential hazards, in practice we have no reliable or convenient carrier detection tests. The effects of biological variability in the general population would tend to make it difficult reliably to distinguish the small proportion (less than $1\frac{1}{2}\%$) of carriers from normals with similar responses.[237] However, our ability to score relatives of patients quite reliably does allow us to give them genetic counselling. Parents with an affected child can be given the probability of another (1 in 4), and brothers and sisters found to be carriers can be warned of the probability of marrying another carrier (1 in *c*. 75), and advised to have their future marriage partner evaluated. This, together with our ability to treat affected patients, makes phenylketonuria a less dreadful prospect than formerly.

6 Other genetic aspects

Since phenylketonuria is associated with brain dysfunction, many people have wondered whether carriers might have some degree of mental impairment. Følling himself[143] found that the aunt of one patient and the great-aunt of another suffered from schizophrenia. Considerable speculation was, however, based on little other solid data, and many of the studies were unsupported by phenylalanine tolerance tests to establish the genetic status of the patients found to be affected by mental disease. Perry *et al.*[342] found no difference in the incidence of psychoses, admissions to mental hospitals, personality disorders, alcoholism, suicide, or crime rate among 1268 relatives of 34 unrelated phenylketonuric patients. Blumenthal[52] examined 108 parents of phenylketonuric children by a standard interview technique, and was unable to detect any increased incidence of mental disease compared with a group of 102 parents of non-phenylketonuric retarded children. Larson and Nyman[253] turned the question round, and compared family data from relatives of phenylketonurics with those from relatives of schizophrenics, and came to similar conclusions as Perry *et al.*,[342] although they felt that the mentally ill relatives of the phenylketonurics were ill in clinically divergent ways from those of the mentally ill relatives of the schizophrenics. Thalhammer *et al.*[425] recently found the IQ of parents of phenylketonurics to be slightly lower than that of controls, and the differences to be chiefly in the verbal parts of the test. Ford and Berman have reported a correlation between ability to handle phenylalanine and intellectual functioning among heterozygotes for phenylketonuria and also fathers, but not mothers, of patients with hyperphenylalaninemia.[147] On the whole, carrying the gene for phenylketonuria does not seem to be associated with any strong risk of mental illness, and indeed one cannot readily see why it should.

If the incidence of phenylketonuria in a population is not declining, in spite of the fact that affected patients are at a severe reproductive disadvantage, this is probably a balanced polymorphism, and is best explained on the basis that the carriers are, statistically speaking, at a reproductive advantage. This has been investigated in Irish and Scottish populations, and the authors concluded that mothers of phenylketonuric children have a lower than normal miscarriage rate, with the possibility that their slightly above average blood phenylalanine levels might, in some not strictly nutritional fashion, mediate this protective effect.[465, 466, 467] It has been proposed that the non-phenylketonuric offspring of phenylketonuria heterozygotes have increased growth compared to normal, and that the reproductive advantage works via this mechanism,[383] but this has not been substantiated.[370, 409] Heterozygote advantage may be based on none of these proposed factors, but could be due to improved protection of carriers against some unidentified disease or deleterious environmental effect.

One way of searching for such an effect is to look at polymorphisms in related enzyme systems, and for genetic linkage with other enzymes. Hsia *et al.*[206] did linkage studies on patients with phenylketonuria (and also Down's syndrome and cystic fibrosis) and found no evidence of linkage with haptoglobins, adenylate kinase (EC 2.7.4.3), phosphoglucomutase (EC 2.7.5.1), acid phosphatase (EC 3.1.3.2), glucose-6-phosphate dehydrogenase (EC 1.1.1.49), and lactate dehydrogenase (EC 1.1.1.2.7). Saugstad[380] found a significantly increased proportion of Rh-negatives among mothers of phenylketonuric children in Norway, and also significant associations with the phsophoglucomutase and Kell systems, while no significant association was found with the ABO, MN, haptoglobin, and Gc systems. These data were reassessed by the lod-score method of Morton,[267] and positive lod-scores were obtained for the PKU–PGM_1 relationship, as well as for PKU–Rh, PKU–Kell, and PKU–Hp.[31] These are interesting pointers on the way to the discovery on which chromosome the gene for phenylalanine hydroxylase occurs.

Fears have been expressed that we are failing to eliminate a deleterious gene when we enable homozygotes for phenylketonuria to reproduce as readily as the rest of the population, and we can apply the methods of population genetics to analyse this. The quantitative effect of increasing the fitness of homozygotes through dietary treatment, and of reducing the hostility of their environment by including screening and treatment, is to add only a very small proportion (less than 0.01%) to the reproducing population. All their children will be heterozygotes, except in the rare case that their parents marry a heterozygote, which is not unknown,[348, 385] in which case half will be heterozygotes and half homozygotes. Consideration of factors such as the proportion not marrying or not reproducing, and forward- and back-mutation rates, leads to the conclusion that it will take 80–100 generations (about 2000 years) to double the carrier frequency.[469] Others estimate that it might take longer,[416] but, on the other hand, heterozygote advantage might accelerate the process.[467] This time scale is so long compared with the rate at which knowledge of inborn errors is advancing that we are more likely to see continued improvement than adverse effects.[391]

G TREATMENT AND MANAGEMENT OF PATIENTS WITH PHENYLKETONURIA AND RELATED DISORDERS

1 Dietary treatment

As we have stressed, the phenylalanine that accumulates in phenylketonuria is that part of the dietary intake that is surplus to the needs of the organism, and so the appropriate reduction in phenylalanine intake should theoretically prevent its accumulation. Bickel, Gerrard, and Hickmans[41] described the first attempt at treating and preventing the brain damage and mental retardation in phenylketonuria with a phenylalanine-depleted diet. They took a commercial casein hydrolysate, and, following a suggestion by Woolf, removed the aromatic amino acids with a charcoal column. Tyrosine, tryptophan, and cystine were then added back in suitable amounts, to give a milk substitute that they fed to their patient, a 2-year-old homozygote for the disease, an idiot, unable to stand, walk, or talk. The patient first received a completely phenylalanine-free diet, and was then carefully 'titrated' with whole milk as a source of dietary phenylalanine sufficient only to achieve satisfactory development. On this regime she started to catch up missed developmental milestones, and that this was due to the diet became evident when a challenge with phenylalanine produced a startling and rapid regression.

Since then, much ingenuity and effort has gone into making phenylalanine-restricted diets for phenylketonurics, and into getting the patients to accept them, since many of them are not very appetising. Dieticians have developed a range of diet sheets and recipes, and to do justice to them would require this section to turn into a 'PKU cook book'. They are all still based on the original principles, and have been reviewed in a symposium held in Bickel's Department in Heidelberg.[2, 388]

Early on, too enthusiastic treatment led to a lowering of blood phenylalanine levels below the patients' requirements for normal growth and development: phenylalanine is after all an essential amino acid. Considerable concern was expressed.[34, 186, 300, 403] Overtreatment is accompanied by characteristic bone changes, which are detectable radiologically, particularly in the wrist.[212, 322] Now that this danger is recognized, phenylketonuric infants are treated to maintain blood phenylalanine between 0.2 and 0.5 mM (3–8 mg %): enough for normal growth and development, but not too much for safety.[34, 186, 408] For some patients, exceeding these limits may be dangerous.[290] Throughout the management of the patients there must be regular monitoring of blood phenylalanine concentrations, to make sure that enough but not too much is being taken. It is also usual to challenge patients with phenylalanine every few months, to see where the concentration levels off, in order to check whether there has been any development in their ability to deal with it. Minor intercurrent infections can be a problem, because they may make it difficult to maintain phenylalanine between the correct limits. Where there is evidence of a rise in phenylalanine excretion and blood levels due to wasting of tissues, the phenylalanine intake must be increased.[434] Throughout dietary treatment the patients must be individually managed, and their dietary requirements, with special regard to phenylalanine, must be established at regular intervals.[138, 139, 239]

2 The results and effectiveness of dietary treatment

The effectiveness of dietary treatment in preventing brain damage has been widely assessed, and patients have been evaluated for physical growth, for mental development by various intelligence, developmental, and social maturity tests, and for behavioural and personality development. There is wide agreement that treatment of patients with classical phenylketonuria prevents brain damage and mental retardation, but some argument about how close to normal the treated patients get. Physical growth has been generally found to be somewhat depressed.[138, 398, 408] When it comes to the assessment of the intelligence of treated patients there are some differences of opinion, which is understandable: the authors have looked at different assortments of individual patients, drawn from different populations, treated by different diets with varying diligence on the part of their parents, and possibly affected by slightly different mutations. Again, in attempts to derive some generally applicable conclusions, the results have been to some extent treated statistically, but the populations of patients have been neither strictly comparable, nor homogeneous. When all the evidence has been reviewed, the precise outcome of treatment in any individual case is still somewhat unpredictable.

The results of treatment have been generally good, in the sense that the treated patient is able to play a normal role in society and is not noticeably affected by being phenylketonuric, although, if sought, minor impairment might be detectable. In some patients there is absence even of minor impairment, with the achievement of superior intelligence. It seems that in many cases the degree of control exercised by the parents, together with the degree of careful monitoring by the medicial team, may not be crucial to the final outcome, provided there are no prolonged episodes of phenylalaninemia or overtreatment. Not every child with phenylketonuria develops completely satisfactorily on the diet; some do not. However, it is clear that the earlier treatment is started, the less brain damage there will be, the higher the patient's ultimate IQ will be, and the more easily they tend to be controlled within the chosen limits of blood phenylalanine. Some feel that the treatment should ideally start before the end of the first month of life:[117, 229, 399] the immediate postnatal period is vital in normal development, and the longer treatment is delayed, the greater the damage that may occur.

Most of the evaluation of dietary treatment has been done with various tests for intelligence, maturity, or development, and the results are expressed as intelligence or developmental quotients. Although somewhat arbitrary yardsticks, they do permit comparisons, provided too much significance is not attached to the loss or gain over a period of a few IQ points. As Hackney *et al.* remarked: '. . . We noted repeatedly that small rises and falls in I.Q. coincided with rises and falls in parental hope . . .'.[186] A trimodal distribution of IQ has been reported,[155, 156, 229] with the upper modes centring on scores of 85–94 and 45–54, and the lowest mode below 45, but there were great variations in the frequency distribution between the modes among these investigations. The factors behind this trimodal response are not understood, but there may be contributions from genetic effects.[155]

Intelligence is not the only criterion of success in treatment, and when considering the behavioural and personality aspects of treatment, some residual impairment of fine motor performance, traces of personality problems, and difficulties with schooling[186] have been reported in comparison with normal children or family members with similar IQs.[361, 405, 414] 'Many treated phenylketonuric children of normal intelligence tend to be somewhat deficient in perceptual function and visual–motor coordination: their major neuropsychological handicap is difficulty in concentrating and in sitting still.'[138] These problems are not unknown in some normal children. . . . A difficulty one comes across in trying to assess these factors is that the presence of an affected child, and the need for it to have a special diet and special attention, impose stresses and distortions on the family whose effect on the child and on the normal children in the family are difficult to separate from the residual effects of the disease.

3 Starting and ending dietary treatment

While dietary treatment should be started as soon as possible after birth, some clinicians have also treated patients with phenylketonuria who were discovered late, particularly in the early days before screening programmes. It is widely felt that children less than 3 years old who are discovered late should be treated. Some benefits can come, especially if treatment is persisted with. This may be true even in older children.[239] There is certainly a feeling that many children with phenylketonuria become less hyperactive and easier to manage when on the diet.[290] Dietary treatment has even been tried on adult phenylketonuric patients. Although in a pilot study subjective assessment was favourable in almost half the patients, a strict double-blind evaluation failed to substantiate this impression.[187]

Considerable discussion has gone into the question of when the dietary treatment can be stopped. Most paediatricians would prefer their patients with phenylketonuria to stay on the diet until they are at least 5 years old. Several studies have suggested that the children whose treatment has been most successful do not deteriorate when their diet is returned to normal, while those whose intelligence did not quite reach normal levels seemed to be harmed by a return to a normal diet.[86, 155, 186, 221, 426, 462] The feeling seems to be growing that it may be preferable to continue with the diet as long as possible. Many patients and their families experience a psychological 'lift' when they can give up the diet, and in other cases family stability may be threatened by continuing. It is obviously a hard decision to stop, as nobody would lightly discard the achievements gained by many years of effort and care. There are indications that some patients establish self-protective dietary preferences.[202] Dietary treatment during pregnancy[16, 129] often appears to be successful in preventing damage to the fetus even if started only half-way. If the mother is breast feeding it should be continued until the baby is weaned.

4 Other attempts at treatment

Lines and Waisman tried to affect the renal tubular reabsorption of phenylalanine in an attempt to help the organism to get rid of some of the excess phenylalanine. The inhibitors β-thienylalanine and cycloleucine were used, and although decreased reabsorption and increased excretion were achieved, the effect on blood phenylalanine was slight,[247, 266] presumably because the renal clearance of phenylalanine is not a large proportion of the total even in phenylketonuria. The inhibition of phenylalanine transport by competition with large neutral amino acids (tryptophan, tyrosine and branched-chain amino acids), singly and in mixtures, was effective in lowering brain phenylalanine levels in rats with a simulated phenylketonuria-like condition.[6, 8] A protective effect was also noted with 5-hydroxytryptophan.[6] Disappointing results were obtained with folic acid administration, possibly because the patients were not deficient.[265] The most recent attempts at a pharmacological treatment was with anabolic steroids.[61] The treatment, which was started with uncontrolled affected patients aged 4–12 years, was successful in lowering blood phenylalanine to quite reasonable levels, and in the younger patients at least, subjective clinical improvement was noted. Interestingly, both with folic acid and anabolic steroids, blood tyrosine went up, suggesting that somehow an increased hydroxylation of phenylalanine was occurring. It was suggested that anabolic steroids were acting by shunting more phenylalanine into protein synthesis.[61]

Attempts have been made to treat patients with phenylalaninemias due to a deficiency of reduced pteridine cofactor. Smith *et al.*[407] suggested trying out the administration of the synthetic 6,7-dimethylpterin (presumably in the reduced form), but this might need amounts of cofactor comparable to those of the substrates, unless it could be recycled.[232, 233] More encouraging are reports that feeding the aromatic amino acids, precursors of the deficient biogenic amines, had a beneficial effect.[25, 70] By giving the patients L-DOPA and 5-hydroxy-tryptophan their inability to hydroxylate phenylalanine, tyrosine, and tryptophan was bypassed.

H FUTURE DEVELOPMENTS

1 Animal models of phenylketonuria

Numerous attempts have been made to render experimental animals 'phenylketonuric'. The obvious approach was to load them with phenylalanine,[168] but, as loading tests in human subjects have shown, in a normal animal it is not possible by feeding phenylalanine to get the blood phenylalanine up to high enough levels for transamination to begin, let alone to levels characteristic of uncontrolled phenylketonuria, because the intact phenylalanine hydroxylase system is active enough to keep pace in converting administered phenylalanine to tyrosine almost as fast as it is absorbed. The relative insolubility of phenylalanine makes it difficult to administer very large loads. The value of this model is also impaired by the increase in the body levels of tyrosine and of tyrosine metabolites, which make the results obtained difficult to interpret.[118]

In 1966 Koe and Weissman[241] found that *p*-chlorophenylalanine (PCPA), a tryptophan hydroxylase inhibitor, also inhibited phenylalanine hydroxylase (these two enzymes are so similar that they were at one time thought to be identical). Lipton *et al.*[270] were the first to exploit this to produce a phenylketonuria-like condition in rats, and this approach has been widely used.[121, 275, 355, 371, 447] PCPA on its own produced only modest increases in blood phenylalanine, and after inhibition of the enzyme it was necessary to administer excess phenylalanine to get up to the necessary concentrations in the blood. Three recent models illustrate the results obtained in this way.[36, 121, 371, 418] Table 3.3 shows the similarity between the animal model and human phenylketonuria that can be achieved. An acute phenylketonuria-like condition can soon be produced which resembles the biochemical and metabolic situation in the uncontrolled phenylketonuric patient fairly closely, and allows metabolic studies to be carried out with reasonably valid extrapolation of the results to the human situation.[36, 121, 122, 371, 417] In spite of the encouraging features of such acute models, they must be carefully set up. The administration of PCPA to experimental animals must be accompanied by well-designed controls[121] to enable one to discount the effects caused by PCPA and its metabolites and of tyrosine and its metabolites (PCPA never completely inhibits phenylalanine hydroxylase), and to put the interpretation of the results on a sound basis.[120, 151, 158, 371, 418] The mechanism for the inhibition of the hydroxylases is thought to involve the incorporation of the inhibitor into the enzyme proteins at some

Table 3.3 A comparison of the biochemical features of uncontrolled human phenylketonuria with those of immature rats with phenylketonuria-like characteristics induced by administration of *p*-chlorophenylalanine and loading with phenylalanine

Biochemical feature	Uncontrolled human PKU	Animal model[121, 122]
1. Blood phenylalanine	Up to 60 mg % (3.6 mM l^{-1})	66±6 mg % (4±0.4 mM l^{-1})
2. Blood phe/tyr ratio	*c.* 20	25
3. Aromatic acid excretion (decreasing amounts)	Phenyllactic Phenylpyruvic *o*-Hydroxyphenylacetic Phenylacetic Mandelic	Phenyllactic Phenylpyruvic *o*-Hydroxyphenylacetic Phenylacetic Mandelic
4. Phenylethylamine turnover	Up to 50 times normal	40 times normal (brain) 25 times normal (liver)
5. Phenylalanine tolerance tests	High levels, slow decline	High levels, slow decline
6. Phenylalanine hydroxylase activity (% of normal)	<1%	4–11%

strategic point along the amino acid sequence, altering the catalytic properties of the active site.[158] It is thought that the *p*-chloro group is ultimately hydrolysed to hydroxyl, and if PCPA is originally incorporated in place of phenylalanine, the net result is substitution of a tyrosine for a phenylalanine residue. This is a more fundamental form of inhibition than that produced by any competitive inhibitor, and the time-course of the changes in phenylalanine hydroxylase activity after PCPA[302] showed a minimum at 24 h., followed by slow recovery to normal values 10 days after administration of a single dose of PCPA.[183, 302] This would be a very attractive animal model of phenylketonuria if it were specific. Unfortunately, as one might expect, PCPA incorporation is quite general: for example acinar cell proteins in pancreas have been found to contain PCPA[148] and no doubt others would be found if looked for. This raises doubts as to whether the effects observed in such model animals are solely due to the phenylketonuria-like characteristics achieved. In acute animal models using PCPA the biochemical and metabolic effects are probably sufficiently marked to overshadow the more subtle effects resulting from widespread alterations to proteins throughout the organism.

Attempts have also been made to produce animal models by inhibition of the dihydropteridine reductase portion of the phenylalanine hydroxylase system. Four substances have been tried: amethopterin;[271, 329] aminopterin;[330] esculetin;[111, 367, 437] and cotrimoxazole;[114] on their own or in combination with PCPA.[271, 437] As we have seen, biopterin is required for other systems, and the increased neurotoxicity of such combinations outweighs the possible advantages of a lower residual ability to hydroxylate phenylalanine.

Attempts have also been made to use PCPA in long-term animal model experiments for the study of learning and behaviour,[197, 443] when the inhibitor was administered as early as the start of pregnancy.[91] These chronic experiments have not been terribly successful because the phenylalanine accumulation has been less than in phenylketonuria patients, and because the toxicity of the drug has led to growth inhibition and high mortality among the immature animals both before and after birth.

The ubiquitous incorporation of PCPA makes it in many ways unsuitable for producing animal models of phenylketonuria, because it elicits biochemical and behavioural side effects which are quite irrelevant. A specific inhibitor is needed, and study of the enzymic mechanism of phenylalanine hydroxylation may lead to the design of an effective 'transition-state analogue' type of inhibitor.[460] The transition-state analogue approach could yield specific inhibitors which are applicable to the design of animal models for inborn errors generally. Another possible approach would be to inject antibodies raised against phenylalanine hydroxylase in order to achieve a specific immunological inactivation of the enzyme *in vivo*.

The phenylketonuria mutation has been looked for in animals: phenylalanine tolerance tests were administered to non-human simian primates, and did uncover some animals with phenylalanine tolerance curves suggestive of the heterozygous state.[222]

It seems doubtful whether animal model experiments, however they may be achieved, will give useful insights into human phenylketonuria as far as learning and behaviour difficulties are concerned, because human and animal intelligence and behaviour are so different that attempts to extend conclusions from the results of animal studies to the human condition are doomed to failure from the outset.

2 Studies on the possibility of prenatal diagnosis of phenylketonuria

Prevention of many inborn errors of metabolism, particularly those where the patients are very severely affected in infancy, has become possible where the biochemical defect has been found to be detectable early enough in pregnancy for safe termination of the pregnancy at the wish of the parents. The appropriate enzyme activity is usually determined in fetal cells cultured from amniotic fluid, obtained by transabdominal amniocentesis at a time between 14 and 18 weeks gestational age. Prenatal diagnosis of phenylketonuria has so far not been described, because amniotic fluid cells have not been found to express phenylalanine hydroxylase activity, though an unconfirmed report has claimed phenylalanine hydroxylase activity in cultured human fibroblasts.[196] It may, in theory, be possible to bring any enzyme to expression in a given cell, since all cell nuclei contain the necessary genetic code for a complete set of proteins, including enzymes. However, in differentiated cells many proteins are synthesized in undetectable amounts (perhaps only a few molecules in each cell cycle), and ways would have to be found of specifically lifting the control on phenylalanine hydroxylase synthesis. It is interesting that phenylalanine hydroxylase activity has been described in cells grown from fetal muscle,[77] although adult muscle tissue does not express this activity. Phenylalanine hydroxylase activity has also been described in human term placenta.[289] Although phenylketonuria is a treatable disorder, many parents who have had an affected child would welcome prenatal diagnosis so that they need not go through this experience again. It has been suggested that prenatal diagnosis of dihydropteridine reductase deficiency should be possible.[105]

3 Looking ahead

Research into the phenylalanine hydroxylase system and its defects is bound to maintain its intrinsic interest because of the possibility that the results might have therapeutic applications, and the hope that experience gained with phenylketonuria may be useful in our understanding of other inborn errors of metabolism and of the causes of brain damage and mental retardation. Attempts at replacing the missing enzyme activity are now being made in many inborn errors.[60, 113, 317] Addition of some normal histidase activity has been achieved via a kidney transplant,[366] an encouraging sign that such an approach might work, although for phenylketonuria it would be necessary to do a liver transplant.[313] The replacement could also in principle be achieved by microencapsulation of the missing enzyme and packing the capsules into an 'accessory organ' that fits into an extracorporeal shunt in the circulation.[80, 194, 313] This would have the advantage of shielding the enzyme from the immune defences of the host. Phenylalanine hydroxylase has been successfully entrapped in a polyacrylamide matrix.[457] There are many problems to be overcome before such experimental procedures become practical, but there is some hope that patients with several inborn errors, possibly including phenylketonuria, might be helped to live more normally as a result of some of the approaches now being investigated.

The prospect of 'gene therapy' is more distant. Although our present expertise scarcely extends beyond microorganisms, clinical attempts have been made to treat arginase deficiency in hyperargininemia by inoculating the patients with Shope papilloma virus[153] which has DNA coding for a viral arginase. Unfortunately the attempt was unsuccessful.[421]

Efforts are now being directed to determination of the amino acid sequences of the normal and mutant enzymes in phenylketonuria and the phenylalaninemias, and into the molecular biology of the mutations. This may help with differentiating these variant forms. There is also bound to be continuing research into the precise pathological mechanisms, both major and minor, for the insights they may give us into the workings of the central nervous system and also their future potential for treating brain damage and mental deficiency. The successes already achieved in preventing the worst of the damage of phenylketonuria encourage us to hope that further efforts may lead to further advances.

4 Conclusion

Why has such a rare disease received so much attention? This was one of the earliest biochemical defects discovered, and required biochemical investigation and biochemically oriented treatment, so that clinically minded biochemists were involved. The idea that lack of a single enzyme could cause mental retardation was a fascinating one, and the mystery of the causal mechanism presented a real challenge. The development of the dietary treatment gave the subject a tremendous boost because of the hopes it raised. Phenylketonuria has been one of the key metabolic disorders, and has provided so much interest and so much stimulus to research in a wide variety of disciplines because it offers the prospect (or is it a mirage?) of answering many difficult questions. And then we were also faced with the decision of the good Samaritan: once we knew what could be done we could not walk by on the other side. Some have argued that screening and treatment of children with phenylketonuria was rushed into prematurely. This is probably so: mistakes were made and some patients were harmed by overtreatment. Nevertheless many more were helped to develop normally, and this happened sooner than if the development of the whole phenylketonuria effort had advanced in a more measured fashion.

I SUMMARY

1. The conversion of phenylalanine to tyrosine is catalysed by the hepatic phenylalanine hydroxylase system. This is a major route for the catabolism of all phenylalanine surplus to the body's requirements for protein synthesis and other metabolic needs. Deficiencies in this enzyme system, particularly the genetically determined complete deficiency of the hydroxylase that is characteristic of classical phenylketonuria, lead to accumulation of phenylalanine in the blood and tissues of the affected patients.
2. As well as deficiencies in phenylalanine hydroxylase, deficiencies in dihydropteridine reductase and also in the metabolic supply of reduced biopterin have been discovered in some patients. A functional lack of reduced biopterin leads to severe neurological disturbances, particularly because of its involvement in the biosynthesis of neurotransmitter monoamines, but phenylalaninemia also occurs in these syndromes.
3. As the blood phenylalanine concentration increases, normal phenylalanine metabolites also increase in concentration, and eventually abnormal metabolites also appear, from the successful competition of phenylalanine as substrate for enzymes that may not normally use it.

4. Toxic effects of phenylalanine and of its metabolites have been described both in clinical and also in model animal investigations, and include inhibition of glycolysis, of amino acid membrane transport, of protein synthesis, of neurotransmitter production, of lipid biosynthesis, and of many other individual enzyme systems. In spite of extensive studies, the exact biochemical mechanisms underlying the brain damage and mental deficiency have not been defined. It is likely that both phenylalanine and phenylpyruvic acid exert important toxic effects, possibly acting synergistically.
5. Patients who are not treated become mentally retarded and may show other signs and symptoms such as microcephaly, seizures, eczema, hypertonicity, and agitated behaviour. An abnormal EEG is often seen. The characteristic mousy odour of phenylacetic acid, and light pigmentation, may also be found. Most of the permanent damage occurs in the first few months of life. Behavioural and learning deficits accompany the biochemical and pathological changes.
6. There may be a moderate reduction in brain weight (related to the microcephaly). There is deficient myelination, the myelin may be immature, and eventually demyelination may occur. It is not certain, though very plausible, that these changes are mainly responsible for the irreversible mental deficiency.
7. These various conditions are all transmitted via the autosomal recessive mode of inheritance, and heterozygotes are clinically unaffected, but may be identified by their responses to a phenylalanine load.
8. The incidence of classical phenylketonuria is not the same in all populations, varying from 1:5000 live births (Ireland) to 1:100 000 (Finland). The genetic defect seems to be associated with populations of Celtic origin, and may be a balanced polymorphism maintained by heterozygote advantage.
9. Early treatment with a phenylalanine-restricted diet has been found to be successful in preventing the biochemical and pathological changes, but attenuated residual learning and behavioural deficits may remain. For greatest effectiveness, treatment should be started as early as possible, ideally within 2–3 weeks of birth. The majority of adequately treated patients are almost entirely protected from mental deficiency. Reports on the effects of ending the treatment before puberty have been contradictory, possibly because of individual differences in susceptibility.
10. Screening programmes for early detection of phenylketonuria are widespread, and have, additionally, revealed atypical states of phenylalaninemia of varying degrees of severity, which present problems in diagnosis and in deciding whether the effected patients need to be treated.
11. Increasingly, homozygous females will become mothers: unless they are treated with a phenylalanine-restricted diet during pregnancy, children of uncontrolled phenylketonurics will be born with brain damage caused by maternal hyperphenylalaninemia.

12. Although these disorders may be treated, no cure for them has yet been found, and prevention of the birth of affected infants through prenatal diagnosis has not yet been achieved.

J REFERENCES

1. Acheson, R. M. (1963). *Publ. Hlth. (London)*, **77**, 261–273.
2. Acosta, P. B., and Wenz, E. (1971). In *Phenylketonuria and Some Other Inborn Errors of Metabolism* (Eds., H. Bickel, F. P. Hudson, and L. I. Woolf), Georg Thieme, Stuttgart, pp. 181–196.
3. Adelman, L. S., Mann, J. D., Caley, D. W., and Bass, N. H. (1973). *J. Neuropathol. Exptl Neurol.*, **32**, 380–393.
4. Adriaenssens, K., Vanheule, R., and van Belle, M. (1967). *Clin. Chim. Acta*, **15**, 362–364.
5. Agrawal, H. C., Bone, A. H., and Davison, A. N. (1970). *Biochem. J.*, **117**, 325–331.
6. Airaksinen, M. M., Leppanen, M.-L., Turakka, H., Marvola, M., and MacDonald, E. J. (1975). *Med. Biol.*, **53**, 481–488.
7. Alvord, E. C., Jr., Stevenson, L. D., Vogel, F. S., and Engle, R. L., Jr. (1950). *J. Neuropathol. Exptl Neurol.*, **9**, 298–310.
8. Andersen, A. E., and Avins, L. (1976). *Arch. Neurol.*, **33**, 684–686.
9. Anderson, J. A., Gravem, H., Ertel, R., and Fisch, R. O. (1962). *J. Pediat.*, **61**, 603–609.
10. Anderson, J. A., Bruhl, H., Michaels, A. J., and Dueden, D. (1967). *Pediat. Res.*, **1**, 372–385.
11. Angeli, E., Denman, A. R., Harris, R. F., Kirman, B. H., and Stern, J. (1974). *Develop. Med. Child Neurol.*, **16**, 800–807.
12. Aoki, K., and Siegel, F. L. (1970). *Science*, **168**, 129–130.
13. Appel, S. (1965). *J. Clin. Invest.*, **44**, 1026.
14. Armstrong, M. D., and Robinson, K. S. (1954). *Arch. Biochem.*, **52**, 287–288.
15. Armstrong, M. D., and Binkley, E. L., Jr. (1956). *Proc. Soc. Exptl Biol. Med.*, **93**, 418–420.
16. Arthur, L. J. H., and Hulme, J. D. (1970). *Pediatrics*, **46**, 235–239.
17. Auerbach, V. H., di George, A. M., and Carpenter, G. G. (1967). In *Amino Acid Metabolism and Genetic Variation* (Ed., W. L. Nyhan), McGraw-Hill, New York, pp. 11–68.
18. Ayling, J. E., Helfand, G. D., and Pirson, W. D. (1975). *Enzyme*, **20**, 6–19.
19. Baños, G., Daniel, P. M., Moorhouse, S. R., and Pratt, O. E. (1974). *Psychol. Med.*, **4**, 262–269.
20. Barbato, L. M. and Barbato, I. M. (1969). *Brain Res.*, **13**, 569–578.
21. Barbesier, J., Boisse, J., Charpentier, C., Lemonnier, A., Mozziconacci, P., and Saudubray, J.-M. (1971). *Presse Med.*, **79**, 395–397.
22. Barbosa, E., Herreros, B., and Ojeda, J. L. (1971). *Experientia*, **27**, 1281–1282.
23. Barranger, J. A., Geiger, P. J., Huzino, A., and Bessman, S. P. (1972). *Science*, **175**, 903–905.
24. Bartholomé, K. (1974). *Lancet ii*, 1580.
25. Bartholomé, K., and Byrd, D. J. (1975). *Lancet ii*, 1046–1047.
26. Bartholomé, K., Byrd, D. J., Kaufman, S., and Milstien, S. (1977). *Pediatrics*, **59**, 757–761.
27. Bechelli, L. M., Gonçalves, R. P., Tanaka, A. M. U., and Pagnano, P. M. G. (1978). *Ann. Dermatol. Venereol. (Paris)*, **105**, 165–173.
28. Behbehani, A. W., Quentin, C.-D., Schulte, F. J., and Neuhoff, V. (1974). *Neuropadiätrie*, **5**, 258–270.

29. Benirschke, K., Brownhill, L., Efron, M. L., and Hoefnagel, D. (1962). *J. Ment. Defic. Res.*, **6**, 44–56.
30. Benuck, M., Stern, F., and Lajtha, A. (1971). *J. Neurochem.*, **18**, 1555–1567; (1972). *J. Neurochem.*, **19**, 949–957.
31. Berg, K., and Saugstad, L. F. (1974). *Clin. Genet.* **6**, 147–152.
32. Berry, H. K., Sutherland, B., and Guest, G. M. (1957). *Amer. J. Hum. Genet.*, **9**, 310–316.
33. Berry, H. K. (1962). *Clin. Chem.*, **8**, 172–173.
34. Berry, H. K., Sutherland, B. S., Umbarger, B., and O'Grady, D. (1967). *Amer. J. Dis. Child.*, **113**, 2–5.
35. Berry, H. K. (1971). *J. Chromatogr.*, **56**, 316–320.
36. Berry, H. K., Butcher, R. E., Kazmaier, K. J., and Poncet, I. B. (1975). *Biol. Neonat.*, **26**, 88–101.
37. Berry, H. K., Poncet, I. B., Sutherland, B. S., and Burkett, R. (1975). *Biol. Neonat.*, **26**, 102–108.
38. Bessman, S. P. (1966). *J. Pediat.*, **69**, 334–338.
39. Bessman, S. P., Wapnir, R., and Due, D. (1967). In *Phenylketonuria and Allied Metabolic Diseases* (Eds., J. A. Anderson, and K. F. Swaiman), US Department of Health, Education, and Welfare, Children's Bureau, Washington DC, pp. 9–17.
40. Bessman, S. P. (1972). *J. Pediat.*, **81**, 834–842.
41. Bickel, H., Gerrard, J., and Hickmans, E. M. (1953). *Lancet ii*, 812–813.
42. Biserte, G. (1974). *Lille Med.*, **19**, 134–140.
43. Blascovics, M. E. (1974). *Clin. Endocrinol. Metab.*, **3**, 87–105.
44. Blascovics, M. E., Schaeffler, G. E., and Hack, S. (1974). *Arch. Dis. Childh.*, **49**, 835–843.
45. Blascovics, M. E., and Shaw, K. N. F. (1974). *Z. Kinderheilk.*, **117**, 265–273.
46. Blau, K. (1970). *Clin. Chim. Acta*, **27**, 5–18.
47. Blau, K., Cameron, H. C., and Summer, G. K. (1970). In *Methods in Medical Research*, Vol. 12 (Ed., R. E. Olson), Yearbook Medical Publishers, Chicago, pp. 100–105.
48. Blau, K., and Edwards, D. J. (1971). *Biochem. Med.*, **5**, 333–341.
49. Blau, K., Summer, G. K., and Edwards, D. J. (1973). In *Advances in Automated Analysis, Technicon International Congress 1972*, Vol. 1, Mediad, Tarrytown, pp. 121–124.
50. Blau, K., Summer, G. K., Newsome, H. C., Edwards, C. M., and Mamer, O. A. (1973). *Clin. Chim. Acta*, **45**, 197–205.
51. Blehova, B., Pazoutova, N., and Subrt, I. (1970). *J. Ment. Defic. Res.*, **14**, 274–275.
52. Blumenthal, M. D. (1967). *J. Psychiat. Res.*, **5**, 59–74.
53. Blyumina, M. G. (1972). *Akush. Ginecol. (Moskva)*, **48**, 52; (1974). *Sov. Genet.*, **8**, 385–390.
54. Boscott, R. J., and Bickel, H. (1953). *Scand. J. Clin. Lab. Invest.*, **5**, 380–382,
55. Boulton, A. A., and Milward, L. (1971). *J. Chromatogr.*, **57**, 287–296.
56. Boulton, A. A., Philips, S. R., and Durden, A. (1973). *J. Chromatogr.*, **82**, 137–142.
57. Boulton, A. A. (1976). In *Trace Amines and the Brain* (Eds., E. Usdin, and M. Sandler), Dekker, New York, pp. 21–39.
58. Bowden, J. A., and McArthur, C. C. (1972). *Nature*, **235**, 230.
59. Boylen, J. B., and Quastel, J. H. (1962). *Nature*, **193**, 376–377.
60. Brady, R. O., Pentchev, P. G., and Gal, A. E. (1976). *Fed. Proc. Fed. Amer. Socs Exptl Biol.*, **34**, 1310–1315.
61. Brambilla, F., Giardini, M., and Russo, R. (1975). *Dis. Nerv. Syst.*, **36**, 257–260.
62. Brand, L. M., and Harper, A. E. (1974). *Proc. Soc. Exptl Biol. Med.*, **147**, 211–215.
63. Brewster, T. G., Abroms, I. F., Kaufman, S., Breslow, J. L., Moskowitz, M. A.,

Villee, D. B., and Snodgrass, R. S. (1976). *Pediat. Res.*, **10**, 446, Abstract No. 872.
64. Brodehl, J., Gellissen, K., and Kaas, W. P. (1970). *Acta Paed. Scand.*, **59**, 241–248.
65. Brouwer, M., de Bree, P. K., van Sprang, F. J., and Wadman, S. K. (1977). *Lancet*, *i*, 1162.
66. Brown, E. S., and Waisman, H. A. (1971). *Pediatrics*, **48**, 401–410.
67. Brown, E. S., Waisman, H. A., Swanson, M. A., Colwell, R. E., Banks, M. E., and Gerritsen, T. (1973). *Clin. Chim. Acta*, **44**, 183–192.
68. Bush, R. T., and Dukes, P. C. (1975). *New Zeal. Med. J.*, **82**, 226–229.
69. Butler, I. J., Holtzman, N. A., Kaufman, S., Koslow, S. H., Krumholz, A., and Milstien, S. (1975). *Pediat. Res.*, **9**, 348, Abstract No. 551.
70. Butler, I. J., Koslow, S. H., Krumholz, A., Holtzman, N. A., and Kaufman, S. (1978). *Ann. Neurol.*, **3**, 224–230.
71. Cabalska, B., and Borzymowska, J. (1975). *Genet. Pol.*, **16**, 371–376.
72. Cahalane, S. F. (1968). *Arch. Dis. Childh.*, **43**, 141–144.
73. Carbonell, J., Feliu, J. E., Marco, J. E., and Solis, A. (1973). *Eur. J. Biochem.*, **37**, 148–156.
74. Carpenter, G. G., Auerbach, V. H., and DiGeorge, A. M. (1968). *Pediat. Clin. North Amer.*, **15**, 313–323.
75. Carter, C. O. (1976). *Irish Med. J.*, **69**, 386–389.
76. Carver, M. J. (1965). *J. Neurochem.*, **12**, 45–50.
77. Cartwright, E. C., and Danks, D. M. (1972). *Biochim. Biophys. Acta*, **264**, 205–209.
78. Cession-Fossion, A., Vandermeulen, R., Dodinval, P., and Chantraine, J.-M. (1966). *Pathol. Biol.*, **14**, 1157–1159.
79. Chalmers, R. A., and Watts, R. W. E. (1974). *Clin. Chim. Acta*, **55**, 281–294.
80. Chang, T. M. S., and Poznansky, M. J. (1968). *Nature*, **218**, 243–245.
81. Chase, H. P., and O'Brien, D. (1970). *Pediat. Res.*, **4**, 96–102.
82. Charpentier, C., Laboureau, J.-P., Saudubray, J.-M., Boisse, J., Lemonnier, A., and Mozziconacci, P. (1973). *Ann. Pédiat.*, **20**, 419–424.
83. Children's Bureau (staff) (1967). *State Laws Pertaining to Phenylketonuria as of November 1966*, US Department of Health, Education, and Welfare, Washington, DC.
84. Christenson, J. G., Dairman, W., and Udenfriend, S. (1970). *Arch. Biochem. Biophys.*, **141**, 356–357.
85. Clarke, J. T. R., and Lowden, J. A. (1969). *Canad. J. Biochem.*, **47**, 291–295.
86. Clayton, B., Moncrieff, A., and Roberts, G. E. (1967). *Brit. Med. J.*, **3**, 133–136.
87. Cohen, B. E., and Szeinberg, A. (1968). In *Proceedings of the First Congress of the International Association for the Scientific Study of Mental Deficiency*, Jackson, Glasgow, pp. 794–800.
88. Colombo, J. P., and Vassella, F. (1968). *Klin. Wschr.*, **46**, 1215–1221.
89. Colombo, J. P. (1971). *Arch. Dis. Childh.*, **46**, 720–721.
90. Cooper, J. D. (1967). In *Phenylketonuria and Allied Metabolic Diseases* (Eds., J. A. Anderson, and K. F. Swaiman), US Department of Health, Education, and Welfare, Children's Bureau, Washington DC, pp. 168–176.
91. Copenhaver, J. H., Carver, M. J., Johnson, E. A., and Saxton, M. J. (1970). *Biochem. Med.*, **4**, 516–530.
92. Coutts, N. A. and Fyfe, W. M. (1971). *Arch. Dis. Childh.*, **46**, 550–552.
93. Cowie, V., and Penrose, L. S. (1951). *Ann. Eugen. (London)*, **15**, 297–301.
94. Cowie, V. A. (1971). In *Phenylketonuria and Some Other Inborn Errors of Metabolism* (Eds., H. Bickel, F. P. Hudson, and L. I. Woolf), Georg Thieme, Stuttgart, pp. 29–39.
95. Creveling, C. R., Kondo, K., and Daly, J. W. (1968). *Clin. Chem.*, **14**, 302–309.
96. Crome, L. (1962). *J. Neurol. Neurosurg. Psychiat.*, **25**, 149–153.

97. Crome, L., Tymms, U., and Woolf, L. I. (1962). *J. Neurol. Neurosurg. Psychiat.*, **25**, 143–148.
98. Crome, L. (1971). In *Phenylketonuria and Some Other Inborn Errors of Metabolism* (Eds., H. Bickel, F. P. Hudson, and L. I. Woolf), Georg Thieme, Stuttgart, pp. 126–131.
99. Crome, L., and Stern, J. (1967). *The Pathology of Mental Retardation*, Churchill, London.
100. Cullen, A. M. and Knox, W. E. (1958). *Proc. Soc. Exptl Biol. Med.*, **99**, 219–222.
101. Culley, W. J. (1969). *Clin. Chem.*, **15**, 902–907.
102. Cumings, J. N., Grundt, I. K., and Yanagihara, T. (1968). *J. Neurol. Neurosurg. Psychiat.*, **31**, 334–337.
103. Curtius, H.-Ch., Baerlocher, K., and Völlmin, J. A. (1972). *Clin. Chim. Acta*, **42**, 235–239.
104. Dancis, J., and Balis, M. E. (1955). *Pediatrics*, **15**, 63–66.
105. Danks, D. M., Cotton, R. G. H., and Schlesinger, P. (1978). *Clin. Genet.*, **13**, 112.
106. Danks, D. M., Bartholomé, K., Clayton, B. E., Curtius, H., Gröbe, H., Kaufman, S., Leeming, R., Pfleiderer, W., Rembold, H., and Rey, F. (1978). *J. Inher. Metab. Dis.*, **1**, 49–53.
107. Davison, A. N. and Sandler, M. (1958). *Nature*, **181**, 186–187.
108. Davison, A. N. and Dobbing, J. (1966). *Brit. Med. Bull.*, **22**, 40–44.
109. Davison, A. N., and Dobbing, J. (1968). In *Applied Neurochemistry* (Eds., A. N. Davison, and J. Dobbing), Blackwell, Oxford, pp. 253–286.
110. Davson, H. (1967). *Physiology of the Cerebrospinal Fluid*, Churchill, London.
111. De Graw, J. I., Cory, M., Skinner, W. A., Theisen, M. C., and Mitoma, C. (1967). *J. Med. Chem.*, **10**, 64–66.
112. Demopoulos, H. B., and Kaley, G. (1963). *J. Nat. Cancer Inst.*, **30**, 611–633.
113. Desnick, R. J., Thorpe, S. R., and Fidler, M. B. (1967). *Physiol. Rev.*, **56**, 57–99.
114. Dhondt, J. L., and Farriaux, J. P. (1977). *Biomedicine*, **27**, 51–52.
115. Dobbing, J. (1968). In *Applied Neurochemistry* (Eds., A. N. Davison, and J. Dobbing), Blackwell, Oxford, pp. 287–316.
116. Dobson, J., and Williamson, M. (1970). *New Engl. J. Med.*, **282**, 1104.
117. Dobson, J. C., Williamson, M. L., Azen, C., and Koch, R. (1976). *Pediatrics*, **60**, 822–827.
118. Dolan, G., and Godin, C. (1967). *Nature*, **213**, 916–917.
119. Doshi, R., Sandry, S. A., Churchill, A. W., and Brownell, B. (1974). *J. Neurol. Neurosurg. Psychiat.*, **37**, 1133–1138.
120. Edwards, D. J., and Blau, K. (1972). *J. Neurochem.*, **19**, 1829–1832.
121. Edwards, D. J., and Blau, K. (1972). *Biochem. J.*, **130**, 495–503.
122. Edwards, D. J., and Blau, K. (1973). *Biochem. J.*, **132**, 95–100.
123. Efron, M. L., Young, D., Moser, H. W., and MacCready, R. A. (1964). *New Engl. J. Med.*, **270**, 1378–1383.
124. Efron, M. L., Kang, E. S., Visakorpi, J., and Fellers, F. X. (1969). *J. Pediat.*, **74**, 399–405.
125. Emery, A. E. H., Farquhar, J. W., and Timson, J. (1972). *Clin. Chim. Acta*, **37**, 544–546.
126. Epps, R. P. (1968). *Clin. Pediat.*, **7**, 607–610.
127. Fahie-Wilson, M. N. (1969). *J. Med. Lab. Technol.*, **26**, 363–370.
128. Farber, M., Knuppel, R. A., Binkiewicz, A., and Kennison, R. D. (1976). *Obstet. Gynecol.*, **47**, 226–228.
129. Farquhar, J. W. (1973). *Arch. Dis. Childh.*, **49**, 205–208.
130. Farrell, D. F., Clark, A. F., Scott, C. R., and Wennberg, R. P. (1975). *Science*, **187**, 1082–1084.

131. Feinberg, S. B., and Fisch, R. O. (1962). *Radiology*, **78**, 394–398.
132. Fellman, J. H. (1956). *Proc. Soc. Exptl Biol. Med.*, **93**, 413–414.
133. Fellman, J. H. (1956). *J. Neurol. Neurosurg. Psychiat.*, **21**, 58–62.
134. Fellman, J. H., Vanbellinghen, P. J., Jones, R. T., and Koler, R. D. (1969). *Biochemistry*, **8**, 615–622.
135. Fellman, J. H., Buist, N. R. M., Kennaway, N. G., and Swanson, R. E. (1972). *Clin. Chim. Acta*, **39**, 243–246.
136. Fisch, R. O., and Horrobin, J. M. (1968). *Clin. Pediat.*, **7**, 226–227.
137. Fisch, R. O., Doeden, D., Lansky, L. L., and Anderson, J. A. (1969). *Amer. J. Dis. Child.*, **118**, 847–858.
138. Fisch, R. O., Torres, F., Gravem, H. J., Greenwood, C. S., and Anderson, J. A. (1969). *Neurology*, **19**, 659–666.
139. Fisch, R. O., Solberg, J. A., and Borud, L. (1971). *J. Amer. Diet. Ass.*, **58**, 32–37.
140. Fischer, E., Heller, B., and Miro, A. H. (1968). *Arzneimittelforschung*, **18**, 1486.
141. Fishman, W. H., Green, S., and Inglis, N. I. (1963). *Nature*, **198**, 685–686.
142. Fois, A., Rosenberg, C., and Gibbs, F. A. (1955). *Electroencephalog. Clin. Neurophysiol.*, **7**, 569–572.
143. Følling, A. (1934). *Hoppe-Seyle's Zeit. Physiol. Chem.*, **227**, 169–176; *Nord. Med. Tidskr.*, **8**, 1054.
144. Følling, A., and Closs, K. (1938). *Hoppe-Seyle's Zeit. Physiol. Chem.*, **254**, 115–116.
145. Følling, A., Closs, K., and Gamnes, T. (1938). *Hoppe-Seyle's Zeit. Physiol. Chem.*, **256**, 1–14.
146. Følling, A. (1971). In *Phenylketonuria and Some Other Inborn Errors of Metabolism* (Eds., H. Bickel, F. P. Hudson, and L. I. Woolf), Georg Thieme, Stuttgart, pp. 1–3.
147. Ford, R. C., and Berman, J. L. (1977). *Lancet i*, 767–771.
148. Forssmann, W. G., and Bieger, W. (1973). *Res. Exptl Med.*, **160**, 1–20.
149. Fox, J. G., Hall, D. L., Haworth, J. C., Maniar, A., and Sekla, L. (1971). *Canad. Med. Ass. J.*, **104**, 1085–1088.
150. Frankenburg, W. K., Duncan, B. R., Coffelt, R. W., Koch, R., Coldwell, J. G., and Son, C. D. (1968). *J. Pediat.*, **73**, 560–570.
151. Friedmann, E., and Maase, C. (1910). *Biochem. Z.*, **27**, 97.
152. Friedman, P. A., Kaufman, S., and Kang, E. S. (1972). *Nature*, **240**, 157–159.
153. Friedman, T. and Roblin, R. (1972). *Science*, **175**, 949–955.
154. Friedman, P. A., Fisher, D. B., Kang, E. S., and Kaufman, S. (1973). *Proc. Nat. Acad. Sci., USA*, **70**, 552–556.
155. Fuller, R. N., and Shuman, J. B. (1969). *Nature*, **221**, 639–642.
156. Fuller R. N., and Shuman, J. B. (1971). *Amer. J. Ment. Defic.*, **75**, 539–545.
157. Fuller, R. W., Snoddy, H. D., Wolen, R. L., Coburn, S. P., and Sirlin, E. M. (1972). *Adv. Enzyme Regulation*, **10**, 153–167.
158. Gál, E. M., Roggeveen, A. E., and Millard, S. A. (1970). *J. Neurochem.*, **17**, 1221–1235.
159. Gallagher, B. B. (1969). *J. Neurochem.*, **16**, 1071–1076.
160. Garelis, E., Young, S. N., Lal, S., and Sourkes, T. L. (1974). *Brain Res.*, **79**, 1–8.
161. Geison, R. L., and Waisman, H. A. (1970). *J. Neurochem.*, **17**, 469–474.
162. Geison, R. L., and Siegel, F. L. (1975). *Brain Res.*, **92**, 431–441.
163. Gibbs, N. K., and Woolf, L. I. (1959). *Brit. Med. J.*, *ii*, 532–535.
164. Gibson, R. (1956). *Canad. Med. Ass. J.*, **74**, 897–900.
165. Gimenez, C., Valdivieso, F., and Mayor, F. (1974). *Biochem. Med.*, **11**, 81–86.
166. Glazer, R. I., and Weber, G. (1971). *Brain Res.*, **33**, 439–450.
167. Glazer, R. I., and Weber, G. (1971). *J. Neurochem.*, **18**, 2371–2382.
168. Goldstein, F. B. (1961). *J. Biol. Chem.*, **236**, 2656–2661.
169. Goldstein, F. B. (1963). *Biochim. Biophys. Acta*, **71**, 204–206.

170. Goldstein, F. B. (1964). In *Proceedings of the International Copenhagen Congress on the Scientific Study of Mental Retardation*, Commun. No. 90, pp. 487–489.
171. Gostomzyk, J. G., and Dressler, F. (1967). *Klin. Wschr.*, **45**, 793–794.
172. Grahame-Smith, D. G. (1964). *Biochem. Biophys. Res. Commun.*, **16**, 586–592.
173. Grahame-Smith, D. G., and Mahoney, L. (1965). *Biochem. J.*, **96**, 66P.
174. Graw, R. G., and Koch, R. (1967). *Amer. J. Dis. Child.*, **114**, 412–418.
175. Grenier, A., and Laberge, C. (1974). *L'Union Med. Canad.*, **103**, 453–456.
176. Griffin, R. F., Humienny, M. E., Hall, E. C., and Elsas, L. J. (1973). *Amer. J. Hum. Genet.*, **25**, 646–654.
177. Grimm, U., Knapp, A., and Teichmann, W. (1971). *Acta Biol. Med. Germ.*, **27**, 443–446.
178. Grimm, U., Knapp, A., Schlenzka, K., and Reddemann, H. (1975). *Clin. Chim. Acta*, **58**, 17–21.
179. Gröbe, H., Bartholomé, K., Milstien, S., and Kaufman, S. (1978). *Eur. J. Pediat.*, **129**, 93–98.
180. Grüttner, R., Sternowsky, H. J., and Rybak, C. (1971). *Mschr. Kinderheilk.*, **119**, 600–604.
181. Guroff, G., and Udenfriend, S. (1962). *J. Biol. Chem.*, **237**, 803–806.
182. Guroff, G., Daly, J. W., Jerina, D. M., Renson, J., Witkop, B., and Udenfriend, S. (1967). *Science*, **157**, 1524–1530.
183. Guroff, G. (1969). *Arch. Biochem. Biophys.*, **134**, 610–611.
184. Guthrie, R., and Susi, A. (1963). *Pediatrics*, **32**, 338–343.
185. Güttler, F., and Rosleff, F. (1973). *Acta Paediat. Scand.*, **62**, 333–336.
186. Hackney, I. M., Hanley, W. B., Davidson, W., and Linsao, L. (1968). *J. Pediat.*, **72**, 646–655.
187. Hambraeus, L., Holmgren, G., and Samuelson, G. (1971). *Nutr. Metabol.*, **13**, 298–317.
188. Hanson, A. (1958). *Naturwissenschaften*, **45**, 423–424.
189. Harper, A. E., Leung, P., and Yoshida, A. (1964). *Fed. Proc. Fed. Amer. Socs Exptl Biol.*, **23**, 1087.
190. Hawcroft, J., and Hudson, F. P. (1973). *Lancet ii*, 702–703.
191. Hill, A., Macaulay, J., and Zaleski, W. A. (1972). *Clin. Biochem.*, **5**, 194–196.
192. Hill, A., and Zaleski, W. A. (1972). *Clin. Biochem.*, **5**, 33–45.
193. Hill, J. B., Summer, G. K., Pender, M. W., and Roszel, N. O. (1965). *Clin. Chem.*, **11**, 541–546.
194. Hitchings, G. H. (1968). *Rev. Franc. Etudes Clin. Biol.*, **13**, 15–16.
195. Ho, C. C. (1971). *Science Progress, Oxford*, **59**, 75–89.
196. Hoffbauer, R. W., and Schrempf, G. (1976). *Lancet ii*, 194
197. Hole, K. (1972). *Develop. Psychobiol.*, **5**, 157–173.
198. Hollerman, C. E., and Calcagno, P. L. (1968). *Amer. J. Dis. Child.*, **121**, 169–178.
199. Holm, V. A., Deering, W. M., and Penn, R. L. (1970). *J. Amer. Med. Ass.*, **212**, 1835–1842.
200. Holton, J. B. (1972). *Ann. Clin. Biochem.*, **9**, 118–122.
201. Holtzman, N. A., Meek, A. G., Mellits, E. D., and Kallman, C. H. (1974). *J. Pediat.*, **85**, 175–181.
202. Holtzman, N. A., Welcher, D. W., and Mellits, E. D. (1978). *New Engl. J. Med.*, **293**, 1121–1124.
203. Hörnchen, H., Stuhlsatz, H. W., Plagemann, L., Eberle, P., and Habedank, M. (1977). *Deut. Med. Wschr.*, **102**, 308–312.
204. Hsia, D. Y. Y., Driscoll, K. W., Troll, W., and Knox, W. E. (1956). *Nature*, **178**, 1239–1240.
205. Hsia, D. Y. Y., O'Flynn, M. E., and Berman, J. L. (1968). *Amer. J. Dis. Child.*, **116**, 143–157.

206. Hsia, D. Y. Y., Shih, L.-Y., Easterberg, S., Farquhar, J., Kim, C.-B., Yeh, S., and Young, A. (1969). *Amer. J. Hum. Genet.*, **21**, 285–289.
207. Hsia, D. Y. Y. (1970). *Progr. Med. Genet.*, **8**, 29–68.
208. Hsia, D. Y. Y., and Dobson, J. (1970). *Lancet i*, 905.
209. Hughes, J. V., and Johnson, T. C. (1977). *Biochem. J.*, **162**, 527–537.
210. Ikeda, M., Levitt, M., and Udenfriend, S. (1967). *Arch. Biochem. Biophys.*, **120**, 420–427.
211. Inwang, E. E., Mosnaim, A. D., and Sabelli, H. C. (1973). *J. Neurochem.*, **20**, 1469–1473.
212. Irtel von Brenndorf, A., and Hagge, W. (1970). *Mschr. Kinderheilk.*, **118**, 433–435.
213. Jackson, D. M., and Temple, D. M. (1970). *Comp. Gen. Pharmacol.*, **1**, 155–159.
214. James, M. O., Smith, R. L., Williams, R. T., and Reidenberg, M. (1972). *Proc. Roy. Soc. Lond. B.*, **182**, 25–35.
215. James, M. O., and Smith, R. L. (1973). *Eur. J. Clin. Pharmacol.*, **5**, 243–246.
216. Jepson, J. B., Lovenberg, W., Zaltzman, P., Oates, J. A., Sjoerdsma, A., and Udenfriend, S. (1960). *Biochem. J.*, **74**, 5P.
217. Jepson, J. B. (1978). In *The Metabolic Basis of Inherited Disease*, 4th ed. (Eds., J. B. Stanbury, J. B. Wyngaarden, and D. S. Fredrickson), McGraw-Hill, New York, pp. 1563–1577.
218. Jervis, G. A. (1953). *Proc. Soc. Exptl Biol. Med.*, **82**, 514–515.
219. Jervis, G. A. (1954). *Proc. Ass. Res. Nerv. Ment. Dis.*, **33**, 259–282.
220. Jervis, G. A. (1960). *Clin. Chim. Acta*, **5**, 471–476.
221. Johnson, C. F. (1972). *Clin. Pediat.*, **11**, 148–156.
222. Jones, T. C., Levy, H. L., MacCready, R. A., Shih, V. E., and Garcia, F. G. (1971). *Proc. Soc. Exptl Biol. Med.*, **136**, 1087–1090.
223. Joanny, P., Natali, J.-P., Hillman, H., and Corriol, J. (1973). *Biochem. J.*, **136**, 77–82.
224. Jonsson, J., Grobecker, H., and Holtz, P. (1966). *Life Sci.*, **5**, 2235–2246.
225. Kääriäinen, R., and Karlsson, R. (1973). *Hereditas (Lund)*, **75**, 109–118.
226. Kamaryt, J., and Mrskos, A. (1974). *Čs. Pediat.*, **29**, 39–41.
227. Kang, E. S., Kennedy, J. L., Gates, L., Burwash, I., and McKinnon, A. (1965). *Pediatrics*, **35**, 932–943.
228. Kang, E. S., Kaufman, S., and Gerald, P. S. (1970). *Pediatrics*, **45**, 83–92.
229. Kang, E. S., Sollee, N. D., and Gerald, P. S. (1970). *Pediatrics*, **46**, 881–890.
230. Kaufman, S. (1958). *Science*, **128**, 1506–1507.
231. Kaufman, S., Milstien, S., and Bartholomé, K. (1975). *Lancet ii*, 708.
232. Kaufman, S. (1975). *Lancet ii*, 767.
233. Kaufman, S., Holtzman, N. A., Milstien, S., Butler, I. J. and Krumholz, A. (1975). *New Engl. J. Med.*, **293**, 785–790.
234. Kaufman, S. (1976). *Biochem. Med.*, **15**, 42–54.
235. Kaufman, S., Berlow, S., Summer, G. K., Milstien, S., Shulman, J. D., Orloff, S., Spielberg, S., and Pueschel, S. (1978). *New Engl. J. Med.*, **299**, 673–679.
236. Kelly, S. (1967). *Public Hlth Rep.*, **82**, 921–924.
237. Kirkman, H. N. (1972). *Progr. Med. Genet.*, **8**, 125–168.
238. Knox, W. E. (1972). In *The Metabolic Basis of Inherited Disease, 3rd ed.* (Eds., J. B. Stanbury, J. B. Wyngaarden, and D. S. Fredrickson), McGraw-Hill, New York, pp. 266–295.
239. Koch, R., Shaw, K. N. F., Acosta, P. B., Fishler, C., Schaeffler, G., Wenz, E., and Wohlers, A. (1970). *J. Pediat.*, **76**, 815–828.
240. Koch, R., Dobson, J., Hsia, D. Y. Y., and Woolf, L. I. (1971). *J. Pediat.*, **78**, 157–159.
241. Koe, B. K., and Weissman, A. (1966). *J. Pharmacol. Exptl Ther.*, **154**, 499–516.
242. Koepp, P., and Hoffmann, B. (1975). *Clin. Chim. Acta*, **58**, 215–221.
243. Koepp, P. (1977). *Lancet ii*, 92–93.

244. Komrower, G. M., and Sardharwalla, I. B. (1974). *Lancet i*, 1047.
245. Kraffczyk, F., Helger, R., and Lang, H. (1969). *Zeit. Klin. Chem.*, **7**, 521–524.
246. Krebs, H. A., and de Gasquet, P. (1964). *Biochem. J.*, **90**, 149–154.
247. Krips, C., and Lines, D. R. (1972). *Austral. Paediat. J.*, **8**, 318–321.
248. Kurtz, D. J., Levy, H., and Kanfer, J. N. (1972). *J. Nutr.*, **102**, 291–298.
249. Kutter, D., and Humbel, R. (1970). *Pharmaceut. Acta Helv.*, **45**, 553–563.
250. Lähdesmaki, P. (1975). *J. Neurobiol.*, **6**, 313–320.
251. Land, J. M., and Clark, J. B. (1973). *Biochem. J.*, **134**, 539–544.
252. Land, J. M., and Clark, J. B. (1973). *Biochem. J.*, **134**, 545–555.
253. Larson, C. A., and Nyman, G. E. (1968). *Psychiat. Clin.*, **1**, 367–374.
254. Lasala, J. M., and Coscia, C. J. (1979). *Science*, **203**, 283–284.
255. Leeming, R. J., Blair, J. A., Melikian, V., and O'Gorman, D. J. (1976). *J. Clin. Pathol.*, **29**, 444–451.
256. Leeming, R. J., Blair, J. A., Green, A., and Raine, D. N. (1976). *Arch. Dis. Childh.*, **51**, 771–777.
257. Leeming, R. J., Blair, J. A., and Rey, F. (1976). *Lancet i*, 99–100.
258. Levine, R. J., Nirenberg, P. Z., Udenfriend, S., and Sjoerdsma, A. (1964). *Life Sci.*, **3**, 651–656.
259. Levitt, M. S., Spector, S., Sjoerdsma, A., and Udenfriend, S. (1965). *J. Pharmacol. Exptl Ther.*, **148**, 1–8.
260. Levy, H. L., Karolkewicz, C., Houghton, S., and MacCready, R. A. (1970). *New Engl. J. Med.*, **282**, 1455–1458.
261. Levy, H., Shih, V. E., Karolkewicz, C., and MacCready, R. A. (1970). *Lancet ii*, 522–523.
262. Levy, H. L., Madigan, P. M., and Shih, V. E. (1972). *Pediatrics*, **49**, 825–836.
263. Lindroos, O. F. C., and Oja, S. S. (1971). *Exptl Brain Res.*, **14**, 48–60.
264. Lines, D. R., and Waisman, H. A. (1971). *J. Pediat.*, **78**, 474–480.
265. Lines, D. R. (1973). *Austral. Paediat.*, **9**, 152–153.
266. Lines, D. R., and Waisman, H. A. (1973). *Austral. New Zeal. J. Med.*, **3**, 169–173.
267. Linneweh, F., and Ehrlich, M. (1960). *Klin. Wschr.*, **38**, 904–910.
268. Linneweh, F., Ehrlich, M., Graul, E. H., and Hundeshagen, H. (1963). *Klin. Wschr.*, **41**, 253–255.
269. Linneweh, F. (1969). *Acta Paed. Scand.*, **10**, 1–12.
270. Lipton, M. A., Gordon, R., Guroff, G., and Udenfriend, S. (1967). *Science*, **156**, 248–250.
271. Longenecker, J. B., Reed, P. B., Lo, G. S., Chang, D. Y., Nasby, M. W., White, M. N., and Ide, S. (1970). *Nutr. Rep. Internat.*, **1**, 105–116.
272. Loo, Y. H., and Ritman, P. (1967). *Nature*, **213**, 914–916.
273. Loo, Y. H. (1974). *J. Neurochem.*, **23**, 139–147.
274. Loo, Y. H. (1967). *J. Neurochem.*, **14**, 813–821.
275. Loo, Y. H., and Mack, K. (1972). *J. Neurochem.*, **19**, 2385–2394.
276. Loo, Y. H., Scotto, L., and Horning, M. G. (1976). *Anal. Biochem.*, **76**, 111–118.
277. Lovenberg, W., Weissbach, H., and Undenfriend, S. (1962). *J. Biol. Chem.*, **237**, 89–92.
278. Lowden, J. A., and LaRamee, M. A. (1969). *Canad. J. Biochem.*, **47**, 883–888.
279. Lowry, R. B., Tischler, B., Cockroft, W. H., and Renwick, J. H. (1972). *Canad. Med. Ass. J.*, **106**, 1299–1302.
280. Lund, E., and Wamberg, E. (1970). *Dan. Med. Bull.*, **17**, 13–18.
281. Lutz, P. (1971). In *Phenylketonuria and Some Other Inborn Errors of Metabolism* (Eds., H. Bickel, F. P. Hudson, and L. I. Woolf), Georg Thieme, Stuttgart, pp. 56–64.
282. Mabry, C. C., Denniston, J. C., Nelson, T. L., and Son, C. D. (1963). *New Engl. J. Med.*, **269**, 1404–1408.

283. Mabry, C. C., Denniston, J. C., and Coldwell, J. G. (1966). *New Engl. J. Med.*, **275**, 1331–1336.
284. MacCready, R. A., and Levy, H. L. (1972). *Amer. J. Obstet. Gynecol.*, **113**, 121–128.
285. MacCready, R. A. (1974). *Pediatrics*, **85**, 383–385.
286. Machill, G., and Knapp, A. (1976). *Humangenetik*, **31**, 107–111.
287. MacInnes, J. W., and Schlesinger, K. (1971). *Brain Res.*, **29**, 101–110.
288. Malamud, N. (1966). *J. Neuropathol. Exptl Neurol.*, **25**, 254–268.
289. Metalon, R., Justice, P., and Deanching, M. N. (1977). *Lancet i*, 854.
290. McBean, M. S. and Stephenson, J. B. P. (1968). *Arch. Dis. Childh.*, **43**, 1–7.
291. McCaman, M. W. and Robins, E. (1962). *J. Lab. Clin. Med.*, **59**, 885–890.
292. McKean, C. M., Boggs, D. E., and Peterson, N. A. (1968). *J. Neurochem.*, **15**, 235–241.
293. McKean, C. M., and Peterson, N. A. (1970). *New Engl. J. Med.*, **283**, 1364–1367.
294. McKean, C. M. (1972). *Brain Res.*, **47**, 469–476.
295. McLean, A., Marwick, M. J., and Clayton, B. E. (1973). *J. Clin. Pathol.*, **26**, 678–683.
296. Medical Research Council, Working Party on Phenylketonuria (1968). *Brit. Med. J. ii*, 7–13.
297. Meister, A., Udenfriend, S., and Bessman, S. P. (1956). *J. Clin. Invest.*, **35**, 619–626.
298. Menkes, J. H. (1968). *Neurology*, **18**, 1003–1008.
299. Menkes, J. H., and Aeberhard, E. (1969). *J. Pediat.*, **74**, 924–931.
300. Mereu, T. (1967). *Amer. J. Dis. Child.*, **113**, 522–523.
301. Miller, A. L., Hawkins, R. A., and Veech, R. L. (1973). *Science*, **173**, 904–906.
302. Miller, M. B., McClure, D., and Shiman, R. (1975). *J. Biol. Chem.*, **250**, 1132–1140.
303. Milstien, S., Holtzman, N. A., O'Flynn, M. E., Thomas, G. H., Butler, I. J., and Kaufman, S. (1976). *J. Pediat.*, **89**, 763–766.
304. Milstien, S., Orloff, S., Speilberg, S., Berlow, S., Shulman, J. D., and Kaufman, S. (1977). *Pediat. Res.*, **11**, 460, Abstract No. 532.
305. Minami, R., Olek, K., and Wardenbach, P. (1975). *Humangenetik*, **28**, 319–324.
306. Mitoma, C., Auld, R. M., and Udenfriend, S. (1957). *Proc. Soc. Exptl Biol. Med.*, **94**, 634–636.
307. Miyamoto, M., and Fitzpatrick, T. B. (1957). *Nature*, **179**, 199–200.
308. Moldave, K., and Meister, A. (1957). *J. Biol. Chem.*, **229**, 463–476.
309. Moldave, K., and Meister, A. (1957). *Biochim. Biophys. Acta*, **25**, 434–435.
310. Moldave, K., and Meister, A. (1957). *Biochim. Biophys. Acta*, **24**, 654–655.
311. Morton, N. E. (1955). *Amer. J. Hum. Genet.*, **7**, 277–318.
312. Mosnaim, A. D., and Wolf, M. E. (1978). *Noncatecholic Phenylethylamines 1. Phenylethylamine, Biological and Clinical Aspects*, Marcel Dekker, New York.
313. Mukherjee, A. B., and Krasner, J. (1973). *Science*, **182**, 68–70.
314. Murthy, L. I., and Berry, H. K. (1974). *Arch. Biochem. Biophys.*, **163**, 225–230.
315. Musacchio, J. M., D'Angelo, G. L., and McQueen, C. A. (1971). *Proc. Nat. Acad. Sci., USA*, **68**, 2087–2091.
316. Nadler, H. L., and Hsia, D. Y. Y. (1961). *Proc. Soc. Exptl Biol. Med.*, **107**, 721–722.
317. Nadler, H. L. (1976). *Birth Defects, Orig. Art. Ser.*, **12**, 177–188.
318. Nakajima, T., Kakimoto, Y., and Sano, I. (1964). *J. Pharmacol. Exptl Ther.*, **143**, 319–325.
319. Neame, K. D. (1968). *Prog. Brain Res.*, **29**, 185–196.
320. Neifakh, S. A., and Šapošnikov, A. M. (1965). *Ž. Nevropathol. Psihiat.*, **65**, 1105–1113.
321. Neriishi, S. (1968). *Jap. J. Hum. Genet.*, **13**, 54–58.
322. Neuhoff, V., Behbehani, A. W., Quentin, C.-D., and Prinz, H. (1974). *Hoppe-Seyler's Zeit. Physiol. Chem.*, **355**, 891–894.
323. Nishikimi, M. (1975). *Biochem. Biophys. Res. Commun.*, **63**, 92–98.
324. Nitz, I., Cobet, G., and Robbe, K. (1971). *Der Radiologe*, **11**, 305–307.

325. Oates, J. A., Nirenberg, P. Z., Jepson, J. B., Sjoerdsma, A., and Udenfriend, S. (1963). *Proc. Soc. Exptl Biol. Med.*, **112**, 1078–1081.
326. Oja, S. S. (1968). *Ann. Med. Exptl Fenn.*, **46**, 541–546.
327. Oja, S. S. (1972). *J. Neurochem.*, **19**, 2057–2069.
328. Olek, K., Oyanagi, K., and Wardenbach, P. (1974). *Humangenetik*, **22**, 85–88.
329. Osborn, M. J., Freeman, M., and Huennekens, F. M. (1958). *Proc. Soc. Exptl Biol. Med.*, **97**, 429–431.
330. Oteruelo, F. T. (1976). *Acta Neuropathol.*, **36**, 295–305.
331. Paine, R. S. (1957). *Pediatrics*, **20**, 290–301.
332. Pare, C. M. B., Sandler, M., and Stacey, R. S. (1957). *Lancet i*, 551–553.
333. Pare, C. M. B., Sandler, M., and Stacey, R. S. (1959). *Arch. Dis. Childh.*, **34**, 422–425.
334. Parker, C. E., Barranger, J., Newhouse, R., and Bessman, S. P. (1977). *Ann. Clin. Biochem.*, **14**, 122–123.
335. Patel, M. S. (1972). *Biochem. J.*, **128**, 677–684.
336. Patel, M. S., Grover, W. D., and Auerbach, V. H. (1973). *J. Neurochem.*, **20**, 289–296.
337. Patel, M. S., and Arinze, I. J. (1975). *Amer. J. Clin. Nutr.*, **28**, 183–188.
338. Partington, M. W., and Vickery, S. K. (1974). *Neuropädiatrie*, **5**, 125–137.
319. Pedersen, H. E. (1974). *Acta Neurol. Scand.*, **50**, 599–610.
340. Pedersen, H. E., and Birket-Smith, E. (1974). *Acta Neurol. Scand.*, **50**, 589–598.
341. Perry, T. L. (1962). *Science*, **136**, 879–880.
342. Perry, T. L., Tischler, B., and Chapple, J. A. (1966). *J. Psychiat. Res.*, **4**, 51–57.
343. Perry, T. L., Tischler, B., Hansen, S., and MacDougall, L. (1967). *Clin. Chim. Acta*, **15**, 47–55.
344. Perry, T. L., Hansen, S., Tischler, B., and Bunting, R. (1967). *Clin. Chim. Acta*, **18**, 51–56.
345. Perry, T. L., and Hansen, S. (1969). *Clin. Chim. Acta*, **25**, 53–58.
346. Perry, T. L., Hansen, S., Tischler, B., Bunting, R., and Diamond, S. (1970). *New Engl. J. Med.*, **282**, 761–766.
347. Perry, T. L., Hansen, S., Tischler, B., Richards, F. M., and Sokol, M. (1973). *New Engl. J. Med.*, **289**, 395–398.
348. Pierson, M., Frentz, R., Marchal, C., Thiriet, M., Didier, F., and Humbel, R. (1968). *Arch. Franc. Pédiat.*, **25**, 439–452.
349. Pitt, D., McFarlane, J., Francis, I., Gaha, T. J., Hill, G., Crotty, J. M., Masters, P., and Cusick, E. (1973). *Med. J. Austral.*, **1**, 170–171.
350. Pitt, D., and Gooch, J. (1974). *Austral. J. Paediat.*, **10**, 337–342.
351. Plöchl, E. (1968). *Clin. Chim. Acta*, **21**, 271–274.
352. Popkin, J. S., Clow, C. L., Scriver, C. L., and Grove, J. (1974). *Lancet i*, 721–722.
353. Poser, C. M., and van Bogaert, L. (1959). *Brain*, **82**, 1–9.
354. Prensky, A. L., Fishman, M. A., and Daftari, B. (1971). *Brain Res.*, **33**, 181–191.
355. Pryor, G. T., and Mitoma, C. (1970). *Neuropharmacology*, **9**, 269–275.
356. Quentin, C.-D., Behbehani, A. W., Schulte, F. J., and Neuhoff, V. (1974). *Neuropädiatrie*, **5**, 138–145.
357. Rama Rao, B. S. S., Narayan, H. S., and Reddy, G. N. N. (1970). *Indian J. Med. Res.*, **58**, 1753–1757.
358. Rampini, S., Anders, P. W., Curtius, H.-Ch., and Marthaler, Th. (1969). *Pediat. Res.*, **3**, 287–297.
359. Rampini, S. (1973). *Schweiz. Med. Wschr.*, **103**, 537–546.
360. Rampini, S., Völlmin, J. A., Bosshard, H. R., Müller, M., and Curtius, H. C. (1974). *Pediat. Res.*, **8**, 704–709.
361. Rapoport, D., Saudubray, J.-M., Hatt, A., Weil-Halpern, F., Depondt, E., Boisse, J., and Mozziconacci, P. (1975). *Ann. Pédiat.*, **22**, 509–516.

362. Rey, F., Pellié, C., Sivy, M., Blandin-Savoja, F., Rey, J., and Frezal, J. (1974). *Pediat. Res.*, **8**, 540–545.
363. Rey, F., Harpey, J. P., Leeming, R. J., Blair, J. A., Aicardi, J., and Rey, J. (1977). *Arch. Franç. Pédiat.*, **34**, Suppl. 2, CIX–CXX.
364. Richards, B. W., Sylvester, P. E., and Hodgson, S. J. (1965). *J. Ment. Defic. Res.*, **9**, 210–218.
365. Rose, D. P., and Cramp, D. G. (1970). *Clin. Chim. Acta*, **29**, 49–53.
366. Rosenblatt, D., Mohyoddin, F., and Scriver, C. S. (1970). *Pediatrics*, **46**, 47–53.
367. Ross, S. B., and Haljasmaa, Ö. (1964). *Life Sci.*, **3**, 579–587.
368. Roth, K. S., Cohn, R. M., Berman, P., and Segal, S. (1976). *J. Pediat.*, **88**, 689–690.
369. Röthler, F., and Karobath, M. (1976). *Clin. Chim. Acta*, **69**, 457–462.
370. Rothman, K. J., and Pueschel, S. W. (1976). *Pediatrics*, **58**, 842–844.
371. Rowe, V. D., Fales, H. M., Pisano, J. J., Andersen, A. E., and Guroff, G. (1975). *Biochem. Med.*, **12**, 123–136.
372. Rozengurt, E., Jimenez de Asua, L., and Carminatti, H. (1970). *FEBS Letters*, **11**, 284–286.
373. Saavedra, J. M., and Axelrod, J. (1973). *Proc. Nat. Acad. Sci., USA*, **70**, 769–772.
374. Sabelli, H. C., and Mosnaim, A. D. (1974). *Amer. J. Psychiat.*, **131**, 695–699.
375. Sabelli, H. C., Borison, R. L., Diamond, B. I., Havdala, H. S., and Narasimhachari, N. (1978). *Biochem. Pharmacol.*, **27**, 1707–1711.
376. Samuels, S., and Ward, S. S. (1966). *J. Lab. Clin. Med.*, **67**, 669–677.
377. Saugstad, L. F., and Wehn, M. (1967). *Electroenceph. Clin. Neurophysiol.*, **23**, 390.
378. Saugstad, L. F., Kelly, S., Korns, R., and Burns, J. (1970). *Lancet ii*, 314–315.
379. Saugstad, L. F. (1973). *Clin. Genet.*, **4**, 105–114.
380. Saugstad, L. F. (1973). *Lancet i*, 1255.
381. Saugstad, L. F. (1975). *Clin. Genet.*, **7**, 40–51.
382. Saugstad, L. F. (1975). *Clin. Genet.*, **7**, 52–61.
383. Saugstad, L. F. (1977). *J. Med. Genet.*, **14**, 20–24.
384. Schaefer, G. J., Barrett, R. J., Sanders-Bush, E., and Vorhees, C. V. (1974). *Pharmacol. Biochem. Behav.*, **2**, 783–789.
385. Scheibenreiter, S. (1972). *Monatsh. Kinderheilk.*, **120**, 189–192.
386. Schoenenberg, H., Radermacher, E. H., Frank, M., and Klemm, W. (1977). *(Klin.) Pädiat.*, **189**, 482–487.
387. Schulte, F. J., Kaiser, H. J., Engelbart, S., Bell, E. F., Castell, R., and Lenard, H. G. (1973). *Pediat. Res.*, **7**, 588–599.
388. Schürrle, L., and Bickel, H. (1971). In *Phenylketonuria and Some Other Inborn Errors of Metabolism* (Eds., H. Bickel, F. P. Hudson, and L. I. Woolf), Georg Thieme, Leipzig, pp. 240–257.
389. Schwark, W. S., Singhal, R. L., and Ling, G. M. (1970). *Life Sci.*, **9**, 939–945.
390. Scriver, C. R., Davies, E., and Cullen, A. M. (1964). *Lancet ii*, 230–232.
391. Scriver, C. R. (1975). *Pediatrics*, **54**, 616–619.
392. Semba, T., and Civen, M. (1970). *J. Neurochem.*, **17**, 795–800.
393. Shah, S. N., Peterson, N. A., and McKean, C. M. (1970). *J. Neurochem.*, **17**, 279–284.
394. Shah, S. N., Peterson, N. A., and McKean, C. M. (1972). *J. Neurochem.*, **19**, 479–485.
395. Shah, S. N., and Johnson, R. C. (1975). *FEBS Letters*, **49**, 404–406.
396. Shizume, K., and Naruse, H. (1958). *J. Ment. Defic. Res.*, **2**, 53.
397. Shrawder, E., and Martinez-Carrion, M. (1972). *J. Biol. Chem.*, **247**, 2486–2492.
398. Sibinga, M. S., Friedman, C. J., Steisel, I. M., and Baker, E. C. (1971). *Dev. Med. Child Neurol.*, **13**, 63–70.
399. Sibinga, M., and Friedman, C. J. (1972). *Dev. Med. Child Neurol.*, **14**, 445–456.
400. Sidransky, H., Bongiorno, M., Sarma, D. S. R., and Verley, E. (1967). *Biochem. Biophys. Res. Commun.*, **27**, 242–248.

401. Siegel, F. L., Aoki, K., and Colwell, R. E. (1971). *J. Neurochem.*, **18**, 537–547.
402. Silberberg, D. H. (1967). *Arch. Neurol.*, **17**, 524–529.
403. Smith, B. A., and Waisman, H. A. (1971). *Amer. J. Clin. Nutr.*, **24**, 423–431.
404. Smith, I., and Lloyd, J. (1974). *Arch. Dis. Childh.*, **49**, 245.
405. Smith, I., and Wolff, O. H. (1974). *Lancet*, *ii*, 540–544.
406. Smith, I., Clayton, B. E., and Wolff, O. H. (1975). *Lancet i*, 328–329.
407. Smith, I., Clayton, B. E., and Wolff, O. H. (1975). *Lancet i*, 1108–1111.
408. Smith, I., Lobascher, M. E., Stevenson, J. E., Wolff, O. H., Schmidt, H., Grubel-Kaiser, S., and Bickel, H. (1978). *Brit. Med. J. ii*, 723–726.
409. Smith, I., Carter, C. O., and Wolff, O. H. (1978). *J. Inher. Metab. Dis.*, **1**, 99–100.
410. Smith, M. E. (1968). *Biochim. Biophys. Acta*, **164**, 285–293.
411. Snyderman, S. E., Norton, P., and Holt, L. E. (1955). *Fed. Proc. Fed. Amer. Socs Exptl Biol.*, **14**, 450–451.
412. Sprinkle, T. J., and Rennert, O. M. (1976). *J. Neurochem.*, **26**, 499–504.
413. Steiner, K. C., and Smith, H. A. (1975). *Public Hlth Rep.*, **90**, 52–54.
414. Steinhausen, H. C. (1974). *Neuropädiatrie*, **5**, 146–156.
415. Stephenson, J. B., and McBean, M. S. (1967). *Brit. Med. J. iii*, 582.
416. Stevenson, R. E., and Howell, R. R. (1972). *Ann. Amer. Acad. Polit. Soc. Sci.*, **399**, 30–37.
417. Stevenson, R. E., and Huntley, C. C. (1967). *Pediatrics*, **40**, 33–45.
418. Szeinberg, A., Shani, M., Crispin, M., Hirshorn, N., Cohen, B. E., and Sheba, Ch. (1970). *Israel J. Med. Sci.*, **6**, 475–478.
419. Tanaka, K., Matsunaga, E., Handa, Y., Murata, T., and Takehara, K. (1961). *Jap. J. Hum. Genet.*, **6**, 65–77.
420. Taniguchi, K., and Armstrong, M. D. (1963). *J. Biol. Chem.*, **238**, 4091–4097.
421. Terheggen, H. G., Lowenthal, A., Lavinha, F., Colombo, J. P., and Rogers, S. (1975). *Kinderheilk.*, **119**, 1–3.
422. Thalhammer, O. (1970). *Wien. Klin. Wschr.*, **82**, 774–777.
423. Thalhammer, O., and Scheiber, V. (1972). *Neuropädiatrie*, **3**, 358–361.
424. Thalhammer, O. (1973). *Mschr. Kinderheilk.*, **121**, 201–204.
425. Thalhammer, O., Havelec, L., Knoll, E., and Wehle, E. (1977). *Wien. Klin. Wschr.*, **89**, Suppl. No. 76, 684–686.
426. Theile, H., Buehrdel, P., and Graustein, I. (1976). *Kinderarztl. Prax.*, **44**, 338–343.
427. Thomas, G. H., Parmley, T. H., Stevenson, R. E., and Howell, R. R. (1971). *Amer. J. Obstet. Gynecol.*, **111**, 38–42.
428. Thompson, R. M., Belanger, B. G., Wappner, R. S., and Brandt, I. K. (1975). *Clin. Chim. Acta*, **61**, 367–374.
429. Ting-Beall, H. P., and Wells, W. W. (1971). *FEBS Letters*, **16**, 352.
430. Tourian, A. Y., and Sidbury, J. B. (1978). In *The Metabolic Basis of Inherited Disease*, 4th ed. (Eds., J. B. Stanbury, J. B. Wyngaarden, and D. S. Fredrickson), McGraw-Hill, New York, pp. 240–255.
431. Trouche, A. M., Gigonnet, J. M., Dorche, C., Nivelon-Chevalier, A., Nivelon, J. L., and Alison, M. (1974). *Pédiatrie*, **29**, 33–50.
432. Tryggvason, K. (1979). (Personal communication.)
433. Tu, J.-B., and Partington, M. W. (1972). *Dev. Med. Child Neurol.*, **14**, 456–466.
434. Umbarger, B., Berry, H. K., and Sutherland, B. S. (1965). *J. Amer. Med. Ass.*, **193**, 784–790.
435. Vahvelainen, M.-L., and Oja, S. S. (1975). *J. Neurochem.*, **24**, 885–892.
436. Valdivieso, F., Maties, M., Ugarte, M., and Mayor, F. (1973). *Biochem. Med.*, **7**, 341–342.
437. Valdivieso, F., Gimenez, C., and Mayor, F. (1975). *Biochem. Med.*, **12**, 72–78.
438. Van Pelt, A., and Levy, H. L. (1974). *New Engl. J. Med.*, **291**, 1414–1416.

439. van Sumere, C. F., Teuchy, H., and Massart, L. (1959). *Clin. Chim. Acta*, **4**, 590–593.
440. Vassella, F., Colombo, J. P., Humbel, R., and Rossi, E. (1968). *Helv. Paed. Acta*, **23**, 22–36.
441. Vijayvargiya, R., Schwark, W. S., and Singhal, R. L. (1969). *Canad. J. Biochem.*, **47**, 895–898.
442. Visakorpi, J. K., Palo, J., and Renkonen, O.-V. (1971). *Acta Paed. Scand.*, **60**, 666–668.
443. Vorhees, C. V., Butcher, R. E., and Berry, H. K. (1972). *Dev. Psychobiol.*, **5**, 175–179.
444. Wadman, S. K., Ketting, D., de Bree, P. K., van der Heiden, C., Grimberg, M. Th., and Kruijswijk, H. (1975). *Clin. Chim. Acta*, **65**, 197–204.
445. Wainer, S. C., and Sideman, L. (1974). *Health Lab. Sci.*, **11**, 306–311.
446. Waisman, H. A. (1967). *Amer. J. Obstet. Gynecol.*, **99**, 431–433.
447. Wapnir, R. A., Hawkins, R. L., Stevenson, J. H., and Bessman, S. P. (1970). *Biochem. Med.*, **3**, 397–403.
448. Wapnir, R. A. and Lifshitz, F. (1974). *Clin. Chim. Acta*, **54**, 349–356.
449. Watson, B. M., Schlesinger, P., and Cotton, R. G. H. (1977). *Clin. Chim. Acta*, **78**, 417–423.
450. Watson, C. W., Nigam, M. P., and Paine, R. S. (1968). *Neurology*, **18**, 203–207.
451. Weber, G. (1969). In *Psychochemical Research in Man* (Eds., A. J. Mandell, and M. P. Mandell), Academic Press, New York, pp. 3–47.
452. Weber, G. (1969). *Adv. Enzyme Regulation*, **7**, 15–40.
453. Weber, G. (1969). *Proc. Nat. Acad. Sci., USA*, **63**, 1365–1369.
454. Weber, H.-P., Brodehl, J., Hensen, S. B., Schröder, M. R., and Shinoda, M. (1971). *Mschr. Kinderheilk.*, **119**, 360–363.
455. Weber, W. W., and Zannoni, V. G. (1966). *J. Biol. Chem.*, **241**, 1345–1349.
456. Weil-Malherbe, H., and Bone, A. D. (1955). *J. Ment. Sci.*, **101**, 733–755.
457. Weiss, B., Hui, M., and Lajtha, A. (1977). *Biochem. Med.*, **18**, 330–343.
458. Weksler, M. E., Bourke, E., and Schreiner, E. (1968). *Clin. Pharmacol. Ther.*, **9**, 647–651.
459. Williams, R. (1968). *Med. J. Austral.*, **2**, 216–219.
460. Wolfenden, R. (1969). *Nature*, **223**, 704–705.
461. Wong, P. K. W., Berman, J. L., Partington, M. W., Vickery, S. K., O'Flynn, M. E., and Hsia, D. Y. Y. (1971). *New Engl. J. Med.*, **285**, 580.
462. Wood, B. (1976). *Dev. Med. Child Neurol.*, **18**, 657–665.
463. Woolf, L. I., Griffiths, R., and Moncrieff, A. (1955). *Brit. Med. J.*, *i*, 57–64.
464. Woolf, L. I., Goodwin, B. L., Cranston, W. I., Wade, D. N., Woolf, F., Hudson, F. P., and McBean, M. S. (1968). *Lancet i*, 114–117.
465. Woolf, L. I., McBean, M. S., Woolf, F. M., and Cahalane, S. F. (1975). *Ann. Hum. Genet.*, **38**, 461–469.
466. Woolf, L. I. (1976). *J. Irish Med. Ass.*, **69**, 398–401.
467. Woolf, L. I. (1978). *J. Inher. Metab. Dis.*, **1**, 101–103.
468. Woolley, D. W., and van der Hoeven, T. (1964). *Science*, **144**, 883–884; 1593–1594.
469. *World Health Organization Technical Report No. 497*, WHO, Geneva (1973).
470. Wu, P. H., and Boulton, A. A. (1974). *Canad. J. Biochem.*, **53**, 42–50.
471. Wunner, W. H., Bell, J., and Munro, H. N. (1966). *Biochem. J.*, **101**, 417–428.
472. Yakymyshyn, L. Y., Reid, D. W. J., and Campbell, D. J. (1972). *Clin. Biochem.*, **5**, 73–81.
473. Yu, J. S., and O'Halloran, M. T. (1970). *Lancet i*, 210–212.
474. Yuwiler, A., and Louttit, R. T. (1961). *Science*, **134**, 831–832.
475. Yuwiler, A., and Geller, E. (1969). *J. Neurochem.*, **16**, 999–1005.
476. Zannoni, V. G., and Weber, W. W. (1966). *J. Biol. Chem.*, **241**, 1340–1345.
477. Zetterström, R., and D'Avignon, M. (1976). *Läkartidningen*, **73**, 1069–1074.

Aromatic Amino Acid Hydroxylases and Mental Disease
Edited by M. B. H. Youdim

CHAPTER 4

Tyrosine-3-monooxygenase (Tyrosine Hydroxylase)

Norman Weiner

A INTRODUCTION AND ENZYMOLOGY OF TYROSINE HYDROXYLASE

Tyrosine-3-monooxygenase (EC 1.14.16.2) (tyrosine hydroxylase) was first demonstrated in bovine adrenal medulla (Nagatsu *et al.*, 1964a; Brenneman and Kaufman, 1964). Because of the availability of this tissue in large amounts and the high activity of the enzyme in this organ, bovine adrenal medulla has been employed by many investigators as a source of this enzyme.

The enzyme is a mixed function oxidase, requiring molecular oxygen (Daly *et al.*, 1968) and a reduced pterin (Nagatsu *et al.*, 1964a; Brenneman and Kaufman, 1964) as cosubstrates. Based on demonstrations of its occurrence in liver (Kaufman, 1963), adrenal medulla (Lloyd and Weiner, 1971), and brain (Gal *et al.*, 1976), the putative natural pterin cosubstrate is believed to be tetrahydrobiopterin (biopterin-H_4). Biopterin can be synthesized from guanosine in mouse neuroblastoma clones (Buff and Dairman, 1974) and from guanosine triphosphate (GTP) in rat brain (Gal and Sherman, 1976). Dihydropteridine reductase, the enzyme that catalyses the reduction of 7,8-dihydropterins to the active 5,6,7,8-tetrahydro form, has been demonstrated in sheep liver (Kaufman, 1964; Nielsen *et al.*, 1969; Craine *et al.*, 1972), beef adrenal medulla (Musacchio, 1969; Musacchio *et al.*, 1971a) and brain (Turner *et al.*, 1974; Spector *et al.*, 1977). The enzyme requires a reduced pyridine nucleotide for activity. Either NADH or NADPH serves as a cofactor for dihydropteridine reductase, but the former pyridine nucleotide exhibits a higher affinity for the enzyme (Nielsen *et al.*, 1969; Craine *et al.*, 1972). Studies in our laboratory (Thoa *et al.*, 1971) and in that of Musacchio (1969) indicate that dihydropteridine reductase is less sensitive to inhibition by methotrexate than dihydrofolate reductase. However, in higher concentrations (greater than 1 μM), methotrexate also inhibits dihydropteridine reductase.

Several synthetic analogues of biopterin-H_4 have been employed as cofactors for the study of tyrosine hydroxylase and other aromatic amino acid hydroxylases. Because of its commercial availability, 6,7-dimethyltetrahydropterin (Me_2PtH_4) has been most frequently used in studies of these enzymes, but this cofactor appears to behave somewhat anomalously in the enzymic reaction (Shiman *et al.*, 1971). Pterin compounds with an alkyl substituent on the 6-position of the pterin ring system are now regarded as the preferred synthetic cosubstrates for use in studies involving aromatic amino acid hydroxylases. Of these, one of the most commonly employed is 6-methyltetrahydropterin (6-$MePtH_4$). Recently, biopterin has become available commercially and, since it can be readily reduced to the tetrahydro form (e.g. Lovenberg *et al.*, 1975), it should be the preferred substrate for studies of this class of enzymes. However, biopterin-H_4 and 6-$MePtH_4$ appear to behave in very similar fashions as cosubstrates in the hydroxylase reactions, and it may be assumed qualitatively similar results would be obtained with the two pterins. Biopterin-H_4 appears to have a somewhat higher affinity for tyrosine hydroxylase than does 6-$MePtH_4$ (Shiman and Kaufman, 1970).

Ferrous ion appears to be required for optimal activity of tyrosine hydroxylase (Nagatsu *et al.*, 1964a; Petrack *et al.*, 1968). There has been considerable debate over the precise role of ferrous ion in the reaction. Since α,α-dipyridyl and other iron chelating reagents completely inhibit the enzyme (Taylor *et al.*, 1969), several workers have assumed that the enzyme may be an iron containing metallo-enzyme (Nagatsu *et al.*, 1964a; Petrack *et al.*, 1968). On the other hand, Shiman *et al.* (1971) and Kaufman and Fisher (1974) observed that catalase could replace iron in the activation of tyrosine hydroxylase. They concluded that ferrous ion may be serving as a catalyst for the breakdown of hydrogen peroxide, which presumably forms during the aromatic hydroxylation reaction and which may inhibit or denature the enzyme. We have confirmed that iron is required for optimal activity in the tyrosine hydroxylation reaction and have demonstrated complete loss of activity in the presence of iron chelators, such as α,α-dipyridyl. Although catalase does increase the activity of tyrosine hydroxylase to some degree, it is unable to replace completely iron in the system. Our results are in agreement with those of Udenfriend and coworkers who claim that iron may be an essential cofactor in the tyrosine hydroxylase reaction (Waymire *et al.*, 1972b).

Recently, Hoeldtke and Kaufman (1977) purified the solubilized bovine adrenal enzyme to near homogeneity and have demonstrated the presence of 0.5–0.75 mol of iron per mole of enzyme monomer. They concluded that, like other hydroxylases, such as phenylalanine hydroxylase, which is an iron-containing enzyme (Fisher *et al.*, 1972), and dopamine β-hydroxylase, which contains copper (Goldstein *et al.*, 1968; Craine *et al.*, 1973), tyrosine hydroxylase is a metallo-enzyme.

Although the detailed mechanism of the tyrosine hydroxylase reaction has not been elucidated definitively, it appears to be analogous to that which occurs with hepatic phenylalanine hydroxylase (Kaufman and Fisher, 1974). On the basis of kinetic studies with both solubilized and particulate bovine adrenal tyrosine hydroxylase, Kaufman and Fisher (1974) proposed that the hydroxylation of tyrosine involves a sequential mechanism in which a quaternary complex, consisting of enzyme, substrate, tetrahydropterin, and oxygen, is an intermediate in the catalysis. These results are not in agreement with earlier studies on soluble bovine adrenal enzyme, which suggested that the hydroxylation of tyrosine involves a ping-pong mechanism and that an intermediate step in the reaction involves removal of two protons from the active site of the enzyme (Ikeda *et al.*, 1966). Results supporting a sequential mechanism were reported by Joh *et al.* (1969) for particulate bovine adrenal enzyme. We have also obtained evidence that the vas deferens enzyme from the guinea pig, in the presence of different concentrations of tyrosine and biopterin-H_4, exhibits kinetics which are consistent with a sequential type reaction (Figure 4.1).

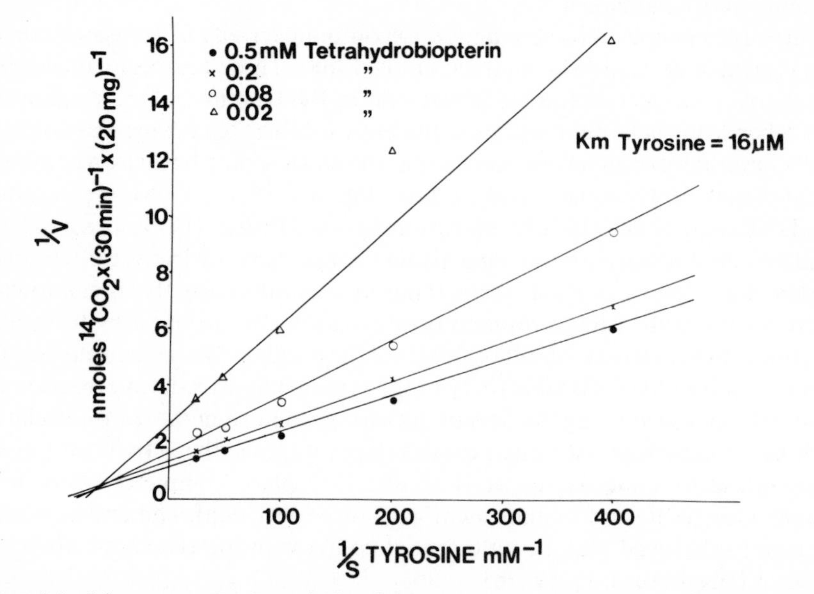

Figure 4.1 Lineweaver–Burk analysis of the activity of guinea pig vas deferens soluble tyrosine hydroxylase in the presence of different concentrations of tetrahydrobiopterin (biopterin-H_4) and tyrosine. The results are consistent with a mechanism of action of the enzyme involving a sequential-type reaction.

Tyrosine hydroxylase activity is expressed as nmol product formed per 30 min. per 20 mg tissue. Crude supernatant enzyme was assayed in these studies

Although bovine adrenal medulla tyrosine hydroxylase has been solubilized by partial proteolytic digestion (Petrack *et al.*, 1968; Shiman *et al.*, 1971), and this enzyme has been purified to near homogeneity (Hoeldtke and Kaufman, 1977), the partially digested enzyme does not behave identically to the native enzyme. For example, it appears to lack much of the sensitivity to putative regulatory factors, such as phospholipids and the cyclic AMP-dependent protein phosphorylating system (Hoeldtke and Kaufman, 1977).

Tyrosine hydroxylase is able to hydroxylate both phenylalanine and tyrosine (Ikeda *et al.*, 1965, 1967). Ikeda *et al.* (1965) suggested, based on the specific activities of tyrosine and dihydroxyphenylalanine (DOPA) formed, that the tyrosine formed from phenylalanine dissociates from the enzyme and equilibrates with the pool of tyrosine in the system. For further hydroxylation to DOPA, the tyrosine must bind again to the active site of the enzyme.

Shiman *et al.* (1971) observed that both phenylalanine and tyrosine were substrates for the bovine adrenal enzyme. Phenylalanine was a relatively better substrate than tyrosine in the presence of biopterin-H_4, the putative natural cofactor, compared with Me_2PtH_4. Furthermore, in the presence of the natural cofactor, high concentrations of tyrosine were associated with substrate inhibition of the enzyme. These workers concluded that, under circumstances which may exist *in vivo* (presence of biopterin-H_4 and relatively high concentrations of tyrosine and phenylalanine), phenylalanine may be the preferred substrate. However, in the intact vas deferens preparation, hydroxylation of tyrosine to DOPA is preferred over the conversion of phenylalanine to DOPA when both are present in the medium in physiological concentrations (Weiner *et al.*, 1974). These *in situ* studies, however, may not reflect the true rate of phynylalanine hydroxylation, since the newly formed labelled tyrosine presumably dissociates from the enzyme and mixes with the much larger pool of unlabelled tyrosine in the neuron. Thus, the specific activity of the tyrosine formed from phenylalanine would be markedly reduced and the rate of conversion of labelled tyrosine to DOPA would be expected to be small in proportion to this dilution.

Katz *et al.* (1976a) compared the rate of hydroxylation of phenylalanine and tyrosine to DOPA by soluble rat striatal enzyme and by the intact P_2 fraction of synaptosomes from this tissue. With soluble enzyme, although there was a brisk conversion of phenylalanine to tyrosine, much more exogenous tyrosine than phenylalanine was converted to DOPA. In contrast, in synaptosomal preparations, a considerable amount of the phenylalanine which was converted to tyrosine was further transformed to DOPA. These results suggest that the synaptosomal pool of tyrosine formed from phenylalanine is preferentially utilized for DOPA formation. This probably reflects the existence of a small pool of the precursor near the enzyme, which is maintained by the structural integrity of the synaptosomal membrane.

The interaction of oxygen with the enzyme is also critically dependent on the cofactor employed. Fisher and Kaufman (1972) observed that with biopterin-H_4, the apparent K_m of the brain and adrenal enzyme for O_2 is approximately 1% (solubilized bovine adrenal enzyme) or less than 1% (bovine brain enzyme). Enzyme inhibition was seen at oxygen concentrations of 40% for the former enzyme and 8% for the latter enzyme. The apparent K_m of O_2 for the enzyme is much higher when Me_2PtH_4 is employed as cofactor, and the enzyme is much more resistant to inhibition by high O_2 concentrations (Fisher and Kaufman, 1972). The physiological relevance of the inhibition of the enzyme by high oxygen concentrations is dubious, however, since Weiner *et al.* (1973) have shown that, with the isolated vas deferens preparation, the hydroxylation of tyrosine to DOPA increases progressively as the oxygen concentration in the medium is altered from 20% to 95%. Similar results have been obtained with brain tyrosine hydroxylase *in vivo* in rats exposed to different concentrations of oxygen in the inspired air (Davis and Carlsson, 1973; Davis, 1976). Of course, it is possible that the high oxygen concentrations in the incubation medium or in the inspired air do not reach the site of the enzyme. Because of diffusion limitations, the oxygen concentration in the isolated organ or in the brain may be very much lower and it may not be possible to reach levels at the site of the enzyme which are inhibitory.

Tyrosine hydroxylase is inhibited by catechol compounds. For most catechol compounds the inhibition of tyrosine hydroxylase appears to be competitive with the pterin cofactor (Nagatsu *et al.*, 1964a; Udenfriend *et al.*, 1965). However, Kaufman has claimed that the inhibition of tyrosine hydroxylase by the immediate product of the hydroxylation reaction, DOPA, is non-competitive with pterin cofactor (Shiman *et al.*, 1971; Kaufman and Fisher, 1974). In their studies, Shiman *et al.* (1971) employed chymotrypsin-solubilized bovine adrenal tyrosine hydroxylase and the native particulate enzyme from that tissue. In contrast, we have employed soluble tyrosine hydroxylase from isolated guinea pig vasa deferentia. With this enzyme preparation, the inhibition of tyrosine hydroxylase by DOPA, like the inhibition of the enzyme by other catechols, appears to be competitive with the pterin cofactor (F. L. Lee, E. Dreyer, and N. Weiner, unpublished). The reason for these differences may be attributable to different properties of the enzyme preparation employed.

The affinity of the enzyme for tyrosine appears to vary with the cofactor employed and the source of the enzyme. Kaufman and coworkers have demonstrated that the affinity of tyrosine for bovine adrenal tyrosine hydroxylase is considerably increased in the presence of biopterin-H_4, exhibits an intermediate value in the presence of 6-$MePtH_4$, and is lowest in the presence of Me_2PtH_4. Furthermore, substrate inhibition by excess tyrosine occurs at lower concentrations of tyrosine in the presence of the putative natural cofactor (Shiman and Kaufman, 1970).

Estimates of the affinity of pterin cofactor for tyrosine hydroxylase vary considerably, depending upon the nature of the cofactor employed, the source of the enzyme, and the method of assay. In general, the affinity of biopterin-H_4 for tyrosine hydroxylase from several sources is higher than that of either 6-$MePtH_4$ or Me_2PtH_4. Stimulation of catecholaminergic neural systems either by administration of receptor blocking agents (Zivković and Guidotti, 1974; Lovenberg and Bruckwick, 1975) or electrical stimulation of the tissue (Morgenroth *et al.*, 1974, 1975a; Weiner *et al.*, 1977) is associated with enhanced affinity of the cofactor for tyrosine hydroxylase. Furthermore, activation of tyrosine hydroxylase by a system which is optimal for cyclic AMP-dependent protein kinase activity is associated with an enhanced affinity of the enzyme for pterin cofactor (Morgenroth *et al.*, 1975c; Lovenberg *et al.*, 1975). The effect of the cyclic AMP-dependent protein kinase system on the affinity of the cofactor for the enzyme is most striking when the enzyme is assayed in the presence of ferrous sulphate and 2-mercaptoethanol. When the pteridine reductase–NADPH catalase system is employed to assay tyrosine hydroxylase, the cyclic AMP-dependent protein kinase system produces an enhancement of cofactor affinity which is of more modest proportions (Weiner *et al.*, 1977).

B ASSAY OF TYROSINE HYDROXYLASE

Many methods for the assay of tyrosine hydroxylase in homogenates and in purified state have been employed by various investigators. Although fluorimetric assays of the measurement of either DOPA or catecholamines from tyrosine or the fluorimetric assay of tyrosine formed from phenylalanine can be employed to estimate tyrosine hydroxylase activity, these assay procedures are generally restricted to purified enzyme of relatively high activity (Shiman *et al.*, 1971). In most studies of tyrosine hydroxylase, radiometric assays have been utilized. The precursor 3,5-ditritiotyrosine has been commonly employed in the assay of tyrosine hydroxylase. During the hydroxylation reaction a tritium atom is displaced from the tyrosine and ultimately is conjugated with OH^- to form ^3HOH as a product of the reaction. The ^3HOH can then be assayed by separation of the radioactive water from precursor and catechol products by Dowex-50 column chromatography (Nagatsu *et al.*, 1964b; Shiman *et al.*, 1971). Alternatively, ^3H-DOPA may be assayed following chromatographic separation of the catechol from tyrosine by absorption on and elution from alumina or alumina plus Dowex-50 columns (Coyle, 1972). In some studies the product has been separated from the precursor by thin layer or paper chromatography. Tyrosine hydroxylase may also be conveniently measured by the use of carboxyl-labelled tyrosine. In the presence of excess L-aromatic amino acid decarboxylase, the DOPA formed is converted to dopamine with the liberation of $^{14}CO_2$, which may then be collected in a basic trapping agent and counted by liquid scintillation spectrometry (Waymire *et al.*, 1971). Other methods to measure the product of the reaction have also been employed, including derivatization and either thin layer or gas chromatographic separation of the derivatized products and mass spectrometry for identification and quantitation of the product (Costa *et al.*, 1972).

A variety of assay systems also has been employed to measure tyrosine hydroxylase activity in various tissues. Most are modifications of the original method of Nagatsu *et al.* (1964a, b) or of that developed by Shiman and Kaufman (1970). In the former procedure, iron is used as an activator metal to maximize tyrosine hydroxylase activity. In addition, the iron appears to serve as a catalyst for the breakdown of hydrogen peroxide which is formed during the reaction and which may inactivate the enzyme during the incubation (Kaufman and Fisher, 1974). 2-Mercaptoethanol is employed in this assay system to maintain the tetrahydropterin in the reduced form throughout the incubation period.

In the assay described by Shiman and Kaufman (1970), catalase is added to the incubation medium in order to catalyse the breakdown of any hydrogen peroxide formed in the reaction system. The pterin cofactor is maintained in the reduced, active form enzymically by addition of optimal amounts of partially purified sheep liver pteridine reductase and either NADH or NADPH. In other respects, the two assay systems are generally the same. Either phosphate, accetate or Tris buffer, or combinations thereof, are included in the assay system to maintain the pH at optimal levels, usually approximately 6.0–6.2.

C LOCALIZATION OF TYROSINE HYDROXYLASE

Tyrosine hydroxylase is localized exclusively either in the adrenal medulla or in catecholaminergic neurons of the central or peripheral nervous system. Sympathetic denervation of various target organs or destruction of sympathetic neurons by 6-hydroxydopamine results in the complete disappearance of tyrosine hydroxylase activity from the target tissue (Potter *et al.*, 1965). It is generally believed that tyrosine hydroxylase is primarily or exclusively localized in the axoplasm of adrenergic neurons in soluble form. A considerable fraction of the tyrosine hydroxylase in bovine adrenal medulla appears to be associated with the chromaffin granules (Petrack *et al.*, 1968). However, it is likely that this enzyme is artefactually adherent to the chromaffin granule membrane, since tyrosine hydroxylase in this particular tissue seems to be highly susceptible to autoaggregation and aggregation with a wide variety of membranes (Wurzburger and Musacchio, 1971). Studies in our laboratory support the results of Musacchio. We have demonstrated that soluble bovine adrenal medulla enzyme will rapidly aggregate and may easily be sedimented by centrifugation. Furthermore, when soluble bovine adrenal medulla enzyme is mixed with membranes from a variety of tissues and the membranes are sedimented by centrifugation, the bulk of the enzyme is found in the sediment. This is in contrast to soluble adrenal enzyme from a variety of other species, including man, guinea-pig, rat, and rabbit. Tyrosine hydroxylase from the adrenals of these species is found mostly or entirely in the cytosol and the enzyme exhibits little tendency to aggregate or to adsorb to membranes in a non-specific fashion (Waymire *et al.*, 1972b). In homogenates of sympathetic ganglia and of bovine splenic nerves, tyrosine hydroxylase appears to be largely in the soluble supernatant fraction (Stjarne, 1966; Mueller *et al.*, 1969a). Some of the brain tyrosine hydroxylase, notably in the striatum, is associated with particles (McGeer *et al.*, 1965; Iyer and McGeer, 1963; Fahn *et al.*, 1969; Patrick and Barchas, 1974a). Although much of the enzyme present in the particulate fraction of brain tissue may actually be present in solution in the axoplasm which is trapped within synaptosomes, a significant fraction does appear to be membrane bound (Kuczenski and Mandell, 1972a). Two types of tyrosine hydroxylase may exist which exhibit different kinetic properties and may be localized in different cell fractions, at least in bovine adrenal medulla (Ikeda *et al.*, 1966; Joh *et al.*, 1969) and in rat striatum (Kuczenski and Mandell, 1972a; Kuczenski, 1973b).

D THE UPTAKE OF TYROSINE OR PHENYLALANINE INTO ADRENERGIC TISSUE

Either phenylalanine or tyrosine must be taken up from the extracellular fluid into the adrenergic neuron prior to aromatic hydroxylation to DOPA. This apparently involves an active transport process since uptake is concentrative, stereospecific, is inhibited by oxygen deprivation and metabolic inhibitors and is competitively antagonized by other aromatic amino acids (Guroff *et al.*, 1961; Neame, 1961; Guroff and Udenfriend, 1962; Joanny *et al.*, 1973). Because adrenergic nervous tissue comprises only a small fraction of most adrenergically innervated organs, the nature and kinetics of tyrosine uptake into neurons are virtually unknown. However, the rate of equilibration of exogenous tyrosine with endogenous pools of this amino acid in isolated tissues is very slow (Weiner and Rabadjija, 1968). Preliminary studies in our laboratory on tyrosine uptake and retention indicate that the rate of uptake and retention of tyrosine in chronically denervated nictitating membranes cannot be distinguished from these processes in innervated membranes (R. A. Bjur and N. Weiner, unpublished observations). However, studies of the specific activity of exogenous tyrosine and that of DOPA formed from this amino acid suggest that uptake must be much more rapid into adrenergic neurons than into the entire tissue (Weiner and Rabadjija, 1968; Costa *et al.*, 1972).

E PURIFICATION OF TYROSINE HYDROXYLASE

Since the simultaneous and independent discovery of tyrosine hydroxylase in bovine adrenal medulla in 1964 by Brenneman and Kaufman and Nagatsu *et al.*, efforts have been made to purify the enzyme to homogeneity. These efforts, however, have not yet met with complete success. Shiman and Kaufman (1970) described the partial purification of bovine adrenal tyrosine hydroxylase, starting with that portion of the adrenal enzyme attached to the chromaffin granules and other particulates of the adrenal gland, and which is solubilized by treatment with chymotrypsin. Following limited digestion with chymotrypsin, the enzyme is fractionated on ammonium sulphate and subjected to Sephadex G-150 chromatography. The enzyme preparation is again carried through ammonium sulphate fractionation. The specific activity of the enzyme prepared in this fashion is approximately 64 nmol min^{-1} (mg protein)$^{-1}$ (Shiman and Kaufman, 1970). Using a similar enzyme preparation, Shiman *et al.* (1971) subjected the enzyme to further fractionation on substituted Sepharose 4B to which was attached 3-iodo-L-tyrosine through the amino group. The enzyme exhibited a specific activity of 90 nmol min^{-1} (mg protein)$^{-1}$ and exhibited a molecular weight of approximately 40 000 on Sephadex G-150. However, on disc gel electrophoresis, the enzyme activity was associated with two major bands. Kuczenski and Mandell (1972a) carried out a limited purification of rat striatal tyrosine hydroxylase, employing ammonium sulphate fractionation and sucrose density gradient centrifugation. The highest activity of the enzyme reported by these workers was approximately 0.1 nmol min^{-1} (mg protein)$^{-1}$.

Musacchio *et al.* (1971b) attempted to purify both soluble bovine adrenal tyrosine hydroxylase and trypsin-treated tyrosine hydroxylase prepared according to the procedure of Petrack *et al.* (1968). Ammonium sulphate fractionation of the soluble supernatant enzyme yielded an enzyme with an activity of 2.5 nmol min^{-1} (mg protein)$^{-1}$. The trypsin-treated enzyme prepared by the procedure of Petrack *et al.* (1968) exhibited a specific activity of 7.3 nmol min^{-1} (mg protein)$^{-1}$.

Joh *et al.* (1973) purified soluble and particulate tyrosine hydroxylase from bovine adrenal medulla for immunotitration and immunocytochemical studies. Their enzyme preparation was precipitated with 80% ammonium sulphate and chromatographed by Sephadex G-200. The active fractions were precipitated with 0.2 M sodium acetate at pH 5.5 and, after centrifugation at 8000**g**, the enzyme was dialysed and chromatography on Sephadex G-200 was repeated. Finally, the enzyme was subjected to disc gel electrophoresis in sodium acetate buffer. The active band was then employed for immunotitration and immunocytochemical studies. No information regarding the specific activity of the enzyme was provided. However, the authors did state that the enzyme does not cross-react with other catecholamine biosynthetic enzymes. Since no indication of the specific activity of the enzyme was provided, it is impossible to ascertain whether a relatively pure enzyme was employed for the production of antiserum for electron and light immunocytochemical microscopy (Pickel *et al.*, 1975a, b, 1976) (see below).

Park and Goldstein (1976) prepared purified tyrosine hydroxylase from human pheochromocytoma tumour cells. The tumour was homogenized in sucrose containing phosphate buffer, pH 6.5, and centrifuged at low speed. After high speed centrifugation, the supernatant enzyme was precipitated by ammonium sulphate fractionation. The 25–35% ammonium sulphate 'cut' was applied to Sepharose 4B and eluted with 20 mM potassium phosphate buffer, pH 6.5. Fractions with highest activity were concentrated and further purified by centrifugation on a linear sucrose density gradient. The enzyme activity was recovered as a single peak. With polyacrylamide disc gel electrophoresis, a single protein band was demonstrable and this band contained the enzymic activity. The enzyme exhibited a specific activity of 129 nmol min^{-1} (mg protein)$^{-1}$.

Hoeldtke and Kaufman (1977) attempted a purification of the enzyme, using bovine adrenal medulla as enzyme source. They employed particulate tyrosine hydroxylase which was solubilized by limited digestion with α-chymotrypsin. The solubilized enzyme was fractionated with ammonium sulphate and the 46–62% ammonium sulphate 'cut' was dissolved and passed over a Sephadex G-25 column to remove residual ammonium sulphate. The enzyme preparation was chromatographed on Sephadex G-150 and subsequently subjected to chromatography on DEAE cellulose and eluted with a potassium chloride gradient from 0 to 0.2 M in 50 mM phosphate buffer, pH 6. The active fractions were concentrated by pressure dialysis and subjected to sucrose density centrifugation. The purified enzyme exhibited a specific activity of approximately 320 nmol min^{-1} (mg protein)$^{-1}$. It contained iron in the amount of approximately 0.5–0.75 mol per mole enzyme. Its subunit molecular weight was estimated to be approximately 34 000. However, this enzyme was found to be unresponsive to activation by either phosphatidyl serine or a cyclic AMP-dependent protein phosphorylating system.

Raese *et al.* (1977) purified tyrosine hydroxylase from bovine striatum by successive chromatography on DEAE Sephadex A50, hydroxylapatite, CM Sephadex G-50, and Sephacryl S-200. They reported an approximately 1600-fold purification of the enzyme by these procedures. The specific activity of the enzyme was reported to be 4.4 nmol min^{-1} (mg protein)$^{-1}$. On sucrose density centrifugation, the enzyme appeared to exist in four catalytically active forms of differing molecular weights. The molecular weights of the several species were roughly multiples of the smallest unit, which is approximately 48 700. If the intrinsic specific activities of tyrosine hydroxylase from all sources are approximately similar, these data would suggest that the preparation of Raese *et al.* (1977) is considerably less pure than would be required to evaluate whether the enzyme is directly phosphorylated (see below).

We have attempted to purify rat pheochromocytoma tyrosine hydroxylase by ammonium sulphate fractionation and chromatography on DEAE cellulose, sucrose density centrifugation, and, finally, preparative polyacrylamide gel electrophoresis (P. R. Vulliet, T. A. Langan and N. Weiner, in preparation). We have obtained an enzyme with a specific activity of 360 nmol min^{-1} (mg protein)$^{-1}$, which exhibits a single band on SDS gels with a molecular weight of approximately 60 000 daltons. Sucrose density gradient centrifugation of the enzyme and chromatography on Sepharose 4B suggests that the enzyme has a molecular weight of approximately 186 000 daltons. In contrast, based upon the behaviour of the enzyme when subjected to molecular sieve chromatography on Sephacryl 200, the enzyme exhibits an apparent molecular weight of 300 000. Using these data, and assuming the enzyme is a non-globular protein, one can calculate a molecular weight for the asymmetric protein of approximately 225 000. These results indicate that the native enzyme from rat pheochromocytoma exists as an oligomer, probably as a tetramer. The results we have obtained thus far with rat pheochromocytoma enzyme indicate that we have obtained an enzyme with a comparable degree of purity to that obtained by Hoeldtke and Kaufman (1977) for the chymotrypsin-solubilized bovine adrenal medulla enzyme. These two enzyme preparations appear to be considerably more pure, assuming similar intrinsic specific activities for the various enzyme preparations, than those reported by other workers.

F THE REGULATION OF NOREPINEPHRINE SYNTHESIS

Tyrosine hydroxylase is generally regarded to be the rate-limiting step in the biosynthesis of catecholamines in brain as well as peripheral adrenergic tissue. Employing the isolated, perfused guinea pig heart, Levitt *et al.* (1965) demonstrated that the rate of norepinephrine synthesis from tyrosine was relatively low and did not increase when physiologic concentrations of tyrosine in the perfusate were exceeded. In contrast, norepinephrine synthesis increased markedly when the perfusate contained either DOPA or dopamine as precursors. The rate of norepinephrine synthesis from either DOPA or dopamine increased in proportion to the concentration of the precursor in the perfusate. Their results suggested that, *in situ*, norepinephrine synthesis is limited by the hydroxylation of tyrosine. Based on the studies of Nagatsu *et al.* (1964a) and Udenfriend *et al.* (1965), which demonstrated inhibition of tyrosine hydroxylase activity by norepinephrine and other catechols, Udenfriend and coworkers suggested that the rate limiting step in the biosynthesis of catecholamines in tissues may be regulated by end-product feedback inhibition.

For many years it had been known that norepinephrine synthesis in adrenergic neurons and in the adrenal medulla is enhanced as a consequence of stimulation of these tissues. This conclusion was based on the observation that acute or chronic stresses to intact animals are associated with increased excretion of catecholamine metabolites in the urine without an associated decline in tissue catecholamine levels, unless either biosynthesis or storage of the biogenic amine is impaired (Euler *et al.*, 1955). In more direct studies, Holland and Schümann (1956), Bygdeman and Euler (1958), and Butterworth and Mann (1957) demonstrated that stimulation of the isolated perfused adrenal gland is associated with the release of amounts of catecholamines from that organ which greatly exceed the difference between the remaining catecholamine content in the stimulated gland and that in the contralateral unstimulated preparation.

Alousi and Weiner (1966) and Roth *et al.* (1967) demonstrated directly that nerve stimulation is associated with an increase in norepinephrine synthesis from tyrosine. In the studies of Alousi and Weiner (1966), the isolated hypogastric nerve–vas deferens preparation of the guinea pig was stimulated in the presence of ^{3}H-tyrosine and the production of ^{3}H-catecholamines in the stimulated preparation was compared with that in the contralateral control. Intermittent nerve stimulation was associated with an increase of approximately 50% in catecholamine synthesis. This effect of nerve stimulation appeared to be at the tyrosine hydroxylation step since no enhanced synthesis of catecholamines was demonstrated when ^{3}H-DOPA was employed as substrate (Weiner and Rabadjija, 1968). Weiner and Rabadjija (1968) observed that the effect of nerve stimulation on catecholamine synthesis could be either reduced or blocked by the addition of catecholamines to the medium. These workers concluded that reduced end-product feedback inhibition was responsible for the accelerated norepinephrine synthesis associated with nerve stimulation. Since no difference in tissue catecholamine content was demonstrable between stimulated and contralateral control vasa deferentia, it was proposed that a small pool of norepinephrine located in solution in the axoplasm of the adrenergic neuron regulated the activity of tyrosine hydroxylase. Presumably, during nerve stimulation this small intracytosolic pool of soluble norepinephrine was reduced in concentration and, as a consequence, the normal tonic suppression of tyrosine hydroxylase activity was diminished and enhanced norepinephrine synthesis ensued.

This attractive theory for the regulation of tyrosine hydroxylase activity during nerve stimulation was supported by a variety of pharmacological studies. Indirectly acting sympathomimetic amines and monoamine oxidase inhibitors which exhibit an ability to release catecholamines from storage sites profoundly reduce the activity of tyrosine hydroxylase in intact tissues (Weiner and Selvaratnam, 1968; Weiner *et al.*, 1972a; Kopin *et al.*, 1969; Bjur and Weiner, 1975). These amines possess no direct inhibitory effect on tyrosine hydroxylase isolated from the same preparations. It was postulated that these agents were able to release stored norepinephrine from intraneuronal vesicles into the axoplasm of the neuron where the catecholamine could interact with tyrosine hydroxylase and inhibit the enzyme. Similarly, acute reserpine administration, which is associated with inhibition of the uptake of norepinephrine and dopamine into adrenergic vesicles, and, by this mechanism, leads to an increase in free cytosolic catecholamine until tissue stores are severely depleted, exerts an inhibitory effect on tyrosine hydroxylase *in situ* (Weiner *et al.*, 1972a; Pfeffer *et al.*, 1975). Again, the drug possesses no direct inhibitory effect on soluble tyrosine hydroxylase. The inhibitory effect of reserpine on tyrosine hydroxylase activity in intact tissue is profoundly enhanced by the concomitant administration of a monoamine oxidase inhibitor (Pfeffer *et al.*, 1975). This result is consistent with the concept that elevated levels of free intraneuronal norepinephrine do indeed regulate tyrosine hydroxylase activity in the intact tissue.

Several years ago, Bjur and Weiner were able to demonstrate that exogenous reduced pterin cofactor could stimulate tyrosine hydroxylase activity in intact tissue (Bjur and Weiner, 1975; Weiner and Bjur, 1972). Presumably, the pterin is able to penetrate the adrenergic nerve terminal and serve as a cofactor for the enzymic hydroxylation of tyrosine. These results suggested that either the concentration of pterin cofactor in the adrenergic nerve terminal is not adequate to saturate the enzyme or free intracytosolic catecholamines are present at levels which are able to antagonize competitively the interaction with the enzyme of otherwise saturating concentrations of cofactor. Bjur and Weiner (1975) and Pfeffer *et al.* (1975) demonstrated that the inhibitory effects of indirectly acting sympathomimetic amines and reserpine on tyrosine hydroxylase activity in either intact vasa deferentia or adrenal slices were overcome competitively by addition of synthetic tetrahydropterin cofactor to the incubation system. In contrast, the inhibition of tyrosine hydroxylase was not antagonized by addition of higher concentrations of tyrosine.

The ability of exogenous pterin cofactor to overcome the inhibitory effects of norepinephrine on tyrosine hydroxylase activity in intact tissue enabled

Cloutier and Weiner (1973) to test the hypothesis that reduced end-product feedback inhibition could account for the accelerated synthesis of norepinephrine associated with nerve stimulation. It was argued that, if the enhancement of tyrosine hydroxylase activity associated with nerve stimulation was due to reduced end-product feedback inhibition, the differential effect of nerve stimulation should be overcome by the presence of high concentrations of pterin cofactor in the medium, since sufficiently high concentrations of exogenous cofactor should eliminate any end-product feedback inhibition in both stimulated and contralateral control preparations. However, Cloutier and Weiner (1973) demonstrated that the enhancement of norepinephrine synthesis associated with nerve stimulation was even further heightened by pterin cofactor and the difference in tyrosine hydroxylase activity between control and stimulated preparations was at least as pronounced as that seen in the absence of exogenous pterin cofactor. This finding was in contrast to the inhibitory effect of exogenous norepinephrine on tyrosine hydroxylase activity in intact vasa deferentia, which could be competitively overcome by added pterin cofactor. These results suggested that the effect of nerve stimulation on tyrosine hydroxylase activity in the intact vas deferens preparation is not due to simple reduction of end-product feedback inhibition by a mechanism involving competitive interaction with reduced pterin.

In their studies of the effects of nerve stimulation on tyrosine hydroxylase activity in intact tissue, Thoa *et al.* (1971) also attempted to determine whether the activity of the soluble enzyme prepared from stimulated organs was modified by nerve stimulation. Tyrosine hydroxylase activity in homogenates of stimulated vasa deferentia and in supernatants prepared from these homogenates was assayed and compared with similar preparations of the contralateral unstimulated organ. Using both saturating and subsaturating concentrations of substrate and cofactor, these workers were unable to elicit changes in the properties or activity of the soluble enzyme (Thoa *et al.*, 1971). It is probably significant that, in these studies, a modification of the assay of Nagatsu *et al.* (1964b) was employed. In this assay system, ferrous ion and 2-mercaptoethanol are included to obtain maximal enzyme activity.

In 1974, Morgenroth *et al.* reported changes in the activity of soluble tyrosine hydroxylase prepared from vasa deferentia after field stimulation, as compared with the enzyme prepared from unstimulated organs. They demonstrated a marked reduction in the K_m for both Me_2PtH_4 and tyrosine and a substantial increase in the K_i for norepinephrine. A modest increase in the maximal velocity of the enzyme reaction was also observed when Me_2PtH_4, but not tyrosine, was employed as the variable substrate. They also demonstrated similar kinetic changes in supernatant enzyme prepared from unstimulated vasa deferentia when 10–50 μM calcium chloride was added to the assay medium. Morgenroth *et al.* concluded that nerve stimulation is associated with an increase of calcium uptake into the nerve terminal and a calcium-mediated activation of tyrosine hydroxylase. In these studies, the enzyme system employed was a modification of that developed by Shiman and Kaufman (1970), and included pteridine reductase, NADPH, and catalase. Ferrous ion and reducing agent were not included in the assay system.

We have attempted to reproduce the results observed by Morgenroth *et al.* (1974), employing a modification of the pteridine reductase–NADPH–catalase assay for tyrosine hydroxylase described by Zivković *et al.* (1974). In these studies tyrosine hydroxylase activity has been assayed in the supernatant enzyme prepared from isolated guinea pig vas deferens preparations. In addition, we have measured changes in tyrosine hydroxylase activity *in situ* during stimulation of the hypogastric nerve to the vas deferens. In both instances, the coupled decarboxylase radiometric assay for tyrosine hydroxylase has been employed (Waymire *et al.*, 1971).

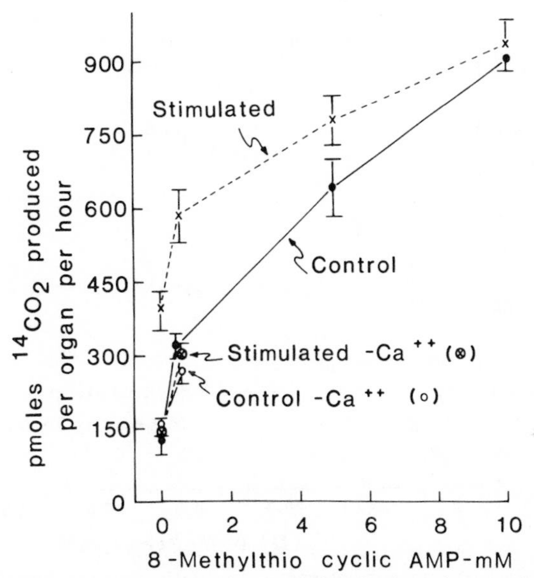

Figure 4.2 Effect of nerve stimulation on tyrosine hydroxylase activity in intact vasa deferentia. Vasa deferentia were stimulated at 25 Hz, supramaximal voltage, 10 s. every 20 s., 5 ms. pulse duration, for 40 min. in the presence of 5×10^{-5} M l-^{14}C-L-tyrosine. The $^{14}CO_2$ produced during stimulation was collected in liquid scintillation fluid containing an organic amine base. The procedure is identical to that described by Weiner *et al.* (1973), except tissues were continuously gassed during the incubation period with 95% O_2–5% CO_2 and the gas was continuously collected. Contralateral control preparations were treated identically, but were not stimulated. Note that nerve stimulation is associated with approximately a threefold increase in tyrosine hydroxylase activity *in situ* (control vs. stimulated, no 8-methylthio cyclic AMP added). Addition of 8-methylthio cyclic AMP to the medium produces a concentration-dependent increase in tyrosine hydroxylase activity in the intact tissue, which is less pronounced during stimulation. The effect of nerve stimulation, but not that of 8-methylthio cyclic AMP, is dependent on the presence of extracellular Ca^{2+}

Supramaximal stimulation of the hypogastric nerve of the vas deferens preparation of the guinea pig at 25 Hz, 5 ms duration, for 10 s every 20 s over a period of 40 min is associated with an approximately three-fold increase in tyrosine hydroxylase activity *in situ* (Figure 4.2). If the vas deferens preparation is stimulated for 30 min in a similar fashion and the tissue is rapidly frozen, homogenized, and centrifuged, and the supernatant assayed for tyrosine hydroxylase activity, a similar enhancement of activity of the soluble enzyme is

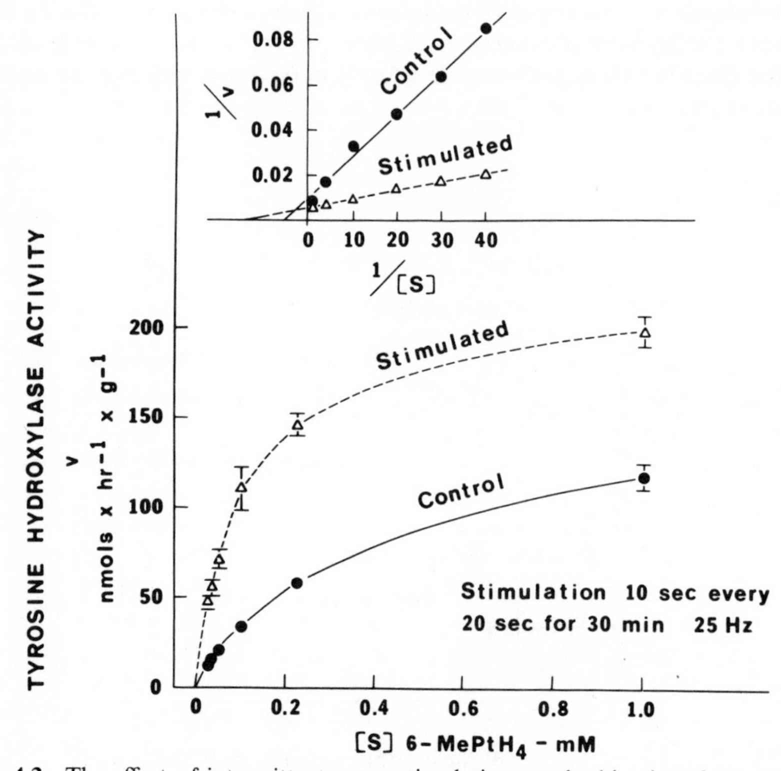

Figure 4.3 The effect of intermittent nerve stimulation on the kinetics of tyrosine hydroxylase activity in the presence of different concentrations of 6-MePtH$_4$. Isolated hypogastric nerve–vas deferens preparations of guinea pig were stimulated at 25 Hz, 10 s. every 20 s. at supramaximal voltage and with a pulse duration of 5 ms. for 30 min. The contralateral control preparation was treated in an identical fashion, but was not stimulated. Tissues were then removed, frozen, homogenized in 15 mM KCl, and centrifuged. The supernatant enzyme was assayed by the coupled decarboxylase assay of Waymire *et al.* (1971) as modified by Zivković *et al.* (1974). Results are presented graphically both as direct enzyme activity versus cofactor concentration and as Lineweaver–Burk reciprocal plots. A considerable activation of the enzyme is seen following nerve stimulation. Note the deviation downward of the control values obtained in the presence of the highest cofactor concentration, when the reciprocal plots are constructed.

Results are expressed as nanomoles of product formed per hour per gram tissue demonstrable. Kinetic analysis of this activation of tyrosine hydroxylase associated with nerve stimulation reveals that the enzyme appears to be activated at both low and high cofactor concentrations, although the activation is greater when cofactor concentration is low (Figure 4.3).

The time course of the activation of tyrosine hydroxylase with nerve stimulation was assessed by varying the duration of intermittent nerve stimulation from 30 s to 30 min. Soluble enzyme prepared from these stimulated tissues revealed similar activities at cofactor concentrations ranging from 0.025 mM to 1.0 mM 6-$MePtH_4$, suggesting that the enzyme was activated very rapidly. However, when the soluble enzyme from the contralateral unstimulated preparation which had been incubated for comparable periods of time was assayed, considerable differences were observed. Incubation of the unstimulated preparation results in a progressive reduction in tyrosine hydroxylase activity, particularly at low cofactor concentrations. The enzyme frequently did not obey Michaelis–Menten kinetics. A Lineweaver–Burk plot of these data revealed a deviation downward of the line at highest cofactor concentrations, suggesting that two enzyme forms may exist, one with a low K_m (high affinity) for pterin cofactor and a second with a high K_m (low affinity) for pterin cofactor (Figure 4.4). Similar enzyme behaviour has been noted when the low and high K_m forms of cyclic AMP phosphodiesterase coexist (Thompson and Appleman, 1971). The proportion of the low K_m (higher affinity) enzyme diminishes with increasing durations of incubation of the tissue.

The observation that nerve stimulation results in a comparable activation of tyrosine hydroxylase regardless of the duration of the stimulation suggested either that nerve stimulation could maintain the enzyme in the active form or that stimulation could rapidly activate an enzyme which was reversibly converted to a less active state. In order to evaluate this, tissues were preincubated for 30–60 min and subsequently stimulated. Soluble supernatant enzyme prepared from these tissues was compared with enzyme from unstimulated organs which were incubated for identical periods of time. Analysis of enzyme activity with different cofactor concentrations revealed that enzyme from unstimulated organs possesses both the low and high K_m enzyme forms, with a greater proportion of the high K_m form present. In contrast, enzyme prepared from stimulated preparations exhibited kinetics suggesting that the bulk of the enzyme was in the low K_m form. These results suggest that tyrosine hydroxylase prepared from either non-incubated or briefly incubated vasa deferentia may be partially in an active state and that, with continuing incubation in the absence of nerve stimulation, partially activated enzyme reverts to a less active form.

The presence of two forms of an enzyme which differ only by an altered affinity for substrate or cosubstrate may be difficult to demonstrate experimentally, particularly if the affinities of the two enzyme forms for the substrate or cosubstrate do not differ markedly. In the instance of vas deferens tyrosine hydroxylase, the enzyme affinities for 6-$MePtH_4$ appear to differ by less than

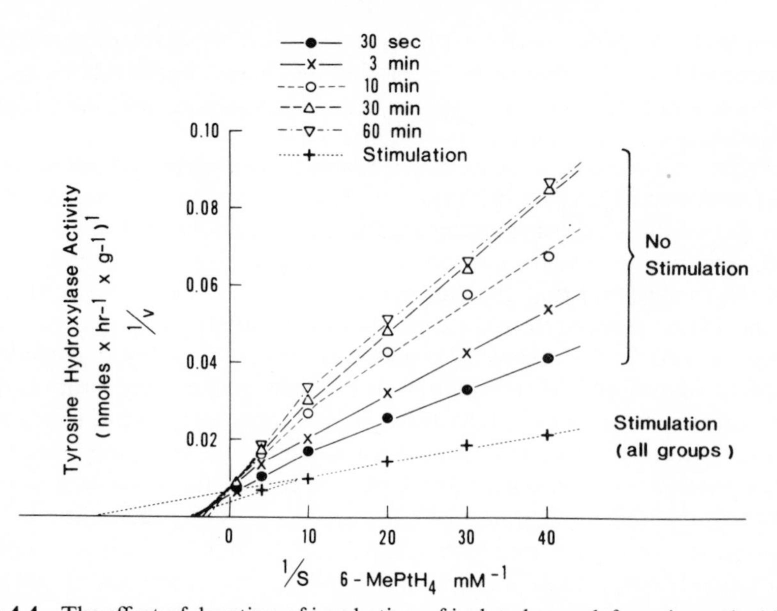

Figure 4.4 The effect of duration of incubation of isolated vasa deferentia on the kinetics of tyrosine hydroxylase subsequently isolated from the tissue. The experimental details are presented in Figure 4.3, legend. Note that, with increasing durations of incubation in the absence of stimulation, there is a progressive reduction in tyrosine hydroxylase activity and increasing downward deviations of the reciprocal Lineweaver–Burk plots in the presence of the highest cofactor concentrations. This downward deviation suggests that, with incubation, a greater proportion of the enzyme is transformed to the less active (high K_m) form (see Figure 4.5). Intermittent stimulation of the hypogastric nerve is, in all instances, associated with a conversion of the less active form of the enzyme to the more active (low K_m) form

tenfold. Under these circumstances, it is relatively easy to demonstrate the two forms of the enzyme when a considerable proportion is in the high K_m (low affinity) form. However, when the low K_m (high affinity) form predominates, it may be difficult to demonstrate the coexistence of the high K_m form, unless several high concentrations of cofactor are employed in the kinetic analysis (Figure 4.5). In general, several cofactor concentrations above the K_m for the low affinity enzyme form have not been examined. The issue is further complicated by the observation that high cofactor concentrations may inhibit the activity of tyrosine hydroxylase (Kaufman, 1973; Kaufman and Fisher, 1974).

It seems conceivable that activation of tyrosine hydroxylase in the freshly removed organ occurs as a consequence of the traumatic method of killing the animal prior to removing the organs for incubation and stimulation. It is highly likely that massive discharge of the sympathetic nervous system occurs consequent to stunning the animal with a blow on the head. In order to test this

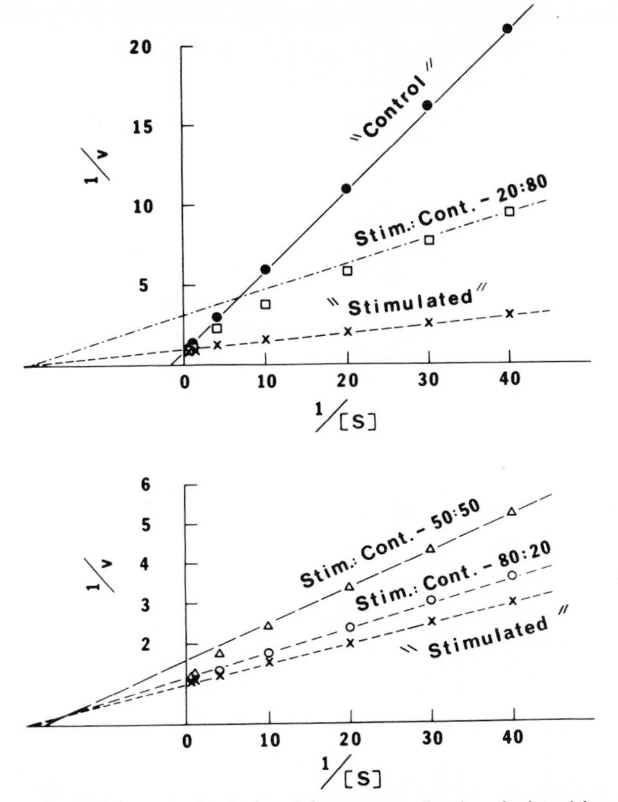

Figure 4.5 Hypothetical curves depicting Lineweaver–Burk relationships of reciprocal enzyme activity versus reciprocal substrate concentration when two forms of the enzyme, which differ only in terms of affinity for substrate, coexist in different proportions. The hypothetical K_ms are 0.05 mM for the low K_m enzyme and 0.5 mM for the high K_m enzyme. Compare these hypothetical curves with the experimentally derived curves seen in Figure 4.4. The downward deviation of the Lineweaver–Burk lines is most apparent when a considerable proportion of the enzyme is in the less active (high K_m; 'control') form.

$1/S$ is presented as $(\text{mM})^{-1}$. Enzyme velocity is presented in arbitrary units with '1'= maximal velocity for the total enzyme present

possibility, animals were pretreated with chlorisondamine, a ganglionic blocking agent, which would be expected to prevent the presumed massive central nervous system stimulation associated with killing the guinea pig from affecting the postganglionic sympathetic neurons. For these experiments, 25 mg chlorisondamine was administered to animals intraperitoneally per kilogram body weight 2 h prior to killing the guinea pigs. Vasa deferentia were removed from chlorisondamine-treated animals and uninjected controls and the enzyme was prepared and assayed for tyrosine hydroxylase activity. The activity of tyrosine hydroxylase prepared from chlorisondamine-treated animals was considerably lower than that of the uninjected controls. Kinetic analysis revealed that the bulk of the enzyme prepared from chlorisondamine-treated animals exhibited a higher K_m than that observed for the corresponding enzyme prepared from uninjected controls (Figure 4.6).

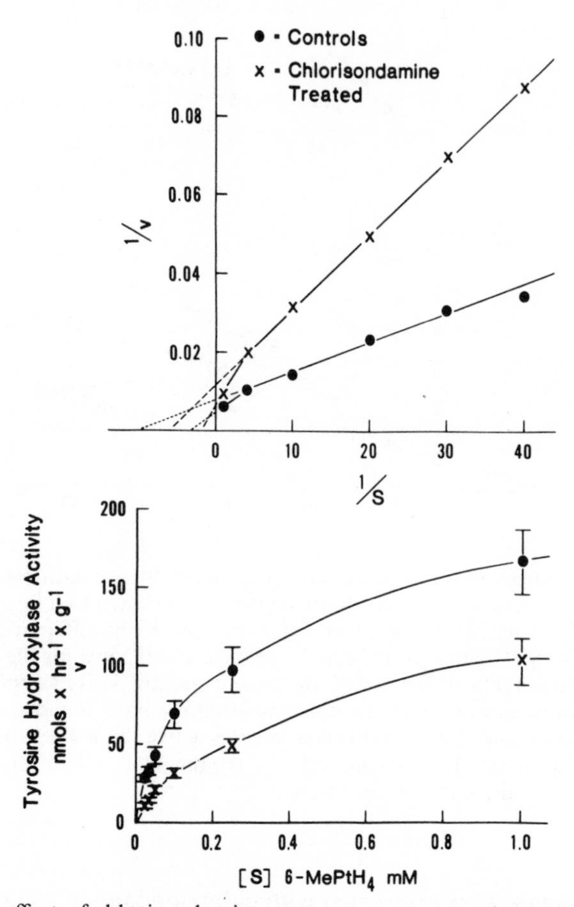

Figure 4.6 The effect of chlorisondamine pretreatment on tyrosine hydroxylase activity in supernatant prepared from isolated vas deferens preparations of guinea pig. Animals were treated with 25 mg chlorisondamine hydrochloride per kilogram body weight 2 h. prior to killing. Vasa deferentia were rapidly removed, frozen and processed as indicated in Figure 4.3, legend. Note that chlorisondamine treatment is associated with reduced enzyme activity, with a greater proportion of the enzyme in the high K_m form. Presumably, during handling and stunning of the animal prior to killing, massive sympathetic discharge occurs and this results in activation of the bulk of the enzyme. This activation is prevented by prior administration of the ganglionic blocking agent

An examination of the stimulation parameters for activation of tyrosine hydroxylase was conducted. Intermittent stimulation at 2 or 3 Hz for 5 min. was associated with a slight activation of tyrosine hydroxylase. A further modest increase in tyrosine hydroxylase activity was seen when the stimulation frequency was increased to 5 Hz. However, the activation of tyrosine hydroxylase which was seen at a stimulation frequency of 25 Hz was considerably greater than that observed at lower frequencies. When intact tissues were preincubated for 30 min., in order to allow the enzyme to revert to the less active form *in situ*, brief bursts of stimulation at 25 Hz led to a rapid activation of the enzyme. Full activation of the enzyme could be achieved in less than 1 min. Significant, but modest, increases in tyrosine hydroxylase activity could be produced with a period of stimulation as short as 10 s. at 25 Hz (Figure 4.7).

The rapid activation of tyrosine hydroxylase with nerve stimulation suggests that enzyme induction is not the mechanism by which this enhanced synthesis of norepinephrine may occur. To evaluate this directly, immunotitration studies were performed on enzyme prepared from tissues stimulated for 30 min. and on enzyme from contralateral control organs. In order to maintain the enzyme in the activated state, it was necessary to shorten the duration of incubation of the antiserum–antigen complex to 4 h. at 4 °C. Under these circumstances, almost complete precipitation of the antigen–antibody complex was attained. Supernatant enzymes from both stimulated and contralateral control vasa deferentia were incubated with different concentrations of anti-rat pheochromocytoma tyrosine hydroxylase serum for 4 h. at 4 °C and the antigen–antibody complex was sedimented by centrifugation. The supernatant was assayed for tyrosine hydroxylase activity. Analysis of these data revealed that identical amounts of enzyme protein were present in supernatant from control and stimulated organs.

Morgenroth *et al.* (1974, 1975a) and Roth *et al.* (1975a, b) reported that the addition of calcium to soluble tyrosine hydroxylase prepared from control vasa deferentia and from rat hippocampi results in an enhanced affinity of the enzyme for both substrate and cofactor and a reduced affinity of the enzyme for norepinephrine. These results are akin to those which they observed following field stimulation of the vas deferens (Morgenroth *et al.*, 1974) or electrical stimulation of the locus coeruleus (Roth *et al.*, 1975a, b; Morgenroth *et al.*, 1975a). They therefore proposed that the mechanism of activation of tyrosine hydroxylase following nerve stimulation involves uptake of calcium during depolarization and direct activation of the enzyme by the divalent cation. Osborne and Neuhoff (1976) also have reported activation of soluble tyrosine hydroxylase from the circumoesophageal ganglion of the snail (*Helix pomatia*) with $CaCl_2$. Definite effects, consisting of approximately a twofold or threefold decrease in K_m for both substrate (tyrosine) and cofactor (6-$MePtH_4$), occurred in the presence of 1 mM $CaCl_2$. At lower concentrations, the effects of calcium were unimpressive or absent.

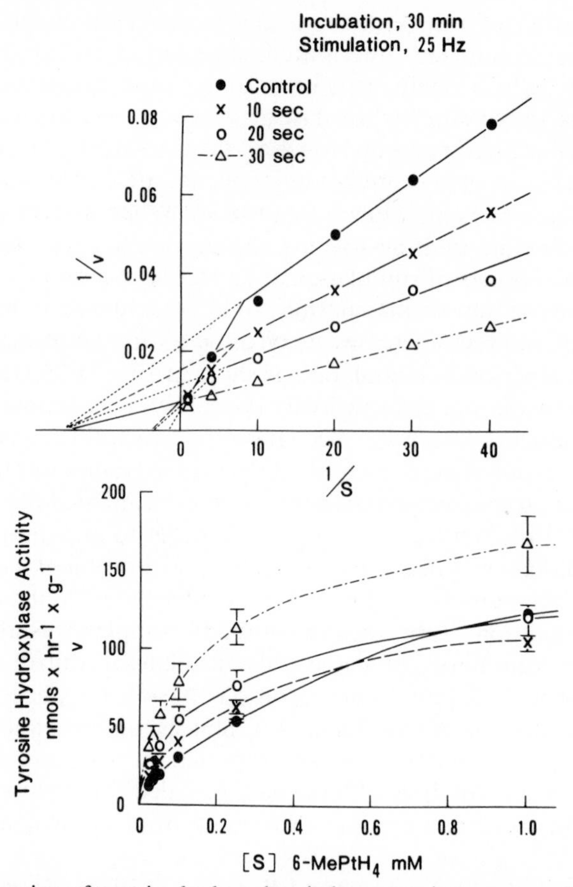

Figure 4.7 Activation of tyrosine hydroxylase in hypogastric nerve–vas deferens preparations of guinea pig following brief periods of stimulation. Vasa deferentia were isolated and incubated for 30 min. without stimulation in order to allow the reversion of most of the enzyme to the less active (high K_m) form. One of each pair was then stimulated for 10, 20, or 30 s. and the contralateral preparation served as the unstimulated control. Tissues were then rapidly removed from the organ bath, frozen, homogenized, and processed as described in Figure 4.3, legend. Note the presence of two enzyme forms in the supernatant prepared from unstimulated tissues. The bulk of the enzyme is in the high K_m (low affinity) state. Note the rapid activation of an increasing proportion of the enzyme following stimulation at 25 Hz for 10, 20, and 30 s. Following 30 s. of stimulation the bulk of the enzyme is in the active form

Several other laboratories, however, have been unable to duplicate these effects of calcium on soluble tyrosine hydroxylase. Knapp *et al.* (1975) and Lerner *et al.* (1977) failed to demonstrate an activation of brain or adrenal tyrosine hydroxylase with calcium. Employing the pteridine reductase–NADPH catalase assay system for tyrosine hydroxylase, we have examined the effect of calcium on the activity of soluble tyrosine hydroxylase prepared from both control and stimulated vasa deferentia. In contrast to the results reported by Morgenroth *et al.* and Roth *et al.*, we have been unable to demonstrate any effect of either calcium or EGTA on the activity of tyrosine hydroxylase prepared from either stimulated or control organs (Figure 4.8). Since, with the conditions of the assay employed, consistent activation of tyrosine hydroxylase was obtained as a consequence of nerve stimulation, a stimulatory effect of calcium would have been expected if the mechanism by which nerve stimulation activates tyrosine hydroxylase involves direct interaction with calcium. It thus appears that the neurally mediated activation of tyrosine hydroxylase is not a consequence of a direct effect of this cation. It is still conceivable, however, that, during nerve stimulation, calcium, which is presumably taken up into the nerve terminal, initiates a series of events which may ultimately be responsible for the enzyme activation (Figure 4.2).

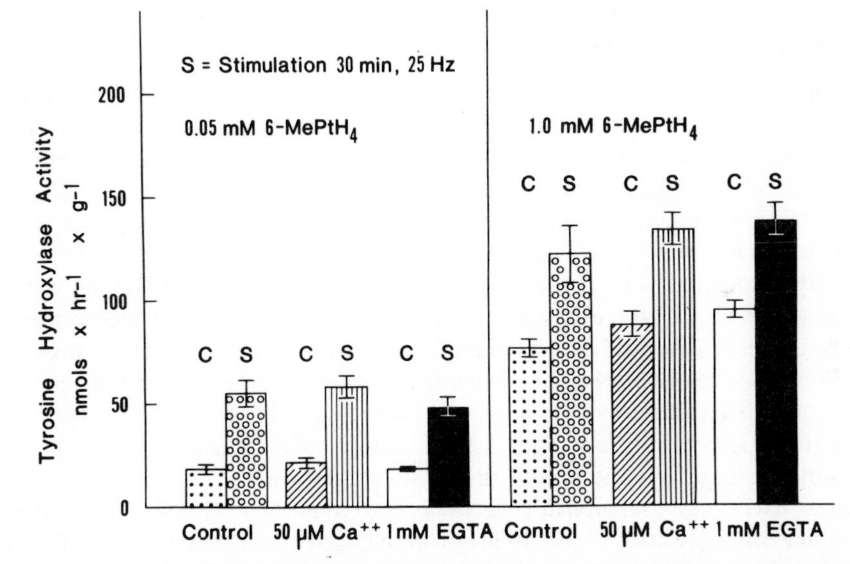

Figure 4.8 Lack of effect of $CaCl_2$ or EGTA on tyrosine hydroxylase supernatant enzyme prepared from guinea pig vas deferens tissue. Vasa deferentia were isolated and the hypogastric nerve of one of each pair was stimulated for 30 min. as outlined in Figure 4.2, legend. Supernatant enzyme was then prepared from each organ and assayed in the presence or absence of 50 μM $CaCl_2$ or EGTA and either 0.05 mM 6-MePtH$_4$ or 1.0 mM 6-MePtH$_4$. Note that, although nerve stimulation is associated with enhanced activity of supernatant tyrosine hydroxylase, neither $CaCl_2$ nor EGTA produced an effect on the enzyme prepared from either unstimulated or stimulated organs

We have also been unable to demonstrate any change in the affinity of the enzyme either for tyrosine or for norepinephrine or DOPA. These results are in marked contrast to results reported by Morgenroth *et al.* (1974, 1975a) or Roth *et al.* (1974a, 1975a), following stimulation of either the vas deferens or the hippocampus. There is no clear-cut explanation for these discrepancies. However, analysis of the Dixon (1953) plots to determine inhibitor affinity constants presented by Morgenroth *et al.* (1974) and Roth *et al.* (1975a) for both vas deferens and hippocampal tyrosine hydroxylase, suggests that these results may not be reliable. Over a 100-fold range of pterin cofactor concentration (10^{-6}–10^{-4} M), they report relatively modest changes in tyrosine hydroxylase activity in both control and stimulated preparations and in the presence and absence of norepinephrine. Based upon their reported affinity constants for pterin cofactor, very much greater changes in tyrosine hydroxylase activity should have been obtained. In a similar study of the kinetics of vas deferens tyrosine hydroxylase, we have observed much greater changes in enzyme activity over a much smaller range of pterin cofactor concentrations, a result that is anticipated assuming the enzyme obeys Michaelis–Menten kinetics.

1 The regulation of tyrosine hydroxylase by the cyclic AMP-dependent protein phosphorylating system

Goldstein *et al.* (1976) and Anagnoste *et al.* (1974) demonstrated that addition of dibutyryl cyclic AMP to incubating slices of rat striatum is associated with a considerable increase in the rate of formation of ^{14}C-catecholamines from ^{14}C-tyrosine. They suggested that cyclic AMP, which is elevated in brain tissue subjected to high potassium or electrical stimulation (Kakiuchi and Rall, 1968; Kakiuchi *et al.*, 1969), may be responsible for the increased tyrosine hydroxylase activity associated with nerve stimulation. Dibutyryl cyclic AMP appears to increase the $\mathbf{V}_{max}$ of the enzyme in striatal slices but does not influence the K_m for tyrosine. Earlier studies suggested that pterin cofactor either is not taken up into synaptosomes or does not stimulate tyrosine hydroxylase in these particles (Patrick and Barchas, 1974b; Kuczenski and Segal, 1974). Based on these observations, it was assumed that kinetic studies with pterin cofactor are not possible in intact synaptosomal preparations. However, more recent studies suggest that exogenous pterin cofactor is able to activate tyrosine hydroxylase in intact brain (Kettler *et al.*, 1974) or in synaptosomal preparations (Patrick and Barchas, 1976).

Harris *et al.* (1974, 1975) reported that dibutyryl cyclic AMP is able to stimulate tyrosine hydroxylase when the cyclic nucleotide is added to rat striatal homogenates. Cyclic AMP was reported to produce a similar effect on soluble striatal tyrosine hydroxylase. According to these investigators, the stimulation of the enzyme by 10 μM cyclic AMP is associated with a reduction in the K_m for Me_2PtH_4 (from 0.62 to 0.08 mM) and an increase in the K_i for dopamine (from 0.09 to 0.64 mM). The K_m for tyrosine was also reduced from 54 to 23 μM. The V_{max} of the enzyme was not altered.

In contrast to these results, Lovenberg and Bruckwick (1975) were unable to demonstrate an effect of cyclic AMP on the kinetic properties of rat striatal tyrosine hydroxylase solubilized by addition of 0.2% Triton X-100. On the other hand, these workers demonstrated a several-fold reduction in the K_m for pterin cofactor when the soluble enzyme was incubated in a system optimal for protein kinase activity (added constituents were ATP, cyclic AMP, Mg^{2+}, EGTA, NaF, and theophylline). Similarly, Goldstein *et al.* (1976) were unable to demonstrate a stimulation of soluble striatal tyrosine hydroxylase activity with cyclic AMP alone. However, they could elicit an activation of the soluble enzyme by addition of ATP and Mg^{2+} along with cyclic AMP. Further enhancement of activity was produced by the addition of protein kinase. The enhancement reported by these workers was associated with a reduced K_m of the enzyme for Me_2PtH_4, an increase in the K_i for dopamine, without any change in the affinity of the enzyme for tyrosine.

Morgenroth *et al.* (1975c) found that cyclic AMP alone is able to activate tyrosine hydroxylase present in a crude high speed supernatant from rat striatum. (This discrepancy with the results of Lovenberg and Bruckwick (1975) may be explained by the use of Triton X-100 in the studies of the latter workers. In the absence of detergent, the brain supernatant may contain sufficient quantities of ATP and protein kinase (and Mg^{2+}) to produce some degree of enzyme activation with addition of only cyclic AMP.) When the enzyme was further purified on a Sephadex G-25 column, cyclic AMP was no longer able to activate the enzyme. However, activation could be produced by addition of ATP, Mg^{2+}, and cyclic AMP and further activation could be obtained by addition of protein kinase. Both Lovenberg and Bruckwick and Morgenroth *et al.* conclude that activation of tyrosine hydroxylase may be mediated by a cyclic AMP-dependent protein kinase.

Recently Morita and Oka (1977) demonstrated that cyclic AMP plus ATP markedly increases tyrosine hydroxylase in preparations of soluble tyrosine hydroxylase from bovine adrenal medulla. The activation was the result of increased affinity of the enzyme for pteridine cofactor without any associated change in the affinity for substrate. Cyclic AMP alone did not activate tyrosine hydroxylase in the supernatant but ATP alone was extremely effective in activating the enzyme. The supernatant enzyme preparation was not examined for adenylate cyclase or cyclic AMP-independent protein kinase activity. It is thus conceivable that some adenylate cyclase or cyclic AMP-independent protein kinase and Mg^{2+} may have been present in the soluble supernatant preparation of the adrenal.

Weiner *et al.* (1977) observed an activation of tyrosine hydroxylase prepared from vasa deferentia in the presence of cyclic AMP, ATP, Mg^{2+}, theophylline, NaF and EGTA. It is thus conceivable that activation of tyrosine hydroxylase by nerve stimulation may be related to a cyclic AMP-dependent protein phosphorylating reaction. In order to test this hypothesis, the effects of cyclic AMP analogues on tyrosine hydroxylase activity in intact vasa deferentia were examined. Several cyclic AMP analogues were able to increase tyrosine hydroxylase in intact vasa deferentia. At low concentrations of cyclic AMP analogues, this increase was demonstrable in both neurally stimulated and unstimulated preparations. When intact vasa deferentia were incubated in the presence of increasing concentrations of 8-methylthio-cyclic AMP, and *in situ* tyrosine hydroxylase was assayed, the incremental changes in enzyme activity were more pronounced in the unstimulated preparation than in the stimulated, contralateral organ. At extremely high concentrations (10 mM 8-methylthio-cyclic AMP), no further increase in tyrosine hydroxylase activity could be elicited with nerve stimulation. At this concentration the difference in *in situ* tyrosine hydroxylase activity between stimulated and unstimulated preparations was no longer apparent. These results are consistent with the notion that the activation of tyrosine hydroxylase associated with nerve stimulation is mediated by a cyclic AMP-dependent phenomenon (Figure 4.2). A similar activation of tyrosine hydroxylase by 8-methylthio-cyclic AMP *in situ* can be produced in a medium lacking calcium (Figure 4.2).

The activation *in situ* by the cyclic AMP analogue appears to involve activation of tyrosine hydroxylase. When carboxyl-labelled DOPA is used as substrate in place of carboxyl-labelled tyrosine, no enhanced production of $^{14}CO_2$ is observed either as a consequence of nerve stimulation or as a consequence of addition of 8-methylthio-cyclic AMP to the medium.

Incubation of intact vasa deferentia with 8-methylthio-cyclic AMP is associated with an alteration of the kinetics of tyrosine hydroxylase prepared from these tissues immediately following the incubation period. Kinetic changes associated with 8-methylthio-cyclic AMP are comparable to those seen following nerve stimulation (Figure 4.9). In addition, when vasa deferentia from animals pretreated with chlorisondamine are incubated in the presence of 8-methylthio-cyclic AMP, full activation of the enzyme can be achieved. These results suggest that the activation of tyrosine hydroxylase by nerve stimulation may be the consequence of a cyclic AMP-mediated series of events.

As noted above, supernatant enzyme prepared from vasa deferentia can be activated by a cyclic AMP-dependent protein phosphorylating system (Figure 4.10). A concentration–effect relationship for the activation of tyrosine hydroxylase by cyclic AMP was determined and it was observed that half-maximal activation of the enzyme can be obtained with approximately 1 μM cyclic AMP (Figure 4.11). In addition to cyclic AMP, the essential ingredients for full activation of the supernatant enzyme are ATP and Mg^{2+}. There appears to be adequate endogenous protein kinase in vas deferens supernatant, since addition of purified bovine heart protein kinase was not associated with further activation of tyrosine hydroxylase. The kinetic changes associated with activation of tyrosine hydroxylase by the cyclic AMP-dependent protein phosphorylating system are very similar to those produced by either nerve stimulation or by incubation of the tissue with 8-methylthio-cyclic AMP.

Several properties of tyrosine hydroxylase from the vas deferens have been assessed following nerve stimulation, following incubation with 8-methylthio-cyclic AMP and during incubation of the soluble enzyme in the presence of the

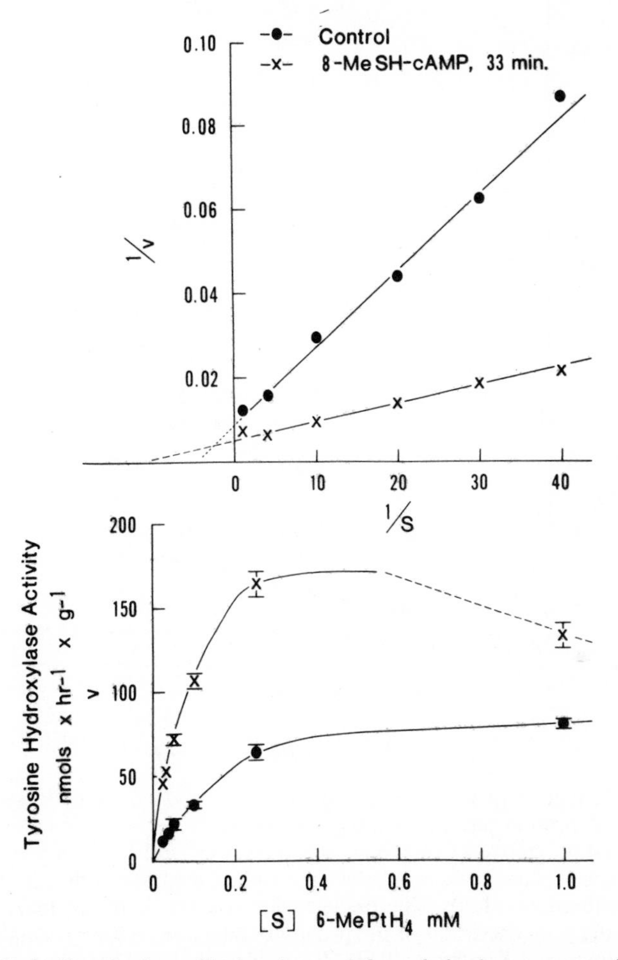

Figure 4.9 Effect of incubation of isolated vasa deferentia in the presence of 8-methylthio cyclic AMP on tyrosine hydroxylase activity. Tissues were incubated for 33 min. in the presence or absence of 5 mM 8-methylthio cyclic AMP (8-MeSH-cAMP) and tissues were subsequently removed, rinsed in saline, frozen, homogenized, and the supernatant enzyme was assayed as described in Figure 4.3, legend. Note that incubation of the tissue in the presence of the stable cyclic AMP analogue is associated with kinetic changes in the enzyme which are similar to those seen following nerve stimulation

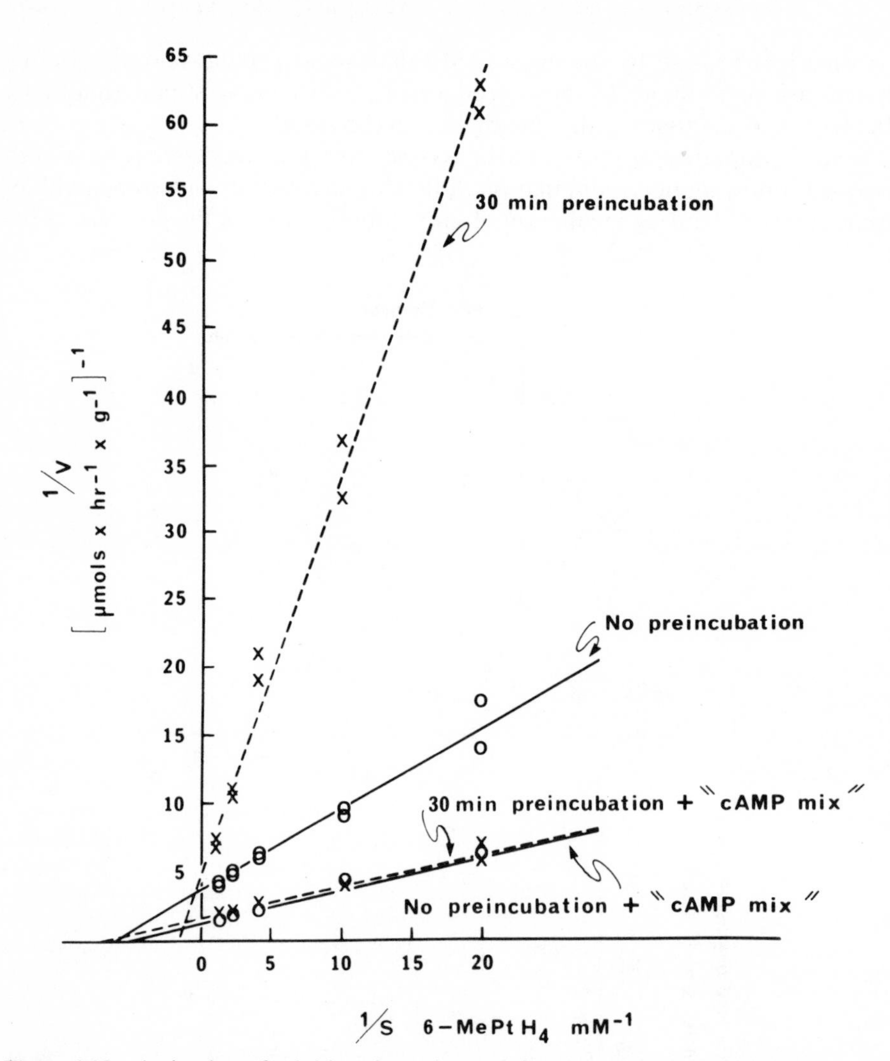

Figure 4.10 Activation of soluble guinea pig vas deferens tyrosine hydroxylase by a cyclic AMP-dependent protein phosphorylating system. Soluble tyrosine hydroxylase was prepared from freshly removed guinea pig vasa deferentia and from vasa deferentia which were preincubated intact for 30 min. in order to allow the bulk of the enzyme to revert to the less active (high K_m) form. The tissues were assayed by the coupled decarboxylase assay, employing a modification of the pteridine reductase–NADPH catalase system developed by Shiman and Kaufman (1970) (Zivković *et al.*, 1974). Assays were performed in the presence and absence of 0.1 mM cyclic AMP, 0.5 mM ATP, 20 mM $Mg(C_2H_3O_2)_2$, 20 mM NaF, 0.8 mM theophylline, and 0.12 mM EGTA ('cAMP mix') (Lovenberg and Bruckwick, 1975). Note that the enzyme preparations from both freshly removed vasa deferentia and vasa deferentia incubated for 30 min. prior to preparation of the enzyme are activated to equal degrees in the presence of the cyclic AMP-dependent protein phosphorylating system

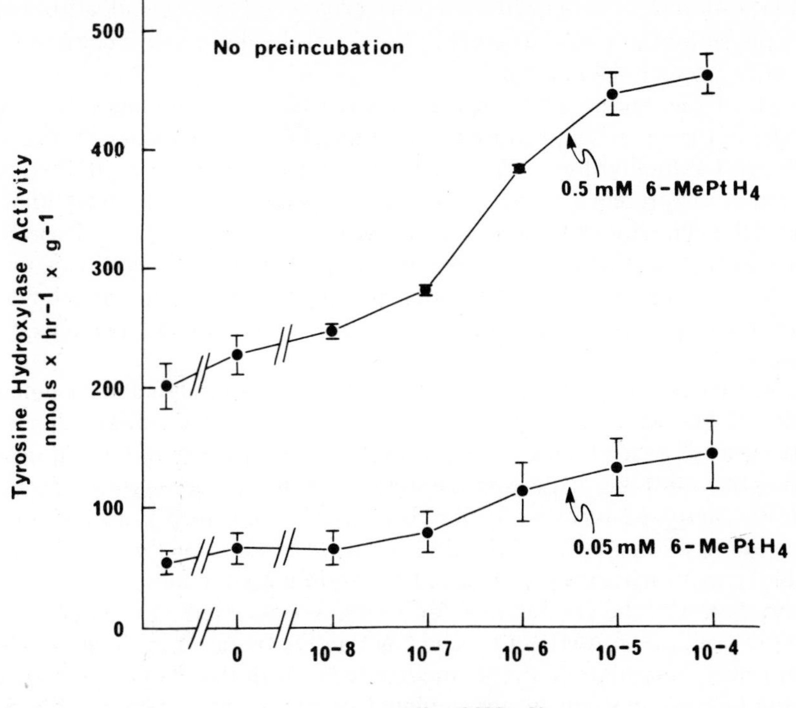

Figure 4.11 Effect of cyclic AMP-dependent protein phosphorylating system on tyrosine hydroxylase activity. Soluble tyrosine hydroxylase prepared from guinea pig vasa deferentia was incubated in the absence (extreme left values) and in the presence of the 'cAMP mix', either minus cyclic AMP ('0'), or in the presence of different concentrations of cyclic AMP. (See legend, Figure 4.10, for details.) The studies were performed in the presence of 0.05 mM and 0.5 mM 6-MePtH$_4$. Note that activation of the enzyme is dependent on the presence of cyclic AMP. The EC50 for enzyme activation is approximately 1 μM

cyclic AMP-dependent protein phosphorylating enzyme system. Under all three circumstances, kinetic analysis revealed that, in the presence of different concentrations of tyrosine the affinity constant for tyrosine is not altered. As expected from these observations, the enzyme obeys Lineweaver–Burk kinetics with different tyrosine concentrations, whether it is primarily in the 'more active form' or 'less active form' or when different proportions of both enzyme forms are present.

Analysis of the pH optimum of the less active and more active forms of the enzyme reveals similar changes following each of the three enzyme activation procedures. For the enzyme from unstimulated organs, the pH optimum is approximately 6.0–6.2. Following stimulation by any of the three procedures, the pH optimum of the enzyme becomes broadened and extended towards the more physiological range. The pH optimum of the more active enzyme form is generally between 6.4 and 6.8.

In all of the studies performed thus far, which were intended to compare changes in the enzyme following nerve stimulation, following incubation of the tissue with 8-methylthio-cyclic AMP and during incubation of the soluble enzyme in the presence of the cyclic AMP-dependent protein phosphorylating system, the behaviour of the enzyme has been found to be similar. These results are consistent with the notion that neurally mediated activation of tyrosine hydroxylase involves a cyclic AMP-dependent process. Presumably, the enzyme activation is a consequence of activation of a cyclic AMP-dependent protein kinase.

In further support of this hypothesis, we have stimulated vasa deferentia at low frequencies (3–5 Hz) to induce submaximal activation of tyrosine hydroxylase. These studies were repeated in the presence and absence of isobutylmethylxanthine, a phosphodiesterase inhibitor. Isobutylmethylxanthine modestly increased tyrosine hydroxylase activity prepared from unstimulated preparations and potentiated the effect of nerve stimulation on the activation of soluble tyrosine hydroxylase prepared from stimulated tissues.

Stimulation of the vas deferens for 5 min. is associated with a significant increase in both cyclic AMP and cyclic GMP in the tissue. After 30 min. of intermittent nerve stimulation, cyclic nucleotide levels in the vas deferens are comparable to those present in unstimulated preparations incubated only 5 min. However, in the control, unstimulated preparation incubated for 30 min., the levels of cyclic nucleotides are reduced approximately 30%. These changes in cyclic nucleotides parallel the changes in the state of activation of tyrosine hydroxylase, when preparations incubated for 5 or 30 min. with and without stimulation are compared. The results suggest that these cyclic nucleotide changes may indeed be occurring in the adrenergic neuron and may be responsible for the changes in tyrosine hydroxylase seen with nerve stimulation. However, this conclusion can only be considered tentative until changes in cyclic nucleotide levels in the adrenergic neurons can be dissected from changes which occur postsynaptically. It is likely that a quantitative immunohistochemical measurement of cyclic AMP in adrenergic neurons will be required to resolve this issue.

It is of interest to note that O'Dea and Zatz (1976) found increases in both cyclic AMP and cyclic GMP in the pineal gland following norepinephrine administration or nerve stimulation. Denervation and pharmacological studies suggested that the changes in cyclic AMP occurred postsynaptically, whereas the changes in cyclic GMP may have been taking place in presynaptic neurons. It is tempting to speculate that, in a tissue where muscle contraction is associated with an increase in cyclic GMP postsynaptically, presynaptic effects may be mediated by alterations in the level of the other cyclic nucleotide. Hsu *et al.* (1976) reported that contraction of the guinea pig vas deferens with methacholine is associated with an increase in tissue cyclic GMP with no significant effect on cyclic AMP.

2 Activation of tyrosine hydroxylase in adrenal glands *in vivo*

We have attempted to ascertain whether the changes in tyrosine hydroxylase activity which have been observed in the vas deferens *in vitro* also can be elicited by acute stress *in vivo*. In these studies we have determined the kinetic changes in adrenal medulla tyrosine hydroxylase associated with several types of acute stress. In the first series of experiments, rats were anaesthetized with pentobarbital, halothane, or ether and the adrenal glands were removed, homogenized, and centrifuged and the supernatants subjected to Sephadex G-25 chromatography to remove inhibitory catecholamines. Another group of animals was killed by decapitation and the adrenals were processed similarly. Tyrosine hydroxylase activity was found to be considerably higher in adrenal glands from animals killed by decapitation. Intermediate levels of tyrosine hydroxylase activity were seen in animals anaesthetized with ether, and lowest levels of tyrosine hydroxylase activity were observed in those rats anaesthetized with either pentobarbital or halothane. Kinetic analyses of the enzyme preparations from each of these groups of animals revealed the existence of two forms of the enzyme, one with a relatively low K_m (high affinity) for pterin cofactor and the second with a relatively high K_m (low affinity) for pterin cofactor. The presence of two forms of the enzyme was much more apparent in preparations from animals treated with either pentobarbital or halothane. After ether, more of the enzyme was in the low K_m form, and following decapitation, virtually all of the enzyme appeared to be in the low K_m form.

Halothane or pentobarbital are anaesthetic agents whose administration presumably is associated with little or no sympathetic discharge, whereas ether anaesthesia is associated with considerable sympathetic discharge. Thus, the known pharmacological actions of these anaesthetic agents are consistent with the degrees of activation of tyrosine hydroxylase which we have obtained in the adrenal glands from these several groups of animals. Further evidence that two forms of the enzyme with different affinities for pterin cofactor coexist in the adrenal gland was obtained by the kinetic analysis of the enzyme using different tyrosine concentrations. As with the guinea pig vas deferens, rat adrenal tyrosine hydroxylase appeared to obey Michaelis–Menten kinetics when this analysis was performed in the presence of different concentrations of tyrosine. The Lineweaver–Burk relationship in the presence of different tyrosine concentrations was linear for enzyme preparations from the several groups of animals. The K_m for tyrosine was found to be the same for the adrenal enzyme from each group. These results are consistent with the notion that, as with the vas deferens, two forms of tyrosine hydroxylase may coexist in the adrenal gland and stress is associated with a conversion of the high pterin cofactor K_m form of the enzyme to a more active, low K_m form.

Rat adrenal tyrosine hydroxylase can also be activated by a cyclic AMP-dependent protein phosphorylating system. Enzyme prepared from animals anaesthetized with ether, pentobarbital, or halothane, or from animals killed by decapitation, could be activated to virtually identical degrees in the presence of a cyclic AMP-dependent protein kinase system. Incubation of the enzyme in the presence of the cyclic AMP-dependent protein phosphorylating system resulated in the conversion of adrenal enzyme virtually completely to the low K_m form. Again, no change in affinity constant for tyrosine was produced by this incubation.

3 The molecular basis of activation of tyrosine hydroxylase by cyclic AMP-dependent protein kinase

Considerable effort has been expended to determine whether tyrosine hydroxylase in the presence of the cyclic AMP-dependent protein kinase system is activated by direct phosphorylation of the enzyme. Lovenberg *et al.* (1975) incubated rat striatal supernatant enzyme with ATP-γ-^{32}P, cyclic AMP, magnesium, and other ingredients to optimize cyclic AMP-dependent activation of tyrosine hydroxylase. After chromatography of the enzyme on a Sephadex G-25 column, the enzyme was incubated with rabbit antityrosine hydroxylase serum for 1 h. at 37 °C and 12 h. at 4 °C. The antibody–antigen complex was centrifuged and counted for radioactivity. No evidence of phosphorylation of the enzyme was detectable. Lloyd and Kaufman (1975) performed similar experiments on highly purified bovine caudate tyrosine hydroxylase. The enzyme could be activated by a cyclic AMP-dependent protein kinase system, but in the presence of ATP-γ-^{32}P no evidence for the phosphorylation of the enzyme was obtained. In these studies the enzyme was isolated following exposure to cyclic AMP-dependent phosphorylating conditions both by immunoprecipitation with antityrosine hydroxylase γ-globulin and counting of the precipitate and by isolation of the enzyme by sucrose density gradient centrifugation. In both of these studies it is conceivable that any phosphorylated enzyme might have been converted back to the dephosphorylated form during incubation of the antigen–antibody complex. We have demonstrated that the activation of tyrosine hydroxylase following nerve stimulation does not persist during a 16-h. period of incubation at 4 °C of the enzyme with antiserum. However, the sucrose density gradient centrifugation studies of Lloyd and Kaufman (1975) do not seem to be subject to this potential pitfall.

In contrast to these studies, Letendre *et al.* (1977a, b) incubated rat adrenal medullae and superior cervical ganglia in organ culture for 16–20 h. in the presence of inorganic $^{32}PO_4^{-3}$. Tyrosine hydroxylase was isolated by immunoprecipitation followed by SDS polyacrylamide gel electrophoresis. The inorganic ^{32}P was incorporated into the enzyme of each tissue and the degree of incorporation of ^{32}P into the enzyme from superior cervical ganglia was increased if the cultures were incubated in the presence of nerve growth factor.

Based on these studies, Letendre *et al.* concluded that almost 1 mol. of phosphate is incorporated per mole of enzyme. They proposed that phosphate may be a constitutive component of the enzyme. They further speculated that the addition of phosphate to, and its removal from, the enzyme may not be involved in the regulation of enzyme activity, since they could not affect the degree of phosphorylation of the enzyme by a variety of procedures, including exposure of the tissue to cyclic AMP and reserpine (Letendre *et al.*, 1977b).

Raese *et al.* (1977), employing bovine striatal tyrosine hydroxylase, were able to purify this enzyme approximately 1600-fold. When they added cyclic AMP, Mg^{2+}, and ATP-γ-^{32}P to the purified enzyme preparation, they obtained evidence which suggested direct phosphorylation of the enzyme. Monitoring the elution patterns of enzyme activity and radioactivity following chromatography on Sephacryl and sucrose density gradient centrifugation, these workers observed several peaks of tyrosine hydroxylase activity, some of which were associated with the radioactive phosphate, suggesting that tyrosine hydroxylase may exist in different molecular forms (oligomers), some of which may be phosphorylated. In the regions of the gradient where both enzyme activity and radioactivity were present, they observed that the enzyme activity and ^{32}P migrated in a similar fashion. However, the patterns of elution for both ^{32}P and tyrosine hydroxylase activity did not superimpose exactly, raising the possibility that some contaminant in the preparation was being phosphorylated, rather than the tyrosine hydroxylase molecule. The specific activity (in nanomoles of product formed per minute per milligram of protein) of the purified tyrosine hydroxylase prepared by Raese *et al.* (1977) was considerably lower than that for either the bovine adrenal medulla enzyme prepared by Hoeldtke and Kaufman (1977) or the purified rat pheochromocytoma enzyme which we have prepared, suggesting that other proteins, some of which may be phosphorylated, may be present in their enzyme preparation.

Vulliet *et al.* (P. R. Vulliet, T. A. Langan, and N. Weiner, manuscript in preparation) have demonstrated direct phosphorylation of purified rat pheochromocytoma tyrosine hydroxylase in the presence of ATP-γ-^{32}P, Mg^{2+} and catalytic subunit of protein kinase. The phosphorylation of the enzyme is highly correlated with enzyme activation. This enzyme, like that from guinea-pig vas deferens and rat adrenal medulla, exhibits both a high K_m and a low K_m form, with respect to pterin cofactor, in the absence of protein phosphorylating factors. The enzyme is converted entirely to the low K_m form by phosphorylation. Approximately 0.75–0.9 mol. of phosphate is incorporated into the enzyme per mol. of subunit monomer of 60 000 daltons. Presumably the failure to attain 1 mol. ^{32}P incorporation per mol. of subunit is due to the fact that a component of the enzyme remains phosphorylated during purification, accounting for the presence of 2 K_m forms in the absence of phosphorylation *in vitro* (Figure 4.12).

Clear-cut evidence also has been provided that phenylalanine hydroxylase can be phosphorylated by a cyclic AMP-dependent protein phosphorylating

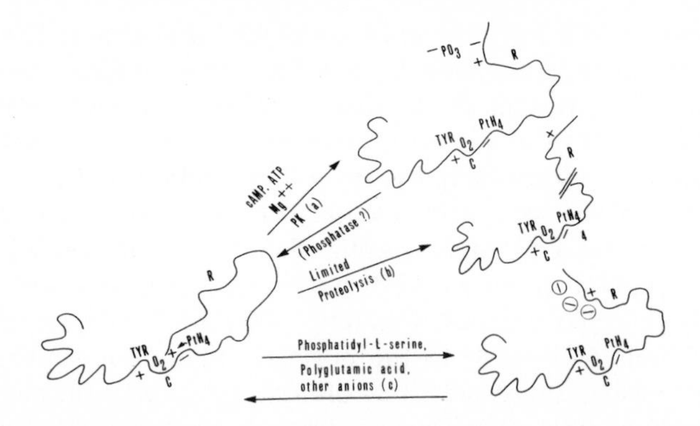

Figure 4.12 Proposed scheme for the activation of tyrosine hydroxylase by either (a) cyclic AMP-dependent protein kinase (PK) system; (b) limited proteolysis; or (c) anionic phospholipids or other anions. The regulatory strand (R) may be either a component of the enzyme, an inhibitory molecule or a subunit of the enzyme which can become juxtaposed, by coulombic and hydrophobic interactions, to an anionic and hydrophobic region at or near the catalytic site of the enzyme. The regulatory component presumably binds in the region where reduced pterin binds and competes with reduced pterin for binding. (a) Phosphorylation of R neutralizes the coulombic interaction between R and the enzyme active site region. (b) Limited proteolysis either releases the regulatory strand or a regulatory subunit from the enzyme or inactivates the regulatory molecule by proteolysis. (c) Polyanions or anionic lipids prevent coulombic interaction between the active site region and the cationic and hydrophobic regions of the R unit

system. Abita *et al.* (1976) and Milstien *et al.* (1976) have demonstrated that approximately 0.7 mol. of ^{32}P are incorporated into each mole of monomer of phenylalanine hydroxylase. They were able to demonstrate that 0·3 mol. of endogenous phosphate was present in each enzyme monomeric unit. These results suggest that, in the presence of a cyclic AMP-dependent protein kinase system, 1 mol. of phosphate is incorporated into each monomeric unit of the enzyme. Phosphorylation of the enzyme was associated with a considerable increase in the activity of the enzyme. However, the major change in enzyme activity was related to an increase in maximal velocity of the enzyme reaction. The apparent K_m values for phenylalanine and biopterin-H_4 were not changed by phosphorylation.

Cyclic nucleotides also may be involved in the release of catecholamines by nerve stimulation (Rasmussen, 1970; Peach, 1972; Wooten *et al.*, 1973; Cubeddu *et al.*, 1975). We and others have observed that both cyclic AMP and cyclic GMP analogues and phosphodiesterase inhibitors enhance release of norepinephrine and dopamine β-hydroxylase during nerve stimulation of the isolated perfused cat spleen. In this preparation, these substances do not appear to initiate the release process, since they do not affect the release of neurotransmitter in the absence of nerve stimulation (Cubeddu *et al.*, 1975). It remains to be determined whether the cyclic nucleotide mediated release of neurotransmitter during nerve stimulation either is responsible for, or is intimately linked to, the concomitant or subsequent changes in biosynthetic enzyme activities or levels.

Dopamine β-hydroxylase levels in adrenergic nervous tissue and in the adrenal medulla are also increased following prolonged sympathetic activity (Molinoff *et al.*, 1970, 1972; Reis *et al.*, 1975). It has been proposed that cyclic AMP may be involved in the induction of this enzyme (Gewirtz *et al.*, 1971). Adrenal cortical hormonal factors also appear to play significant roles in the determination of the synthesis and levels of norepinephrine biosynthetic enzymes, especially in the adrenal gland (Weiner, 1975).

4 Other possible allosteric regulators of tyrosine hydroxylase activity

A considerable number of putative allosteric modulators of rat striatal tyrosine hydroxylase has been proposed and one or more of these substances may be responsible for the enhanced activity of this enzyme which is associated with neural activity (Figure 4.12). Mandell *et al.* (1972) noted that administration of either amphetamine, reserpine or α-methyl-*p*-tyrosine to rats is associated with a shift of caudate tyrosine hydroxylase from the soluble to the particulate state, presumably as a consequence of membrane binding. This was associated with a modest activation of the enzyme. However, stresses that presumably enhance central neural activity did not reproduce the effects of the drugs on this enzyme.

Kuczenski and Mandell (1972b) observed that sulphate ions or heparin activated rat caudate soluble tyrosine hydroxylase *in vitro*. Both the maximal velocity and the affinity of the enzyme for cofactor were enhanced by heparin. In addition, the K_i for dopamine was increased by this sulphated polysaccharide. The changes produced by heparin resulted in an enzyme whose kinetic properties resembled closely those of the particulate enzyme. Kuczenski and Mandell (1972a) proposed that physiological stimuli may modify the soluble enzyme or its environment in a manner which allows the molecule to become membrane bound. As a consequence, the enzyme is activated. However, the relevance of this effect of heparin to neurally induced enzyme activation is questionable (Weiner *et al.*, 1974).

Other anions, such as phosphatidyl-L-serine and polyglutamic acid, also are able to activate rat caudate tyrosine hydroxylase by enhancing the affinity of the enzyme for cofactor and shifting the optimal pH of the enzyme towards the more physiological pH region (Lloyd and Kaufman, 1974; Katz *et al.*, 1976b).

Raese *et al.* (1976) have also demonstrated activation of rat striatal tyrosine hydroxylase by lysolecithin and phosphatidylserine. The activation involves a three- or fourfold reduction in the K_m for 6-MePtH$_4$ or biopterin-H$_4$. No change in the apparent affinity constant for L-tyrosine or in the maximal velocity of the reaction is demonstrable. The K_i for dopamine inhibition is increased approximately fourfold. It is of interest that both heparin and phosphatidyl-L-serine have been reported to be without effect on the activity of tyrosine hydroxylase from peripheral adrenergic neurons (Numata and Nagatsu, 1975).

In contrast to the purported stimulatory effects of calcium on tyrosine hydroxylase prepared from central and peripheral noradrenergic neurons (Morgenroth *et al.*, 1974, 1975a; Roth *et al.*, 1975a, b; Roth and Salzman, 1977), Goldstein *et al.* (1970) found that calcium ions inhibited tyrosine hydroxylase of rat striatal slices. Roth *et al.* (1974b, 1975b) obtained analogous results with a high-speed supernatant preparation of striatal enzyme. Furthermore, Roth *et al.* (1974b) and Morgenroth *et al.* (1975b) observed that ethylene glycol-*bis*-(γ-aminoethyl ether)-*N*,*N'*-tetraacetic acid (EGTA) activates striatal tyrosine hydroxylase by enhancing its affinity for substrate and cofactor and reducing its affinity for dopamine.

The stimulatory effects of EGTA and the inhibitory effects of Ca^{2+} on striatal tyrosine hydroxylase have been used to explain the peculiar observation that reducing impulse flow in the dopamine neurons of the striatum, either by administration of γ-hydroxybutyrolactone or by lesioning the nigrostriatal tract, is associated with increased tyrosine hydroxylase activity in the striatum (Roth *et al.*, 1974b). This enhanced activity of tyrosine hydroxylase can be antagonized by administration of dopamine agonists, such as apomorphine or piribedil (Walters and Roth, 1974). These agonists, like γ-hydroxybutyrolactone, have been reported to suppress firing of dopamine neurons of the nigrostriatal tract (Walters *et al.*, 1975). The tentative explanation proposed for these observations is that there exist dopamine presynaptic receptors which, when activated by dopamine or dopamine agonists in the presynaptic space, initiate a negative feedback effect on presynaptic tyrosine hydroxylase. Dopamine receptor blockers are able to prevent this effect of dopamine agonists. Presumably following lesioning of the nigrostriatal tract, or after acute administration of γ-hydroxybutyrolactone, there is no dopamine released into the synaptic cleft, and the presynaptic dopamine receptors therefore are not activated. Thus, the presynaptic tyrosine hydroxylase is not subject to the dopamine presynaptic receptor mediated inhibitory mechanism.

Administration of haloperidol, chlorpromazine, or other neuroleptics results in activation of tyrosine hydroxylase in the striatum (Zivković *et al.*, 1974, 1975; Costa *et al.*, 1974; Lovenberg and Bruckwick, 1975; Roth *et al.*, 1975b). Zivković *et al.* (1974, 1975), Costa *et al.* (1974), and Lovenberg and Bruckwick (1975) have observed that administration of neuroleptics to rats results in an activation of soluble striatal tyrosine hydroxylase which is characterized by a reduced K_m for pterin cofactor and increased K_i for dopamine. In contrast to these studies, in which only an increase in cofactor affinity was reported, Roth *et al.* (1975b) also found that the enzyme exhibited a reduced K_m for tyrosine. The reason for this discrepancy is unclear. It seems paradoxical that enhanced nigrostriatal nervous activity, which presumably occurs following neuroleptic administration, and which is assumed to be associated with enhanced firing of nigrostriatal neurons and consequent calcium uptake into nerve terminals, results in activation of striatal tyrosine hydroxylase, in view of the reported effects of EGTA and calcium which are summarized above.

5 Protein kinases and the regulation of tyrosine hydroxylase activity

Our recent studies and those of Lovenberg *et al.* (1975), Lloyd and Kaufman (1975), and Morgenroth *et al.* (1975b) suggest that tyrosine hydroxylase is activated by a cyclic AMP-dependent protein kinase reaction. The existence of cyclic AMP-dependent protein kinase was first reported by Walsh *et al.* (1968) and Langan (1968). Since that time, many investigators have demonstrated the virtually ubiquitous presence of this enzyme system in mammalian tissues, including brain (Kuo and Greengard, 1969). The enzyme has been shown to exist as a complex between a regulatory subunit and a catalytic subunit. In the presence of cyclic AMP, a complex between the cyclic nucleotide and the regulatory subunit is formed and the catalytic and regulatory subunits dissociate. In the dissociated form, the catalytic subunit is enzymically active (Tao *et al.*, 1970; Reimann *et al.*, 1971).

There are at least two forms of cyclic AMP-dependent protein kinase in tissues, which have been designated 'Type I' and 'Type II'. The terminology refers to the order in which they are eluted from DEAE–cellulose columns. Type I enzyme is the major type of cyclic AMP-dependent protein kinase in rat skeletal muscle (Beavo *et al.*, 1975), whereas type II enzyme predominates in bovine heart muscle (Erlichman *et al.*, 1973). There is evidence to suggest that the catalytic subunits of these two types of protein kinase are identical, but the regulatory subunits differ (Bechtel *et al.*, 1977).

Type I protein kinase, in addition to eluting from DEAE–cellulose columns at lower concentrations of NaCl (approximately 0.05 M), is readily dissociated into regulatory and catalytic subunits by 0.5 M NaCl. In contrast, the Type II protein kinase holoenzyme is not readily dissociated by 0.5 M NaCl.

Although protein kinases from many tissues, including rat brain and adipose tissue, appear to be of either one or the other type (Corbin *et al.*, 1975) it is possible that there are other protein kinases which are tissue-specific, with unique substrate specificities or other features. Apparently the Type II protein kinases from adipose tissue and heart are not identical, based on chromatographic characteristics (Corbin *et al.*, 1975). Different proportions of the two types of enzymes have been found in different tissues. In addition, particulate forms of protein kinase have been demonstrated (Corbin and Keely, 1977). Corbin *et al.* (1977) showed that, in the presence of cyclic AMP, the catalytic subunit is translocated from the particulate to the soluble fraction. Corbin *et al.* (1977) proposed that the holoenzyme may be, in part, particulate in rabbit heart. In the presence of cyclic AMP, the holoenzyme dissociates and the catalytic subunit is released from the membrane into the supernatant. In contrast, Costa and coworkers propose that, in the adrenal medulla, cyclic AMP-dependent activation of protein kinase is associated with translocation of the active catalytic subunit of soluble cytosolic protein kinase into the nucleus, where phosphorylation of nuclear proteins is enhanced (Chuang *et al.*, 1977). In the presence of cyclic AMP, the catalytic subunits of both beef adrenal and beef heart cyclic AMP-dependent protein kinase holoenzyme were translocated into beef adrenal nuclei, suggesting the protein kinases are not tissue specific.

In developing rat brown adipose tissue, eight distinct cyclic AMP-dependent protein kinases are demonstrable on polyacrylamide disc gel electrophoresis. Changes in the electrophoretic properties of these bands were seen with increasing age of the animals (Knight and Skala, 1977).

Ueda *et al.* (1973) observed that at least two proteins in synaptic membrane fractions were phosphorylated by a cyclic AMP-dependent protein kinase system. Synaptic membrane fractions also contain high levels of cyclic AMP-dependent protein kinase (Johnson *et al.*, 1971) and phosphoprotein phosphatase (Maeno and Greengard, 1972). Greengard and coworkers have proposed that nervous system function and the function of other tissues may be regulated by protein phosphorylation by a cyclic AMP-dependent system (Greengard, 1975).

The role of calcium in cyclic AMP-dependent protein phosphorylation reactions is by no means clear. Indeed, calcium appears to produce opposite effects in different systems. Brostrom *et al.* (1975) demonstrated the presence of calcium-binding protein in pig cerebral cortex which activates adenylate cyclase in this tissue. The activation of pig cerebral cortex adenylate cyclase is inhibited by EGTA. The calcium-binding protein appears to be identical to a phosphodiesterase-activating factor, which is also Ca^{2+} dependent. These workers have suggested that activation of this system leads to enhanced synthesis of cyclic AMP and enhanced hydrolysis of cyclic GMP.

Krueger *et al.* (1977) showed that either veratridine or depolarization of synaptosomes in the presence of calcium in the medium is associated with phosphorylation of specific proteins in the synaptosomes. Calcium ionophores produced analogous actions. The effects of each of these treatments were inhibited by EGTA. However, none of these procedures was associated with an increase in cyclic AMP in the synaptosomes. Krueger *et al.* (1977) speculate that postsynaptic protein phosphorylation may be mediated by a cyclic AMP-dependent protein kinase reaction, whereas presynaptic (synaptosomal) protein phosphorylation may be mediated by Ca^{2+} and may not involve cyclic AMP.

Ferrendelli *et al.* (1976) reported that depolarizing concentrations of K^+ increased cyclic AMP and cyclic GMP levels in cerebellar slices. Only the increase of cyclic GMP was dependent on the presence of extracellular Ca^{2+}. Incubation of cerebellar slices in the presence of the divalent ionophore,

A23187, and extracellular Ca^{2+} also was associated with an increase in cerebellar content of cyclic GMP, but not cyclic AMP.

A specific cytosolic protein from mouse brain is phosphorylated in the presence of a cyclic GMP-dependent protein kinase and Ca^{2+}. Low concentrations of cyclic AMP also stimulated the phosphorylation of this protein in the presence of Ca^{2+}, whereas the phosphorylation was inhibited by higher concentrations of cyclic AMP in the presence of Ca^{2+} (Malkinson, 1975).

Other regulators of protein kinase mediated reactions which are likely to be of physiological importance are: phosphoprotein phosphatases (Maeno *et al.*, 1975; Chou *et al.*, 1977); endogenous protein kinase inhibitors (Walsh *et al.*, 1971; Roskoski and Frederick, 1977), and autophosphorylation of the regulatory subunit of protein kinase (Rangel-Aldao and Rosen, 1977). Much additional study is required before the importance of these various protein phosphorylating modulators is appreciated for each system whose function is regulated by protein kinase, including the tyrosine hydroxylase system.

6 Comparison of the regulation of tryptophan and phenylalanine hydroxylase activities with that of tyrosine hydroxylase

Both phenylalanine and tryptophan hydroxylases exhibit many regulatory sensitivities which are similar, although not identical, to those of tyrosine hydroxylase (Kaufman and Fisher, 1974). All three hydroxylases are activated by ATP and Mg^{2+}. With phenylalanine hydroxylase, the activation is dependent on cyclic AMP and involves direct enzyme phosphorylation and an increase in the intrinsic specific activity (V_{max}) of the enzyme. Unlike tyrosine hydroxylase, no change in the affinity of the enzyme for either phenylalanine or pterin cofactor is apparent (Milstien *et al.*, 1976). In contrast, activation of tryptophan hydroxylase by ATP and Mg^{2+} is not cyclic AMP dependent, although it appears to involve protein phosphorylation, since the non-metabolizable analogue of ATP, adenylylimidodiphosphate, cannot replace ATP. The kinetic changes in tryptophan hydroxylase associated with exposure to ATP plus Mg^{2+} are an enhanced affinity for pterin cofactor and a shift in the pH optimum to a more alkaline range (Hamon *et al.*, 1977a).

In contrast to most reported studies for tyrosine hydroxylase, tryptophan hydroxylase is activated by Ca^{2+} (Boadle-Biber, 1975; Knapp, 1975; Hamon *et al.*, 1977a, b). The effect of Ca^{2+} on tryptophan hydroxylase appears to be mediated by activation of a calcium-dependent protease. Presumably the enzyme is activated by partial proteolysis (Hamon *et al.*, 1977a, b). Similar activation of the enzyme can be achieved by exposure of the enzyme to trypsin for a brief period (Hamon *et al.*, 1977a, b). Both tyrosine hydroxylase (Petrack *et al.*, 1968; Shiman *et al.*, 1971; Kuczenski, 1973a, b) and phenylalanine hydroxylase (Fisher and Kaufman, 1973; Kaufman and Fisher, 1974) also can be activated by limited proteolysis (Figure 4.12).

It is of interest that Inoue *et al.* (1977) and Kishimoto *et al.* (1977) have demonstrated a calcium-activated, cyclic nucleotide-independent protein kinase in rat brain and other tissues. The activation of the protein kinase is mediated by a calcium-dependent protease which converts a proenzyme to active protein kinase. The protein kinase is capable of phosphorylating the subunits of rabbit skeletal muscle glycogen phosphorylase kinase. It is conceivable that the mechanism of activation of tryptophan hydroxylase may not involve proteolysis of tryptophan hydroxylase, but, rather, of a proenzyme of protein kinase which, in turn, activates the hydroxylase by phosphorylation.

Tyrosine and phenylalanine hydroxylases can be activated by lysolecithin or phosphatidyl-L-serine (Fisher and Kaufman, 1973; Lloyd and Kaufman, 1974; Lloyd, 1977). Tryptophan hydroxylase is also activated by phospholipids (Hamon *et al.*, 1977a). The kinetic changes in tryptophan hydroxylase following exposure to phospholipids are similar to those produced by ATP plus Mg^{2+}.

The observation that either phospholipids or limited proteolysis can activate tryptophan, tyrosine, and phenylalanine hydroxylases and that a protein phosphorylating system can activate these enzymes is of interest in view of the recent observation of Hoeldtke and Kaufman (1977) that bovine adrenal tyrosine hydroxylase subject to limited proteolysis is no longer sensitive to activation by phosphatidyl serine or a cyclic AMP-dependent protein phosphorylating system. It is conceivable that the regulatory component of the enzyme is inhibitory to the catalytic site. In some fashion, phosphorylation of the regulatory component (or another molecule that interacts with the regulatory subunit), and interaction of the enzyme with phospholipid or anions of several types (Katz *et al.*, 1976b) result in a dissociation of the regulatory component of the enzyme from the catalytic site, thus leading to enzyme activation. With limited proteolysis, the regulatory unit may be cleaved from the enzyme and the catalytic component may then be permanently activated and no longer responsive to the reversible activators of the enzyme (Figure 4.12).

7 The induction of norepinephrine biosynthetic enzymes following prolonged stimulation

In addition to the enhanced activity of tyrosine hydroxylase which occurs during and shortly following nerve stimulation, other mechanisms for the regulation of the biosynthesis of this transmitter exist. Increased sympathetic activity for many hours or days is associated with an increase in the content of tyrosine hydroxylase in the adrenal gland, in sympathetic nervous tissue and in the central nervous system (Dairman *et al.*, 1968; Mueller *et al.*, 1969a, b; Viveros *et al.*, 1969). Immunochemical studies with specific antibody indicate that the amount of enzyme protein increases both in the central nervous system after reserpine treatment (Joh *et al.*, 1973) and in the adrenal gland after either immobilization stress, cold exposure, or 6-hydroxydopamine administration (Hoeldtke *et al.*, 1974).

A large number of pharmacological agents which affect adrenergic nervous function either directly or indirectly can influence the level of tyrosine hydroxylase in peripheral or central adrenergic neurons. Insulin administration and the consequent hypoglycemia is associated with a reflex increase in adrenergic stimulation and elevated levels of tyrosine hydroxylase (Viveros *et al.*, 1969; Weiner and Mosimann, 1970) and dopamine β-hydroxylase (Viveros *et al.*, 1969) in the adrenal medulla. The administration of either reserpine, which impairs sympathetic function by depletion of catecholamine stores, or phenoxybenzamine, which produces blockade of sympathetic function at the postsynaptic receptor site, presumably leads to a compensatory reflex increase in adrenergic nervous activity and a consequent increase in the levels of the catecholamine biosynthetic enzymes. These effects can be inhibited at least in part by administration of inhibitors of protein synthesis or actinomycin D (Thoenen *et al.*, 1969, 1970; Weiner and Mosimann, 1970). Conversely, after chronic denervation of the adrenal gland, the level of tyrosine hydroxylase is significantly less than that in the contralateral, innervated gland (Weiner *et al.*, 1972b).

The mechanism by which tyrosine hydroxylase is induced in adrenergic nervous tissue following chronic adrenergic nerve stimulation has been examined in a variety of systems. Waymire *et al.* (1972a) demonstrated that the addition of dibutyryl cyclic AMP, cyclic AMP analogues, or phosphodiesterase inhibitors to growing neuroblastoma C1300 cells in culture is associated with arrest of cell division, cell differentiation, and increased levels of tyrosine hydroxylase after 48–72 h. Immunotitration studies reveal that the level of enzyme protein in neuroblastoma cells is increased. Kinetic properties of the induced enzyme, however, are not different from those of the enzyme present in cells from control cultures. 4-(3-butoxy-4-methoxybenzyl)-2-imidazolidinone (Ro 20-1724), a phosphodiesterase inhibitor, and PGE_1, a known activator of adenylate cyclase in neuroblastoma cells in culture (Gilman and Nirenberg, 1971; Matsuzawa and Nirenberg, 1975; Blume and Foster, 1976), also produce increases in tyrosine hydroxylase activity in neuroblastoma in 24–48h. These effects can be inhibited by carbachol. The inhibitory action of carbachol is blocked by atropine (A. W. Tank, M. Posiviata, and N. Weiner, in preparation). Incubation of neuroblastoma cells with either PGE_1 or Ro 20-1724 is associated with an increase in cyclic AMP in the cells within a matter of minutes. This increase is sustained for several hours. Carbachol is able to inhibit partially the increase in cyclic AMP associated with these enzyme inducers and the effect of carbachol on cyclic AMP levels is blocked by atropine. These results tend to suggest that induction of tyrosine hydroxylase in neuroblastoma cells in culture is indeed mediated by a cyclic AMP-dependent process (A. W. Tank, M. Posiviata, and N. Weiner, in preparation). Carbachol does not inhibit the increase in tyrosine hydroxylase associated with incubation of the neuroblastoma cells in the presence of 8-methylthio cyclic AMP. Mackay and Iversen (1972) observed a similar increase in tyrosine hydroxylase in sympathetic ganglia cultured in the presence of dibutyryl cyclic AMP.

Costa and coworkers have presented a series of elegant experiments which support their hypothesis that stimulation of the adrenal medulla is associated with a cyclic AMP-dependent activation of protein kinase; translocation of the activated cytosolic protein kinase to the nucleus; increase in RNA synthesis as a consequence of this series of events; and, ultimately, induction of tyrosine hydroxylase (Costa *et al.*, 1975, 1976, 1977; Guidotti *et al.*, 1975; Chuang *et al.*, 1976; Kurosawa *et al.*, 1976a, b; Chuang and Costa, 1975). Guidotti *et al.* (1973) and Guidotti and Costa (1973) demonstrated a rapid and transient increase in adrenal medulla cyclic AMP after administration of carbachol or reserpine to rats. This increase was followed by an elevation in adrenal tyrosine hydroxylase 24 h. later. A similar series of events occurs following cold exposure (Chuang and Costa, 1974; Guidotti *et al.*, 1973). The rise in cyclic AMP following each of these stresses is transient. It may last only 60–120 min. In some instances, the absolute rise in cyclic AMP seems to be less important than an elevation in the ratio of the concentrations of cyclic AMP to cyclic GMP (Costa *et al.*, 1975). Following the rise in cyclic AMP, an activation of protein kinase in the cytosol is observed. The proportion of cyclic AMP-independent protein kinase rises markedly in the period following the rise in cyclic AMP and this persists for several hours (Guidotti *et al.*, 1975; Costa *et al.*, 1975). Following this, the total cytosolic protein kinase activity decreases and there is an increase in nuclear protein kinase. Presumably these events indicate a shift of the catalytic unit of cytosolic protein kinase into the nucleus (Costa *et al.*, 1975; Kurosawa *et al.*, 1976a, b). The active protein kinase in the nucleus presumably catalyses the phosphorylation of nuclear proteins which then results in enhanced template activity (Chuang *et al.*, 1976, 1977). At least one consequence of the enhanced template activity and increased ribonucleic acid synthesis (Chuang and Costa, 1975) is an increased synthesis of tyrosine hydroxylase (Chuang *et al.*, 1975). Chuang *et al.* (1975) have demonstrated that the elevated level of tyrosine hydroxylase in the rat adrenal gland after cold stress results from enhanced synthesis of the enzyme rather than reduced degradation.

However, several studies tend to discount the association between increased levels of cyclic AMP in adrenal medulla and the subsequent induction of tyrosine hydroxylase. For example, Paul *et al.* (1971) demonstrated a rise in rat adrenal cortex cyclic AMP after immobilization stress, but these workers were unable to show an elevation of the cyclic nucleotide in the adrenal medulla. Immobilization stress is associated with elevated levels of tyrosine hydroxylase after one or more days (Kvetňanský *et al.*, 1970). Otten *et al.* (1973) administered reserpine to rats or subjected groups of these animals either to intermittent swimming or cold stress and measured cyclic AMP and tyrosine hydroxylase levels in the adrenal medulla and superior cervical ganglia at intervals thereafter. Early increases in adrenal cyclic AMP and later elevations in tyrosine hydroxylase were detectable after each of these procedures. However, a good correlation between the extent of the rise in adrenal medulla cyclic AMP and the subsequent induction of tyrosine hydroxylase was not apparent. Furthermore, no elevation in cyclic AMP was demonstrable in the superior cervical ganglion after swimming stress, although a subsequent elevation in tyrosine hydroxylase was noted. Further studies by Otten *et al.* (1974) also suggest that there may not be a causal relationship between elevations in cyclic AMP and induction of tyrosine hydroxylase. These workers showed that isoproterenol administration, which results in considerable increases in cyclic AMP levels in the superior cervical ganglion, is not associated with elevated tyrosine hydroxylase levels subsequently. Furthermore, propranolol administration to rats markedly diminishes the rise in adrenal medulla cyclic AMP after reserpine treatment, but does not modify the subsequent rise in tyrosine hydroxylase.

Otten and Thoenen (1976) also showed that blockade of the release of ACTH from the pituitary, either by administration of dexamethasone or by hypophysectomy, abolishes the reserpine-mediated increase in cyclic AMP and the increase in the cyclic AMP:cyclic GMP ratio in the adrenal medulla. However, the reserpine-mediated induction of tyrosine hydroxylase was unaffected. Thoenen and Otten (1977) have recently summarized the various points of view regarding this controversy. The validity of some of the studies of Thoenen and coworkers have been challenged by Costa's group (Guidotti *et al.*, 1976).

Costa *et al.* (1975) suggest that the failure to obtain perfect correlations between elevations of adrenal medulla cyclic AMP and tyrosine hydroxylase induction may result from the very transient elevation of cyclic AMP, which may persist for only 60–90 min. This is in part due to the activation of a cyclic AMP phosphodiesterase which appears to be cyclic AMP-dependent (Uzunov *et al.*, 1975). Nevertheless, the transient elevation in cyclic AMP may suffice to activate cytosolic protein kinase, converting the cyclic AMP-dependent species (and thus the inactive form) into a cyclic AMP-independent form.

G ULTRASTRUCTURAL LOCALIZATION OF TYROSINE HYDROXYLASE

In view of the multiple regulatory factors which have been implicated in enhancing tyrosine hydroxylase activity with nerve stimulation and in view of the more favourable kinetic properties that rat striatal tyrosine hydroxylase which is associated with membranes appears to exhibit (Kuczenski and Mandell, 1972a), it has been proposed that nerve stimulation may be associated with a physical transformation or translocation of the enzyme which results in enhanced activity. In fact, Udenfriend (1966) suggested several years ago that tyrosine hydroxylase and the other enzymes involved in norepinephrine biosynthesis may be an integral component of the synaptic vesicle. Considerable evidence has accumulated to indicate that tyrosine hydroxylase is not within the synaptic vesicle (Weiner, 1970), but, during stimulation, the enzyme may become loosely associated with the outer membrane of the vesicle in some manner. There is much evidence to suggest that nervous activity is associated with enhanced phosphorylation of both lipids (Burt and Larrabee, 1973; White *et al.*, 1974) and proteins (Johnson *et al.*, 1971, 1972; Sloboda *et al.*, 1975) and that membrane and microtubule protein phosphorylation involves a cyclic AMP-dependent protein kinase (Johnson *et al.*, 1972; Sloboda *et al.*, 1975). Cyclic AMP-dependent phosphorylation of tyrosine hydroxylase has been suggested as a mechanism for enzyme activation during nerve stimulation (see above). It is possible that these processes result in interaction of the enzyme with membranous structures in the neuron.

Employing antiserum raised against purified bovine adrenal tyrosine hydroxylase, Pickel *et al.* (1975a, b, 1976) have examined the cellular and subcellular distribution of tyrosine hydroxylase in brain by immunocytochemical techniques, employing both light and electron microscopy. They observed rather extensive localization of the enzyme and suggest that the enzyme may be associated both with neurotubules in axon processes and with endoplasmic reticulum and the Golgi apparatus in the cell bodies. However, because the sections, which were treated by the peroxidase–antiperoxidase technique, were 20 μm thick and because the peroxidase–antiperoxidase–immunoglobulin complex has a molecular weight of approximately 340 000, precise determination of the ultrastructural localization of the enzyme would be extremely difficult. Furthermore, the homogeneity and purity of the enzyme which Pickel *et al.* used has not been reported in full detail and the monospecificity of the antibody employed by these workers was not conclusively established in the reports describing their preparation (Joh *et al.*, 1973).

Hokfelt and coworkers (1976, 1977) have described the immunohistochemical localization of tyrosine hydroxylase in brain, employing antibody against purified human pheochromocytoma tumour. Since the specific activity of the enzyme employed for developing antibody for these immunohistochemical studies is approximately one third that obtained by Hoeldtke and Kaufman (1977) and in our laboratories, the purity of their enzyme and the specificity of their antiserum also may be challenged. In fact, these workers report that tyrosine hydroxylase positive fibres can be demonstrated in the hippocampal formation and the distribution of these fibres appears to be identical to that of mossy fibres. Since there is no evidence for catecholamine or 5-hydroxytryptamine-containing nerve terminals in this area, they presume that the reaction is due to a cross-reacting protein. An equally plausible explanation might be that the antiserum is not entirely monospecific.

H REFERENCES

Abita, J.-P., Milstien, S., Chang, N., and Kaufman, S. (1976). *J. Biol. Chem.*, **251**, 5310–5314.

Alousi, A., and Weiner, N. (1966). *Proc. Nat. Acad. Sci., USA*, **56**, 1491–1496.

Anagnoste, B., Shirron, C., Friedman, E., and Goldstein, M. (1974). *J. Pharmacol. Exptl Ther.*, **191**, 370–376.

Beavo, J. A., Bechtel, P. J., and Krebs, E. G. (1975). *Adv. Cyclic Nucleotide Res.*, **5**, 241–251.

Bechtel, P. J., Beavo, J. A., and Krebs, E. G. (1977). *J. Biol. Chem.*, **252**, 2691–2697.

Bjur, R. A., and Weiner, N. (1975). *J. Pharmacol. Exptl Ther.*, **194**, 9–26.

Blume, A. J., and Foster, C. J. (1976). *J. Neurochem.*, **26**, 305–311.

Boadle-Biber, M. C. (1975). *Biochem. Pharmacol.*, **24**, 1455–1460.

Brenneman, A. R., and Kaufman, S. (1964). *Biochem. Biophys. Res. Commun.*, **17**, 177–183.

Brostrom, C. O., Huang, Y.-C., Breckenridge, B. McL., and Wolff, D. J. (1975). *Proc. Nat. Acad. Sci.*, **72**, 64–68.

Buff, K., and Dairman, W. (1974). *Mol. Pharmacol.*, **11**, 87–93.

Burt, D. R., and Larabee, M. G. (1973). *J. Neurochem.*, **21**, 255–272.

Butterworth, K. R., and Mann, M. (1957). *Brit. J. Pharmacol. Chemother.*, **12**, 422–426.

Bygdeman, S., and Euler, U. S. von (1958). *Acta. Physiol. Scand.*, **44**, 375–383.

Chou, C.-K., Alfano, J., and Rosen, O. M. (1977). *J. Biol. Chem.*, **252**, 2855–2859.

Chuang, D. M., and Costa, E. (1974). *Proc. Nat. Acad. Sci., USA*, **71**, 4570–4574.

Chuang, D. M., and Costa, E. (1975). *Mol. Pharm.*, **12**, 514–518.

Chuang, D. M., Hollenbeck, R. A., and Costa, E. (1976). *Science*, **193**, 60–62.

Chuang, D. M., Hollenbeck, R. A., and Costa, E. (1977). *J. Biol. Chem.*, **252**, 8365–8373.

Chuang, D., Zsilla, G., and Costa, E. (1975). *Mol. Pharmacol.*, **11**, 784–794.

Cloutier, G., and Weiner, N. (1973). *J. Pharmacol. Exptl Ther.*, **186**, 75–85.

Corbin, J. D., and Keely, S. L. (1977). *J. Biol. Chem.*, **252**, 910–918.

Corbin, J. D., Keely, S. L., and Park, C. R. (1975). *J. Biol. Chem.*, **250**, 218–225.

Corbin, J. D., Sugden, P. H., Lincoln, T. M., and Keely, S. L. (1977). *J. Biol. Chem.*, **252**, 3854–3861.

Costa, E., Chuang, D. M., and Guidotti, A. (1977). In *Structure and Function of Monoamine Enzymes* (Eds., E. Usdin, N. Weiner, and M. B. H. Youdim), Marcel Dekker, New York, pp. 279–310.

Costa, E., Guidotti, A., and Kurosawa, A. (1975). In *Biological Membranes. Neurochemistry*, Vol. 41, *Proceedings of the Tenth FEBS Meeting*, North Holland/American Elsevier, Amsterdam, pp. 137–149.

Costa, E., Guidotti, A., and Zivković, B. (1974). In *Neuropsychopharmacology of Monoamines and their Regulatory Enzymes* (Ed., E. Usdin), Raven Press, New York, pp. 161–175.

Costa, E., Kurosawa, A., and Guidotti, A. (1976). *Proc. Nat. Acad. Sci., USA*, **73**, 1058–1062.

Costa, E., Green, A. R., Koslow, S. H., LeFevre, H. F., Revuelta, A. V., and Wang, C. (1972). *Pharmacol. Rev.*, **24**, 167–190.

Coyle, J. T. (1972). *Biochem. Pharmacol.*, **21**, 1935–1944.

Craine, J. E., Daniels, G. H., and Kaufman, S. (1973). *J. Biol. Chem.*, **248**, 7838–7844.

Craine, J. E., Hall, E. S., and Kaufman, S. (1972). *J. Biol. Chem.*, **247**, 6082–6091.

Cubeddu, L. X., Barnes, E., and Weiner, N. (1975). *J. Pharmacol. Exptl Ther.*, **193**, 105–127.

Daly, J., Levitt, M., Guroff, G., and Udenfriend, S. (1968). *Arch. Biochem. Biophys.*, **126**, 593–598.

Dairman, W., Gordon, R., Spector, S., Sjoerdsma, A., and Udenfriend, S. (1968). *Mol. Pharmacol.*, **4**, 457–464.

Davis, J. N. (1976). *J. Neurochem.*, **27**, 211–215.

Davis, J. N., and Carlsson, A. (1973). *J. Neurochem.*, **20**, 913–915.
Dixon, M. (1953). *Biochem. J.*, **55**, 170–171.
Erlichman, J., Rubin, C. S., and Rosen, O. M. (1973). *J. Biol. Chem.*, **248**, 7607–7609.
Euler, U. S. von, Luft, R., and Sundin, T. (1955). *Acta. Physiol. Scand.*, **34**, 169–174.
Fahn, S., Rodman, J. S., and Coté, L. J. (1969). *J. Neurochem.*, **16**, 1293–1300.
Ferrendelli, J. A., Rubin, E. H., and Kinscherf, D. A. (1976). *J. Neurochem.*, **26**, 741–748.
Fisher, D. B., and Kaufman, S. (1972). *J. Neurochem.*, **19**, 1359–1365.
Fisher, D. B., and Kaufman, S. (1973). *J. Biol. Chem.*, **248**, 4345–4353.
Fisher, D. B., Kirkwood, R., and Kaufman, S. (1972). *J. Biol. Chem.*, **16**, 5161–5167.
Gal, E. M. and Sherman, A. D. (1976). *Neurochem. Res.*, **1**, 627–639.
Gal, E. M., Hanson, G., and Sherman, A. (1976). *Neurochem. Res.*, **1**, 511–523.
Gewirtz, G. P., Kvetňanský, R., Weise, V. K., and Kopin, I. J. (1971). *Nature*, **230**, 462–464.
Gilman, A. G., and Nirenberg, M. (1971). *Nature*, **234**, 356–358.
Goldstein, M., Joh, T. H., and Garvey, T. Q., III (1968). *Biochemistry*, **7**, 2724–2730.
Goldstein, M., Backstrom, T., Ohi, Y., and Frenkel, R. (1970). *Life Sci.*, **9**, 919–924.
Goldstein, M., Bronaugh, R. L., Ebstein, B., and Roberge, C. (1976). *Brain Res.*, **109**, 563–574.
Greengard, P. (1975). *Cyclic Nucleotide Res.*, **5**, 585–601.
Guidotti, A., and Costa, E. (1973). *Science*, **179**, 902–904.
Guidotti, A., Kurosawa, A., and Costa, E. (1976). *Naunyn-Schmiedeberg's Arch. Pharmacol.*, **295**, 135–140.
Guidotti, A., Kurosawa, A., Chuang, D. M., and Costa, E. (1975). *Proc. Nat. Acad. Sci., USA*, **72**, 1152–1156.
Guidotti, A., Zivković, B., Pfeiffer, R., and Costa, E. (1973). *Naunyn-Schmiedeberg's Arch. Pharmacol.*, **278**, 195–206.
Guroff, G., King, W., and Udenfriend, S. (1961). *J. Biol. Chem.*, **236**, 1773–1777.
Guroff, G., and Udenfriend, S. (1962). *J. Biol. Chem.*, **237**, 803–806.
Hamon, M., Burgoin, S., Artaud, H., and Héry, F. (1977a). *J. Neurochem.*, **28**, 811–818.
Hamon, M., Burgoin, S., Héry, F., and Glowinski, J. (1977b). In *Structure and Function of Monoamine Enzymes* (Eds., E. Usdin, N. Weiner, and M. B. H. Youdim), Marcel Dekker, New York, pp. 59–90.
Harris, J. E., Baldessarini, R. J., Morgenroth, V. H., III, and Roth, R. H. (1975). *Proc. Nat. Acad. Sci., USA*, **72**, 789–793.
Harris, J. E., Morgenroth, V. H., III, Roth, R. H., and Baldessarini, R. J. (1974). *Nature*, **252**, 156–158.
Hoeldtke, R., and Kaufman, S. (1977). *J. Biol. Chem.*, **252**, 3160–3169.
Hoeldtke, R., Lloyd, T., and Kaufman, S. (1974). *Biochem. Biophys. Res. Commun.*, **57**, 1045–1053.
Hokfelt, T., Johansson, O., Fuxe, K., Goldstein, M., and Park, D. (1976). *Med. Biol.*, **54**, 427–453.
Hokfelt, T., Johansson, O., Fuxe, K., Goldstein, M., and Park, D. (1977). *Med. Biol.*, **55**, 21–40.
Holland, W. C., and Schümann, H. J. (1956). *Brit. J. Pharmacol. Chemother.*, **11**, 449–453.
Hsu, C.-Y., Leighton, H. J., Westfall, T. C., and Brooker, G. (1976). *J. Cyclic Nucleotide Res.*, **2**, 359–363.
Ikeda, M., Fahien, L. A., and Udenfriend, S. (1966). *J. Biol. Chem.*, **241**, 4452–4456.
Ikeda, M., Levitt, M., and Udenfriend, S. (1965). *Biochem. Biophys. Res. Commun.*, **18**, 482–488.
Ikeda, M., Levitt, M., and Udenfriend, S. (1967). *Arch. Biochem. Biophys.*, **120**, 420–427.
Inoue, M., Kishimoto, A., Takai, Y., and Nishizuka, Y. (1977). *J. Biol. Chem.*, **252**, 7610–7616.
Iyer, N. T., and McGeer, E. G. (1963). *Canad. J. Biochem. Physiol.*, **41**, 1565–1570.

Joanny, P., Natali, J.-P., Hillman, H., and Corriol, J. (1973). *Biochem. J.*, **136**, 77–82.
Joh, T. H., and Reis, D. J. (1975). *Brain Res.*, **85**, 146–151.
Joh, T. H., Geghman, C., and Reis, D. (1973). *Proc. Nat. Acad. Sci., USA*, **70**, 2667–2771.
Joh, T. H., Kapit, R., and Goldstein, M. (1969). *Biochem. Biophys. Acta.*, **171**, 378–380.
Johnson, E. M., Maeno, H., and Greengard, P. (1971). *J. Biol. Chem.*, **246**, 7731–7739.
Johnson, E. M., Ueda, T., Maeno, H., and Greengard, P. (1972). *J. Biol. Chem.*, **247**, 5650–5652.
Kakiuchi, S., and Rall, T. W. (1968). *Mol. Pharmacol.*, **4**, 379–388.
Kakiuchi, S., Rall, T. W., and McIlwain, H. (1969). *J. Neurochem.*, **16**, 485–491.
Katz, I., Lloyd, T., and Kaufman, S. (1976a). *Biochim. Biophys. Acta*, **444**, 567–578.
Katz, I. R., Yamauchi, T., and Kaufman, S. (1976b). *Biochim. Biophys. Acta*, **429**, 84–95.
Kaufman, S. (1963). Proc. Nat. Acad. Sci., *USA*, **50**, 1085–1093.
Kaufman, S. (1964). *J. Biol. Chem.*, **239**, 332–338.
Kaufman, S. (1973). In *Ciba Foundation Symposium: Aromatic Amino Acids in the Brain*, Elsevier North-Holland Publ., Amsterdam, pp. 85–115.
Kaufman, S., and Fisher, D. B. (1974). In *Molecular Mechanisms of Oxygen Activation* (Ed., O. Hayaishi), Academic Press, New York, pp. 285–369.
Kettler, R., Bartholini, G., and Pletscher, A. (1974). *Nature*, **249**, 476–478.
Kishimoto, A., Taki, Y., and Yasutomi, N. (1977). *J. Biol. Chem.*, **252**, 7449–7452.
Knapp, S., Mandell, A. J., and Bullard, W. P. (1975). *Life Sci.*, **10**, 1583–1594.
Knight, B. L., and Skala, J. P. (1977). *J. Biol. Chem.*, **252**, 5356–5362.
Kopin, I. J., Weise, V. K., and Sedvall, G. C. (1969). *J. Pharmacol. Exptl Ther.*, **170**, 246–252.
Krueger, B. K., Forn, J., and Greengard, P. (1977). *J. Biol. Chem.*, **252**, 2764–2773.
Kuczenski, R. (1973a). *J. Biol. Chem.*, **248**, 2261–2265.
Kuczenski, R. (1973b). *Life Sci.*, **13**, 247–255.
Kuczenski, R. T., and Mandell, A. J. (1972a). *J. Biol. Chem.*, **247**, 3114–3122.
Kuczenski, R. T., and Mandell, A. J. (1972b). *J. Neurochem.*, **19**, 131–137.
Kuczenski, R., and Segal, D. S. (1974). *J. Neurochem.*, **22**, 1039–1044.
Kuo, J. F., and Greengard, P. (1969). *Proc. Nat. Acad. Sci. USA*, **64**, 1349–1353.
Kurosawa, A., Guidotti, A., and Costa, E. (1976a). *Science*, **193**, 691–693.
Kurosawa, A., Guidotti, A., and Costa, E. (1976b). *Mol. Pharmacol.*, **12**, 420–432.
Kvetňanský, R., Weise, V. K., and Kopin, I. J. (1970). *Endocrinology*, **87**, 744–749.
Langan, T. A. (1968). *Science*, **162**, 579–580.
Lerner, P., Ames, M. M., and Lovenberg, W. (1977). *Mol. Pharmacol.*, **13**, 44–49.
Letendre, C. H., MacDonnell, P. C., and Guroff, G. (1977a). *Biochem. Biophys. Res. Commun.*, **74**, 891–897.
Letendre, C. H., MacDonnell, P. C., and Guroff, G. (1977b). *Biochem. Biophys. Res. Commun.*, **76**, 615–617.
Levitt, M., Spector, S., Sjoerdsma, A., and Udenfriend, S. (1965). *J. Pharmacol. Exptl Ther.*, **148**, 1–8.
Lloyd, T. (1977). In *Structure and Function of Monoamine Enzymes* (Eds., E. Usdin, N. Weiner, and M. B. H. Youdim), Marcel Dekker, New York, pp. 211–230.
Lloyd, T., and Kaufman, S. (1973). *Mol. Pharmacol.*, **9**, 438–444.
Lloyd, T., and Kaufman, S. (1974). *Biochem. Biophys. Res. Commun.*, **59**, 1262–1269.
Lloyd, T., and Kaufman, S. (1975). *Biochem. Biophys. Res. Commun.*, **66**, 907–913.
Lloyd, T., and Weiner, N. (1971). *Mol. Pharmacol.*, **7**, 569–580.
Lovenberg, W., and Bruckwick, E. A. (1975). In *Pre- and Post-synaptic Receptors* (Eds., E. Usdin and W. E. Bunney, Jr), Marcel Dekker, New York, pp. 149–168.
Lovenberg, W., Bruckwick, E. A., and Hanbauer, I. (1975). *Proc. Nat. Acad. Sci., USA*, **72**, 2955–2958.
Mackay, A. V. P., and Iversen, L. L. (1972). *Brain Res.*, **48**, 424–426.

McGeer, P. L., Bagchi, S. P., and McGeer, E. G. (1965). *Life Sci.*, **4**, 1859–1867.
Maeno, H., and Greengard, P. (1972). *J. Biol. Chem.*, **247**, 3269–3277.
Maeno, H., Ueda, T., and Greengard, P. (1975). *J. Cyclic Nucleotide Res.*, **1**, 37–48.
Malkinson, A. M. (1975). *Biochem. Biophys. Res. Commun.*, **67**, 752–759.
Mandell, A. J., Knapp, S., Kuczenski, R. T., and Segal, D. S. (1972). *Biochem. Pharmacol.*, **21**, 2737–2750.
Matsuzawa, H., and Nirenberg, M. (1975). *Proc. Nat. Acad. Sci., USA*, **72**, 3472–3476.
Milstien, S., Abita, J.-P., Chang, N., and Kaufman, S. (1976). *Proc. Nat. Acad. Sci., USA*, **73**, 1591–1593.
Molinoff, P. B., Brimijoin, S., and Axelrod, J. (1972). *J. Pharmacol. Exptl Ther.*, **182**, 116–129.
Molinoff, P. B., Brimijoin, S., Weinshilboum, R., and Axelrod, J. (1970). *Proc. Nat. Acad. Sci., USA*, **66**, 453–458.
Morgenroth, V. H., III, Boadle-Biber, M., and Roth, R. H. (1974). *Proc. Nat. Acad. Sci., USA*, **71**, 4283–4287.
Morgenroth, V. H., III, Boadle-Biber, M. C., and Roth, R. H. (1975a), *Mol. Pharmacol.*, **11**, 427–435.
Morgenroth, V. H., III, Boadle-Biber, M. C., and Roth, R. H. (1975b). *Mol. Pharmacol.*, **12**, 41–48.
Morgenroth, V. H., III, Hegstrand, L. R., Roth, R. H., and Greengard, P. (1975c). *J. Biol. Chem.*, **250**, 1946–1948.
Morita, K., and Oka, M. (1977). *FEBS Letters*, **76**, 148–150.
Mueller, R. A., Thoenen, H., and Axelrod, J. (1969a). *J. Pharmacol. Exptl Ther.*, **169**, 74–79.
Mueller, R. A., Thoenen, H., and Axelrod, J. (1969b). *Science*, **163**, 468–469.
Musacchio, J. M. (1969). *Biochem. Biophys. Acta*, **191**, 485–487.
Musacchio, J. M., D'Angelo, G. L., and McQueen, C. A. (1971a). *Proc. Nat. Acad. Sci., USA*, **68**, 2087–2091.
Musacchio, J. M., Wurzburger, R. J., and D'Angelo, G. L. (1971b). *Mol. Pharmacol.*, **7**, 136–146.
Nagatsu, T., Levitt, M., and Udenfriend, S. (1964a). *J. Biol. Chem.*, **238**, 2910–2917.
Nagatsu, T., Levitt, M., and Udenfriend, S. (1964b). *Anal. Biochem.*, **9**, 122–126.
Neame, K. D. (1961). *J. Neurochem.*, **6**, 358–366.
Nielsen, K. H., Simonsen, V., and Lind, K. E. (1969). *Eur. J. Biochem.*, **9**, 497–502.
Numata (Sudo), Y., and Nagatsu, T. (1975). *J. Neurochem.*, **24**, 317–322.
O'Dea, R. F., and Zatz, M. (1976). *Proc. Nat. Acad. Sci., USA*, **73**, 3398–3402.
Osborne, N. N., and Neuhoff, V. (1976). *Hoppe-Seyle's J. Physiol. Chem.*, **357**, 1271–1275.
Otten, U., Mueller, R. A., Oesch, F., and Thoenen, H. (1974). *Proc. Nat. Acad. Sci.*, **71**, 2217–2221.
Otten, U., Oesch, F., and Thoenen, H. (1973). *Naunyn-Schmiedeberg's Arch. Pharmacol.*, **280**, 129–140.
Otten, U., and Thoenen, H. (1976). *Naunyn-Schmiedeberg's Arch. Pharmacol.*, **293**, 105–108.
Park, D. H., and Goldstein, M. (1976). *Life Sci.*, **18**, 55–60.
Patrick, R. L., and Barchas, J. D. (1974a). *Nature*, **250**, 737–739.
Patrick, R. L., and Barchas, J. D. (1974b). *J. Neurochem.*, **23**, 7–15.
Patrick, R. L., and Barchas, J. D. (1976). *J. Pharmacol. Exptl Ther.*, **197**, 97–104.
Paul, M. I., Kvetnansky, R., Cramer, R., Silbergeld, S., and Kopin, I. J. (1971). *Endocrinology*, **88**, 338–344.
Peach, M. J. (1972). *Proc. Nat. Acad. Sci., USA*, **69**, 834–836.
Petrack, B., Sheppy, F., and Fetzer, V. (1968). *J. Biol. Chem.*, **243**, 743–748.

Pfeffer, R. I., Mosimann, W., and Weiner, N. (1975). *J. Pharmacol. Exptl Ther.*, **193**, 533–548.

Pickel, V. M., Joh, T. H., and Reis, D. J. (1975a). *Proc. Nat. Acad. Sci., USA*, **72**, 659–663.

Pickel, V. M., Joh, T. H., and Reis, D. J. (1975b). *Brain. Res.*, **85**, 295–300.

Pickel, V. M., Joh, T. H., and Reis, D. J. (1976). *J. Histochem. Cytochem.*, **24**, 792–806.

Potter, L. T., Cooper, T., Willman, V. L., and Wolfe, D. E. (1965). *Circulation Res.*, **16**, 468–481.

Raese, J. D., Edelman, A. M., Lazar, M. A., and Barchas, J. D. (1977). In *Structure and Function of Monoamine Enzymes* (Eds., E. Usdin, N. Weiner, and M. B. H. Youdim), Marcel Dekker, New York, pp, 383–421.

Raese, J., Patrick, R. L., and Barchas, J. D. (1976). *Biochem. Pharmacol.* **25**, 2245–2250.

Rangel-Aldao, R., and Rosen, O. M. (1977). *J. Biol. Chem.*, **252**, 7140–7145.

Rasmussen, H. (1970). *Science*, **170**, 404–412.

Reimann, E. M., Brostrom, C. O., Corbin, J. D., King, C. A., and Krebs, E. G. (1971). *Biochem. Biophys. Res. Commun.*, **42**, 187–194.

Reis, D. J., Joh, T. H., and Ross, R. A. (1975). *J. Pharmacol. Exptl Ther.*, **193**, 775–784.

Roskoski, R., Jr., and Frederick, C. E. (1977). *J. Neurochem.*, **28**, 543–547.

Roth, R. H., and Salzman, P. M. (1977). In *Structure and Function of Monoamine Enzymes* (Eds., E. Usdin, N. Weiner, and M. B. H. Youdim), Marcel Dekker, New York, pp. 149–168.

Roth, R. H., Morgenroth, V. H., III, and Salzman, P. M. (1975a), *Naunyn-Schmiedeberg's Arch. Pharmacol.*, **289**, 327–343.

Roth, R. H., Salzman, P. M., and Morgenroth, V. H., III (1974a). *Biochem. Pharmacol.*, **23**, 2779–2784.

Roth, R. H., Stjarne, L., and Euler, U. S. von. (1967). *J. Pharmacol. Exptl Ther.*, **158**, 373–377.

Roth, R. H., Walters, J. R., and Morgenroth, V. H., III (1974b). In *Neuropsychopharmacology of Monoamines and Their Regulatory Enzymes* (Ed., E. Usdin), Raven Press, New York, pp. 369–384.

Roth, R. H., Walters, J. R., Murrin, L. C., and Morgenroth, V. H., III (1975b). In *Pre- and Postsynaptic Receptors* (Eds., E. Usdin and W. E. Bunney, Jr.), Marcel Dekker, New York, pp. 5–46.

Shiman, R., and Kaufman, S. (1970). In *Methods in Enzymology*, Vol. XVII, *Metabolism of Amino Acids and Amines*, Part A (Eds., H. Tabor and C. W. Tabor), Academic Press, New York, pp. 609–615.

Shiman, R., Akino, M., and Kaufman, S. (1971). *J. Biol. Chem.*, **246**, 1330–1340.

Sloboda, R. D., Rudolph, S. A., Rosenbaum., J. L., and Greengard, P. (1975). *Proc. Nat. Acad. Sci., USA*, **72**, 177–181.

Spector, R., Levy, P., and Abelson, H. T. (1977). *Biochem. Pharmacol.*, **26**, 1507–1511.

Stjarne, L. (1966). *Acta. Physiol. Scand.*, **67**, 441–454.

Tao, M., Salas, M. L., and Lipmann, F. (1970). *Proc. Nat. Acad. Sci.*, **67**, 408–414.

Taylor, R. J., Jr., Stubbs, C. S., and Ellenbogen, L. (1969). *Biochem. Pharmacol.*, **18**, 587–594.

Thoa, N. B., Johnson, D. G., Kopin, I. J., and Weiner, N. (1971). *J. Pharmacol. Exptl Ther.*, **178**, 442–449.

Thoenen, H., and Otten, U. (1977). In *Essays in Neurochemistry and Neuropharmacology*, Vol. 1 (Eds., M. B. H. Youdim, W. Lovenberg, D. F. Sharman, and J. R. Lagnado), John Wiley and Sons, London, pp. 73–101.

Thoenen, H., Mueller, R. A., and Axelrod, J. (1969). *J. Pharmacol. Exptl Ther.*, **169**, 249–254.

Thoenen, H., Mueller, R. A., and Axelrod, J. (1970). *Proc. Nat. Acad. Sci., USA*, **65**, 58–62.

Thompson, W. J., and Appleman, M. M. (1971). *Biochemistry*, **10**, 311–316.

Turner, A. J., Ponzio, F., and Algeri, S. (1974). *Brain Res.*, **70**, 553–558.

Udenfriend, S. (1966). In *The Harvey Lectures*, Vol. 60, Academic Press, New York, pp. 57–83.

Udenfriend, S., Zaltzman-Nirenberg, P., and Nagatsu, T. (1965), *Biochem. Pharmacol.*, **14**, 837–845.

Ueda, T., Maeno, H., and Greengard, P. (1973). *J. Biol. Chem.*, **248**, 8295–8305.

Uzunov, P., Revuelta, A., and Costa, E. (1975). *Mol. Pharmacol.*, **11**, 506–510.

Viveros, O. H., Arqueros, L., Connett, R. J., and Kirshner, N. (1969). *Mol. Pharmacol.*, **5**, 69–82.

Walsh, D. A., Perkins, J. P., and Krebs, E. G. (1968). *J. Biol. Chem.*, **243**, 3763–3774.

Walsh, D. A., Ashby, C. D., Gonzales, C., Calkins, D., Fischer, E. H., and Krebs, E. G. (1971). *J. Biol. Chem.*, **246**, 1977–1985.

Walters, J. R., and Roth, R. H. (1974). *J. Pharmacol. Exptl Ther.*, **191**, 82–91.

Walters, J. R., Bunney, B. S., and Roth, R. H. (1975). *Adv. Neurol.*, **9**, 273–284.

Waymire, J. C., Bjur, R., and Weiner, N. (1971). *Anal. Biochem.*, **43**, 588–600.

Waymire, J. C., Weiner, N., and Prasad, K. N. (1972a). *Proc. Nat. Acad. Sci., USA*, **69**, 2241–2245.

Waymire, J. C., Weiner, N., Schneider, F. H., Goldstein, M., and Freedman, L. S. (1972b). *J. Clin. Invest.*, **51**, 1798–1804.

Weiner, N. (1970). *Ann. Rev. Pharmacol.*, **10**, 273–290.

Weiner, N. (1975). In *Handbook of Physiology*, Section 7: *Endocrinology*, Vol. VI, *Adrenal Gland* (Eds., H. Blaschko, G. Sayers, and A. D. Smith), American Physiological Society, Washington, DC, pp. 357–366.

Weiner, N., and Bjur, R. (1972). *Adv. Biochem. Psychopharmacol.*, **5**, 409–415.

Weiner, N., and Mosimann, W. F. (1970). *Biochem. Pharmacol.*, **19**, 1189–1199.

Weiner, N., and Rabadjija, M. (1968). *J. Pharmacol. Exptl Ther.*, **160**, 61–71.

Weiner, N., and Selvaratnam, I. (1968). *J. Pharmacol. Exptl Ther.*, **161**, 21–33.

Weiner, N., Waymire, J. C., and Prasad, K. N. (1972b). In *Cell Interactions, Third Lepetit Colloquin* (Ed., L. Silvestri), North Holland Publishing Co., Amsterdam, pp. 54–70.

Weiner, N., Cloutier, G., Bjur, R., and Pfeffer, R. I. (1972a). *Pharmacol. Rev.*, **24**, 203–221.

Weiner, N., Lee, F.-L., Barnes, E., and Dreyer, E. (1977). In *Structure and Function of Monoamine Enzymes* (Eds., E. Usdin, N. Weiner, and M. B. H. Youdim), Marcel Dekker, New York, pp. 109–148.

Weiner, N., Lee, F.-L., Waymire, J. C., and Posiviata, M. (1974). In *Ciba Foundation Symposium* 22, *Aromatic Amino Acids in the Brain* (Ed., R. J. Wurtman), Elsevier, Amsterdam, pp. 135–147.

Weiner, N., Bjur, R., Lee, F.-L., Becker, G., and Mosimann, W. F. (1973). In *Frontiers in Catecholamine Research* (Eds., E. Usdin, and S. H. Snyder), Pergamon Press, New York, pp. 211–221.

White, G. L., Schellhase, H. U., and Hawthorne, J. N. (1974). *J. Neurochem.*, **22**, 149–158.

Wooten, F. G., Thoa, N. B., Kopin, I. J., and Axelrod, J. (1973). *Mol. Pharmacol.*, **9**, 178–183.

Wurzburger, R. J., and Musacchio, J. M. (1971). *J. Pharmacol. Exptl Ther.*, **177**, 155–167.

Zivković, B., and Guidotti, A. (1974). *Brain. Res.*, **79**, 505–509.

Zivković, B., Guidotti, A., and Costa, E. (1974). *Mol. Pharmacol.*, **10**, 727–735.

Zivković, B., Guidotti, A., Revuelta, A., and Costa, E. (1975). *J. Pharmacol. Exptl Ther.*, **194**, 37–46.

Aromatic Amino Acid Hydroxylases and Mental Disease
Edited by M. B. H. Youdim

CHAPTER 5

Catecholamines and the Affective Disorders

James W. Maas

A INTRODUCTION

In the 1950s two accidential observations occurred which were to have important implications for the development of biological psychiatry. It was noticed that patients being treated with ipromiazid for tuberculosis in some cases developed mood elevations which were beyond those which might be expected to occur with amelioration of their physical illness. In separate studies it was found that a small but significant number of patients having hypertension who were being treated with reserpine developed severe depressive states which were clinically indistinguishable from endogenous depressions. When it was subsequently discovered that reserpine produced depletions of three important biogenic amines within brain, i.e. dopamine (DA), norepinephrine (NE), and serotonin (5-HT) whereas ipromiazid, via inhibition of monoamine oxidase (MAO), produced an elevation of these amines, the idea that levels of these amines within the central nervous system might have to do with the regulation of mood began to develop. This so called 'amine hypothesis of the affective disorders' was initially explicated by Jacobsen.[1] Following these observations came Kuhn's report that treatment of schizophrenic patients with a tricyclic compound which was an analogue of the phenothiazines resulted in a lifting of the depressive component of the patient's illness. Kuhn concluded that this phenothiazine analogue imipramine, was without effect in altering the schizophrenic process but that it might have important antidepressant properties.[2] Numerous clinical trials have proved the validity of his observation.

Further, in 1964 it was reported that this drug was able to block the reuptake of norepinephrine in brain and, as reuptake is the principal mode of inactivation of the neurotransmitters dopamine, norepinephrine, and serotonin, it was suggested that such a blockade of reuptake would produce more functional amine at the synapse and that this might explain its antidepressant properties[3] in a way which was consistent with the earlier observations as to the actions of reserpine and the monoamine oxidase inhibitors. These early observations, as well as many other reports suggesting that biogenic amine metabolism and/or disposition within the brain might be integrally related to affective illness, were summarized in a series of three review articles appearing in 1965 and 1966.[4–6] However, as will be noted, these earlier unitary hypotheses relating a single amine to affective illness *per se* are certainly too simplistic. Rather, the present picture suggests that affective disorders are a biochemically heterogenous group of illnesses. In addition it would appear that alterations in two or more brain amine systems may be present in different types of affective disease.

In the decade following 1965 a huge basic and clinical research literature dealing with the biogenic amines and the affective disorders has appeared and has been the subject of several exhaustive reviews.[7–15] Because this area has been so well surveyed the present essay will be quite selective in scope and will be aimed at the difficult task of attempting to synthesize and criticize some of the basic and clinical data relevant to brain NE, DA, and 5-HT systems as they relate to affective illness which have emerged during the past several years. Less attention will be given to 5-HT in that indoleamines and the affective disorders are reviewed in a separate essay in this volume. In the section dealing with basic neuropharmacology prime emphasis will be placed on experimental data relevant to the acute and chronic modes of action of the tricyclic antidepressants. This selective emphasis has been chosen because of the relevance of these data to our understanding of the neurobiological processes which may be altered in the affective disorders. Further, a careful review of this area is needed in order to begin best to understand emerging new data which suggests that affective disease is a biochemically heterogenous illness. Finally, an effort will be made in this essay to emphasize points of congruence as well as areas of uncertainty or disagreement.

B BASIC NEUROPSYCHOPHARMACOLOGICAL STUDIES RELEVANT TO BIOGENIC AMINES AND THE AFFECTIVE DISORDERS

1 The tricyclic antidepressants—acute effects on the blockade of the amine pump

(a) The experimental data

In 1964 Glowinski and Axelrod demonstrated that desipramine and imipramine, but not chlorpromazine, decreased the uptake of intraventricularly administered NE.[3] Subsequent studies, in which a variety of techniques have been used, have in general supported these original observations, but it has also been found that there are differences in the degree to which the various tricyclic antidepressants are able to block the uptake of NE within brain and further that there are differential effects upon specific amines and aminergic systems. Some of the published reports bearing upon this specificity of drug action may be summarized as follows.

Carlsson *et al.* pretreated rats with reserpine to deplete brain stores of biogenic amines and administered a monoamine oxidase inhibitor, nialamide, and then gave L-DOPA and studied the reappearance of catecholamine fluorescence. It was found that the reappearance of fluorescence was blocked in both brain and heart by desipramine and protriptyline and that these effects were restricted to NE fibres in that the drugs did not alter the reappearance of fluorescence in DA fibres. Similar results as to differential drug effects on DA and NE neurons were obtained with the use of an *in vitro* brain slice technique.[16] Glowinski *et al.* also found that while desipramine decreased the uptake of intraventricular administered ^{3}H-NE into several areas of brain, this drug was without effect on DA uptake.[17] Ross and Renyi, however, noted that the differential effects of the tricyclic drugs on blockade of uptake on DA and NE systems was not absolute. For example, desipramine in the incubating media produced a 50% inhibition of uptake of NE in brain slices at a concentration of 3×10^{-8} M and 50% inhibition of uptake of DA into striatal slices at a concentration of 5×10^{-5} M, i.e. a blockade of DA uptake equal to that of NE required an approximate 1000-fold increase in drug concentration. Similar differences in the concentration of imipramine required to produce a 50% inhibition of uptake in noradrenergic and dopaminergic systems were found. These workers also made the interesting discovery that there were marked differences in the concentrations of desipramine and imipramine which were needed to block uptake of NE into cerebral slices by 50%, whereas there was relatively little difference in the quantities of these two drugs required to give a 50% blockade of uptake of DA into striatal slices. They calculated, for example, that a 50% inhibition of uptake of NE by brain slices could be produced by incubating the slices with

3×10^{-7} M imipramine or 3×10^{-8} M desipramine (to produce the same degree of blockade of uptake by pretreatment of the animals it was necessary to give 6 mg of imipramine per kilogram body weight but only 2 mg of desipramine per kilogram body weight). These *in vivo* and *in vitro* biochemical findings were buttressed by physiological data, i.e. the amount of imipramine required to give significant inhibition of a reserpine-produced ptosis was 7 mg (kg body weight)$^{-1}$, and for hypothermia 5 mg (kg body weight)$^{-1}$, whereas desipramine in doses of 0.8 and 0.5 mg (kg body weight)$^{-1}$ respectively produced the same antagonistic effects.[18] Haggendal and Hamberger also produced data which is in essential agreement with the preceding work. They pretreated rats with reserpine and nialamide, prepared slices from cerebral cortex and neostriatum, and examined the uptake of NE in both of these areas with and without desipramine. They demonstrated that the amine pump in the striatum was quite active for NE, i.e. NE was concentrated against a gradient, as it was in cerebral cortex, but that the blockade of the uptake in these two areas by desipramine and chlorpromazine was rather different, i.e. at a concentration of 1×10^{-5} M there was a significant blockade of uptake of NE by cortical slices, but this effect was much reduced in slices of striatum.[19] That these actions of tricyclic drugs on the uptake of NE also occur *in vivo* was established by Sulser *et al.* This group pretreated cats with reserpine and desipramine and, with the use of push–pull cannulae implanted in the hypothalamus, assayed labelled NE and NM in the perfusate. As expected, desipramine produced an increase in both NE and NM in the perfusate.[20]

Bickel and Brodie noted that treatment of animals with a benzequinolizine produced a reserpine-like syndrome and that desipramine blocked all signs of this syndrome and, in higher doses, desipramine was even able to produce a hyperactive animal. Taking this as a model situation, these investigators tested a large number of drugs as well as a number of structurally altered analogues of desipramine in terms of their potency in reversing the benzoquinalizine syndrome. They noted the following. Activity was restricted to compounds having two or three carbons on the side chain whereas compounds with branch chains or chains containing more than four carbons tended to be inactive or toxic. In terms of *N*-substitution it was noted that activity was confined to methyl-substituted or unsubstituted amines, whereas ethyl or higher alkyl groups on the side chain nitrogen resulted in compounds which were either inactive or toxic. A number of ring-substituted compounds were active, i.e. the 3-chloro, 10-methyl, or the 10,11-dimethyl compounds. Changes in the bridge between the two phenyl groups from CH_2-CH_2 to $CH=CH$ did not change activity. Removal of the ring nitrogen and substitution with a carbon made little difference in terms of activity. Given works which will be reviewed later as to the differential effects between tertiary and secondary tricyclic amines on uptake of 5-HT and NE, it is of interest that these workers noted that 'almost all antidepressant compounds are primary and secondary amines . . .' and 'The possibility exists that the antidepressant action of the two active tertiary amines are mediated through the rapid formation of their secondary analogues in the body'.[21]

Maxwell *et al.*[22–25] explored in some detail the molecular features of the tricyclic antidepressants as they may affect the inhibition of the uptake of NE. Although most of the work of this group has been done with rabbit aortic strips, there is one publication which suggests that the general conclusions reached by these workers can be applied to brain.[25] Their data agree with other work, which has been or will be cited, as to the greater potency of desipramine, as compared to imipramine, in blocking the uptake of NE, and they also found that at lower concentrations of NE the inhibition of NE uptake by desipramine departs markedly from linearity. For example, at a concentration of NE of 1×10^{-7} M there was a much more marked inhibition of uptake of NE than at a concentration of 4×10^{-7} M. This deviation from linearity was not observed for imipramine, nortriptyline, or a primary amine derivative. They note that at lower concentrations of NE the potency of desimipramine in blocking NE uptake relative to imipramine may be therefore much greater than the factor of 10 which is usually quoted (q.v.). This group has also made some interesting and important observations about mechanisms by which structural differences may alter the inhibition of uptake. They compared systems in which the bridge between the two phenyl groups was either absent or was formed by sulphur, a $-CH_2-CH_2-$, a $-CH=CH-$, an oxygen, or a bond between the two carbons of the phenyl groups. They found that high potency (in blocking uptake of NE) occurs with tricyclic compounds in which the phenyl groups are held at considerable angles to one another (examples of this would be, imipramine, amitriptyline, phenothiazines, or protriptyline); intermediate potency occurs with tricyclic drugs in which the two phenyl groups are held at slight angles or in which there is no bridge between the diphenyl systems. Tricyclics in which the phenyl rings are coplanar, such as carbazole, are only weakly active. They suggest that if the assumption is made that the receptor site is best fit by the extended confirmation of phenylethylamine, then the presence of two phenyl rings which are not in the same plane will allow the side chain amine group of the drug to be inserted into the receptor site and further than the phenyl ring which is above the plane can occupy a position somewhat analogous to that of the hydroxyl group on the β carbon of NE. Theoretical considerations led to the suggestion that the fit of the secondary amine into the receptor site is a tight one of the lock-and-key type. They calculated the difference in total free energy of binding for desipramine and its primary amine analogue and note that the difference is of the order of -1.4 kcal, which is quite close to the sum of -700 cal which occurs with the transfer of a methyl group from water to a non-aqueous phase, and -600 cal, which is the maximal increment for van der Waals interactions of a methyl group with methylene groups in an enzyme. It is also apparent that since these drugs are not capable of blocking the uptake of DA (q.v.), which lacks a hydroxyl group on the β carbon, the interaction of the raised phenyl ring with another portion of the receptor must be of considerable importance.

In general, the cited studies as a group are consistent in that they indicate that the tricyclic drugs desipramine and imipramine markedly block the uptake of NE by neural tissues and that these effects are either absent or much less marked in DA systems, viz. the concentration of desipramine or imipramine required to give a 50% inhibition of uptake in striatum is 100–1000 times greater than that for cerebral cortex. In addition to the differential effects on DA versus NE neurons there is also agreement that desipramine is a more potent inhibitor of NE uptake than is imipramine, i.e. depending upon the experimental conditions the amount of desipramine required to give the same effect is 3–10 times less than that of imipramine.

Interestingly, in most of these studies the effects of these drugs upon the uptake of 5-HT was ignored. This omission, however, was soon rectified with some interesting results. Blackburn *et al.* using rat brain slices found that at concentrations of 1×10^{-4} and 1×10^{-5} M, imipramine, desipramine, and chlorpromazine were all inhibitory of 5-HT uptake. However, at the 1×10^{-5} M concentration imipramine blocked uptake by 38 and 30%.[26] In partial contradiction to this work, Palaic *et al.*[27] found that desipramine in doses of 10 mg (kg body weight)$^{-1}$, when given intraperitoneally 20 min. before killing, did not affect the uptake of labelled 5-HT in an experimental situation in which the brain was perfused with labelled 5-HT and later assayed for both 5-HT and 5-hydroxyindoleacetic acid (5-HIAA). Other published reports by different groups of investigators gave information relevant to the apparent discrepancy between the work of Blackburn *et al.* and Palaic *et al.* For example, Alpers and Himwich, using slices from rat brainstems, estimated the concentrations of imipramine, amitriptyline, and desipramine required to give a 50% inhibition of 5-HT uptake.[28] The concentrations required for imipramine and amitriptyline were essentially the same, i.e. $3–4.5 \times 10^{-5}$ M, whereas for desipramine the concentration required was 10 times greater, i.e. 3×10^{-4} M. Ross and Renyi[18] and Carlsson *et al.*[29] also found that the tertiary tricyclic amines, imipramine and amitriptyline, were more potent in blocking 5-HT uptake than were their demethylated derivatives desipramine and nortriptyline. For example, the latter authors found that a 50% inhibition of uptake of 5-HT by brain slices occurred at a concentration of imipramine of 6×10^{-7} M whereas the concentration required for desipramine was 4×10^{-6} M. A similar difference, albeit not as marked, was found between amitriptyline and nortriptyline. Both Ross and Renyi[18] and Carlsson *et al.*[29] noted that the difference in the inhibitory potency for the uptake of 5-HT by brain slices was exactly opposite that chlorimipramine was somewhat more potent that imipramine.[30] Glowinski and coworkers have used a somewhat different but important technique to assess the differential effects of the secondary and tertiary tricyclic amines on serotonergic, noradrenergic, and dopaminergic systems. Their approach has been to dissect out specific structures such as the hypothalamus, medulla oblongata, or striatum, to make slice preparations from these areas, and incubate the tissues for 15 min. with labelled tyrosine or tryptophan. Given the short period of incubation, they feel that they may be specifically focusing on the release and reuptake of newly synthesized amines which may be preferentially released during nerve stimulation. The data obtained with this technique are essentially in agreement with the work which has been reviewed above, i.e. desipramine markedly increased the content of labelled NE in the media in which slices of medulla oblongata were incubated, whereas there were no effects on the quantities of labelled DA found under similar conditions with slices of striatum. Similarly, with slices of hypothalamus, imipramine resulted in marked increases in the quantities of labelled 5-HT found in the media. In general, these results were obtained whether the drugs were administered *in vivo* or were added to the incubating media.[31] Schildkraut compared the effects of amitriptyline, nortriptyline, imipramine, and desipramine upon the uptake of NE injected intraventricularly (in rats). The last three drugs, but not amitriptyline, significantly decreased NE uptake, but in all cases there was a decrement in oxidatively deaminated metabolites and an increase in *o*-methylated products.[32]

Given the agreement among the studies cited above as to the differential effects of the tricyclic amines on blockade of uptake of the three amines DA, NE, and 5-HT, a logical inference is that these different drugs might have specific effects on aminergic systems as morphologically defined. As expected, this issue has been explored with the use of histochemical fluorescence techniques. Fuxe and Ungerstedt pretreated experimental animals with reserpine and then gave intraventricular injections of DA, 5-HT, or NE. In some cases prior to the injection of the amines the animals were pretreated with desipramine or imipramine. They found that with the control animals, i.e. those which had been pretreated with reserpine but not with the tricyclic drug, there was a partial to marked increase in fluorescence following injection of DA, NE, or α-methyl-NE in areas which were close to the ventricle. They further found that pretreatment with desipramine or protriptyline prevented the increase in fluorescence in NE terminals but not in DA terminals. This effect was dose dependent. The blockade was greatest in those terminals just beneath the fourth ventricle and in the subarachnoid space of the medulla and pons, but little blockade was observed in NE nerve terminals of the hippocampal formation or septal area, where the concentration of injected amine would have been expected to be high. In contrast to the findings with desipramine or protriptyline they found that pretreatment with imipramine only slightly decreased the return of fluorescence following the injection of the DA or NE. Further, pretreatment with desipramine did not block the accumulation of fluorescence following the injection of 5-HT in any of the areas examined, whereas there were partial effects following pretreatment with imipramine.[33] While these findings agree with and extend those obtained by biochemical or pharmacological approaches, there remains the problem of the unphysiological route of administration of the amines. To avoid this potentially confounding problem Carlsson *et al.*[34] injected rats and mice intraperitoneally with 4-α-dimethyl-*m*-tyramine (H77-77), which causes the depletion of NE and DA in both central and peripheral pools, and by the use of biochemical and histochemical fluorescence techniques examined the effects of imipramine, desipramine, or protriptyline on the drug-induced amine depletions. The H77-77-induced depletion of NE, but not DA, was prevented by pretreatment with desipramine and protriptyline, and this blocking effect appeared to be dose dependent. In contrast, pretreatment with imipramine, amitriptyline, nortriptyline, or chlorimipramine blocked NE nerve terminal depletion by H77-77 only at the highest doses. Data obtained by biochemical analysis of brains for NE and DA content were, in general, in agreement with those obtained with the histochemical fluorescence technique. Carlsson *et al.* also found[35] that an analogue of H77-77, 4-methyl-α-ethyl-*m*-tyramine (H75-12), was capable of causing depletion not only of catecholamines but also of 5-HT stores in brain and, as before, used this agent to examine the action of the tricyclic drugs on serotonergic systems in brain. In these investigations the 5-HT nerve terminals examined were mainly in the mesencephalon and diencephalon, particularly the nucleus suprachiasmaticus. It was found that chlorimipramine and amitryptiline were the most potent drugs in blocking the H75-12-induced amine depletion in 5-HT *nerve* terminals, whereas drugs such as protriptyline, desipramine, and nortriptyline had little blocking activity in the doses studied. The biochemical data was supportive of and consistent with the histochemical fluorescence data.

Lidbrink *et al.*[36] approached the problem as to the effects of the tricyclic drugs on 5-HT and NE uptake by a variety of methods. First they examined the effects of the drugs on uptake of amines by synaptosomal preparations and found that for 5-HT systems the order of potency was chlorimipramine, imipramine, and desipramine with the latter compound having only negligible effects. In contrast, for the uptake of NE, desipramine was the most potent with imipramine and chlorimipramine being about equal. Next these investigators pretreated animals with reserpine and a monoamine oxidase inhibitor and then injected either 5-HT or NE intraventricularly and examined the effects of pretreatment of tricyclic drugs on reappearance of fluorescence in monoamine nerve terminals. Chlorimipramine was quite potent in blocking the reappearance of fluorescence. Amitriptiline was also noted to be effective in the 5-HT preparation. In contrast neither chlorimipramine or amitryptiline had significant effects on the reappearance of the fluorescence following NE injection. Finally these same authors, using 5-hydroxytryptophan (5-HTP), induced increase in the hindlimb extension reflex of animals which had been pretreated with reserpine and nialamide; they found that amitryptiline and chlorimipramine increased reflex activity as compared with controls. In contrast the DOPA-induced increase in the reflex activity was not increased by amitryptiline and only a slight increase at the higher dose was found with chlorimipramine.

Other investigators[37, 38] have also examined the uptake of biogenic amines into synaptosomal preparations as influenced by tricyclic drugs and while for the most part the results are the same as those obtained in other experimental preparations there are some exceptions, viz. in one of their studies, but not in the other, amitryptiline was found to inhibit NE uptake to a greater extent than nortryptiline.

Sheard *et al.*[39] examined the effects of some of the tricyclic antidepressants on the firing rate of 5-HT neurons in the dorsal raphe and found that amitryptiline and imipramine, but not desipramine, decreased firing rates. Further, this effect was abolished by pretreatment with *p*-chlorophenylalamine. Nyback *et al.*[40] examined the neurophysiological effects of the tricyclic class of drugs by recording the firing rate of NE neurons in the locus coeruleus and estimating the dosage of drug required to produce a 50% decrement in firing rate. The order of potency was desipramine > chlordesipramine > imipramine ≃ nortriptyline > amitriptyline > chlorimipramine.

(b) A summary statement

In summary the following may be said as to the acute effects of tricyclic drugs in blocking uptake of NE, DA, and 5-HT by the neuronal membrane pump.

1. There is good agreement among almost all investigators that, except at the highest doses, which are unlikely to be achieved in a clinical situation, the tricyclic drugs are ineffective in blocking the uptake of DA. Unless some other type of action of these drugs upon DA systems can be demonstrated it would appear that this finding is strong evidence against alterations in brain DA systems having an important role in depressive states. (As will be noted lated in this review, however, it does appear that DA systems may be altered in manic states.)
2. In general it appears that demethylation from a tertiary to a secondary tricyclic amine results in a compound which is more potent in blocking NE uptake and less potent in blocking 5-HT uptake. This general finding, however, has led to the *incorrect* generalization that tertiary tricyclic amines as a class block the uptake of 5-HT, whereas secondary tricyclic amines block the uptake of NE. For example, both amitryptiline and imipramine are tertiary

amines but amitryptiline is without effect upon NE systems whereas imipramine in most studies has been found to be a reasonably potent blocker of NE uptake. Similarly nortryptiline, desmethylimipramine, and chlordesmethylimipramine are all secondary tricyclic amines but it would appear that there are 10-fold differences in their abilities to block the uptake of NE, and further there is some possibility that nortryptiline may have actions on 5-HT systems.

3. In addition to the above general comments the following conclusions regarding specific drugs appear warranted from the available data. (i) In almost all studies desmethylimipramine, chlordesmethylimipramine, and protryptiline in concentrations which are likely to be found *in vivo* have been found to be potent blockers of the uptake of NE. These same drugs have little effect upon 5-HT systems. (ii) There is good agreement that amitryptiline and chlorimipramine, at least in doses akin to those given to humans, do not block the uptake of NE but do block the uptake of 5-HT. (iii) Imipramine blocks the uptake of 5-HT and most studies indicate that imipramine also blocks the uptake of NE, but there are some exceptions such as the report of Carlsson.[34] (iv) Nortryptiline has not been as extensively studied as the other drugs but on balance the evidence indicates that this drug blocks NE uptake. The most persuasive argument is the work of Nyback *et al.*[40] Until this report, the data were contradictory. This report indicates that imipramine and nortryptiline are approximately equipotent in their actions on NE systems *in vivo*. Further, it would appear that nortryptiline is less potent than desmethylimipramine or chlordesmethylimipramine in blocking the uptake of norepinephrine.

These effects of the drugs on the blockade of amine uptake are summarized in Table 5.1.

Table 5.1 Summary of effects of various antidepressant or mood altering drugs on blockade of uptake of biogenic amines

Drug	Biogenic amine		
	Serotonin	Norepinephrine	Dopamine
Amitriptyline	++++[a]	0[b]	0
Nortriptyline	++	++	0
Imipramine	+++	++	0
Desipramine	0	++++	0
Dextroamphetamine	0	+++	++++

[a] Indicates most active.

[b] Indicates probable lack of activity *in vivo* at tissue levels clinically achievable.

2 The tricyclic antidepressants—acute effects on synthesis and/or turnover of brain amines

(a) *The experimental data*

The effects of the tricyclic antidepressants on the turnover of DA, NE, and 5-HT have been less well studied, probably because of technical problems, viz. the effects of drug treatment on the uptake of AMPT or the problem of measuring the effects of a drug on the rate of disappearance of a labelled amine whose uptake is also influenced by the same drug. There are some data in this area, however, which are summarized below. In 1966 Neff and Costa published data which indicated that if rats were given protriptyline (29 mg (kg body weight)$^{-1}$) or desipramine (10 mg (kg body weight)$^{-1}$) $\times$ 5 over a period of 3 days, an increase of turnover of brain NE but not DA was produced,[41] and in contrast to the lack of an effect of the tricyclic drugs on DA turnover, chlorpromazine in doses of 5 mg (kg body weight)$^{-1}$ did increase the turnover of DA. Corrodi and Fuxe pretreated experimental animals with imipramine and inhibitors of tyrosine hydroxylase or tryptophan hydroxylase and estimated the rates of disappearance of catecholamines and indoleamines by biochemical and histochemical fluorescence techniques. They found that, even with high doses of imipramine, there were no changes in the rates of depletion of NE or DA, whereas there was a significant slowing of the decrease in the disappearance of 5-HT.[42] This finding as to an effect of imipramine on the turnover of 5-HT and the lack of an effect of NE or DA systems, of course, 'fits' with studies cited earlier which indicate that the tertiary tricyclic amines act primarily on serotonergic systems whereas the secondary amines have a more specific action on noradrenergic systems. In later work Corrodi and Fuxe[43] examined the effects of amitriptyline, chlorimipramine, and nortriptyline on brain 5-HT turnover by assessing the effects of these agents on the depletion of brain 5-HT and the rates of disappearance of fluorescence in serotonergic areas following the administration of tryptophan hydroxylase inhibitor. They found that it was necessary to use extremely high doses to obtain significant slowing of turnover of 5-HT, and even then the results were very modest. (These findings are in contrast to the rather marked effects of some of these drugs in blocking 5-HT uptake as noted in studies cited elsewhere in this paper.) Using a somewhat different technique and approach Meek and Werdinius[44] found that following the administration of chlorimipramine and probenecid there was a decrease in the accumulation of 5-HIAA in brain which is consistent with and supportive of the earlier suggestion that the imipramine-like drugs slow the turnover of 5-HT.

Schubert *et al.*[45] approached the problem of turnover time as influenced by the tricyclic drugs by giving labelled tryptophan or tyrosine either as a single pulse or as an infusion and then measuring the amount of 5-HT, DA, or NE which accumulated during the infusion, or by the amount of labelled amine which was found in brain at some point after the pulse was given. Drugs evaluated were imipramine, desipramine, amitriptyline, and nortriptyline. It was very clear from their data that none of these drugs affected DA accumulation or disappearance. For 5-HT the disappearance was decreased by imipramine and amitriptyline but was unaltered by desipramine or nortriptyline. The accumulation of 5-HT was decreased by imipramine but not by the other drugs. In general these findings may be considered to be consistent with the notion that imipramine slows the turnover of 5-HT whereas desipramine or nortriptyline are without an effect. The accumulation of NE was not increased by any of the drugs, but the disappearance was increased by desipramine and nortriptyline, which would suggest that these two drugs may increase the turnover of this amine. Since only single time points are available, however, these conclusions as to effects of these four agents on turnover must be quite tentative and it can only be firmly concluded that these drugs appear to be without effect on DA systems but do have effects on the quantities of labelled 5-HT and NE found in brain.

In an investigation of the effects of psychoactive drugs on 5-HT metabolism, Schildkraut *et al.*[46] noted that when ^{14}C-5-HT was administered by intracisternal injection and imipramine was subsequently injected intraperitoneally (the animals were killed 2 h. following the injection of the labelled amine), there was a significant increase in levels of ^{14}C-5-HT in those animals which had been treated with imipramine. This finding is consistent with the work noted above which indicates that imipramine results in a slowing of the turnover of brain 5-HT.

Glowinski and coworkers, using a technique which was described earlier,[31] examined the effects of some tricyclic antidepressants on the synthesis of NE, DA, and 5-HT from labelled precursors. They found that the total labelled 5-HT found (media and tissue) was decreased if imipramine was present in the media or if tissue was obtained from animals which had been pretreated with this drug. Desipramine produced marked increases in NE synthesis in slices of medulla but no effects on the synthesis of DA in striatum were noted.

(b) *A summary statement*

In the aggregate, the available data indicate that the tricyclic drugs, whether of the tertiary or secondary amine type, when given acutely do not alter the turnover or disposition of DA. Acute administration of imipramine or amitryptiline produce effects which are consistent with a slowing of the turnover of 5-HT. The tricyclic drugs may alter the turnover of NE but the available data is in conflict and is difficult to interpret clearly. The differential acute effects (or lack thereof) of the tertiary and secondary tricyclic amines on the turnover of DA, and 5-HT are generally consistent with the demonstrated differential blockade of uptake of these two biogenic amines but more work is needed.

3 The tricyclic drugs—acute actions other than those on the membrane pump or turnover

As is indicated by the above, there has been a great deal of attention paid to the effects of tricyclic drugs on the neuronal membrane pump for CNS biogenic amines as well as to drug produced changes in synthesis or turnover. There are other modes of actions of these drugs, however, which are important, and these should be noted. Brodie *et al.*[47] found that desipramine blocked the uptake of small doses of tyramine by rat heart, but if the tyramine was given in doses of 20 mg (kg body weight)$^{-1}$ desipramine did not alter the intracellular concentrations of tyramine but it was able to prevent the depletion of NE by tyramine. Similarly, Leitz[48] showed that while desipramine was able to prevent the efflux of NE from heart slices following treatment with metaraminol this effect was much greater than the blockade of the uptake of metaraminol *per se*. He interpreted this data to indicate that while desipramine blocked the uptake of metaraminol at the neuronal membrane it also blocked the entrance of metaraminol into the granules, and thus was in agreement with the findings of Brodie which suggested that desipramine has an intraneuronal action, i.e. it can block the uptake of amines into NE-containing granules.

Roth and Gillis[49] have found that oxidative deamination of tyramine, 5-HT, and β-phenylethylamine (PEA) by mitochondrial preparations from rabbit lung and brain was inhibited by imipramine. At concentrations of 1×10^{-4} M of imipramine the deamination of PEA, tyramine, and 5-HT was inhibited approximately 70, 45, and 45%. The desmethyl and didesmethyl derivatives of imipramine were equally as effective as imipramine in inhibiting the deamination of PEA. Kodama *et al.*[50] found that imipramine, desipramine, and amitryptiline elevated the level of cyclic AMP (cAMP) and increased the relative formation of cAMP in slices more than fivefold. This effect was maximal at 0.5 mM. In these latter experiments and in those dealing with inhibition of monoamine oxidase the high concentrations which were necessary to achieve significant effects raises questions as to whether or not such mechanisms might be operative *in vivo* in a clinical situation.

Taylor and Randall[51] made the most interesting observation that a single dose of imipramine (10 mg (kg body weight)$^{-1}$) resulted in a 50% depletion in brain *S*-adenosylmethionine (SAMe) but that the values quickly returned to normal within a period of 4–8 h. After chronic treatment with imipramine (20 mg (kg body weight)$^{-1}$ d^{-1} for 3 weeks) the level of SAMe had been depleted to the same extent as occurred 1 h. after a single injection of the drug, but in contrast to the acutely treated animals the brain SAMe levels remained at the level of maximal depletion as long as the mice received chronic imipramine treatment. The authors note that it may be, particularly in the chronic situation, that a decrease in *o*-methylation of NE which would occur as a function of the decrement in SAMe might possibly result in more functional norepinephrine being available and hence would contribute to the neurochemical effects of the antidepressive drugs. They also note, in terms of an issue to be discussed later, that such a mechanism might help to explain the delayed onset of the therapeutic action of these drugs.

Iprindole is a tricyclic type drug which has been reported to have antidepressive effects, although it has subsequently been withdrawn from the market. It is of interest that this drug, however, has not been found to have significant effects in blocking the reuptake of DA, 5-HT, or NE. It has also been found not to alter the turnover of the biogenic amines significantly, nor in the earlier noted study did it produce an alteration in the SAMe levels in brain.[51,52] This lack of an effect of iprindole on biogenic amines, coupled with its antidepressant properties, poses unresolved problems regarding the biogenic amines and affective illness.

Finally the anticholinergic properties of the tricyclic drugs should not be overlooked. It is common clinical experience that patients experience various degrees of blockade of cholinergic function while receiving the tricyclic antidepressant drugs. This blockade is sometimes of sufficient severity to require discontinuation of the drug. Snyder[53] has evidence that most of the tricyclic drugs possess a significant capacity to bind to muscarinic cholinergic receptors within brain. It is of interest that the most potent of these drugs appears to be amitryptiline.

From the above it is apparent that the tricyclic drugs possess actions other than those which might be attributable to blockade of reuptake of amines, but in some cases questions as to the doses which are required to produce the effects suggest that these actions are not operative *in vivo* in human patients. The effects on SAMe in a chronic treatment situation and the anticholinergic properties, however, are most interesting and may have relevance to the clinical efficacy of these drugs. It is interesting to speculate that it may be some combination of effects such as the anticholinergic properties and effects upon the neuronal pump which may have to do with the therapeutic actions of these drugs.

4 The tricyclic drugs—chronic effects or problems in extrapolating from the acute experimental paradigm to the chronic treatment situation

The effects of chronic tricyclic drug administration on the synthesis, metabolism, disposition, and turnover of biogenic amines assumes importance because of the fact that the therapeutic effect of these drugs in most instances is not observed for 2–3 weeks. As such, easy extrapolation of the above acute drug data to therapeutic effects which occur in the chronic situation is not possible, and this issue will be reviewed and discussed in this section. In reviewing the literature on the chronic effects of tricyclic drug administration the first thing one notes is the relative paucity of data, relative to that for acute experiments. In fact, in this particular area, there is probably more data dealing with the pharmacological actions of these drugs which have been obtained from human subjects than from experimental animals. For this reason both the basic and clinical studies relevant to this issue will be reviewed together. Following this review, some technical and theoretical issues having to do with the interpretation of data obtained in the chronic treatment situation will be noted.

(a) Animal studies

Schildkraut[54] examined the effects of imipramine, protryptiline, and desmethylimipramine in both the acute and chronic situations, using the following paradigm. Animals were given either a single injection of one of the drugs or were treated chronically for 3 weeks. In both the acute and chronic situations the control groups received saline injections. At the termination of the experiments the animals were given intracisternal injections of ^{3}H-NE and then were killed at 6 min. or 270 min. after the injection of the labelled NE. It was found that with both the acute and chronic experimental conditions all drugs tested produced a decrement in the uptake of NE, indicating that chronic treatment is also associated with a blockade of the neuronal pump. In addition, by comparing the amount of labelled NE present at 270 min. to that present at 6 min. it was possible to make some relative statements regarding rates of disappearance. In the acute experiments it appeared that all drugs produced a slowing of disappearance of NE, whereas in the chronic situation the disappearance of the labeled NE appeared either not to differ from or to be increased over controls. Finally, in contrast to the findings of other investigators,[55] it was noted that chronic treatment with the noted drugs produced a 10–15% decrement in endogenous NE present in brain.

In another approach to the problem Rofler-Torlov[57] found that long-term treatment with desmethylimipramine enhanced the release of rat-brain NE, DA, and 5-HT after reserpine. In agreement with Schildkraut she also found that chronic treatment with desmethylimipramine produced a significant decrement in the endogenous NE content of brain.

Taking earlier data from the periphery suggesting the presence of transynaptic induction, Mandell and associates examined the effects of chronic administration of a tricyclic antidepressant, desmethylimipramine, on tyrosine hydroxylase activities in the locus coeruleus and in the hippocampal cortex. They found that after 8 days of drug administration the tyrosine hydroxylase activity in the locus is decreased by about 30% whereas in the hippocampal cortex it is reduced by 50%. They suggest that this change in enzyme activity may occur as a consequence of an alteration in impulse flow in noradrenergic neurons produced via a neuronal feedback system as a consequence of drug administration.[56]

As has been previously noted, it has also been found that chronic treatment of mice with imipramine produces a 50% decrement in brain levels of SAMe which persists as long as the drug is administered. This finding has led to the suggestion that this long-term effect may result in more functional NE being available.

(b) Human studies

In several studies[58–60] it has been found that patients chronically treated with imipramine excrete more normetanephrine (NM) and less vanillylmandelic acid (4-hydroxy-3-methoxymandelic acid) (VMA) into urine relative to pre-drug periods. In general, the increase in NM excretion is not equal to the decrement for VMA excretion. These findings, in agreement with the data obtained in animal experiments, suggest that chronic drug treatment with imipramine may result in a decreased synthesis of NE. Urinary VMA and NM, however, probably reflect peripheral catecholamine metabolism and the relationship of such changes to events in the central nervous system are unclear.[61,62] It was found in one study,[60] but not another,[63] that changes in urinary 3-methoxy-4-hydroxyphenethyleneglycol (MHPG) and NM following imipramine treatment may differ as a function of whether or not the patient shows a therapeutic response to the drug. Changes in urinary NM did occur as a function of treatment response whereas the decrements in VMA occurred without regard to treatment outcome.

Bowers *et al.*[64] found that patients treated with amitryptiline have significant reductions in cerebrospinal fluid (CSF) 5-hydroxyindoleacetic acid (5-HIAA), whereas there were no significant differences in homovanillic acid (4-hydroxy-3-methoxylphenylacetic acid) (HVA) concentration between baseline and treatment periods. Papeschi and McClure[65] also found significant changes in CSF 5-HIAA but not HVA following 2 weeks of imipramine treatment. Asburg *et al.*[66] found that 5-HIAA as well as indoleacetic acid levels in the CSF of depressed patients were decreased as a function of chronic treatment with nortryptiline. Goodwin and coworkers examined the effects of either imipramine or amitryptiline in depressed patients on CSF 5-HIAA, HVA, and MHPG before and after the administration of probenecid. It was found that treatment with either of these tricyclic drugs produced a significant decrement in 5-HIAA accumulation following probenecid administration, whereas the probenecid-produced increments in HVA were not altered, suggesting that the changes in 5-HIAA were unlikely to be due to nonspecific transport effects or reduction in probenecid effectiveness during imipramine or amitryptiline administration. It was of interest that the changes observed in 5-HIAA accumulation did not appear to be related to treatment outcome. In contrast to the effects on 5-HIAA neither amitryptiline or imipramine when given chronically to patients produced significant alterations in CSF MHPG. However, in agreement with earlier findings as to urinary MHPG,[60] it was found that those patients who improved on the tricyclic drugs had slight increases in CSF MHPG whereas those who did not improve had significant decreases.[67]

Bertilsson[68] examined the effects of chlorimipramine and nortryptiline on 5-HIAA and MHPG in the CSF of 18 depressed patients. They found that the MHPG levels in CSF decreased significantly with either drug while a change in 5-HIAA did not occur with nortryptiline treatment. This finding regarding 5-HIAA is in disagreement with the earlier noted work of Asburg.[64] However, the potentially confounding effect of baseline values should be noted. For example Asburg noted that if one looked at the patients in terms of those having low versus high CSF 5-HIAA levels, for the seven patients who had pretreatment 5-HIAA levels below 50 mg (ml CSF)$^{-1}$ the 5-HIAA concentration increased in five patients and decreased in two. In contrast, for the 13 patients with an initial 5-HIAA concentration higher than 50 mg (ml CSF)$^{-1}$, the 5-HIAA levels decreased in all.

(c) A summary statement regarding chronic treatment effects of the tricyclic antidepressants

1. Perhaps the clearest point of agreement among different groups is that chronic treatment with the tricyclic drugs seems to be without effect on CSF HVA and hence by inference on the metabolism and/or turnover of DA.
2. Most animal and human investigations would appear to agree that chronic treatment with amitryptiline or imipramine is associated with either a reduction in baseline 5-HIAA levels and/or that the rate of accumulation of this serotonin metabolite is decreased following probenecid suggesting that the synthesis of 5-HT is slowed.
3. The situation with NE is less clear in that the animal data suggests that chronic treatment may be associated with a small ($\sim$10–15%) decrement in production of endogenous NE without an alteration in turnover time. In one instance,[68] data obtained from human studies indicates that there is a decrement in NE synthesis with chronic treatment with tricyclic drugs. In other studies in which other antidepressant drugs were used no evidence for a decrement in synthesis for the total patient group was found.[60,67]

(d) Some problems in evaluating chronic tricyclic treatment, changes in amine systems, and therapeutic outcome

In evaluating data obtained from these chronic experiments the following should be noted. In those situations where tertiary tricyclic amines have been administered chronically, viz. amitryptiline, chlorimipramine, or imipramine, it is certain that a significant fraction of these tertiary amines have been metabolically converted, *in vivo*, to the secondary amines nortryptiline, chlordesipramine, or desipramine respectively, and these products have different effects on amine systems than do the parent compounds (see Table 5.1). In addition to there being possible large individual differences in the rates at which these conversions occur it has also been found that the ratio of the tertiary to secondary amine differs between classes of drugs. For example, it would appear that a much larger fraction of imipramine is converted to desmethylimipramine than is the case for the conversion of amitryptiline to nortryptiline.[69, 70] As a consequence, in those studies in which the tertiary amine was given and changes in amine metabolites observed one may infer but not know with certainty whether the parent compound, its metabolite, or both were responsible.

Further, however difficult it may make the experimental situation, the variable of treatment response should not be ignored when assessing the effects of drugs on amine systems. Perhaps 30% of depressed patients will not respond to a given drug, a third have modest responses, and only about one third have marked improvement. The available data suggest that in some cases the effect that a drug may have on a given amine system is a function of both the treatment response of the subject and the pretreatment functioning of the particular system. For example, there is data as noted to suggest that depressed patients who have decrements in urinary or CSF MHPG will have different *pharmacological and treatment* responses than subjects having normal MHPG levels when both groups are treated with the same drug. This issue is of importance in terms of data analysis and because it raises the possibility that a further specification of the biochemical lesion(s) which may underlie some affective illnesses cannot be completely detailed via studies on experimental animals or subjects which do not have this lesion. In this regard Mandell[56] notes a series of studies from his laboratory suggesting that there may be neurobiological constraints on the limits within which mood may vary. He then applies this data directly to speculations as to the biological genesis of depression in man. However, it should be remembered that depressed patients, in particular patients with manic and depressive episodes, in fact have an illness which is characterized by a lack of constraints on mood swings. It is entirely conceivable that these patients will have a different response to drug treatment than will normal subjects or normal rats.

Finally, in a situation in which a drug produces more functionally available neurotransmitter at a synapse, alters metabolism, changes turnover time, and reduces synthesis there must be a question as to what the net overall effect is. For example, in the noradrenergic neuron it would appear that the chronic administration of desmethylimipramine may reduce tyrosine hydroxylase levels in the locus coeruleus by 30% and endogenous levels of NE in whole brain by 10–15%. At the same time, significant effects on metabolism and reuptake processes are present in the chronic situation. The question then arises, 'Is the amount of neurotransmitter which is released during neuronal firing more or less as a consequence of drug administration?' The answer to this question is of course unknown.

5 Lithium and electroshock therapy (EST)

Investigation of the actions of these therapeutic agents, i.e. Li^+ and EST, has been, and continues to be, an active area. Both have been the subjects of symposia and reviews and will not be covered in any depth here. The reader who is interested in the details of these studies should consult references 71 and 72. The following few comments, however, which are relevant to these forms of treatment, biogenic amines, and the affective disorders will be made.

Lithium at reasonably low doses, somewhat akin to those that might be expected to be used in the clinical treatment situation, is able to promote increased turnover of both 5-HT and NE and yet produce a decrement in the release of both of these amines.[72,73] There is also some evidence that Li^+ inhibits the activating effect of NE upon adenyl cyclase in brain homogenates.[74] While these modes of action can be considered as generally consistent with the postulated association between mania and NE or 5-HT, it must also be noted that the magnitude of these effects are disappointingly small. For example, at concentrations of 2 mequiv in media, the decrement in release of either 5-HT or NE from brain slices is of the order of 15%. A similar small percentage change is seen with the inhibition of activation of adenyl cyclase. It might be argued, of course, that the overall magnitude of the effect may not be so important as action at critical sites, and while this may be true, caution as to a direct extrapolation of these findings with Li^+ to amine hypotheses of affective disorders is in order. Similarly, the role of EST in altering amine metabolism is somewhat problematic in that the therapeutic efficacy of this form of treatment contrasts with the magnitude of its effects upon brain amine systems.[72,75]

In summary, while the known modes of action as well as the magnitude of the effects of lithium and EST on amine systems in no way negate a role for biogenic amines in the affective disorders, research data obtained with these agents serve as warning signs that a full appreciation of the neurobiological processes which underlie the affective disorders will probably include amine systems, and neuronal functions other than those involving NE and 5-HT.

C CLINICAL STUDIES RELEVANT TO CATECHOLAMINES AND THE AFFECTIVE DISORDERS

1 Enzymes and the affective disorders

(a) *Catechol-*o*-methyl transferase*

In 1970 Cohn *et al*. reported that women having primary affective disorders (depressed type) had significantly lower catechol-*o*-methyl transferase (COMT) activity in red blood cells than did a comparison group of women. In contrast depressed males categorized as having primary affective disorders did not differ significantly from comparison subjects. It was also found that female and male schizophrenic patients had values which were similar to those found for comparison subjects. In addition it was noted that histamine-*N*-methyl-transferase was elevated in the depressed women in comparison to the normal subjects, whereas no differences were found for males. The authors also reported that clinical improvement in the depressed patients did not change the activity of COMT and further that values obtained during a manic episode were not different from those found during depression.[76]

In another study done by Dunner *et al.*[77] it was again found that the activity of red blood cell COMT was significantly lower in women with primary affective disorders (depressed type) as compared to control women, schizophrenic women, or men with primary affective disorders (depressed type). Further, when the depressed women were divided into unipolar and bipolar groups it was found that the unipolar women had significantly lower COMT activity than did the bipolar women. Finally, in this report it was noted that 10 patients had been followed through an episode of illness and into the recovery phase and that COMT activity varied relatively little over time.

In support of these earlier findings Briggs and Briggs reported that women with depressions had mean values of COMT which were significantly lower than those found for normal subjects, whereas COMT activity was normal in both men and women with schizophrenia and in men with depression. They also noted that they were unable to find any effects of age or stage of menstrual cycle. However, they did find significant effects of hormones, i.e. COMT was significantly reduced in women taking oral contraceptives and also in a group of 12 women who had received a 21 day course of oral oestrogens and another smaller group of women who had taken a 21 day course of oral progestagin. Finally they noted significant reductions in COMT in the second and third trimesters of pregnancy.[78]

In contrast to these earlier reports, Gershon and Jonas found that patients with primary affective disorders had higher red blood cell COMT activity than did normal controls even after controlling for differences between men and women. Further, no differences between unipolar and bipolar affective disorder-type patients were found.[79]

In addition to the above data suggesting the COMT activity is a relatively constant characteristic of an individual, it has been noted that there is a good correlation between siblings as to COMT levels,[80] and Gershon and Jonas[79] have reported that there is a positive correlation for COMT between relatives and that within families increased COMT activity distinguishes healthy relatives from patients and ill relatives. These authors suggest that COMT levels may provide a genetic marker for vulnerability to affective illness.

(b) Monoamine oxidase

Robinson *et al.* reported that MAO activity in plasma, platelets, and human hindbrain increased with increasing age. In each of the tissues studied the correlations were significant. Further they noted that as a group women had significantly higher platelet and plasma MAO activity than did men.[81] The same group of investigators also found that with age there was a significant decrease in NE, little change in 5-HT, and an increase in 5-HIAA after the age of 65 in human hindbrain.[82] These findings were of interest not only because of the suggested relationships between biogenic amines and depression but also because of the well-known correlations between increasing age and vulnerability to depression, as well as the increased incidence of depression in female versus male subjects.

In addition to these sex and age factors which may govern MAO activity it has also been reported that levels of this enzyme may be under genetic control.[83,84]

Finally, there is persuasive evidence that in addition to the genetic factor MAO activity varies as a function of the menstrual cycle.[85,86] There is some discrepancy in these data, however, in that one group of investigators have found relationships between menstrual cycle, oestrogen therapy, and plasma MAO activities in human subjects, whereas another group which studied Rhesus monkeys under strictly controlled conditions found similar relationships for platelet MAO activity but not the plasma enzyme. Again, as with the age and sex determinants of MAO, these observed relationships between menstrual cycle and MAO activity remind one of the well-known observation that mood shifts occur with great frequency as a function of the time of the menstrual cycle.

Results obtained from clinical studies are discrepant, but do suggest that there may be some relationship between MAO activity and depression. For example, Nies *et al.*[87] found that both platelet and plasma monoamine oxidase levels were significantly higher in depressed patients as compared to normals at each decade. Klaiber *et al.*[88] also found that patients having endogenous depressions had levels of plasma MAO activity which were significantly higher than those of nondepressed women. In contrast to the work of Nies *et al.*,[87] Murphy and Weiss[89] found that bipolar depressed patients had significantly lower platelet monoamine oxidase activity than did comparison subjects. Further, they noted that the nonbipolar depressed patients had platelet enzyme activities which were not significantly different from those of normal subjects. They also found that when four bipolar patients were studied through a series of depressive and manic episodes there did not appear to be a consistent direction of change in enzyme activity.

(c) A summary statement

1. COMT activity seems to be reasonably constant over time, i.e. an individual who has a low or high COMT level at one point in time, if studied again several months later, is likely to have again a similar value. The COMT activity also seems to be unaffected by a patient's clinical status. For example, it would appear from the reports noted above that, if a patient has a low COMT value while depressed, the same patient will have a low value whether in a manic or euthymic phase. Presently available evidence indicates that the COMT levels for any given individual are heritable and it has been suggested that the amount of enzyme present may serve as a marker for vulnerability to affective illness.[79]
2. The available data is discrepant as to issue of high versus low amounts of COMT activity in depressed subjects. It would appear that estrogens and oral progestagins can influence enzyme activity, but the mechanisms by which this occurs are not certain.
3. It seems likely that MAO activity in plasma platelets changes as a function of age, sex, and genetic factors. In addition, within a given individual there are changes which occur as a function of the subject's position in her menstrual cycle.
4. The relationship of MAO activity to depressive states is less clear and the available findings are discrepant. These discrepancies may have occurred because of differing diagnostic criteria or because of technical differences in the assays as they were performed in different laboratories. For example the possible contamination of plasma with platelets may be a confounding variable. Also, different results may occur as a function of the use of assays which utilize different substrates. Until some of these issues have been dealt with experimentally, the question as to whether or not there are indeed changes in monoamine oxidase activity in depressed patients relative to comparison subjects, and whether these changes are specific to one or more types of depression, must remain open.

2 L-DOPA, depression, and mania

(a) A survey of clinical studies

Given the suggested role of the catecholamines in the genesis of the affective disorders a number of investigating groups became interested in giving L-dihydroxyphenylalanine (L-DOPA), which can enter brain and serve as a precursor to DA and NE. It was originally hoped that L-DOPA administration might therefore produce an amelioration in the patient's depressed state. In the earliest of such studies no effects on the patient's mood were noted, but the doses of L-DOPA given to patients were small and/or were given for only brief periods.[90, 91] Because of subsequent work which indicated that large doses of L-DOPA were necessary to reverse a Parkinsonism syndrome where a DA deficiency had been demonstrated, the issue was reinvestigated by Goodwin *et al.* by the administration of large doses of L-DOPA with and without an inhibitor of DOPA decarboxylase, which acts peripherally but not centrally.[92] In addition to this study, there are other studies[93, 94] in which doses of L-DOPA were employed which were intermediate between the small doses given in the initial studies and those used in the experiments done by Goodwin *et al.* These previous studies have been critically reviewed by Mendels *et al.*[95] with particular emphasis upon numbers of subjects, types of design, and L-DOPA dosage. In addition to this survey, Mendels *et al.* also experimentally administered large doses of L-DOPA (up to 8 g d^{-1}) on a double blind basis to six depressed patients, half of whom were bipolar and half of whom were unipolar.

In general, the results of these studies with L-DOPA have varied from findings of no change in the clinical state to an improvement in the depression sufficient to warrant discharge in ~20% of patients. Goodwin's report[92] is of particular interest in that there were 21 patients in this study, the design was double blind with placebo substitution, there were both bipolar and unipolar patients within the group, and finally, prior to being given the L-DOPA or placebo, patients were classified as retarded or agitated and the response of the patient was also evaluated in terms of this pretreatment dimension. They found that of the 21 depressed patients, six showed a consistent improvement in depression and in seven of the eight trials the responders relapsed following placebo substitution. It was of interest that none of the patients who responded had been classified as agitated prior to initiation of treatment, and all had been classified as retarded. (This is not to say that all retarded patients responded, in that 14 patients had been classified as retarded, and of these 14 only six were considered to have shown a therapeutic response.) The authors suggest that in those patients who did respond the level of psychomotor activation might have been more central to the depressive illness than in those who did not respond. They further speculated that it may be that in some of these patients an increase in activity might have been sufficient to trigger off the subsequent overall improvement.

Another interesting finding emerged from this NIMH group of investigators and was published by Murphy *et al.*[96] It was noted that in six patients the L-DOPA administration produced a hypomania or manic state characterized by increased motor activity, pressured speech, increased social involvement, intrusiveness, etc., without euphoria. Further, of these six subjects who developed the hypomanic or manic states, five had been previously diagnosed as bipolar type 1, i.e. they had had a documented history of manic episodes.

(b) Summary and critique

In evaluating the above data the following points should be noted.

1. L-DOPA when given in small doses is probably converted to DA, which is sequestered about small blood vessels and hence is essentially outside the central nervous system. When given in larger doses it would appear that a real elevation of brain DA above normal does occur. This had been found both with normal experimental animals, experimental animals depleted of DA by various means,[97] in patients having Parkinson's disease who were being treated with L-DOPA,[98] and in depressed patients treated with L-DOPA.[92] In contrast, even with large doses of L-DOPA it would appear that, in normal animals and in patients with Parkinson's disease who were being treated with L-DOPA, increases in brain NE did not occur.[97,98] (If animals are experimentally depleted of brain NE and then given L-DOPA there is an increase of NE levels relative to those which are normally found.[99]) It should be remembered, however, that change in endogenous content of an amine does not necessarily mean that a functional change in the neurotransmitter has been produced *per se*. Further, when large doses of L-DOPA are given this amine may be taken up by neurons which do not ordinarily contain DOPA, DA, or NE and displace normal amine transmitters such as 5-HT.
2. Despite the caveats, given the above data, it seems reasonable to assume that L-DOPA administration does increase pools of brain DA and, given the general failure of L-DOPA to alleviate depression, it may be tentatively concluded that brain DA is not integrally related to depressed mood. On the other hand, it would appear that brain DA does have a role in modulating one aspect of the depressed state, i.e. psychomotor activity. Further, in predisposed individuals, i.e. those with bipolar-type illnesses, L-DOPA may trigger off a manic episode, presumably as a consequence of the increase in brain DA.
3. The role for NE in depression and mania as deduced from the studies with L-DOPA is less clear. If one takes at face value the findings from patients with Parkinson's disease that CSF MHPG levels are not increased following administration of large amounts of L-DOPA, one might tentatively suggest that alterations in NE systems are not central to manic episodes, whereas the role of NE in depressed mood has not been adequately examined via the L-DOPA-type studies.
4. From a clinical standpoint most investigators who have worked with L-DOPA have concluded that this amino acid is not particularly effective in the treatment of depression.

3 CSF amine metabolites in the affective disorders

(a) A summary statement as to origins

Before reviewing some of the literature dealing with alteration in metabolites in CSF in patients having affective illness, brief comments should be made regarding the origin of such metabolites in that this information is rather crucial to the interpretation of the published data. HVA is a major metabolite of DA. Concentrations of this acid within the ventricular system reflect the concentrations of the metabolite in those structures which are particularly rich in DA.[100] Further, all of the available data would suggest that there is a ventriculo-cisternal-lumbar space gradient for HVA.[101] The level of HVA in lumbar CSF is not altered by transections of the cord but is decreased by blocks of the flow of CSF which occur above the lumbar space.[101-103] Finally it would appear that HVA in plasma or blood will not penetrate to any appreciable extent into CSF.[100] In the aggregate these data indicate that assays of HVA in lumbar CSF will provide some reflection of HVA production in higher central nervous system structures. It is of course unclear, because of the time gradient, the degree to which transient changes in functional activity and metabolite production in higher centres will be reflected in CSF obtained from the lumbar space. In addition to these data which indicate that essays of HVA and CSF may reflect metabolism of DA in brain structures, it is also established that administration of probenecid blocks the efflux of HVA from CSF and, from the accumulation of HVA, rough calculations as to the rate of production of HVA may be made.[111]

In contrast to the situation with HVA, the data regarding lumbar CSF MHPG being representative of MHPG in higher centres is less convincing. Since it would appear that MHPG in the periphery will not penetrate into CSF,[104,105] and since MHPG is present in the CSF, it appears that MHPG in the lumbar space is of central origin. However, it has also been shown that that section of the cord produces a marked decrement in lumbar CSF MHPG, whereas blockade of CSF flow above the lumbar space is not associated with any change in CSF MHPG, indicating that CSF MHPG has its origins within the cord rather than in higher centres.[103] This is of particular importance in that it has been repetitively demonstrated that the rates of metabolism, and consequently tissue levels of amine metabolites, change markedly as a consequence of shifts in functional activities of amine neuronal systems. Shifts in functional activities of noradrenergic neurons in brain will probably not be reflected in CSF MHPG obtained from the lumbar space given the above data. Further, it would appear that although significant increases in MHPG may be produced by probenecid these are modest in comparison with HVA and 5-HIAA. As such, the use of this technique for estimating turnover of NE or rates of accumulation of MHPG is questionable. Since it is such functional activity which is postulated to be involved in affective illness the use of CSF MHPG assays in studies of affective illness may not give the desired information.

(b) Clinical studies

Denker *et al.*[106] studied HVA levels in patients who were both depressed and manic. Their primary emphasis in this initial study was on 5-HIAA and the number, *N*, of patients with effective disease whose CSF was assayed for HVA was relatively small. They found that the range for HVA for healthy volunteers was approximately 0.02–0.05 μg ml,$^{-1}$ whereas in five patients with early depression the HVA values appeared to be low in two and in the range of 0.03–0.05 μg ml^{-1} in the remaining three subjects. In the earlier stages of mania there appeared to be no marked differences in HVA levels. The authors concluded in this study that the CSF 5-HIAA levels were more closely related to mental changes than were those of HVA. In another study from this group, levels of HVA before and after probenecid were examined in six depressed patients, seven manic patients, and seven control patients and/or volunteers. Although the number of patients was small, inspection of the data suggests that there was relatively little difference in baseline levels of HVA between groups, but following probenecid both the manic patients and the subjects in the comparison groups had an approximate doubling of HVA whereas there was only a very modest change in HVA in the depressed patients.[107]

In another study from Göteborg, comparisons of HVA levels in lumbar CSF were noted before and after electroshock treatment in a group of 20 patients with endogenous depressions. In this study no comparison groups were available, but the authors did note that if they compared the values of HVA obtained in this study with those previously obtained from normal subjects than the value of HVA appeared somewhat low. There appeared to be little change in CSF HVA after EST despite clinical improvement.[105]

Bowers *et al.*[64] examined levels of HVA in lumbar CSF in eight depressives, six schizophrenics, and seven manics both before and after treatment. (The depressed patients were treated with amitryptiline, the manic patients with lithium, and the schizophrenics with phenothiazines equivalent to 600 mg of chlorpromiazine per day.) The depressive subjects had significantly lower CSF HVA than did the schizophrenics during treatment, but it should be noted that the schizophrenics had received some brief phenoziazine treatment prior to admission. It is of interest that in none of the treatment groups was there a significant change in lumbar CSF HVA as a consequence of treatment. In another study dealing with this problem, Bowers[109] examined the levels of HVA following probenecid in unipolar depressives treated with amitryptiline. For this study a group of inmate comparison subjects was used. In contrast to other studies it was found that following probenecid there were significantly higher levels of HVA in the depressed subjects than in the comparison sample. During amitryptiline treatment there tended to be a decrease in HVA, but this was not found in all subjects.

Papeschi and McClure[65] studied 18 patients having diagnoses of endogenous depression and used as their comparison groups 18 subjects from the Montreal General Hospital who had various neurological diseases other than extra-pyramidal disease or epilepsy. Patients and comparison subjects were free of medication which were likely to have interfered with HVA levels in CSF. These investigators found a rather marked and statistically significant decrement of HVA in the depressed subjects relative to the control patients. Finally, it was noted that HVA did not change significantly after 2 weeks of treatment with imipramine. They note that their findings as to decreased HVA were unlikely to be a function of decreased physical activity because of the lack of correlation between the amount of psychomotor activity and the decrease of HVA and because both the depressed and control patients were kept in bed for 12 h. prior to the lumbar puncture. Negative correlations between a variety of mood or behavioural parameters and HVA levels were searched for but not found, and the authors note in the discussion, 'This finding will strengthen the view that other neurotransmitters, and not dopamine, are involved in the control of mood and of higher psychic activities in human beings, whereas it leaves open the question of which symptom results from a dopamine deficiency in depression.'

Two other groups examined the relationship of HVA to the diagnosis of bipolar versus unipolar illness. Mendels *et al.*[110] compared six subjects having manic depressive or bipolar illness, four subjects having a manic episode, and six subjects having unipolar depressions. It would appear from their data that there are no significant differences in the HVA levels between the unipolar and bipolar depressives. Two of the manic patients had elevated levels of HVA, but these were clearly within the range that was found for the depressive group. A normal comparison group was not examined, but the values as noted for both the bipolar and unipolar patients groups were lower than results obtained from a neurological patient population by other investigators using the same general technique.

Ashcroft *et al.*[112] studied 31 psychiatric patients and 31 neurological patients. The psychiatric patients were diagnosed as having a unipolar or bipolar affective disorder (either depression or hypomania) of sufficient severity to warrant hospital admission. It was found that the unipolar depressives had significantly less HVA in CSF than did the bipolar depressives, the bipolar manic patients, or the normal subjects. Further, after treatment the HVA levels in the unipolar group did not change appreciably whereas in the bipolar depressed group there was a decrement in all cases. The relationship between the type of treatment for a specific patient, degree of improvement, etc., was not specified and cannot be evaluated. Wilk *et al.* noted that the CSF HVA levels had a statistically nonsignificant tendency to be low in depressed patients as compared to normal subjects.[113] Subrahmanyan[114] found no differences between CSF HVA levels in manic-depressives (? depressed) and normal or schizophrenic subjects.

The number of reports dealing with CSF MHPG levels in psychiatric patients and normal subjects is smaller than that for HVA, but no less conflicted. Two groups find that CSF MHPG levels in depressed patients are significantly lower than in comparison subjects,[114,115] whereas two other groups find that they are the same.[116,117] As was noted earlier[67] it has been reported that in those patients who did not respond to treatment there were decrements in CSF MHPG, whereas in those subjects who did respond the CSF MHPG either increased slightly or did not change.

(c) A summary and critique of studies of CSF amine metabolites in patients with affective illness

The available data regarding HVA and MHPG in CSF in normal and depressed subjects is discrepant. The points about which there is agreement mostly have to do with issues other than psychopathology, and as such a true summary or synthesis relevant to this essay is not possible. However, the following statements and impressions may be noted.

1. Both CSF HVA and MHPG seem to have their origins in the central nervous system. Most, if not all, of the HVA in lumbar space CSF is derived from DA-containing brain structures. Lumbar space CSF MHPG probably has its principle origins in the spinal cord and not the brain. Because of the origin of lumbar CSF MHPG, this metabolite (in the lumbar space) probably does not reflect the functional activity of brain NE neurons. Because of the ventriculo-lumbar gradient for HVA, this metabolite may or may not reflect the functional activity of brain DA neurons.
2. There is a nonsignificant *tendency* in many studies for CSF HVA to be low in depressed patients versus normal subjects. Identifying characteristics of those patients who have the low HVA levels have not been found. CSF HVA is not clearly related to mood, nor do the levels seem to change when patients are treated with antidepressant drugs (with or without significant improvement).
3. Some investigators have found lower than normal amounts of CSF MHPG in patients with depression, whereas others have not. The number of studies thus far published are too few to make definite statements regarding differences in CSF MHPG as a function of diagnosis or state variables. In one study, however, which was cited earlier,[67] it was found that those patients who did not respond to drug treatment had decrements in CSF MHPG, whereas those patients who did respond to treatment had essentially no change or small increments in this metabolite.

4. In those studies dealing with CSF HVA and MHPG there are methodological issues which are important. Some investigators keep patients at bed rest prior to the lumbar puncture whereas others do not; diagnostic classifications and criteria differ; some investigators use probenecid whereas others do not; and activity, a variable which is notably difficult to measure, has not been quantitatively assessed in most studies.

4 Urinary MHPG excretion in relationship to affective illness

(a) Comments as to the origins of urinary catecholamine metabolites

The avilable data indicate that there is a barrier to the movement of NE and NM into and out of the brain[61,62,105,118] and that in man essentially all of the urinary NE and NM can be accounted for by the disposition of circulating NE.[119,120–122] VMA is a minor metabolite of NE in brain,[61,123,124,] and essentially all of the urinary VMA is due to catecholamine (CA) metabolism outside the central nervous system (CNS).[61,62] MHPG is the major metabolite of CNS NE[62,123–126] and it has been shown that MHPG in brain increases when impulse flow is increased.[127,128] In the dog approximately 25% of urinary MHPG is derived from CNS NE metabolism.[62] In the rat the estimates as to the urinary MHPG derived from the CNS vary from a few percent to 30%.[129–131] In the monkey *(Macaca arctoides)* 35–65% of urinary MHPG originates in brain.[132] In man it has been found that a large fraction (~75%) of urinary MHPG is derived from CA pools which, relative to their size and/or rates of synthesis, are poorly penetrated by circulating NE.[122] One such possible pool is the brain. Other less direct evidence also suggests that in man urinary MHPG may serve, directly or indirectly, as an index of CNS NE metabolism, i.e. in one study it was shown that there is a relationship between urinary MHPG and human growth hormone.[133]

In the aggregate these data suggest that urinary epinephrine, metanephrine, NE, NM, and VMA will reflect catecholamine metabolism in tissues other than the CNS. Urinary MHPG may give some index of CNS NE metabolism, but it is not known with certainty, in man, how much urinary MHPG is derived from central pools.

(b) A survey of clinical studies

In 1968 it was reported that a heterogeneous group of depressed patients excreted significantly less MHPG into urine than did a healthy comparison group, whereas the quantities of urinary NM and metanephrine (M) excreted by these two groups were similar.[134] Subsequently, five separate groups have investigated the excretion of MHPG by cycling manic-depressive patients; in four of these studies it has been found that the patients excreted less MHPG during periods of depression than they did during periods of either euthymia or hypomania.[135–139] In two of these reports it was suggested that the increments in urinary MHPG preceded the shift in the behavioural state of patients. It has been noted that recovered, unipolar, depressed patients excrete greater amounts of MHPG relative to the period of depression.[117] It has also been found that patients taking amphetamines who were clinically hypomanic excreted greater than normal amounts of MHPG, but that the withdrawal of the amphetamines and the development of a depression urinary MHPG was decreased. Subsequently there was a lifting of mood and an increase in MHPG.[140]

In another study, in which 68 patients were compared with 40 comparison subjects, it was found that the quantities of urinary MHPG excreted by the patient groups were significantly less than that by the healthy subjects, whereas the quantities of NM, M, and VMA excreted by the two groups were similar. Furthermore, it was noted that the differences between the patient and comparison groups could not be explained in terms of age, body weight, creatinine excretion, or urine volume.[141–143]

One report has appeared which indicates that decrements in urinary MHPG may be a function of activity,[144] but it has also been found that strenuous exercise by healthy subjects which is sufficient to produce a 3–10-fold increase in urinary NE does not alter MHPG.[145] Further, no relationships between agitation or retardation in depressed patients and MHPG excretion have been found.[143,146] There are two reports, however, which indicate that stressful situations in healthy and depressed subjects may result in increases in urinary MHPG, but here there are also increases in other catecholamine metabolits.[147,148] It seems clear that further studies, with better measures of state variables as to activity–stress–mood interactions as they may relate to urinary MHPG excretion in healthy and depressed subjects are needed.

These studies in the aggregate indicate that depressed patients *as a group* excrete significantly less MHPG than do healthy subjects, but inspection of the means, variances, and levels of statistical significance leads one to conclude that many depressed patients excrete normal or even greater than normal quantities of MHPG. It would thus appear that while some depressed patients excrete less than normal quantities of MHPG, this is by no means true for every patient. These findings raise the possibility that those patients who excrete less than normal amounts of MHPG may represent a particular subgroup which has special clinical, biochemical, and pharmacological characteristics.

Diagonostically, four separate groups have reported data suggesting that it is the bipolar depressed patients who are particularly likely to have low urinary MHPG values.[114,142,143,149,150] In another study it has been found that when patients are rigorously classified by the criteria of Feighner *et al.*[151] it is those patients who are identified as having primary affective disorders who excrete less than normal amounts of MHPG.[143,152] This relationship between low MHPG and primary affective illness was not found in an alternative study.[63]

The urinary excretion of MHPG by patients with psychopathological conditions other than affective illness has not been intensively studied, although there is one report which suggests that it is low in schizophrenia,[114] whereas another group have found it to be normal for a schizophrenic group.[152]

In addition to these clinical classifications there are also data suggesting that there are subgroups of depressed patients who may be identified biochemically and pharmacologically. The initial findings bearing upon this issue were first reported in 1968.[153] In that report it was noted that those patients who excreted less than normal quantities of MHPG prior to treatment with imipramine or desmethylimipramine had favourable responses to treatment with these drugs, whereas those patients who excreted normal amounts of MHPG did not respond well to treatment with either desmethylimipramine or imipramine. Furthermore, it was noted that those patients who did respond to treatment had either no change or modest increments in urinary MHPG following 4 weeks of drug treatment, whereas those patients who had normal pretreatment urinary MHPGs and subsequent failure of response had decrements in the excretion of urinary MHPG. Subsequent to this original observation, which was made on a series of 12 patients, a second series of 16 patients was studied in more detail. This group of patients was chosen in terms of having depression of sufficient severity to warrant hospitalization and organic treatment, and they were maintained free of all medication for 3 weeks. Nurses' depression ratings were taken daily, and any patient who showed a spontaneous change in his depression during these 3 weeks was excluded from the study. Again, in this replication study, it was found that there was a direct correlation between the pretreatment MHPG level and subsequent response to imipramine, i.e. a low pretreatment MHPG predicted a favourable response to treatment with this particular antidepressant drug. Furthermore, it was found that pretreatment levels of NM, M, or VMA were not correlated with subsequent treatment response. As before, it was also found that those patients who responded to treatment with imipramine had increments in urinary NM and either no change or modest increments in urinary MHPG, whereas those patients who did not respond had rather marked decrements in both metabolites. These findings led to the suggestion that there are two populations of depressed patients who may be identified in terms of MHPG excretion.[60]

In support of the possibility that there are two distinct groups of patients are findings obtained with *d*-amphetamine.[154,155] In these studies patients were given, on a double blind basis, alternately either placebo or *d*-amphetamine (15 mg b.i.d.), and the effects of the *d*-amphetamine on mood elevation were noted. These studies were done originally to see if one could predict subsequent response to tricyclics by alterations in mood as induced by *d*-amphetamine. It was found that those patients who did respond with an elevation of mood to *d*-amphetamine also had favourable responses to treatment with tricyclic drugs. Further, it was noted that those patients who had elevations in mood following *d*-amphetamine were those patients who excreted less than normal quantities of MHPG, whereas there was a tendency for those patients who excreted normal amounts of MHPG to have either no elevation of mood with *d*-amphetamine, or a worsening. It is of interest that, as with the use of imipramine or desmethylimipramine, those patients who responded to *d*-amphetamine with a brightening of mood had increments in urinary MHPG excretion, whereas those patients who did not had decrements.

Schildkraut has reported that patients who excrete normal to higher than normal amounts of urinary MHPG have a favourable treatment response to amitriptyline, whereas those who excrete lower amounts of MHPG fail to respond to this drug. He also noted that one of the non-responders had a history of a favourable response to imipramine treatment.[156] More recently, a study has emerged from the NIMH group which confirmed and extended the above findings. Beckmann and Goodwin[63] examined a group of patients who were *unequivocal* responders or non-responders to either amitryptiline or imipramine in terms of their pretreatment MHPG levels. They found that those patients who had low pretreatment MHPGs responded well to imipramine, whereas those patients who had high MHPGs did not respond to treatment with this drug. The high pretreatment MHPG patients responded well to amitriptyline, and those who excreted low quantities of this metabolite did not respond to amitriptyline. Although the number of patients per group in this study was small, it is of particular interest that there were no crossovers in values and all four cells were covered.

(c) *A summary statement regarding urinary MHPG excretion by patients with affective illness*

1. There is general agreement among different investigators that MHPG is a major metabolic produce of NE in brain. In man the available data suggests that urinary MHPGs may give, directly or indirectly, some index of CNS NE metabolism, but this issue is far from settled and is clearly open for more definitive types of research.
2. There is agreement between several groups of investigators that when manic-depressive patients are followed longitudinally urinary MHPG is low during depressive phases and is normal or greater than normal during periods of euthymia and/or hypomania.
3. There is reasonably good consensus that not all depressed patients excrete less than normal amounts of MHPG but that the diagnosis of bipolar illness, depressed type, significantly identifies those patients who excrete less than normal amounts of MHPG. In two studies, but not in a third, the diagnosis of primary affective illness depressed type also identified the 'low MHPG' excretions.

4. There is emerging agreement that pretreatment urinary MHPG provides information relevant to treatment response with imipramine or amitryptiline, i.e. depressed patients who excrete less than normal amounts of MHPG respond best to imipramine and not amitryptiline, whereas for those patients who excrete normal or greater than normal amounts of MHPG the reverse is found.
5. Relationships between activity, anxiety, and urinary MHPG need further investigation with the use of good quantitative measures of these state variables. Similarly, more studies of urinary MHPG excretion by patients having psychiatric conditions other than mania or depression are needed.

(d) Depression—a biochemically heterogeneous illness—hypotheses as to mechanisms

In the aggregate the studies which were cited above under section C.4(b) suggest that there are two biochemically and pharmacologically identifiable subgroups of depressed patients. The first has been called Group A[15] and these patients are characterized by: (i) a low pretreatment MHPG; (ii) a favourable response to treatment with imipramine or desmethylimipramine; (iii) a brightening of mood following a trial of *d*-amphetamine; (iv) modest increments or no change in urinary MHPG following treatment with imipramine, desmethylimipramine, or a brief trial on *d*-amphetamine; (v) a failure to respond to amitryptiline. The second group of patients has been called Group B, and is characterized by: (i) a normal or high urinary MHPG; (ii) a favourable treatment response to amitryptiline; (iii) a lack of mood change during a trial of *d*-amphetamine; (iv) decrements in urinary MHPG following treatment with imipramine, desmethylimipramine, or a brief trial of *d*-amphetamine; (v) a failure to respond to imipramine.

Given these groups as defined, it is of use to apply the classical pharmacological stratagem of examining some of the modes of action of imipramine, desmethylimipramine, *d*-amphetamine, and amitryptiline in order to develop hypotheses as to biological mechanisms which may be involved in these two groups of depressive disorders. A summary of some of the available data as to the effects of these and other drugs on amine systems was presented earlier in this essay (see Table 5.1). It may be noted from Table 5.1 that in contrast to *d*-amphetamine, imipramine, desmethylimipramine, amitryptiline, and the other tricyclic antidepressants have little or no effect on dopamine reuptake or turnover in animals or man. Further, the studies dealing with the effects of L-DOPA in depressed subjects which were reviewed earlier suggest that only a small percentage of depressed patients improve when treated with this drug, despite the fact that the L-DOPA treatment produced an elevation in brain DA and CSF HVA. Finally, although there is a tendency for CSF HVA to be reduced in some depressed patients, these subjects cannot be identified clinically, by treatment response, nor do the antidepressant drugs increase changes in CSF HVA with or without concomitant improvement. In the aggregate these findings lead to the conclusion that it is unlikely that alterations in brain DA systems are integrally associated with mood disturbances in the majority of depressions. This conclusion, however, is not to be taken as ruling out a role for DA systems in regulating activity levels in mania and perhaps depression as it has been noted that L-DOPA, when given to bipolar patients, regularly produces some hypomania symptoms, although the dysphoric mood of the patients was not altered.

The fact that desmethylimipramine is effective in the treatment of Group A patients, that *d*-amphetamine produces a mood elevation in these subjects, and that both these agents have relatively weak effects upon blocking reuptake in 5-HT systems suggests that 5-HT systems are not altered in the Group A patients. In contrast, all agents which are effective in treating Group A patients, or result in a mood elevation, block NE uptake. Further, depressed patients who respond to treatment with imipramine, desmethylimipramine, or *d*-amphetamine excrete less than normal amounts of MHPG into urine prior to treatment. It is, therefore, suggested that Group A patients have an alteration in noradrenergic systems, probably central, which is integrally related to their depressive state. In contrast, the failure of a mood change with *d*-amphetamine lead to the conclusion that Group B patients do not have a deficit in NE systems, although 5-HT appears to be a worthwhile candidate for investigation (see Chapter 7 of this book by H. M. van Praag).

If the above conclusions are correct, one may ask why Group A patients who have the postulated deficit in central NE systems do not respond to treatment with amitryptiline, i.e. amitryptiline itself is not an effective agent in blocking the reuptake of NE, but it is demethylated *in vivo* to nortryptiline, which does block NE uptake (see Table 5.1). Two explanations for this failure of Type A patients to respond to amitryptiline are as follows. First, amitryptiline is demethylated *in vivo* to nortryptiline, as is imipramine to desmethylimipramine, but the conversion of amitryptiline to the secondary amine proceeds at a slower rate than nortryptiline hydroxylation, with the result that it has been found that plasma levels of amitryptiline are about the same, or higher, than those of nortryptiline.[69] In contrast, when imipramine is given to patients the plasma levels of desipramine are consistently higher than those of imipramine.[70] As a consequence of the above it is possible, even likely, that when Group A patients are given amitryptiline then therapeutically effective concentrations of nortryptiline are not achieved, i.e. the nortryptiline levels are not sufficient to correct the deficit in noradrenergic systems. Parenthetically, if this line of reasoning is correct it would strengthen the possibility that Group B patients have a disorder in 5-HT systems. These suggestions are testable by examining the therapeutic effects of adequate doses of nortryptiline on depressions of the A and B types as defined in this paper, i.e. Group A patients should respond to nortryptiline. Because of the possible effects of nortryptiline on 5-HT systems (see Table 5.1) it is also possible, however, that some Group B patients may respond.

Neither the basic studies nor clinical investigations can be used at this point to specify the nature of the suggested alteration in noradrenergic systems in Group A patients. For example, increased or decreased NE release during neuronal firing, or an increase in receptor sensitivity, could all produce the same final end result as follows. An alteration in NE neurons which would lead to increased NE release will be expected via actions on presynaptic receptors and neuronal feedback loops to dampen excessively the activity of the NE neuron. Similar results could occur as a function of increased receptor sensitivity. In the case of decreased NE release a failure of compensating mechanisms could exist. In any of these situations the end result in terms of MHPG formation, mood elevation with *d*-amphetamine, or the response to imipramine or desipramine might be the same.

In addition to the above considerations, it is possible, even likely, that other brain amine systems than those already discussed are also involved in the genesis of depression in the Group A patients. For example, the tricyclic antidepressants have been demonstrated to have significant central peripheral anticholinergic properties, and it is possible that these drug actions upon central cholinergic systems when combined with their effects upon NE systems may account for their beneficial actions. Parenthetically, this could also possibly explain why *d*-amphetamine is able to produce a mood elevation in patients but does not have a long-term therapeutic value in the treatment of depression, i.e. amphetamine has an effect on noradrenergic systems, but it does not possess significant anticholinergic properties. In this regard it is of interest that it has recently been reported that physostigmine, which is an inhibitor of cholinesterase, is able to reverse manic states, and in patients with an affective component to their illness can produce a depressive-like syndrome. These data have led to the suggestion that manic depressive illness is a function of adrenergic–cholinergic balance.[157]

Inferences as to the aminergic systems involved in Group B depressions are more difficult to make, but since these patients are characterized by no change in mood with *d*-amphetamine, have normal to high urinary MHPG values, and respond to amitryptiline (see Table 5.1), the possibility of an alteration in 5-HT systems must be seriously considered. In this respect the following observations are of interest. Praag *et al.*[158] have reported preliminary data which indicates that depressed patients who have less than normal increments in 5-HIAA following probenecid will have a favourable therapeutic response to 5-hydroxytryptophan (see also Chapter 7 of this volume). Asberg *et al.*[66] have noted that those patients who had CSF 5-HIAA levels below 15 mg ml^{-1} failed to respond to nortryptiline, whereas those patients with CSF 5-HIAA concentrations above 15 mg ml^{-1} had favourable responses to treatment with nortryptiline.

D REFERENCES

1. Jacobsen, E. (1964). *Depression: Proceedings of the Symposium held at Cambridge, September 22–26, 1959* (Ed., E. B. Davis), Cambridge University Press, London.
2. Kuhn, R. (1958). *Amer. J. Psychiat.*, **115**, 459.
3. Glowinski, J., and J. Axelrod (1964). *Nature*, **204**, 1318.
4. Bunney, W. E., Jr., and Davis, J. M. (1965). *Arch. Gen. Psychiat.*, **13**, 483.
5. Schildkraut, J. (1965). *Amer. J. Psychiat.*, **122**, 508.
6. Schildkraut, J., and Kety, S. S. (1967). *Science*, **156**, 21.
7. Coppen, A. (1967). *Brit. J. Psychiat.*, **113**, 1237.
8. Davis, J. M. (1970). *Int. Rev. Neurobiol.*, **12**, 145.
9. Himwich, H. E. (1970). *Biochemistry of Schizophrenia and Affective Illnesses* (Ed., H. E. Himwich), Williams and Wilkins, Baltimore, p. 230.
10. Praag, H. M. van (1969). *Pharmakopsychiat. Neuro-Psychopharmakol.*, **2**, 151.
11. Schildkraut, J. J. (1970). *Neuropsychopharmacology and the Affective Disorders*, Little Brown, Boston.
12. Weil-Malherbe, H. (1967). *Adv. Enzymol.*, **29**, 479.
13. Schildkraut, J. J. (1975). In *American Handbook of Psychiatry*, Vol. VI (Ed., D. Hamburg), Basic Books, New York.
14. Baldessarini, R. J. (1975). *Arch. Gen. Psychiat.*, **32**, 1087.
15. Maas, J. W. (1975). *Arch. Gen. Psychiat.*, **32**, 1357.
16. Carlsson, A., Fuxe, K., Hamberger, B., and Lindquist, M. (1966). *Acta Physiol. Scand*, **67**, 481.
17. Glowinski, J., Axelrod, J., and Iverson, L. L. (1966). *J. Pharmacol. Exptl Ther.*, **153**, 30.
18. Ross, S. B., and Renyi, A. L. (1969). *Eur. J. Pharmacol.*, **7**, 270.
19. Haggendal, T., and Hamberger, B. (1967). *Acta Physiol. Scand.*, **70**, 277.
20. Sulser, F., Owens, M. L., Strada, S. J., and Dingell, J. V. (1969). *J. Pharmacol. Exptl Ther.*, **168**, 272.
21. Bickel, M. H., and Brodie, B. B. (1964). *Int. J. Neuropharmacol.*, **3**, 611.
22. Maxwell, R. A., Keenan, P. D., Chaplin, E., Roth, B., and Eckhardt, S. B. (1969). *J. Pharmacol. Exptl Ther.*, **166**, 320.
23. Maxwell, R. A., Eckhardt, S. B., and Hite, G. (1970). *J. Pharmacol. Exptl Ther.*, **171**, 62.
24. Maxwell, R. A., Chaplin, E., Eckhardt, S. B., Soares, J. R., and Hite, G. (1970). *J. Pharmacol. Exptl Ther.*, **173**, 158.
25. Salama, A. I., Insalaco, J. R., and Maxwell, R. A. (1971). *J. Pharmacol. Exptl Ther.*, **173**, 474.
26. Blackburn, K. T., Funch, D. C., and Merrills, R. J. (1967). *Life Sci.*, **6**, 1653.
27. Palaic, D., Page, I., and Khairallah, P. (1967). *J. Neurochem.*, **14**, 63.
28. Alpers, H. S., and Himwich, H. (1969). *Biol. Psychiat.*, **1**, 81.
29. Carlsson, A., Corrodi, H., Fuxe, K., and Hokfelt, T. (1969) *Eur. J. Pharmacol.*, **5**, 357.
30. Carlsson, A. (1970). *J. Pharm. Pharmacol.*, **22**, 729.
31. Glowinski, J. (1970). *New Aspects of Storage and Release Mechanisms of Catecholamine*, Springer, Berlin, p. 301.
32. Schildkraut, J. J., Dodge, G. A., and Logue, M. A. (1969). *J. Psychiat. Res.*, **7**, 29.
33. Fuxe, K., and Ungerstedt, U. (1968). *Eur. J. Pharmacol.*, **4**, 135.
34. Carlsson, A., Corrodi, H., Fuxe, K., and Hokfelt, T. (1969). *Eur. J. Pharmacol.*, **5**, 357.

35. Carlsson, A., Corrodi, H., Fuxe, K., and Hokfelt, T. (1969). *Eur. J. Pharmacol.*, **5**, 357.
36. Lidbrink, P., Jonsson, G., and Fuxe, K. (1971). *Neuropharmacology*, **10**, 521.
37. Ross, S. B., and Renyi, A. L. (1975). *Acta Pharmacol. Toxicol.*, **36**, 382.
38. Horn, A. S., Coyle, J. T., and Snyder, S. H. (1971). *Mol. Pharmacol.*, **7**, 66.
40. Nyback, H. V., Walters, J. R., Aghajanian, G. K., and Roth, R. H. (1975). *Eur. J. Pharmacol.*, **32**, 302.
41. Neff, N. H., and Costa, E. (1966). *Antidepressant Drugs*, Excerpta Medica Foundation, Milan.
42. Corrodi, H., and Fuxe, K. (1968). *J. Pharm. Pharmacol.*, **20**, 230.
43. Corrodi, H., and Fuxe, K. (1969). *Eur. J. Pharmacol.*, **7**, 56.
44. Meek, J., and Werdinius, B. (1970). *J. Pharm. Pharmacol.*, **22**, 136.
45. Schubert, J., Nyback, H., and Sedvall, G. (1970). *J. Pharm. Pharmacol.*, **22**, 136.
46. Schildkraut, J., Schanberg, S., Breese, G. R., and Kopin, I. J. (1969). *Biochem. Pharmacol.*, **18**, 1971.
47. Brodie, B. B., Costa, E., Groppetti, A., and Matsumoto, C. (1968). *Brit. J. Pharmacol.*, **34**, 648.
48. Leitz, F. H. (1970). *J. Pharmacol. Exptl Ther.*, **173**, 152.
49. Roth, J. A., and Gillis, C. N. (1974). *Biochem. Pharmacol.*, **23**, 2537.
50. Kodama, T., Matsukado, Y., Suzuki, T., Tanaka, S., and Shimizu, H. (1971). *Biochim. Biophys. Acta*, **252**, 165.
51. Taylor, K. M., and Randall, P. K. (1975). *J. Pharmacol. Exptl Ther.*, **194**, 303.
52. Gluckman, M. I., and Baum, T. (1969). *Psychopharmacologia*, **15**, 169.
53. Snyder, S. (1975). Oral communication.
54. Schildkraut, J. (1975). In *Neurobiological Mechanisms of Adaptation and Behavior* (Ed., A. J. Mandell), Raven Press, New York.
55. Alpers, H. S., and Himwich, H. E. (1972). *J. Pharmacol. Exptl Ther.*, **180**, 531.
56. Mandell, A. J., Segal, D. S., and Kuczenski, R. (1975). In *Catecholamines and Behavior*, Vol. 2 (Ed., A. Friedhoff), Plenum Press, New York.
57. Roffler-Tarlov, S. (1975). *Biochem. Pharmacol.*, **24**, 1321.
58. Schildkraut, J., Gordon, E. K., and Durell, J. (1965). *J. Psychiat. Res.*, **3**, 213.
59. Prange, A. J., Wilson, I. C., and Knox, A. E. (1971). In *Symposium: Brain Chemistry and Mental Disease* (Eds., B. T. Ho, and W. McIsaac), Plenum Press, New York.
60. Maas, J. W., Fawcett, J. A., and Dekirmenjian, H. (1972). *Arch. Gen. Psychiat.*, **26**, 252.
61. Maas, J. W., and Landis, H. (1966). *Psychom. Med.*, **28**, 247.
62. Maas, J. W., and Landis, H. (1968). *J. Pharmacol. Exptl Ther.*, **163**, 147.
63. Beckmann, H., and Goodwin, F. K. (1975). *Arch. Gen. Psychiat.*, **32**, 17.
64. Bowers, M. B., Heninger, G. R., and Gerbode, F. (1969). *Int. J. Neuropharm.*, **8**, 255.
65. Papeschi, R., and McClure, D. J. (1971). *Arch. Gen. Psychiat.*, **25**, 354.
66. Asburg, M., Bertilsson, L., Tuck, D., Cronholm, B., and Sjoqvist, F. (1973). *Clin. Pharm. Ther.*, **14**, 277.
67. Goodwin, F. K., Post, R. M., and Sack, R. L. (1975). In *Neurobiological Mechanisms of Adaptation and Behavior* (Ed., A. J. Mandell), Raven Press, New York.
68. Bertilsson, L., Asburg, M., and Thorenp, P. (1974). *Eur. J. Clin. Pharmacol.*, **7**, 365.
69. Braithwaite, R. A., and Widdop, B. (1971). *Clin. Chim. Acta*, **35**, 461.
70. Moody, J. P., Tait, A. C., and Todrick, A. (1967). *Brit. J. Psychiat.*, **113**, 183.
71. Gershon, S., and Shopsin, B. (Eds.) (1973). *Lithium: Its Role in Psychiatric Research and Treatment*, Plenum Press, New Yok.
72. Sheard, M. H., and Aghajanian, G. K. (1970). *Life Sci.*, **9**, 285.
73. Katz., R. I., Chase, T. N., and Kopin, I. J. (1968) *Science*, **162**, 466.
74. Forn, J., and Valdecasa, F. (1971). *Biochem. Pharm.*, **23**, 2773.

75. Ebert, M. H., Baldessarini, R. J., Lipinski, J. F., and Beav, K. (1973). *Arch. Gen. Psychiat.*, **29**, 397.
76. Cohn, C. K., Dunner, D. L., and Axelrod, J. (1970). *Science*, **170**, 1323.
77. Dunner, D. L., Cohn, C. K., Gershon, E. S., and Goodwin, F. K., *Arch. Gen. Psychiat.*, **25**, 348.
78. Briggs, M. H., and Briggs, M. (1973). *Experientia.*, **29**, 278.
80. Weinshilboum, R. M., Raymond, F. A., Elveback, L. R. (1974). *Nature*, **252**, 490.
81. Robinson, D. S., Davis, J. M., Nies, A., Ravaris, C. L., and Sylwester, B. (1971). *Arch. Gen. Psychiat.*, **24**, 536.
82. Robinson, D. S., Nies, A., Davis, J. N., Bunney, W. E., Davis, J. M., Colburn, R. W., Bourne, H. R., Shaw, D. M., and Coppen, A. J. (1972). *Lancet i*, 290.
83. Nies, A., Robinson, D. S., Lamborn, K. R., and Lampert, R. P. (1973). *Arch. Gen. Psychiat.*, **28**, 834.
84. Murphy, D. L., Belmaker, R., and Wyatt, R. J. (1974). *J. Psychiat. Res.*, **11**, 221.
85. Klaiber, E. L., Kobayashi, Y., Broverman, D. M., and Hall, F. (1971). *J. Clin. Endocrinol.*, **33**, 630.
86. Redmond, D. E., Murphy, D. L., Baulu, J., Ziegler, M. G., and Lake, C. R. (1975). *Psychosomat. Med.*, **37**, 417.
87. Nies, A., Robinson, D. S., Riveris, C. L., and Davis, J. M. (1971). *Psychosomat. Med.*, **33**, 470.
88. Klaiber, E. L., Broverman, D. M., Vogel, W., Kobayashi, Y., and Moriarity, D. (1972). *Amer. J. Psychiat.*, **128**, 1492.
89. Murphy, D. L., and Weiss, R. (1972). *Amer. J. Psychiat.*, **128**, 1351.
90. Pare, C. N. B., and Sandler, M. (1959). *J. Neurol., Neurosurg., Psychiat.*, **22**, 247.
91. Klerman, G. L., Schildkraut, J. J., Hasenbush, L. L., Greenblatt, M., and Friend, G. G. (1963). *J. Psychiat. Res.*, **1**, 289.
92. Goodwin, F. K., Murphy, D. L., Brody, H. K. H., and Bunney, W. E. (1971). *J. Pharm. Ther.*, **12**, 383.
93. Matussek, N. (1971). In *Advances in Neuropharmacology* (Eds., O. Liner, Z. Votava, and P. B. Bradley), North Holland Publishing Co., Amsterdam.
94. Persson, T., and Walinder, J. (1971). *Brit. J. Psychiat.*, **119**, 277.
95. Mendels, J., Stinnett, J. L., Burns, D., and Frazer, A. (1975). *Arch. Gen. Psychiat.*, **32**, 22.
96. Murphy, D. L., Brodie, H. K. H., Goodwin, F. K., and Bunney, W. E. (1971). *Nature*, **229**, 135.
97. Everett, G. M., and Borcherding, J. W. (1970). *Science*, **168**, 849.
98. Wilk, S., and Mones, R. (1971). *J. Neurochem.*, **18**, 1771.
99. Seiden, L. S., and Peterson, D. D. (1968). *J. Pharmacol. Exptl Ther.*, **163**, 84.
100. Goldberg, H. C. M. (1968). *Brit. J. Pharmacol.*, **33**, 457.
101. Garelis, E., and Sourkes, T. L. (1973). *J. Neurol. Neurosurg. Psychiat.*, **36**, 625.
102. Curzon, G., Gumpert, E. J., and Sharpe, D. M. (1971). *Nature New Biol.*, **231**, 189.
103. Post, R. M., Goodwin, F. K., Gordon, E., and Watkin, D. M. (1973). *Science*, **179**, 897.
104. Chase, T. N., Gordon, E. K., and Ng, L. K. (1973). *J. Neurochem.*, **21**, 581.
105. Maas, J. W., and Landis, D. H. (1965). *Psychosomat. Med.*, **26**, 399.
106. Dencker, S. J., Malm, U., Roos, B. E., and Werdinius, B. (1966). *J. Neurochem.*, **13**, 1545.
107. Roos, B. E., and Sjostrom, R. (1969). *Pharmacol. Clin.*, **1**, 153.
108. Nordin, G., Ottosson, J. O., and Roos, B. E. (1971). *Psychopharmacologia*, **20**, 315.
109. Bowers, M. B. (1972). *Psychopharmacologia*, **23**, 26.
110. Mendels, J., Frazer, A., Fitzgerald, R. G., Ramsey, T. A., and Stokes, J. W. (1972). *Science*, **175**, 1380.

111. Bowers, M. B. (1972). *Neuropharmacology*, **11**, 101.
112. Ashcroft, G. W., Blackburn, I. M., Eccleston, O., Glen, A. I. M., Hartley, W., Kinloch, N. E., Lonergan, M., Murray, L. G., and Pullar, I. A. (1973). *Psychol. Med.*, **3**, 319.
113. Wilk, S., Shopsin, B., Gershon, S., and Suhl, M. (1972): *Nature*, **235**, 440.
114. Subrahmanyam, S. (1975). *Brain Res.*, **87**, 355.
115. Post, R. M., Gordon, E. K. Goodwin, F. K., and Bunney, W. E. (1973). *Science*, **179**, 1002.
116. Shopsin, B., Wilk, S., Gershon, S., and Suhl, M. (1973). *Arch. Gen. Psychiat.*, **28**, 230.
117. Shaw, D. M., O'Keefe, R., MacSweeney, D. A., Brooksbank, B. W. L., Nogvera, R., Coppen, A. (1973). *Psychol. Med.*, **3**, 333.
118. Weil-Malherbe, H., Woodby, L., Axelrod, J. (1961). *J. Neurochem.*, **8**, 55.
119. Maas, J. W. (1970). *J. Pharmacol. Exptl Ther.*, **174**, 369.
120. Patel, M., and Maas, J. W. (1970). *J. Pharmacol. Exptl Ther.*, **174**, 379.
121. Maas, J. W., Benensohn, H., and Landis, D. H. (1970). *J. Pharmacol. Exptl Ther.*, **174**, 381.
122. Maas, J. W., and Landis, D. H. (1971). *J. Pharmacol. Exptl Ther.*, **177**, 600.
123. Sjoquist, B. (1975) *J. Neurochem.*, **24**, 199.
124. Karoum, F., Gillin, J. C., and Wyatt, R. J. (1975). *J. Neurochem.*, **25**, 653.
125. Mannarino, E., Kirshner, N., and Nashold, B. S. (1963). *J. Neurochem.*, **10**, 373.
126. Schanberg, S. M., Schildkraut, J. J., Breese, G. R., and Kopin, I. J. (1968). *Biochem. Pharmacol.*, **17**, 247.
127. Korf, J., Roth, R. H., and Aghajanian, G. K. (1973). *Eur. J. Pharmacol.*, **23**, 276.
128. Walter, D. S., and Eccleston, D. (1973). *J. Neurochem.*, **21**, 281.
129. Breese, G. R., Prange, A. J., and Howard, J. L. (1972). *Nature New Biol.*, **240**, 286.
130. Karoum, F., Wyatt, R., and Costa, E. (1974). *Neuropharmacology*, **13**, 165.
131. Bareggi, S. R., Marc, V., and Morselli, P. L. (1974). *Brain Res.*, **75**, 177.
132. Maas, J. W., Dekirmenjian, H., Garver, D., Redmond, D. E., and Landis, D. H. (1973). *Eur. J. Pharmacol.*, **23**, 121.
133. Garver, D. L., Pandey, G. N., Dekirmenjian, H., and DeLeon-Jones, F. (1975). *Amer. J. Psychiat.*, **132**, 1149.
134. Maas, J. W., Fawcett, J. A., and Dekirmenjian, H. (1968). *Arch. Gen. Psychiat.*, **19**, 129
135. Greenspan, J., Schildkraut, J. J., Gordon, E. K., Baer, L., Aronoff, M. S., and Durell, J. (1970). *J. Psychiat. Res.*, **7**, 171.
136. Bond, P. A., Jenner, F. A., and Sampson, G. A. (1972). *Psychol. Med.*, **2**, 81.
137. Jones, F., Maas, J. W., Dekirmenjian, H., and Fawcett, J. A. (1973). *Science*, **179**, 300.
138. Shopsin, B., Wilk, S., Gershon, S., Roffman, M., and Goldstein, M. (1974). In *Frontiers in Catecholamine Research* (Eds., E. Usdin and S. Snyder), Pergamon Press, New York, p. 1173.
139. Post, R., and Bunney, W. (1974). Paper read before the World Psychiatric Association, Biological Psychiatric Section, Munich, Germany.
140. Watson, R., Hartmann, E., and Schildkraut, J. J. (1972). *Amer. J. Psychiat.*, **129**, 263.
141. Dekirmenjian, H., Maas, J., and Fawcett, J. (1973). Paper read before the annual meeting of the American Psychiatric Association, Honolulu, Hawaii.
142. Maas, J. W., Dekirmenjian, H., and DeLeon-Jones (1974). In *Frontiers in Catecholamine Research* (Eds., E. Usdin and S. Snyder), Pergamon Press, New York, p. 1091.
143. DeLeon-Jones, F., Maas, J. W., Dekirmenjian, H., and Sanchez, J. (1975). *Amer. J. Psychiat.*, **132**, 1149.
144. Ebert, M. H., Post, R. M., and Goodman, F. K. (1972). *Lancett*, *ii*, 766.

145. Goode, D. J., Dekirmenjian, H., Meltzer, H. Y., and Maas, J. W. (1973). *Arch. Gen. Psychiat.*, **29**, 396.
146. Schildkraut, J. J. (1974). In *Frontiers of Catecholamine Research* (Eds., E. Usdin and S. Snyder), Pergamon Press, New York, p. 1165.
147. Rubin, R. T., Miller, R. G., Clark, B. R., Poland, R. E., and Arthur, R. J. (1970). *Psychom. Med.*, **32**, 589.
148. Maas, J. W., Dekirmenjian, H., and Fawcett, J. A. (1971). *Nature*, **230**, 330.
149. Schildkraut, J. J., Keeler, B. A., Papousek, M., and Hartmann, E. (1973). *Science*, **181**, 762.
150. Goodwin, F. K., and Beckmann, H. (1975). In *The Biology of the Major Psychosis: A Comparative Analysis* (Ed., D. Freedman), Raven Press, New York.
151. Feighner, J. P., Robins, E., Guze, S. B., Woodruff, R. A., Winokur, G., and Munoz, R. (1972). *Arch. Gen. Psychiat.*, **26**, 57.
152. Taube, S. L., Kirstein, L. S., Sweeney, D. R., and Maas, J. W. (1978). *Amer. J. Psychiat.*, **135**, 78–82.
153. Maas, J. W., Fawcett, J. A., and Dekirmenjian, H. (1968). Paper read before the annual meeting of the American Psychiatric Association, Boston.
154. Fawcett, J., Maas, J. W., and Dekirmenjian, H. (1972). *Arch. Gen. Psychiat.*, **26**, 246.
155. Fawcett, J., and Siomopoulous, V. (1971). *Arch. Gen. Psychiat.*, **25**, 247.
156. Schildkraut, J. J. (1973). *Amer. J. Psychiat.*, **130**, 695.
157. Janowsky, D. S., El-Yousef, M. K., Davis, J. M., and Sekerke, H. J. (1972). *Lancet ii*, 632.
158. Praag, H. M. van, and Korf, J. (1974). *Int. Pharmacopsychiat.*, **9**, 35.

Aromatic Amino Acid Hydroxylases and Mental Disease
Edited by M. B. H. Youdim

CHAPTER 6

Tryptophan Hydroxylation in the Central Nervous System and Other Tissues

M. Hamon, S. Bourgoin, and M. B. H. Youdim

A INTRODUCTION

Although 5-hydroxytryptamine (serotonin) (5-HT) was discovered in the central nervous system (CNS) in 1953 by Twarog and Page, the synthetic pathway of this neurotransmitter was only recently established. Gál *et al.* (1963) were the first to demonstrate hydroxylation of tryptophan by brain tissue in rats and pigeons. In 1964, Grahame-Smith reported the occurrence of a tryptophan hydroxylating enzyme in brain homogenates. Since this date, much effort has been devoted to the purification and characterization of tryptophan hydroxylase (EC 1.14.16.4). In contrast to tyrosine hydroxylase (EC 1.14.16.2), the first enzyme in the biosynthetic pathway of catecholamines in nervous tissue, tryptophan hydroxylase is not present in neurons or related cells outside the brain. Most of our knowledge on tyrosine hydroxylase in brain derives from previous extensive studies on the adrenal gland enzyme; the lack of a serotoninergic peripheral organ analogous to the adrenal medulla for catecholaminergic neurons explains why considerably less is known about tryptophan hydroxylase. However, several other organs contain tryptophan hydroxylating enzyme, e.g. the liver, the kidney, and the intestine. No activity can be detected in either heart or spleen (Lovenberg *et al.*, 1968). In liver and kidney, hydroxylation of tryptophan is very likely due to phenylalanine hydroxylase (EC 1.14.16.1) (Lovenberg *et al.*, 1968). Thus much interest has focused on the brain enzyme which catalyses the first step of biosynthesis of the neurotransmitter serotonin. Other non-neuronal cells like pinealocytes, mast cells, and, to a lesser degree, platelets contain significant amount of tryptophan hydroxylase. Although the present review deals mainly with the brain enzyme, references to the biochemistry and pharmacology of tryptophan hydroxylase in other tissues are also given when possible (see Lovenberg, 1977; Hamon *et al.*, 1977).

B ASSAYS OF TRYPTOPHAN HYDROXYLASE

As for all other hydroxylating enzymes, the development of a reliable, sensitive assay of tryptophan hydroxylase (L-tryptophan, tetrahydropterin: oxygen oxidoreductase (5-hydroxylating) (EC 1.14.16.4)) is a *sine qua non* condition to further studies on its physicochemical characteristics and regulatory properties. This is particularly true for brain tryptophan hydroxylase. Indeed, the activity of this enzyme is low and the enzyme itself is rather unstable. Thus, until the introduction of a very potent stimulating pterin cofactor of the enzyme, the 6-methyltetrahydropterin (6-MPH_4) by Friedman *et al.* (1972), most assay conditions required a radioactive substrate in order to measure the activity.

The stoichiometry of the enzymic conversion of tryptophan into 5-hydroxytryptophan was established by Kaufman (1974) (Figure 6.1):

$$\text{L-Tryptophan} + BH_4 + O_2 \rightarrow \text{5-Hydroxy-L-tryptophan} + BH_2 + H_2O$$

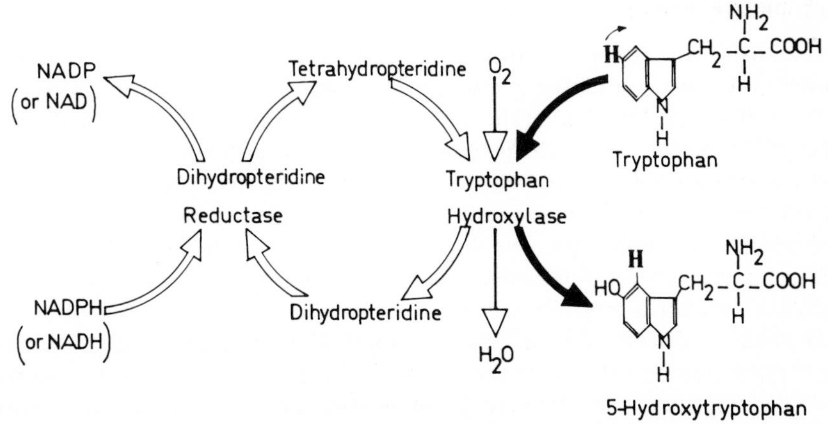

Figure 6.1 Mechanism of enzymic tryptophan hydroxylation in serotoninergic neurons

where BH_4 stands for tetrahydrobiopterin and BH_2 for the quinonoid dihydrobiopterin. In most assay conditions described, tetrahydrobiopterin is replaced by the commercially available synthetic cofactors: 6-methyl- and 6,7-dimethyl-5,6,7,8-tetrahydropterins (6-MPH_4 and $DMPH_4$ respectively). Efforts to determine the stoichiometry with $DMPH_4$ have generally been unsuccessful because the high rate of non-enzymic oxidation of the pterin made an accurate estimation of the tryptophan-dependent rate of $DMPH_4$ oxidation impossible (Friedman *et al.*, 1972). The uncoupling induced by the synthetic cofactor leads to the formation of an abnormal product in the hydroxylation, namely hydrogen peroxide, H_2O_2 (Kaufman, 1974), which could alter the characteristics of the enzyme. However, with extensively purified tryptophan hydroxylase, the coupling between tryptophan hydroxylation and pterin oxidation becomes perfect (the ratio of 5-hydroxytryptophan formed over the cofactor oxidized is not significantly different from 1) whichever cofactor is used: $DMPH_4$, 6-MPH_4, or BH_4 (Tong and Kaufman, 1975).

In vivo, the cofactor is regenerated by a NADH-dependent enzyme: dihydropteridine reductase (EC 1.6.99.7) (Musacchio *et al.*, 1972) according to the following reaction:

$$BH_2 + NADH + H^+ \rightarrow BH_4 + NAD^+$$

NADPH can replace NADH in the reduction process to BH_4, but both the affinity and maximal velocity of dihydropteridine reductase are then lower (Turner *et al.*, 1974). To determine the tryptophan hydroxylase activity, most authors have looked for rapid and sensitive procedures to measure the reaction product: L-5-hydroxytryptophan. On one occasion (Friedman *et al.*, 1972) the estimation of the oxidation rate of NADPH (spectrophotometric reading at 340 nm) was used.

1 Standard assay

Usually brain tissue is homogenized in 0.05 M Tris-HCl or Tris-acetate buffer (3:1 v/w) containing 1 mM 2-mercaptoethanol or dithiothreitol. Phosphate buffer must be avoided since it markedly inhibits tryptophan hydroxylase (Gál, 1974; Kizer *et al.*, 1975). Very often authors speak about 'tryptophan hydroxylase activity' when they measure the rate of conversion of labelled tryptophan into labelled serotonin in homogenates made in 0.32 M sucrose. Although these studies raise the important question of the possible existence of a particulate form of the enzyme, they concern mainly the synthesis of serotonin in intact synaptosomes. Therefore, it is incorrect to call this kind of assay tryptophan hydroxylase measurement, since what is determined is the sum of a complex interaction between the uptake of the labelled amino acid, tryptophan hydroxylase, and L-5-hydroxytryptophan decarboxylase in the cytoplasm included in synaptosomes.

After homogenization, the source of the enzyme is the 20 000–30 000 **g** supernatant of brainstem homogenate (which contains 80–90% of the activity). The assay mixture includes L-tryptophan (0.01–1.0 mM), 6-MPH_4 or $DMPH_4$ (0.1–1 mM). The regenerating system allowing the reduction of the quinonoid dihydropterin in the course of the assay consists of the addition of dihydropteridine reductase and NADH or NADPH, or more simply of a thiol reagent (2-mercaptoethanol or dithiothreitol). The inclusion of catalase in the assay mixture was reported to stimulate the tryptophan hydroxylase activity. According to Friedman *et al.* (1972) the formation of hydrogen peroxide during the conversion of tryptophan into 5-hydroxy-L-tryptophan in the presence of exogenous pterin cofactors would inhibit the latter reaction. Indeed, these authors could replace catalase by Fe^{2+}, another agent for decomposition of H_2O_2, in their assay mixture with only a small reduction in the tryptophan hydroxylase activity. The reaction proceeds linearly for 15–60 min. at 37 °C under 1 atm of air. However, the conversion of tryptophan into 5-hydroxytryptophan is greatly enhanced in the presence of pure O_2 at 1 atm barometric pressure (Green and Sawyer, 1966; Youdim *et al.*, 1975*a*).

Although assay conditions do not vary very much from one laboratory to another, methods used for the determination of the 5-hydroxytryptophan formed are numerous.

2 Direct measurement of 5-hydroxytryptophan

For this purpose it is necessary to include in the assay mixture an inhibitor of 5-hydroxytryptophan decarboxylase (EC 4.1.1.28) usually NSD 1034 (*N*-methyl-*N*-3-hydroxyphenylhydrazine) or NSD 1015 (*m*-hydroxybenzylhydrazine). If the inhibitor is not included, even in purified preparations of tryptophan hydroxylase (Youdim *et al.*, 1975a), the occurrence of small amounts of 5-hydroxytryptophan decarboxylase activity is sufficient quantitatively to convert 5-hydroxytryptophan into 5-hydroxytryptamine (5-HT). This contaminating activity is in fact involved in some assay procedures of tryptophan hydroxylase (see below).

Spectrofluorimetric determinations of 5-hydroxytryptophan are based on various methods. Friedman *et al.* (1972) measured 5-hydroxytryptophan by its native fluorescence at 538 nm in 3 N HCl under excitation light of 295 nm wavelength (Bogdanski *et al.*, 1956). A much more sensitive method was developed by Gál and Patterson (1973), who measured the hydroxylated product after condensation with *o*-phthaldialdehyde (Maickel *et al.*, 1968). The highly fluorescent derivative formed allows this rapid method to estimate tryptophan hydroxylase activity in 0.1–3 mg of protein homogenate from rat brainstem.

Recently, Saavedra *et al.* (1973, 1974) have developed a new micromethod to determine 5-HT or 5-hydroxytryptophan in small punches of brain tissue. This procedure consists in the conversion of 5-HT into ^3H-melatonin by two enzymic steps: the *N*-acetylation of the indoleamine by rat liver *N*-acetyltransferase (EC 2.3.1.5) with acetyl-CoA and the *O*-methylation of *N*-acetylserotonin formed into ^{3}H-melatonin by the beef pineal hydroxyindole-*O*-methyltransferase (EC 2.1.1.4), with (^{3}H-methyl)-*S*-adenosylmethionine as the ^{3}H-methyl donor. This method can also be used for the determination of serotonin in biological fluids (Boireau *et al.*, 1976). Its high sensitivity allows one to measure as little as 10–50 pg of the indoleamine (Saavedra *et al.*, 1973; Boireau *et al.*, 1976). When applied to the determination of 5-hydroxytryptophan formed in the course of tryptophan hydroxylase assay, this method permits the measurement of the hydroxylase activity in as little as 5 μg of a crude brainstem homogenate (Kizer *et al.*, 1975). This microassay has been successfully used for the determination of tryptophan hydroxylase activity in isolated nuclei of the rat hypothalamus (Kizer *et al.*, 1975).

A large improvement in the detection of 5-hydroxytryptophan owing to its native fluorescence in strong acidic medium allowed Meek and Neckers (1975) to measure tryptophan hydroxylase activity in single nuclei of the rat brain. After performing the enzymic assay in similar conditions to those described by Gál and Patterson (1973), these authors selectively separated the 5-hydroxytryptophan formed by high pressure liquid chromatography on a strong cation exchanger. The fluorescence of the hydroxylated amino acid in strong acid was then quantitatively recorded. According to Meek and Neckers (1975) this automatized assay requires only 5 min. per sample. Its limit of detection (0.5 ng of 5-hydroxytryptophan) is in the same range as that of the radioenzymic procedure (Kizer *et al.*, 1975).

In radiometric assays, the direct measurement of the ^{14}C-5-hydroxytryptophan formed can be performed by thin layer chromatography: Håkanson and Hoffman (1967) have described a method to separate the product (^{14}C-5-hydroxytryptophan) from the substrate (^{14}C-tryptophan) on silica gel layer using 14% NaCl in water as the solvent. The migration was achieved in less than 2 h. The R_f values of the two amino acids were sufficiently different to result in a low blank value in the ^{14}C-5-hydroxytryptophan spot. This method has been successfully used to estimate the conversion rate of ^{3}H-tryptophan into ^{3}H-5-hydroxytryptophan in intact tissues (Hamon *et al.*, 1973). As in the cases of the microassays developed by Kizer *et al.* (1975) and Meek and Neckers (1975), this procedure has been employed to measure tryptophan hydroxylase activity in single nuclei of the rat brain. Kan *et al.* (1975) were thus able to estimate the enzymic activity in the various raphe nuclei containing 5-HT cell bodies in the rat brainstem. Renson *et al.* (1966) have demonstrated that with 5-^{3}H-tryptophan, the ^{3}H atom in the 5-position is transferred to the 4-position in the course of enzymic hydroxylation (see Figure 6.1). The exact mechanism of this transposition (NIH shift) is not clearly established. Some authors (Jerina *et al.*, 1970) have postulated the formation of an intermediary arene oxide in this process. Whatever it may be, the proton in the 4-position of the hydroxyindoles is readily exchangeable with water under acid conditions (Guroff *et al.*, 1967a, b). Thus, the following procedure has been designed to determine tryptophan hydroxylase assay:

$$5\text{-}^3\text{H-Tryptophan} + O_2 + BH_4 \rightarrow BH_2 + H_2O + 5\text{-Hydroxy-4-}^3\text{H-tryptophan}$$

$$5\text{-Hydroxy-4-}^3\text{H-tryptophan} + H_2O \xrightarrow{+H} {}^3\text{H}-\text{OH} + 5\text{-Hydroxytryptophan}$$

However, the blank values using commercially prepared 5-^{3}H-tryptophan were too high to get reliable data with tryptophan hydroxylase from brain homogenates. Recently Lovenberg *et al.* (1971) have described a new method for the preparation of 5-^{3}H-tryptophan which results in much lower blank values. After the conversion of 5-^{3}H-tryptophan into 5-hydroxy-4-^{3}H-tryptophan in usual conditions for the measurement of tryptophan hydroxylase activity (see above), the assay mixture is brought to 1.5 N with perchloric acid and incubated for 20 min. at 37 °C. Tritium in the 4-position of the labelled 5-hydroxytryptophan molecule exchanges totally with water during this step. Thus ^{3}H—OH formed can be selectively extracted by retention of other radioactive molecules on strong anionic and cationic exchange resins (Dowex 1-X2 and Dowex 50-X4 respectively). Using this method, Lovenberg *et al.* (1971) measured tryptophan hydroxylase activity in homogenates of various tissues. More recently this technique was also applied to estimate the rate of conversion of ^{3}H-tryptophan into ^{3}H-5-hydroxytryptophan in intact cultured pineal glands (Bensinger *et al.*, 1974) and in brain slices (Renson, 1973). The main objection against using this procedure is that until now only the racemic substrate, D,L-^{3}H-tryptophan has been used. However, the commercial availability of both the active synthetic pterin cofactor, 6-MPH_4, and L-5-^{3}H-tryptophan of good quality should allow the extensive use of this method.

3 Indirect measurement of 5-hydroxytryptophan

The separation of serotonin from tryptophan is much easier than that of two amino acids such 5-hydroxytryptophan from tryptophan. Hence several assays of tryptophan hydroxylase including immediate decarboxylation of the newly formed 5-hydroxytryptophan were designed. Such procedures have been used by Grahame-Smith (1964, 1967) to demonstrate the presence of tryptophan hydroxylase in brain tissue. After isolating ^{14}C-5-hydroxytryptophan formed in the course of the enzymic assay from ^{14}C-tryptophan by paper chromatography, it was further decarboxylated into ^{14}C-5-HT using partially purified aromatic L-amino acid decarboxylase from guinea pig kidney. The final product was then extracted by paper chromatography. A simpler procedure was devised by Green and Sawyer (1966). Since crude homogenates contain large amounts of 5-hydroxytryptophan decarboxylase activity, in absence of a specific inhibitor (see above), the product of tryptophan hydroxylation, ^{14}C-5-hydroxytryptophan, is totally converted into ^{14}C-5-HT. When further catabolism of the latter compound is inhibited (by tranylcypromine, a potent monoamine oxidase inhibitor), its determination can be considered as a measurement of tryptophan hydroxylation. The selective extraction of ^{14}C-5-HT was achieved on ion-exchange resin (Amberlite CG 50). The overall method is easy and rather rapid, and many laboratories have developed this kind of assay with minor modifications (Figure 6.2) (Peters *et al.*, 1968; Lovenberg *et al.*, 1967; Sanders-Bush *et al.*, 1972a; Youdim *et al.*, 1974, 1975a; Hulme *et al.*, 1974).

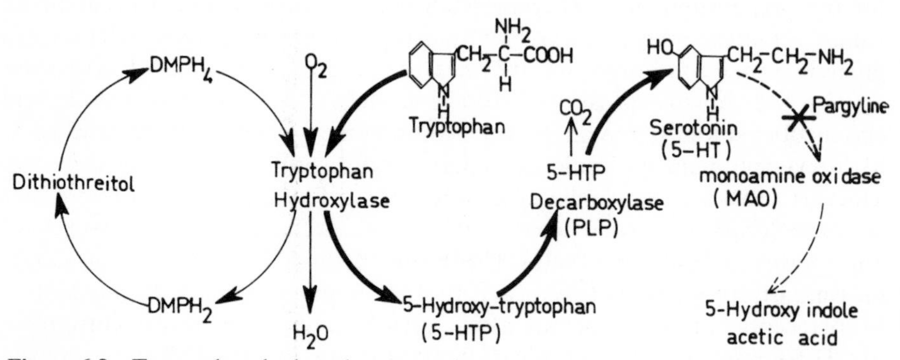

Figure 6.2 Tryptophan hydroxylase assay by coupling with 5-hydroxytryptophan decarboxylase activity (Youdim *et al.*, 1975a)

Another method for measuring tryptophan hydroxylase activity which involves subsequent conversion of 5-hydroxytryptophan into serotonin consists of determining the amount of CO_2 released during 5-hydroxytryptophan decarboxylation. The substrate is labelled on the first carbon of the side chain so that $^{14}CO_2$ is finally formed. The carbon dioxide is then trapped into organic alkali (phenethylamine or hydroxide of hyamine for instance) after acidifying the reaction mixture with perchloric acid. This method, first described by Ichiyama *et al.* (1968, 1970), is widely used owing to its rapidity and simplicity (Knapp and Mandell, 1972a, b, 1973, 1974; Kuhar *et al.*, 1972; Azmitia and McEwen, 1974).

Ichiyama *et al.* (1970) have pointed out that the K_m of L-aromatic amino acid decarboxylase for L-5-hydroxytryptophan is at least two orders of magnitude lower than for L-tryptophan, so that in a mixture of both amino acids the hydroxylated molecule is preferentially decarboxylated. However, when the tryptophan concentration exceeds 20 μM, this amino acid is significantly decarboxylated, leading to the formation of tryptamine and CO_2 (Figure 6.3).

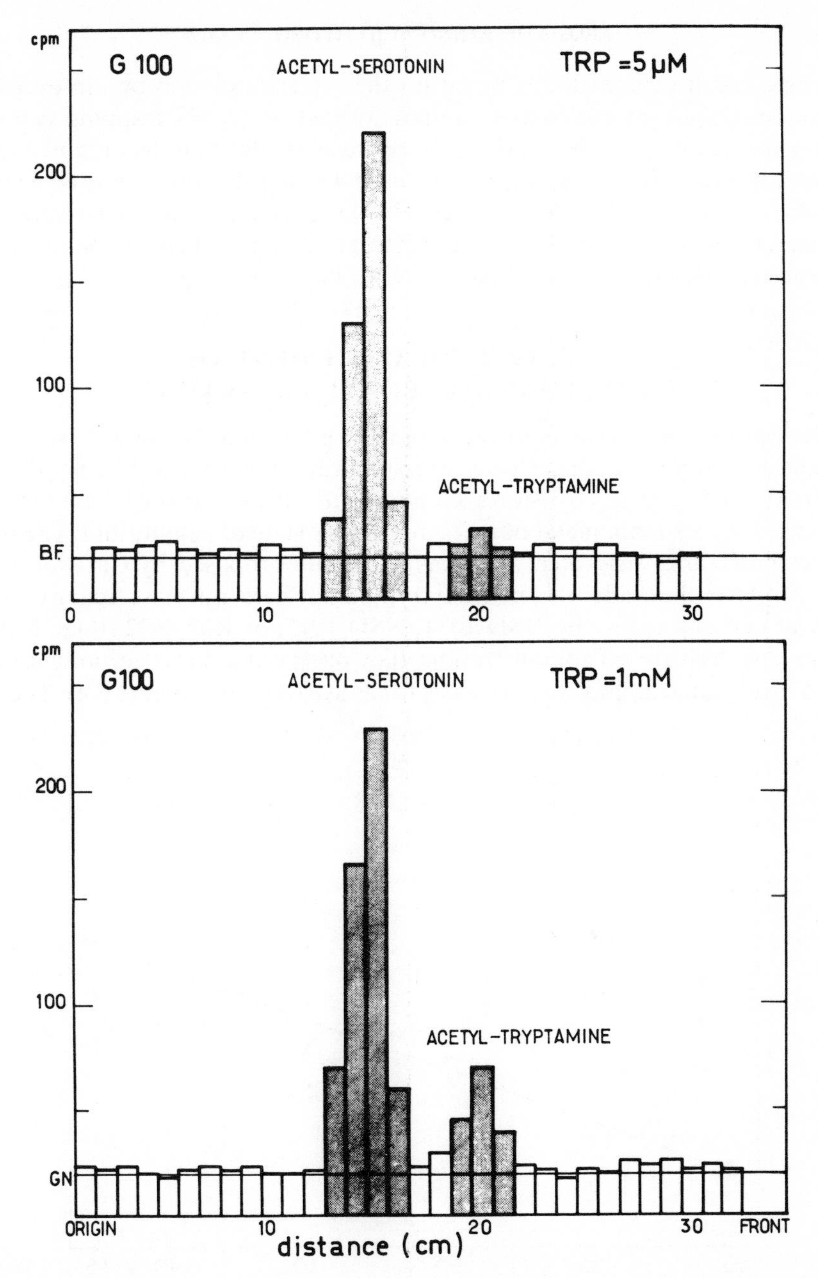

Figure 6.3 Limits in the use of the tryptophan hydroxylase assay coupled with 5-hydroxytryptophan decarboxylase and formation of ^{3}H-tryptamine in presence of high concentrations of ^{3}H-tryptophan.

^{3}H-5-HT and ^{3}H-tryptamine formed in the course of tryptophan hydroxylase assay (Youdim *et al.*, 1975a) were identified by paper chromatography of their acetylated derivatives (Hamon *et al.*, 1973; M. Hamon and S. Bourgoin, unpublished observations)

In these conditions, methods based on the crude extraction by ion exchange chromatography of radioactive amines formed or $^{14}CO_2$ trapping can give erroneous results. In particular, only approximate data can be obtained when large concentrations of substrate are used for the determination of enzyme kinetics, the K_m and V_{max} for instance. However, in numerous pharmacological studies, this kind of assay is adapted to detect any change in tryptophanhydroxylase activity (Knapp and Mandell, 1972a, b, 1973, 1974).

C *IN VIVO* MEASUREMENT OF TRYPTOPHAN HYDROXYLASE ACTIVITY

Several methods have been proposed for estimating the rate of serotonin synthesis in brain, i.e. the velocity of its limiting step catalysed by tryptophan hydroxylase (cf. review by Morot-Gaudry *et al.*, 1974). Most of these methods require that serotonin metabolism is in a steady state of equilibrium. The more direct procedure very often used was originally proposed by Carlsson *et al.* (1972). After blockade of 5-hydroxytryptophan decarboxylase activity in the central nervous system by NSD 1034, NSD 1015, or Ro4-4602 (three benzylhydrazine derivatives) administration, they observed a linear accumulation of the hydroxylated amino acid in tissues for at least 45 min. (Figure 6.4). The rate

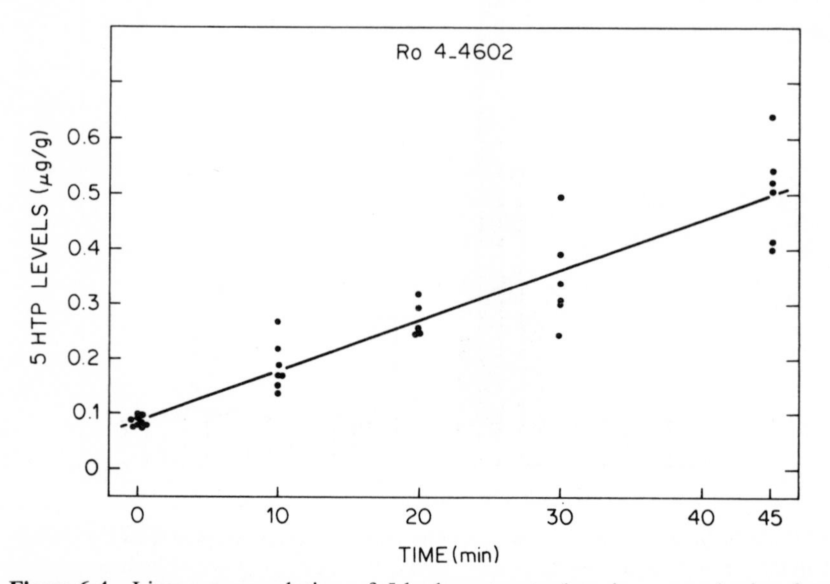

Figure 6.4 Linear accumulation of 5-hydroxytryptophan in mouse brain after the inhibition of central 5-hydroxytryptophan decarboxylase activity by Ro 4-4602 (800 mg (kg body weight)$^{-1}$) (Morot-Gaudry *et al.*, *Naunyn-Schmiedeberg's Arch. Pharmacol.*, 1974, **282**, 223–238). Reproduced by permission of Springer-Verlag

of this accumulation, although slightly lower than that of 5-HT after monoamine oxidase inhibition or that of 5-HT synthesis calculated by radiometric methods (Morot-Gaudry *et al.*, 1974), can be considered as an estimation of tryptophan hydroxylation in brain. *In vivo*, changes in the rate of conversion of tryptophan into 5-hydroxytryptophan can result from fluctuations in the concentration of tryptophan in the vicinity of tryptophan hydroxylase (since the enzyme is not saturated by its substrate in physiological conditions) and/or in the actual activity of this enzyme. One way to predict if tryptophan hydroxylase is altered *in vivo* consists of looking for the persistence of a given effect on tryptophan hydroxylation (i.e. the rate of 5-hydroxytryptophan accumulation) after the administration of a large dose of tryptophan. In these conditions, alterations induced by changes in the substrate concentration are totally 'flooded'. Thus, it could be shown that following spinal cord transection (Carlsson *et al.*, 1973) or halothane anaesthesia (Figure 6.5) (Bourgoin *et al.*, 1975) the *in vivo* activity of the hydroxylating enzyme was actually reduced in the distal part of the spinal cord or in the brain respectively.

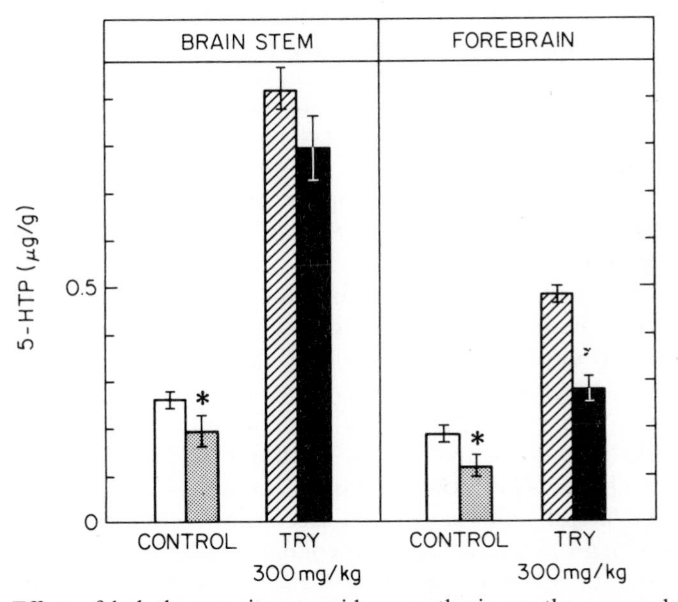

Figure 6.5 Effect of halothane + nitrous oxide anaesthesia on the accumulation of 5-hydroxytryptophan in the brainstem and forebrain after 5-hydroxytryptophan decarboxylase inhibition.

Since the inhibitory effect of anaesthesia (solid bars) was still detected after tryptophan loading (300 mg (kg body weight)$^{-1}$ intraperitoneally), it is very likely mediated by a decreased tryptophan hydroxylase activity (Bourgoin *et al.*, *Naunyn-Schmiedeberg's Arch. Pharmacol.*, 1975, **288**, 109–121). Reproduced by permission of Springer-Verlag

D LOCALIZATION OF TRYPTOPHAN HYDROXYLASE

The map of serotoninergic neurons in the central nervous system was established owing to the histofluorescence method developed by Carlsson, Falck, and Hillarp (1962). Thus, Dahlström and Fuxe (1964) have shown that the cell bodies of these neurones are distributed in the medial part of the brainstem; they have distinguished several groups of serotoninergic cell bodies, called B1–B9 from the posterior to the anterior brainstem, corresponding in first approximation to the raphe nuclei described by Brodal *et al.* (1960). The distribution of tryptophan hydroxylase in the cat brain closely resembles that of yellow fluorescence in the Falck–Hillarp technique and superimposes with regional changes in 5-HT concentration (Peters *et al.*, 1968). This is also true for the rat brain, although data available in the literature do not agree completely. The procedures used to measure tryptophan hydroxylase activity appear to be critical in this respect (Knapp and Mandell, 1972a; Renson, 1973; Lovenberg and Victor, 1974a, b; Sanders-Bush *et al.*, 1975). The highest level of activity (25–30 nmol (mg protein)$^{-1}$ h^{-1}) is generally found in the mesencephalic tegmentum, an area rich in serotoninergic cell bodies. Precise dissection of this region has demonstrated that most tryptophan hydroxylase activity occurs in the raphe nuclei (Deguchi *et al.*, 1973). Fairly high activity is also found in the tectum, the hypothalamus, the septum, the pons medulla, and the spinal cord (all areas containing 5-HT terminals). The cerebellum exhibits the lowest level of activity (Table 6.1). In addition, the cerebellar enzyme does not share all the properties of tryptophan hydroxylase in other brain areas (Renson, 1973; Gál *et al.*, 1975a). Selective destruction of median raphe (nucleus raphe dorsalis and nucleus raphe centralis superior, the two nuclei sending most of the 5-HT projections into the forebrain; Dahlström and Fuxe, 1964) results in the impairment of ^{3}H-5-HT synthesis from ^{3}H-tryptophan in cat brain slices (Pujol *et al.*, 1971). Similarly, in the rat selective midbrain raphe lesions induce a large reduction in the conversion of ^{3}H-tryptophan to ^{3}H-5-HT in brain sliçes (Herr and Roth, 1972) and the progressive disappearance of tryptophan hydroxylase activity in the forebrain (Kuhar *et al.*, 1971).

Similar results have been obtained by injecting 5,6- or 5,7-dihydroxytryptamine into the lateral ventricles of rats. According to Baumgarten *et al.* (1972, 1973), these drugs induce, in particular conditions, the selective degeneration of 5-HT terminals mainly in the spinal cord and partly in various structures in the forebrain. This toxic effect induces simultaneously the decrease of both the yellow fluorescence (by the Falck–Hillarp technique) and 5-HT levels in brain tissues. Tryptophan hydroxylase activity is also largely reduced by these treatments (Lovenberg and Victor, 1974). The intraperitoneal administration of *p*-chloroamphetamine, which selectively destroys the B9 group of serotoninergic cell bodies, i.e. the nucleus linearis (Harvey *et al.*, 1975), also induced a large

Table 6.1 Regional variations of ^{3}H-5-HT synthesis from ^{3}H-tryptophan in brain slices. Comparison with the distribution of soluble tryptophan hydroxylase activity in the same areas

Structure	^{3}H-5-Hydroxyindoles (μCi g^{-1})	^{3}H-5-Hydroxyindoles / ^{3}H-Tryptophan (%)	Tryptophan hydroxylase activity (nmol 5-hydroxytryptophan mg^{-1} h^{-1})
Raphe	3.801 ± 0.219	100	—
Brainstem	2.483 ± 0.159	53	0.980
Hypothalamus	4.254 ± 0.298	54	0.481
Spinal cord	0.912 ± 0.102	46	—
Striatum	1.877 ± 0.165	32	0.192
Hippocampus	1.486 ± 0.094	28	0.162
Cerebral cortex	0.595 ± 0.036	11	0.081
Cerebellum	0.205 ± 0.014	4	—

Soluble tryptophan hydroxylase was measured according to Gál and Patterson (1973) with 20 µM tryptophan. Hydroxylation of ^{3}H-tryptophan in brain slices largely depends on the accumulation of the labelled substrate in tissues. When expressed as the ratio of ^{3}H-5-hydroxyindoles synthesized over the ^{3}H-tryptophan accumulated in tissues, regional changes of tryptophan hydroxylation in slices resemble those of soluble tryptophan hydroxylase activity (correlation coefficient $r = 0.823$; M. Hamon and S. Bourgoin, unpublished observations).

reduction of tryptophan hydroxylase activity in most forebrain areas (Sanders-Bush *et al.*, 1975).

All these data were recently confirmed by immunohistochemistry: Joh *et al.* (1975), using a specific antibody against rat midbrain tryptophan hydroxylase, were able to visualize directly the enzyme in 5-HT neurons by the peroxidase–antiperoxidase method. Therefore, it may be definitely established that next to the high affinity serotonin uptake process, tryptophan hydroxylase is the most selective marker of serotoninergic neurons in the rat brain. However, a recent report by Harvey and Gál (1974) has challenged this view. According to these authors, midbrain raphe lesions, which induced a large reduction (-72%) in 5-HT levels in the septum 10 days after surgery, did not significantly affect tryptophan hydroxylase activity in this area. On the other hand, Victor *et al.* (1974) observed a similar decrease of tryptophan hydroxylase activity in the septum as compared to other areas such as the striatum, hypothalamus, and pons medulla after the intraventricular injection of 5,6-dihydroxytryptamine. At present, no clear explanation can be given for these contradictory results. However, it must be pointed out that both groups did not use the same assay conditions to estimate tryptophan hydroxylase activity. Victor *et al.* (1974) measured as final products 5-hydroxytryptophan and 5-HT formed in the absence of a 5-hydroxytryptophan decarboxylase inhibitor by their native fluorescence in 3 N HCl, whereas Harvey and Gál (1974) used the '*o*-phthaldialdehyde' method of Maickel *et al.* (1968). Since there are conflicting results about the relative enzymic activity in a given area such as the striatum, the poorest area according to Lovenberg and Victor (1974a, b), a finding in contrast with other data (Peters *et al.*, 1968; Knapp and Mandell, 1972a; M. Hamon and S. Bourgoin, S. Bourgoin, unpublished observations, see Table 6.1), one possible explanation may be that the technical aspects of enzyme assay are critical for the measurement of this particular enzyme.

Inside 5-HT neurons, tryptophan hydroxylase exists both in terminals and perikaryons. Thus peripheral injection of tryptophan in rats induces a large increase in the yellow fluorescence of 5-HT raphe cells (Aghajanian and Asher, 1971). Since the indoleamine cannot cross the blood–brain barrier, its local synthesis from tryptophan is highly probable. Indeed, Knapp and Mandell (1972a), studying the regional localization of tryptophan hydroxylase in the rat brain, have shown that a 'soluble' form is mainly present in areas containing 5-HT cell bodies (midbrain, median raphe, mesencephalic tegmentum) whereas a 'particulate' form occurs in 5-HT terminal regions (cerebellum, caudate nucleus, septum, frontal cortex). Although the exact significance of the two forms of tryptophan hydroxylase is rather questionable, data reported by Knapp and Mandell (1972a) indicate that the enzymic activity in areas containing cell bodies is at least in the same range as that occurring in structures containing only 5-HT terminals.

New microdissection techniques developed by Palkovits and coworkers

(Palkovits, 1973; Saavedra *et al.*, 1974) allow one to measure tryptophan hydroxylase activity in discrete nuclei of the rat brain. Thus Kizer *et al.* (1975) have demonstrated the occurrence of the enzyme in median eminence, confirming previous work showing the synthesis of ^{3}H-5-HT from ^{3}H-tryptophan by this structure *in vitro* (Hamon *et al.*, 1970). Microdissection techniques were also used by Kan *et al.* (1975) to study the distribution of tryptophan hydroxylase activity in various raphe nuclei of the rat. According to these authors, the raphe nucleus dorsalis (B7) has the highest enzymic activity (2636.8 ± 151.7 pmol of 5-hydroxytryptophan formed per hour and per milligram of protein homogenate in presence of 40 μM ^{14}C-tryptophan). In the other nucleus, B8 (raphe centralis superior), which also contains the 5-HT cell bodies sending fibres to the forebrain, tryptophan hydroxylase activity is half that of the B7 group. In other B nuclei, the enzymic activity is much lower: for instance in B5 (raphe pontis) it is equal to only 7.7% that of the dorsal raphe. Similar studies were recently made by Meek and Neckers (1975) and Brownstein *et al.* (1975). The latter authors have measured tryptophan hydroxylase activity in numerous nuclei in the brain, and, concerning the raphe nuclei, reported findings in close agreement with those of Kan *et al.* (1975). Thus, maximal tryptophan hydroxylase activity was measured in the dorsal raphe nucleus (43 times that found in the cerebellar cortex). In contrast, working at saturable concentrations of both tryptophan and 6-methyltetrahydropterin, Meek and Neckers (1975) report an enzymic activity in B7 nucleus 46 times higher than Kan *et al.* (1975). Since the latter authors use a saturable concentration of the pterin cofactor and a tryptophan concentration slightly lower than the respective K_m of the enzyme, this large difference is quite surprising. Discrepancies can be mainly explained by the respective dissection procedures and assay conditions used.

Joh *et al.* (1975), using a specific antibody against tryptophan hydroxylase, have described the intracellular localization of the enzyme in 5-HT raphe cells. According to these authors, tryptophan hydroxylase is associated with endoplasmic reticulum in perikaryons and neurotubules in processes.

The mammalian nervous system is not the only one which can synthesize 5-hydroxytryptamine. The giant C1 cell of the nervous system of *Aplysia* is able to synthesize ^{3}H-serotonin from ^{3}H-tryptophan (Eisenstadt *et al.*, 1973). Similarly, Cardot (1972) has demonstrated that the central nervous system of *Helix pomatia* contains a significant amount of tryptophan hydroxylase activity. In both cases, recent studies are in favour of the presence of more than one specific transmitter-synthesizing enzyme in the molluscan neurons. Thus tryptophan hydroxylase activity can be detected in cholinergic R2 cells of *Aplysia* (see Brownstein *et al.*, 1974) and choline acetyltransferase is found in serotoninergic neurons of *Helix* (Hanley *et al.*, 1974). Since the destruction of the serotoninergic neurons in the rat brain does not induce any significant decrease of septal tryptophan hydroxylase activity (Harvey and Gál, 1974) the possible occurrence of this enzyme in other kind of neurons may also be proposed for the mammalian brain. However, an alternative explanation has also been suggested by Harvey and Gál (1974).

Outside the brain, tryptophan hydroxylase exists in various types of cells. In the pineal gland, tryptophan hydroxylase is the first enzyme involved in the synthesis of melatonin in pinealocytes. Although the identity of this enzyme with that of serotoninergic neurons is doubtful (Deguchi and Barchas, 1972a), its very high specific activity has lead some authors to use pineal extracts as the starting point for attempting the purification of tryptophan hydroxylase (Lovenberg *et al.*, 1973). Mast cells also contain high amounts of tryptophan hydroxylase (Lovenberg *et al.*, 1968), more than 200 times that of a crude brainstem homogenate. Some carcinoid tumours were also reported to have a high tryptophan hydroxylase activity (Grahame-Smith, 1972; Lovenberg *et al.*, 1968). Peripheral organ homogenates (liver, kidney, intestinal mucosa) are very often able to convert tryptophan into 5-hydroxytryptophan. In most cases, with the exception of the gut (Noguchi *et al.*, 1973), tryptophan is hydroxylated by phenylalanine hydroxylase (EC 1.14.16.1). In particular, Renson *et al.* (1962) have clearly shown that this enzyme is responsible for the synthesis of 5-HTP in the rat liver. However, since the catalytic site for tryptophan on phenylalanine hydroxylase is not shared in the hydroxylation of phenylalanine (Coulson *et al.*, 1968) one can call the liver enzyme tryptophan hydroxylase (Sullivan *et al.*, 1973). Although one report mentioned a slight but significant activity in platelets (Lovenberg *et al.*, 1968), more recent studies definitively established that tryptophan hydroxylase is absent in these microcells (Morrissey *et al.*, 1977).

E EFFECT OF AGE ON TRYPTOPHAN HYDROXYLASE

At birth, the activity of tryptophan hydroxylase extracted from the whole brain is about a quarter that found in adult rats (Schmidt and Sanders-Bush, 1971). The enzymic activity increases rapidly during the postnatal period to reach the adult level around the third week after birth. As already mentioned in preceding sections, quite different results can be obtained by authors using various kinds of assays to measure tryptophan hydroxylase activity. This is also true concerning studies on tryptophan hydroxylase activity in developing rat brain. Thus, using different homogenization conditions (0.15 M KCl instead of 50 mM Tris-acetate pH 7.4; Schmidt and Sanders-Bush, 1971) and enzymic assays (fluorimetric determination of 5-hydroxyindoles formed instead of measurement of the conversion of ^{14}C-tryptophan into ^{14}C-5-HT; Schmidt and Sanders-Bush, 1971), Wapnir *et al.* (1971) found that whole brain tryptophan hydroxylase activity at birth was in the same range as that of adult rats. Between the fourth and sixth days after birth, the enzymic activity may be even greater than in the whole brain of adults. More recently, Renson (1973) reported findings which are in agreement with Wapnir *et al.* (1971). Using another method, ^{3}H-release from ^{3}H-5-hydroxytryptophan formed in brain slices, he could observe that tryptophan hydroxylase activity in the brain of a 21-day-old rat foetus had already reached the adult level. In the same report, Renson (1973) mentioned that tryptophan hydroxylase activity was detectable on the 16th day of foetal life, the level being equal to 20% that of adults. In an effort to reconcile these contradictory results, we have measured tryptophan hydroxylase activity in both newborn and adult rats by two different methods. Thus, we observed that the rate of tryptophan hydroxylation in brainstem slices of newborn rats was about 80% that estimated in adult tissues (Bourgoin *et al.*, 1974), although the activity of soluble tryptophan hydroxylase (assayed in the 30 000 **g** supernatant of brainstem homogenate according to Gál and Patterson, 1973) of newborns was less than half that of adults (S. Bourgoin and M. Hamon, unpublished observations). This difference may be explained by the higher capacity of newborn slices to accumulate tryptophan (Bourgoin *et al.*, 1974), thus leading to a substrate concentration in the vicinity of tryptophan hydroxylase much greater than in adult slices. Therefore, tryptophan hydroxylation in slices is the result of a complex interaction of several mechanisms and cannot be considered as an index of tryptophan hydroxylase activity alone (see Table 6.1).

The growth of serotoninergic neurons during development may vary considerably from one area to the other. Thus Loizou (1972) observed that at birth the yellow fluorescence (specific for serotonin) closely resembles that of adults only in the brainstem. In forebrain areas, which contain 5-HT terminals, the density of the yellow fluorescence is much lower in the newborn. In agreement with these observations, several authors (Deguchi and Barchas, 1972b; Bourgoin *et al.*, 1974) have reported that tryptophan hydroxylase activity is higher in the brainstem than anywhere else in the brain of very young rats. In particular, in the forebrain ('cortex') tryptophan hydroxylase activity is very low at birth (about 10% that of adult rats). The large increase in the enzymic activity up to the adult level occurs between the 10th and the 30th day after birth.

Similar findings were recently reported by Baumgarten *et al.* (1975). Whereas, tryptophan hydroxylase activity is in the same range in the pons medulla (containing 5-HT cell bodies) of newborn and 40-day-old rats, in the hypothalamus (rich in 5-HT terminals) the enzymic activity at birth is only about 15% that of mature animals. The postnatal phase of rapid maturation of tryptophan hydroxylase occurs mainly between the sixth and the 20th day following birth. Slightly different data were described by Rastogi and Singhal (1974a). Thus, according to these authors, tryptophan hydroxylase activity increases linearly during the first 2 weeks of postnatal life and then it levels off between the 14th and 28th days after birth. Moreover, in 1-day-old rats they detected an enzymic activity equal to 40% of that occurring at 60 days of age, although the brain region studied corresponded in first approximation to the 'cortex' (containing 5-HT terminals) dissected by Deguchi and Barchas (1972b). In spite of these conflicting data (partly explained by species differences, dissections, and enzymic assay conditions) it is very likely that the development of tryptophan hydroxylase follows a similar pattern to that of tyrosine hydroxylase in the brain (Coyle and Axelrod, 1972). Indeed, the decrease in its apparent K_m value for tryptophan which is observed in rat brainstem slices in the course of ontogenesis (Figure 6.6) (Bourgoin *et al.*, 1974) might be related to a shift in the subcellular distribution of tryptophan hydroxylase mainly in the soluble fraction in the foetal brain to the particulate (i.e. synaptosomal) fraction in the adult brain. In contrast, the determination of kinetic properties of soluble (30 000 **g** supernatant) tryptophan hydroxylase extracted from the brainstem of newborn or adult rats indicate that the affinities of the enzyme for its substrate amino acid and 6-MPH_4 do not change as a function of age (S. Bourgoin and M. Hamon, unpublished observations).

The maturation of tryptophan hydroxylase in the brain of mice closely follows the pattern already described for the rat: the adult level of enzymic

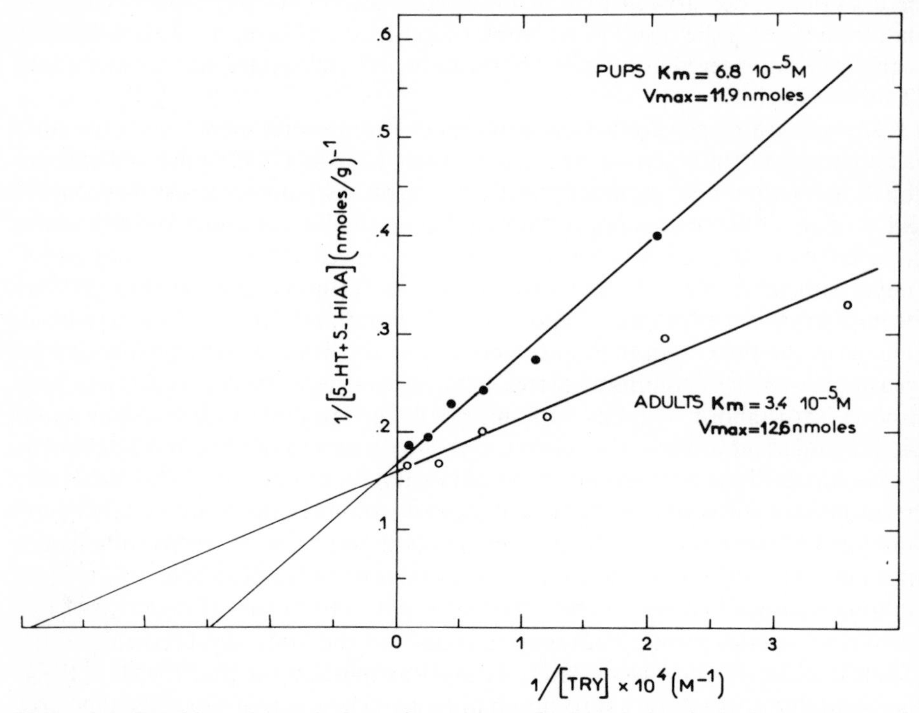

Figure 6.6 Double reciprocal plots of tryptophan hydroxylation in brainstem slices of newborn or adult rats.

Slices were incubated in the presence of various concentrations of ^{3}H-tryptophan for 30 min. Each point is the mean of six determinations made with newborn (less than 24 h. old) or adult tissues. V_{max} is expressed in nanomoles of 5-HT synthesized per hour in 1 g of fresh tissue (Bourgoin *et al.*, 1974). Reproduced by permission of Pergamon Press activity is reached earlier in the brainstem (1 week *post partum*) than in forebrain areas (4 weeks *post partum*; Hoff and Goodrich, 1975). In man, tryptophan hydroxylase is present and functioning in subcortical areas by the end of the first trimester of foetal life (Millard and Gál, 1972). At the end of the second trimester its activity is in the range of that found in various areas of the adult brain (Millard and Gál, 1972).

F PROPERTIES OF TRYPTOPHAN HYDROXYLASE

1 Purification

First attempts to purify tryptophan hydroxylase were made by Jequier *et al.* (1969). The enzyme source was either the beef pineal or the rat brainstem. In both cases, tissues were first homogenized in 0.05 M Tris-acetate buffer (pH 7.4). The supernatant obtained by centrifugation at 30 000 **g** for 15–20 min. was dialysed overnight against 0.1 M 2-mercaptoethanol in 0.05 M Tris-acetate (pH 7.4). In the case of tryptophan hydroxylase from beef pineal, dialysis resulted in doubling its specific activity as compared to that of the crude supernatant. At this stage, the enzyme was already purified 4.5 times when starting with the non-dialysed whole homogenate in Tris-acetate buffer. Finally, tryptophan hydroxylase was precipitated from the dialysed supernatant between 50 and 60% saturation with ammonium sulphate. The precipitate was then suspended in 2 ml of Tris buffer and dialysed for 4 h. against 0.1 M 2-mercaptoethanol in the same buffer (0.05 M Tris-acetate, pH 7.4). The specific activity of tryptophan hydroxylase in the dialysed solution was 12.7 times higher than in a crude pineal homogenate (Jequier *et al.*, 1969). With the rat brainstem as the starting material, Jequier *et al.* (1969) observed that the enzyme was precipitated between 30 and 60% ammonium sulphate saturation. A twofold increase in enzymic specific activity was reached by adsorption and elution from calcium phosphate gel (Jequier *et al.*, 1969). The overall procedure allowed the purification of the enzyme by only 3.5-fold when starting with the dialysed 30 000 **g** supernatant of brainstem homogenate. In contrast to this procedure, Ichiyama *et al.* (1970) precipitated tryptophan hydroxylase by ammonium sulphate between 0 and 40% saturation in an attempt to partially purify the enzyme from the guinea pig brainstem. More recently, Friedman *et al.* (1972) have used a narrow range of ammonium sulphate concentration to precipitate tryptophan hydroxylase from the rabbit hindbrain (between 18 and 28% saturation). Although the procedure also involved calcium phosphate gel adsorption, the final enzyme preparation was only 10 times more active than the initial crude extract.

During the last 2 years, several groups have made attempts to purify tryptophan hydroxylase. Thus, Lovenberg *et al.* (1973) first succeeded in obtaining a protein solution with an enzymic specific activity 20–30 times higher than that of a crude extract of bovine pineal glands. The procedure used involved ammonium sulphate fractionation (the enzyme precipitated between 37 and 57% saturation), DEAE Sephadex chromatography, and finally Sephadex G-100 gel fitration. About the same extent of purification was obtained by Gál and Moses (1974) starting with the rat brainstem as the enzyme source. The final step also consisted of a Sephadex (G-200) gel filtration. Combination of several procedures already used by various groups enabled us to obtain 30–40-fold purification of tryptophan hydroxylase from the pig brainstem (Table 6.2, Youdim

Table 6.2 Partial purification of pig brain tryptophan hydroxylase

Step	Volume (ml)	Protein (mg)	Units	Specific activity (unit (mg protein)$^{-1}$)	Yield (%)
Extract (homogenate)	450	7200	93.6	0.013	100
Ammonium sulphate (25–60%)	80	1420	21.3	0.025	23
Sephadex G-25	100	350	34.6	0.099	37
Calcium phosphate	100	125	22.9	0.183	24.5
Sephadex G-100	8	17	6.8	0.400	7
Sephadex G-200	8	15	7.5	0.499	8

One unit of enzyme is that amount which catalyses the formation of 1 nmol of 5-hydroxytryptophan per minute under standard assay conditions at 37 °C (Youdim *et al.*, 1975a).
Reproduced by permission of Pergamon Press.

et al., 1974, 1975a). However, in most studies, the instability of the enzyme has never been totally solved, so that the final recovery of these methods is generally low (15% Lovenberg *et al.*, 1973; 7–8 per cent, Youdim *et al.*, 1974, 1975a). The recent discovery by Rogawski *et al.* (1974) of the occurrence of an hydrophobic site on the enzyme molecule would suggest that use of glycerol or ethylene glycol in the course of the purification procedure might result in an improved stability of tryptophan hydroxylase. Indeed, Tong and Kaufman (1975) observed no loss of activity for a period of 3 weeks when tryptophan hydroxylase solution containing trace amounts of polyethylene glycol was kept at −80 °C. Large improvements in the purification of tryptophan hydroxylase were recently reported by Tong and Kaufman (1975) and Joh *et al.* (1975). The method described by Tong and Kaufman (1975) for the enzyme extracted from the rabbit hindbrain involves adsorption and elution from calcium phosphate gel, fractionation with polyethylene glycol, gel filtration on Sepharose 6B, and DEAE cellulose chromatography. Finally, tryptophan hydroxylase was purified 58 times with a final recovery of 25%. The purity of the enzyme was established by sucrose gradient centrifugation and disc gel electrophoresis: the final enzyme solution gave only one major band which coincided with tryptophan hydroxylase activity. The enzyme was estimated 85–90% pure when eluted from DEAE cellulose. The procedure devised by Joh *et al.* (1975) for tryptophan hydroxylase extracted from the region of raphe nucleus of rat midbrain consists of precipitation with 80% $(NH_4)_2SO_4$, chromatography over Sepharose 4B, hydroxylapatite, and Sephadex G-200 columns, and finally disc gel electrophoresis. The final enzyme solution was pure enough to obtain a specific antibody after injection into rabbits (Joh *et al.*, 1975).

2 Subcellular distribution

Gál *et al.* (1966) first reported that the crude mitochondrial fraction contained a significant proportion of brain tryptophan hydroxylase (Figure 6.7).

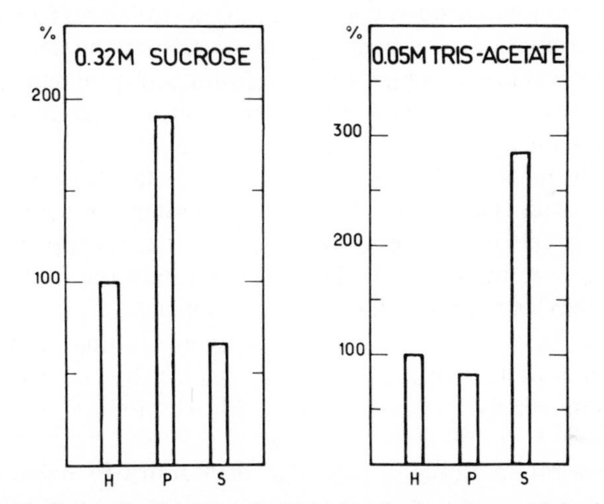

Figure 6.7 Subcellular distribution of tryptophan hydroxylase activity in the pig brain. Pig brainstems and caudate nuclei were homogenized either in 0.32 M sucrose or 0.05 M Tris-acetate, pH 7.4. They were then centrifuged at 10 000 **g** for 30 min. and tryptophan hydroxylase activity was measured in the pellet (P) and the supernatant (S). The enzymic activity (per milligram of proteins in each fraction) was expressed as percentage of that found in the corresponding homogenate. In isotonic conditions the enzyme is mainly included in synaptosomes (P); in hypotonic medium the enzyme is released in the soluble fraction. (M. Hamon *et al.*, unpublished observations)

In this fraction, i.e. when the enzyme is included in synaptosomes, it did not exhibit the same properties as the soluble enzyme. Whereas the latter was completely dependent on 5,6,7,8-tetrahydro-6,7-dimethylpterin for activity, the particulate enzyme was already fully active in absence of the cofactor (Gál *et al.*, 1966; Robinson *et al.*, 1968; Ichiyama *et al.*, 1970; Knapp and Mandell,

1972a, b; Youdim *et al.*, 1974, 1975a). On occasions high concentrations (around 1 mM) of the dimethyl cofactor were reported to inhibit even particulate tryptophan hydroxylase activity (Robinson *et al.*, 1968; Ichiyama *et al.*, 1970; Youdim *et al.*, 1975a). One explanation may be the presence of a significant proportion of the cofactor molecule in an oxidised form which directly inhibits the enzyme (see Kizer *et al.*, 1975).

The different sensitivities of the soluble and the synaptosomal tryptophan hydroxylase to the addition of $DMPH_4$ have led Ichiyama *et al.* (1970) to propose the existence of two different enzymes. Subsequent studies by Knapp and Mandell (1972a, b, 1973, 1974) have shown that these two measurable tryptophan hydroxylase activities could be differentially affected by various pharmacological treatments. As postulated by Ichiyama *et al.* (1970) and demonstrated by several authors (Héry *et al.*, 1970, 1972, 1974; Hamon *et al.*, 1973, 1974a, b, 1975), Knapp and coworkers have shown that the uptake of tryptophan occurring in particles containing the so-called 'particulate' tryptophan hydroxylase could be a critical step in studies on the enzymic activity. Indeed, in assay conditions which do not saturate tryptophan hydroxylase, fluctuations in the accumulation of tryptophan in serotoninergic synaptosomes can directly affect the hydroxylation rate of the amino acid in these particles. The inhibition by *p*-chlorophenylalanine of tryptophan hydroxylase activity may illustrate this fact. The K_i of *p*-chlorophenylalanine competitive inhibition of tryptophan uptake in synaptosomes ($K_i = 9.8$ μM, Knapp and Mandell, 1972a) is much lower than that of the drug inhibiting the midbrain soluble tryptophan hydroxylase activity ($K_i = 0.3$ mM, Jequier *et al.*, 1967; Knapp and Mandell, 1972a). As a result, the particulate hydroxylating activity is much more sensitive to inhibition by *p*-chlorophenylalanine than the soluble enzyme (Knapp and Mandell, 1972a; Peters, 1972; M. Hamon *et al.*, unpublished observations).

Although these observations suggest that occlusion of soluble tryptophan hydroxylase in synaptosomes induces alterations in its activity mainly by the role of tryptophan transport in the particles, the possible occurrence of an actual particulate enzyme in synaptosomes is still an open question. The lack of stimulating effect of $DMPH_4$ on tryptophan hydroxylase activity in crude mitochondrial fraction might result from the inability of the reduced cofactor to enter the particles (Grahame-Smith, 1967; Knapp and Mandell, 1972a). However, *in vitro* addition of the exogenous cofactor to vas deferens (Cloutier and Weiner, 1973) or rat striatal slices (T. C. Westfall, personal communication) induced a large increase in the synthesis of norepinephrine or dopamine from tyrosine respectively. Similarly, Kettler *et al.* (1974) have shown that *in vivo* intraventricular administration of tetrahydrobiopterin resulted in an enhancement of tyrosine hydroxylation in dopaminergic terminals of the rat striatum. These observations show that, at least in the case of catecholaminergic neurons, the exogenous cofactor is taken up. Although similar studies on serotoninergic neurons have not been reported yet, the hypothesis suggesting that the exogenous cofactor does not stimulate particulate tryptophan hydroxylase since it does not cross the synaptosomal membrane is rather doubtful.

Hypo-osmotic shock of synaptosomes results in the appearance of soluble tryptophan hydroxylase totally dependent on pterin cofactor (Grahame-Smith, 1967). However, homogenization of rat brainstem in hypo-osmotic conditions (0.05 M Tris-acetate buffer, pH 7.4) does not totally solubilize the enzyme since a large proportion (70%) of tryptophan hydroxylase remains in the pellet after centrifugation at 30 000 **g** for 15 min. (Robinson *et al.*, 1968). Similarly, using the same hypo-osmotic buffer, Friedman *et al.* (1972) and Youdim *et al.* (1975a) did not recover all tryptophan hydroxylase activity in the high speed supernatant of rabbit hindbrain and pig brainstem homogenates respectively. In addition, subsequent treatments of the 'particulate' form of the enzyme prepared in isotonic 0.25–0.33 M sucrose, with hypotonic buffer solutions, failed to solubilize it (Mandell *et al.*, 1972; Gál, 1974). Although no

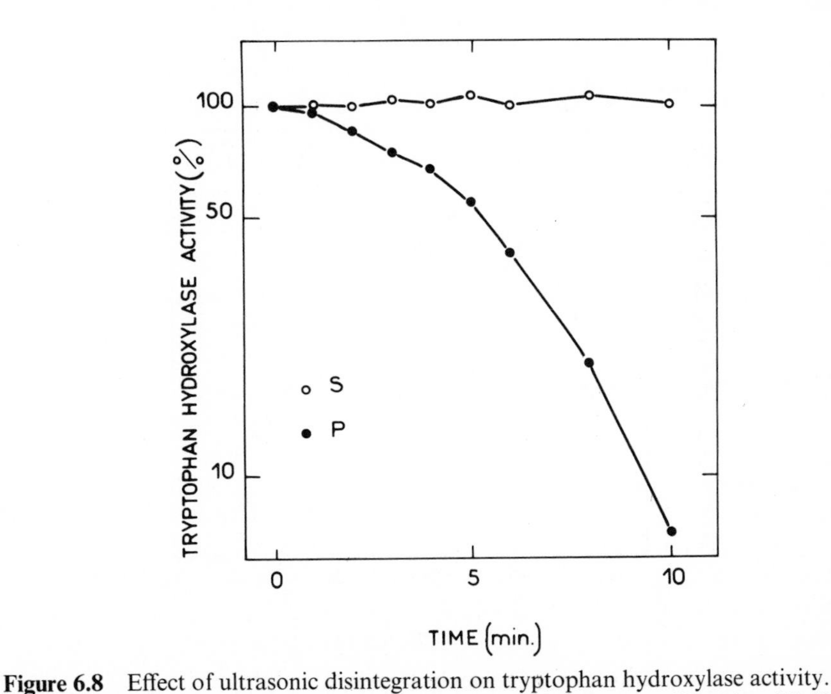

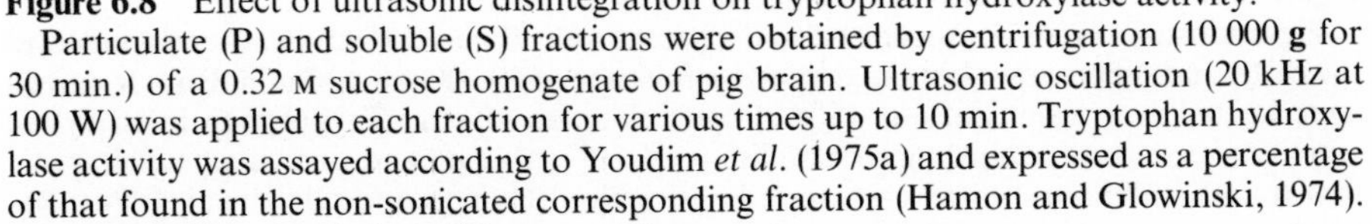

Figure 6.8 Effect of ultrasonic disintegration on tryptophan hydroxylase activity.
Particulate (P) and soluble (S) fractions were obtained by centrifugation (10 000 **g** for 30 min.) of a 0.32 M sucrose homogenate of pig brain. Ultrasonic oscillation (20 kHz at 100 W) was applied to each fraction for various times up to 10 min. Tryptophan hydroxylase activity was assayed according to Youdim *et al.* (1975a) and expressed as a percentage of that found in the non-sonicated corresponding fraction (Hamon and Glowinski, 1974). Reproduced by permission of Pergamon Press

definite answer can be given at present, tryptophan hydroxylase is very likely more or less bound to membranes in synaptosomes. Elegant studies on striatal tyrosine hydroxylase (Kuczenski and Mandell, 1972; Kuczenski, 1973a, b) have demonstrated the existence of a membrane-bound enzyme having properties different from that of the soluble one. Since the sensitivities of soluble and synaptosomal tryptophan hydroxylase to ultrasonic oscillations (Figure 6.8) and heat inactivation (Figure 6.9) are different (Hamon and Glowinski, 1974; Youdim *et al.*, 1975a) the actual existence of two different forms of the enzyme may be suggested. Whether this hypothesis is true or not, immunological studies indicate that the same antibody reacts with the (soluble) form of 5-HT perikaryons and the (particulate) form of 5-HT terminals in the septum (Gál *et al.*, 1975b; Joh *et al.*, 1975). Owing to the availability of a specific antibody, great progress about the subcellular localization was recently made by immunocytochemistry: Joh *et al.* (1975) observed that tryptophan hydroxylase was associated with endoplasmic reticulum in perikaryons and with neuro-

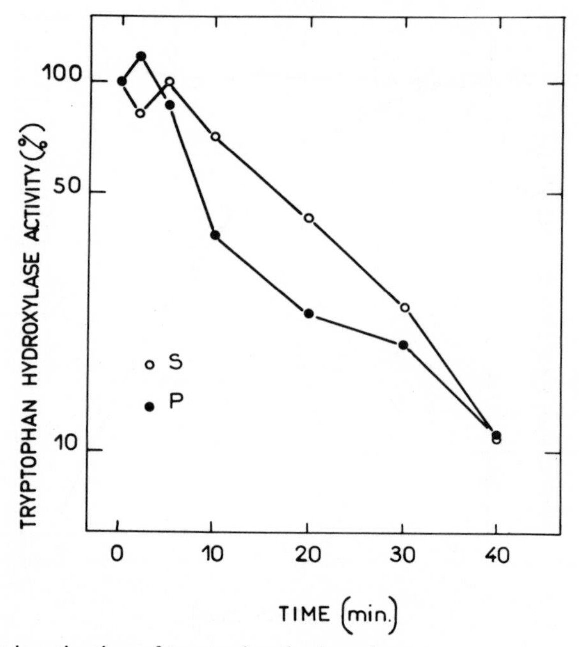

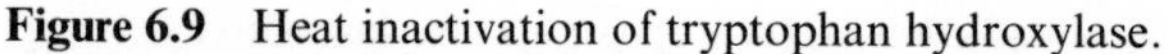

Figure 6.9 Heat inactivation of tryptophan hydroxylase.

Pig brainstems were homogenized in 0.32 M sucrose; P and S fractions corresponded to the pellet and supernatant obtained by centrifugation at 10 000 **g** for 30 min. Both fractions were heated in a water bath at 45 °C for times up to 40 min. Tryptophan hydroxylase activity was assayed according to Youdim *et al.* (1975a) and expressed as a percentage of that found in the non-heated corresponding fraction (Hamon and Glowinski, 1974). Reproduced by permission of Pergamon Press

tubules in processes; therefore, the soluble form of the enzyme may be only the result of biochemical techniques used for extraction. This is a critical point since the properties of tryptophan hydroxylase, mainly studied on the soluble form of the enzyme, may be different from those exhibited by the enzyme in its natural environment.

3 Molecular weight—composition

Gel filtration on Sephadex G-100 of soluble tryptophan hydroxylase prepared from pig brainstem (Figures 6.10, 6.11) allowed determination of the enzyme's molecular weight: 55–60 000 daltons (Youdim *et al.*, 1974, 1975a). The same procedure applied to the enzyme extracted from beef pineals indicated a molecular weight of about 50 000 daltons (Lovenberg *et al.*, 1973; Lovenberg, 1977). From Sepharose 6B gel filtration, polyacrylamide gel electrophoresis, and sucrose density gradient centrifugation of soluble tryptophan hydroxylase extracted from the rabbit hindbrain, Tong and Kaufman (1975) estimated a molecular weight around 230 000 daltons. The electrophoresis pattern of the purified enzyme on sodium dodecyl sulphate polyacrylamide gel suggested that tryptophan hydroxylase is in fact a tetramer composed of two subunits very close in molecular weight (57 500 and 60 900 daltons). Thus, the value obtained

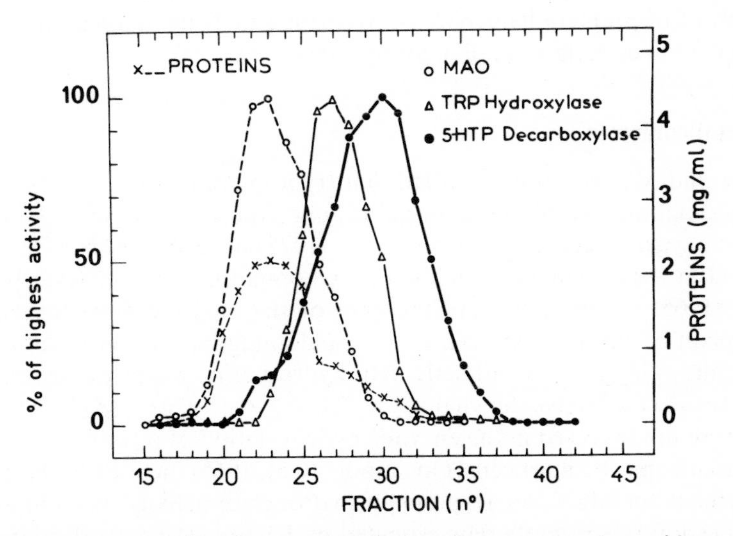

Figure 6.10 Elution profile of tryptophan hydroxylase, monoamine oxidase, and 5-hydroxytryptophan decarboxylase from Sephadex G-100. In each fraction, the enzymic activities were expressed as a percentage of the highest value (elution peak) (Youdim *et al.*, 1975a). Reproduced by permission of Pergamon Press

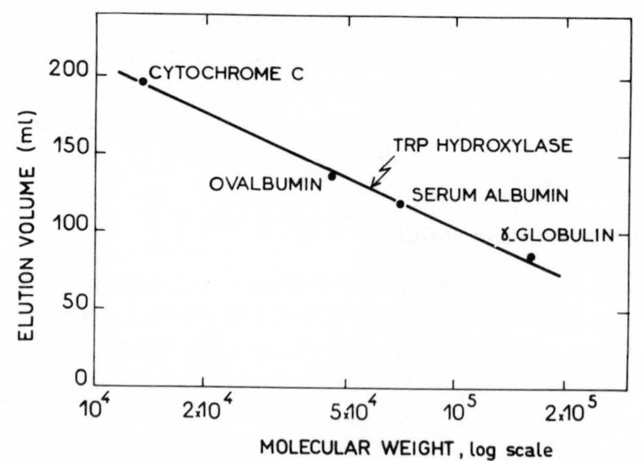

Figure 6.11 Determination of the molecular weight of tryptophan hydroxylase by gel filtration on Sephadex G-100 (Youdim *et al.*, 1975a). Reproduced by permission of Pergamon Press

by Youdim *et al.* (1974, 1975a) very probably corresponds to an active subunit of the enzyme after ultrasonic disintegration. This quaternary structure of tryptophan hydroxylase might account for the allosteric properties of the enzyme (see below). At present, not much is known about the biochemical composition of tryptophan hydroxylase. According to Tong and Kaufman (1975), tryptophan hydroxylase is not a glycoprotein.

4 Metal cofactors

Cooper and Melcer (1961) reported that tryptophan hydroxylation activity in the intestinal mucosa of the rat or guinea pig required ascorbic acid and Cu^{2+} but not oxygen. Recently, Noguchi *et al.* (1973) have shown that Cu^{2+} is not involved in the enzymic conversion of tryptophan into 5-hydroxytryptophan. Fuller (1965) reported that, in the case of the enzymic 5-hydroxylation of tryptophan by rat liver extracts, Cu^{2+} could antagonize the stimulatory effect of 2-amino-4-hydroxy-6,7-dimethyltetrahydropterin. More extensive studies by Sato *et al.* (1967) have shown that Ca^{2+}, Co^{2+}, Cu^{+}, Cu^{2+}, Mg^{2+}, Mn^{2+}, and Ni^{2+} were not involved in the enzymic hydroxylation of tryptophan occurring in mouse neoplastic mast cells. One report (Gál, 1965) mentioned the absolute requirement for Mg^{2+} for tryptophan hydroxylase activity in brain extracts. This question was recently reinvestigated by Knapp *et al.* (1975), who did not confirm the finding of Gál (1965). Indeed, they mentioned that not only Mg^{2+} but also Ba^{2+} and Mn^{2+} have no activating effects on tryptophan hydroxylase. In contrast, they observed that Ca^{2+} activates the midbrain soluble enzyme, in agreement with others (Boadle-Biber *et al.*, 1975; Hamon *et al.*, 1977; Table 6.3). This activation results from an increased affinity of the enzyme for both its substrate and the pterin cofactor.

Table 6.3 Effect of Ca^{2+} on the soluble tryptophan hydroxylase activity

	Tryptophan hydroxylase activity (nmol 5-hydroxytryptophan formed mg^{-1} h^{-1})	%
Control	1.574	100
2 mM Ca^{2+}	2.356	149
3 mM Ca^{2+}	2.690	171
4.5 mM Ca^{2+}	2.956	188
1.5 mM EGTA	1.228	78

Tryptophan hydroxylase activity in the 30 000 **g** supernatant of a rat brainstem homogenate was estimated according to Gál and Patterson (1973) with 22 μM tryptophan and 0.16 mM 6-MPH_4. The addition of Ca^{2+} to the assay mixture activated the enzyme whereas the cation chelating agent sodium ethylene glycol tetraacetate (EGTA) slightly inhibited its activity (M. Hamon and S. Bourgoin, unpublished observations).

The possible role of Fe^{2+} in the enzymic hydroxylation process was also extensively studied. Omission of Fe^{2+} in the assay resulted in a large reduction in tryptophan hydroxylase activity from mast cells (Lovenberg, *et al.*, 1965; Sato *et al.*, 1967). Addition of Fe^{2+} to the enzyme extracted from beef pineals stimulated its activity about twofold. The calculated K_m value of the pineal enzyme for iron is about 25 μM (Jequier *et al.*, 1969). The exclusion of Fe^{2+} from the assay mixture resulted in a 35% reduction of the activity of partially purified tryptophan 5-hydroxylase from the rat intestine (Noguchi *et al.*, 1973). In the brain, the effect of Fe^{2+} on the enzymic conversion of tryptophan into 5-hydroxytryptophan is much less obvious. Whereas Friedman *et al.* (1972) found Fe^{2+} to have a stimulatory effect when rabbit hindbrain tryptophan hydroxylase was assayed in absence of catalase, Jequier *et al.* (1969), using the same conditions, failed to detect any effect of Fe^{2+} on the activity of tryptophan hydroxylase from the rat brainstem. According to Friedman *et al.* (1972) Fe^{2+} protects a sensitive component of the tryptophan hydroxylation system from inactivation by H_2O_2. However, in the assay conditions used by Jequier *et al.* (1969), when Fe^{2+} was removed no H_2O_2 decomposing agent was added to replace it (2-mercaptoethanol is not active in this respect), therefore the hypothesis of Friedman *et al.* (1972) is very unlikely. According to Kaufman (1974) the formation of hydrogen peroxide in the course of assays of aromatic amino acids hydroxylases results from oxidation of pterin cofactor uncoupled with the hydroxylation. In the case of tryptophan hydroxylase, the ratio of tetrahydropterin oxidized to 5-hydroxytryptophan formed is very close to one, whichever cofactor is used—BH_4 (Friedman *et al.*, 1972), 6-MPH_4, or $DMPH_4$ (Tong and Kaufman, 1975)—indicating that formation of H_2O_2 during the enzymic assay is doubtful. However, Fe^{2+} may act as a cofactor and is very likely responsible for the higher rate of enzymic tryptophan hydroxylation observed in the presence of the cation. In agreement with this hypothesis, stimulation of soluble tryptophan hydroxylase from guinea pig brainstem by Fe^{2+} was detected even in presence of catalase (Ichiyama *et al.*, 1970). However, the stimulatory effect of added Fe^{2+} was not observed on the rat brainstem enzyme (Jequier *et al.*, 1969) or on that extracted from pig brainstem (Youdim *et al.*, 1974, 1975a). On the contrary, the inhibitory effects of iron chelating agents (Table 6.4) were reported by most authors. Thus, α,α'-bipyridyl

Table 6.4 Effect of iron and of some chelating agents on soluble tryptophan hydroxylase activity

Addition	Tryptophan hydroxylase activity
None	100
+ 0.5 mM Fe^{2+}	88
+ 1 mM 1, 10-phenanthroline	65
+ 1 mM EDTA	59
+ 1 mM 8-hydroxyquinoline	53

Pig brain tryptophan hydroxylase activity was expressed as a percentage of the standard value (Youdim *et al.*, 1975a). Reproduced by permission of Pergamon Press.

inhibited the enzymic activity of the rat intestine (Noguchi *et al.*, 1973). Similarly the pineal enzyme was almost completely inactivated by addition of α,α'-bipyridyl or *O*-phenanthroline to the assay mixture (Jequier *et al.*, 1969). With the brain enzyme, several authors (Jequier *et al.*, 1969; Ichiyama *et al.*, 1970; Youdim *et al.*, 1974, 1975a) mentioned a significant inhibition exerted by these chelating agents. Therefore the reduction of tryptophan hydroxylase activity in phosphate buffer (see Gál, 1974) might be the consequence of the mobilization of Fe^{2+} (in the prosthetic group?) to form a microprecipitate of insoluble ferrous phosphate. Finally the role played by Fe^{2+} in the hydroxylation reaction is unclear at present. Studies on the purified enzyme would decide whether tryptophan hydroxylase is a metalloenzyme like phenylalanine hydroxylase (Fisher *et al.*, 1972).

5 pH dependence

In standard conditions the pH optimum of tryptophan hydroxylation by the brainstem and beef pineal enzymes is 7.5 (Jequier *et al.*, 1969). The pH effect is most significant since the activity is halved at pH 7.0. The upper limit which results in similar reduction of activity is pH 8.5 (Jequier *et al.*, 1969). A close dependence on pH is also exhibited by the enzyme prepared from mast cells (Sato *et al.*, 1967). In this case, the optimal value is pH 6.75 and the activity is at least equal to 50% of its maximum provided that the assay is performed between pH 6.0 and 7.5. According to Noguchi *et al.* (1973) tryptophan hydroxylase activity from the rat intestine is maximum at pH 8.0. A similar value was also determined for the soluble enzyme of guinea pig brainstem (Ichiyama *et al.*, 1970). When tryptophan hydroxylase activity is estimated by the $^{14}CO_2$ trapping method the pH of this assay mixture is critical. Increasing the pH results not only in a higher rate of decarboxylation of 5-hydroxytryptophan (Ichiyama *et al.*, 1970) but also in direct decarboxylation of ^{14}C-tryptophan. Indeed the optimal pH determined by Ichiyama *et al.* (1970) may not represent only that of tryptophan hydroxylase activity (equal to 7.5; Jequier *et al.*, 1969), but that combined with hydroxylation of tryptophan *and* decarboxylation of 5-hydroxytryptophan.

6 Substrate specificity

First studies on tryptophan hydroxylase were in favour of a highly specific enzyme, catalysing only the hydroxylation of tryptophan. Thus, Hosada and Glick (1966) did not detect any significant phenylalanine hydroxylating activity of tryptophan hydroxylase from neoplastic murine mast cells. Similarly, the brain enzyme was said to be highly specific for tryptophan (Jequier *et al.*, 1969; Lovenberg and Victor, 1974). Recent reports have challenged this view since tryptophan hydroxylase can catalyse the hydroxylation of other amino acids whatever the enzyme source. Thus, Sato *et al.* (1967) demonstrated that the neoplastic mast cells do hydroxylate tryptophan as well as phenylalanine. However, since the enzyme exhibits the same affinity for tryptophan and phenylalanine, it is not identical with the rat liver phenylalanine hydroxylase. In the latter case, the affinity is about 10 times higher for phenylalanine than for tryptophan (Sato *et al.*, 1967). Similarly, the pineal enzyme can hydroxylate tryptophan and phenylalanine to the same extent (Jequier *et al.*, 1969; Lovenberg *et al.*, 1973). More recently, Tong and Kaufman (1975) have reported hydroxylation of L-phenylalanine by extensively purified rabbit hindbrain tryptophan hydroxylase. The rate of phenylalanine hydroxylation could reach 85.9% that of tryptophan when the enzymic assay was performed in the presence of BH_4 as the pterin cofactor. Although rat liver phenylalanine hydroxylase and brain tryptophan hydroxylase used both phenylalanine and tryptophan as substrates, immunological studies demonstrate that these enzymes were structurally unrelated (Gál *et al.*, 1975b). Owing to the unspecificity of tryptophan hydroxylase for its substrate amino acid, Breakefield and Nirenberg (1974) recently obtained a serotoninergic neuroblastoma line (N-T 16) from the uncloned neuroblastoma C1300. The method they used for the selection of neuroblastoma cells consisted of culturing uncloned neuroblastoma in the absence of tyrosine, an essential amino acid for most mammalian cells, so that only the cells able to synthesize tyrosine from phenylalanine could survive. The occurrence of serotoninergic neuroblastoma among the living cells can be considered as an elegant demonstration of phenylalanine hydroxylating capacity of tryptophan hydroxylase in these cells. The enzyme extracted from the rat intestine cannot hydroxylate L-phenylalanine and L-tyrosine (Noguchi *et al.*, 1973). Another peculiarity of this enzyme is that, in contrast to tryptophan hydroxylase from the rat, pigeon, and rabbit brains (Gál, 1965; Friedman *et al.*, 1972), D-tryptophan can serve as a substrate (Noguchi *et al.*, 1973). The hydroxylation rate is one third of that occurring with L-tryptophan. With the enzyme extracted from the rat pineal, cerebellum, and septum, L-tryptophan was reported not to be the best substrate since the enzymic conversion of α-methyltryptophan into α-methyl-5-hydroxytryptophan occurred at a faster rate than that of the natural substrate (Gál and Christiansen, 1975). In all other areas, L-tryptophan was a better substrate. Indeed, when tryptophan hydroxylase was extracted from subcortical areas of the rat brain, its V_{max} with L-α-methyltryptophan was about half that with L-tryptophan whichever the cofactor was used (6-methyltetrahydropterin or 6,7-dimethyltetrahydropterin). These regional differences could reveal the existence of 'isoenzymes' in particular nuclei of the rat brain (Gál and Christiansen, 1975).

7 Cofactor requirements

It has already been mentioned that the requirement for exogenous reduced pterin cofactor is absolute for the activity of the soluble enzyme. On the contrary, the synaptosomal conversion of tryptophan into 5-hydroxytryptophan is not stimulated by such addition. An inhibitory effect of $DMPH_4$ has even been reported in a few cases (Robinson *et al.*, 1968; Ichiyama *et al.*, 1970; Youdim *et al.*, 1975a). Similarly, concentrations above 1 mM of tetrahydrobiopterin were reported to inhibit rat liver phenylalanine hydroxylase (Stone and Townsley, 1973). The exact nature of the endogenous pterin cofactor of tryptophan hydroxylase in brain is still unknown. Recently, Gál and Roggeveen (1973) partially purified a stimulating factor of the enzyme from the rat brain. Although they reported that this molecule has properties indicative of a pterin-like structure, its identification is still an open question. Therefore, the activity of tryptophan hydroxylase is always determined in the presence of an exogenous cofactor. The three molecules used are 6,7-dimethyl-5,6,7,8-tetrahydropterin ($DMPH_4$), its 6-monomethyl derivative ($6\text{-}MPH_4$) and tetrahydrobiopterin (BH_4) (Figure 6.12). The respective K_m values of the rabbit enzyme for these cofactors are 130 μM, 67 μM, and 31 μM (Friedman *et al.*, 1972). With the tryptophan hydroxylase from the rat brainstem, Jequier *et al.* (1969) obtained much lower values: 30 μM for $DMPH_4$ and 5 μM for BH_4 as the cofactors. In the case of $DMPH_4$, Ichiyama *et al.* (1970) reported a similar value: 60 μM for the soluble enzyme extracted from the guinea pig brainstem. A very recent study by Kizer *et al.* (1975) on soluble tryptophan hydroxylase from the rat brainstem mentioned a K_m value for $6\text{-}MPH_4$ equal to 65 μM. Recently, using the same enzyme source but assaying tryptophan hydroxylase activity according to Gál and Patterson (1973), we obtained a K_m equal to 200 μM for $6\text{-}MPH_4$ (M. Hamon and S. Bourgoin, unpublished observations). Knapp *et al.* (1974) reported similar values: 500 ± 50 μM and 200 ± 25 μM for $DMPH_4$ and $6\text{-}MPH_4$ respectively with the rat midbrain soluble tryptophan hydroxylase Differences in assay conditions were very likely responsible for these discrepancies. Among the pterin molecules, and by analogy with the rat liver phenylalanine hydroxylase, tetrahydrobiopterin is a good candidate to be the endogenous cofactor of tryptophan hydroxylase in brain. Indeed, with the endogenous cofactor in synaptosomes, the reported K_m of tryptophan hydroxylase for tryptophan (20 μM: Ichiyama *et al.*, 1968; Youdim *et al.*, 1974, 1975a; 22 μM: Peters *et al.*, 1968) is in close agreement with that obtained for the soluble enzyme in the presence of BH_4. A similar value was determined recently for tryptophan hydroxylation in rat brainstem slices (34 μM: Bourgoin *et al.*, 1974). Finally, by measuring the rate of 5-hydroxytryptophan accumulation in brain after both peripheral administration of various amounts of L-tryptophan and a 5-hydroxytryptophan decarboxylase inhibitor (NSD 1015), Carlsson and Lindqvist (1972) have estimated an *in vivo* K_m value of tryptophan hydroxylation for tryptophan. This value ($K_m = 60$ μM) closely resembled that obtained *in vitro* with BH_4 as the pterin cofactor. It is not surprising that the K_m for the naturally occurring cofactor is lower than those for the two other synthetic compounds. A much more interesting and somewhat unexpected finding is that the apparent K_m's of tryptophan hydroxylase (as for phenylalanine and tyrosine hydroxylases) for its substrates, the amino acid and oxygen, also vary with the pterin used (Kaufman, 1970). Thus, whereas the K_m for tryptophan is relatively high when the soluble enzyme is working in the presence of $DMPH_4$ (300 μM: Jequier *et al.*, 1969; Ichiyama *et al.*, 1970; 290 μM: Friedman *et al.*, 1972; 400 μM: Youdim *et al.*, 1974, 1975a), it comes in the range of the endogenous amino acid concentration in brain tissues when the cofactor is either $6\text{-}MPH_4$ (78 μM: Friedman *et al.*, 1972; 89–104 μM: Gál, 1974; Gál *et al.*, 1975a; 80 μM: Lovenberg and Victor, 1974; Kizer *et al.*, 1975, 200 μM: M. Hamon and S. Bourgoin, unpublished observations), or BH_4 (50 μM: Friedman *et al.*, 1972; 25–38 μM: Gál, 1974).

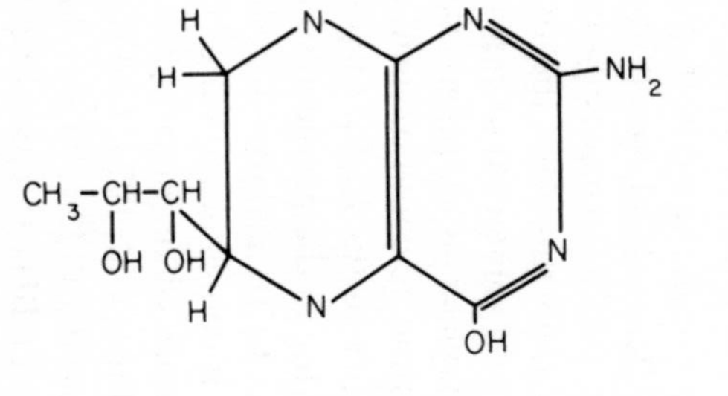

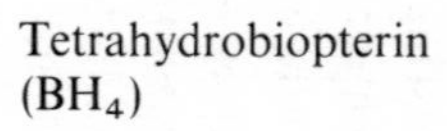

Tetrahydrobiopterin
(BH_4)

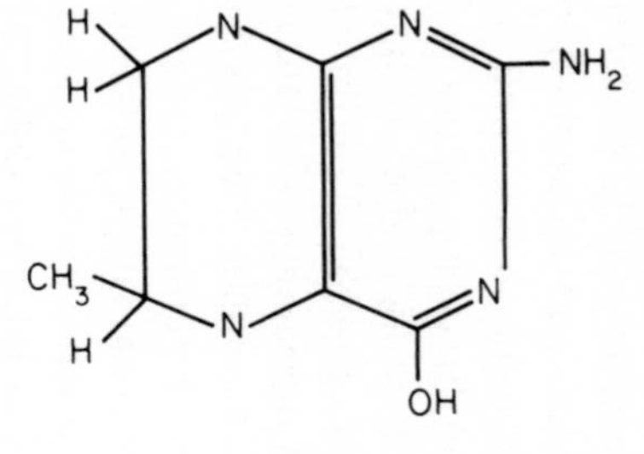

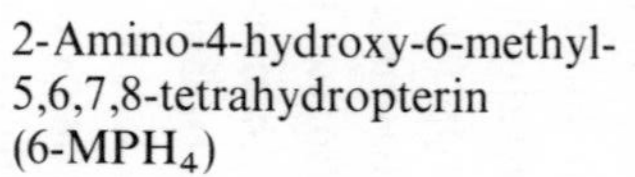

2-Amino-4-hydroxy-6-methyl-
5,6,7,8-tetrahydropterin
(6-MPH_4)

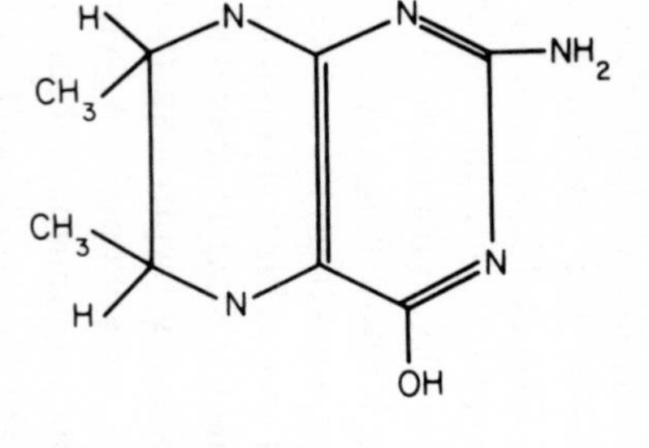

2-Amino-4-hydroxy-6,7-dimethyl-
5,6,7,8-tetrahydropterin
($DMPH_4$)

Figure 6.12 Pterin cofactors used in the assay of tryptophan hydroxylase activity

The apparent K_m of tryptophan hydroxylase for molecular oxygen with $DMPH_4$ as cofactor is equal to 20% (Friedman *et al.*, 1972). In the presence of tetrahydrobiopterin this value decreases to 2.5%, a value in the range of the oxygen concentration in the brain of animals breathing air (Jamieson and Van den Brenck, 1965).

In the case of tyrosine hydroxylase, the influence of the pterin cofactor on the affinity of the enzyme for the amino acid substrate is so high that the substrate specificity of the enzyme can be altered. Thus whereas L-phenylalanine is a very poor substrate of the adrenal tyrosine hydroxylase as compared to L-tyrosine in the presence of $DMPH_4$ as cofactor (Ikeda *et al.*, 1965), with BH_4 the rate of phenylalanine hydroxylation is equal to or even greater than the rate of tyrosine hydroxylation (Shiman *et al.*, 1971). Indeed, recently Karobath and Baldessarini (1972) have demonstrated that *in vivo* phenylalanine could serve as an important precursor of noradrenaline. Although less marked, such effects were also observed with purified rabbit hindbrain tryptophan hydroxylase. Thus, Tong and Kaufman (1975) found that the rate of hydroxylation of L-phenylalanine was 28.8%, 44.3%, and 85.9% of those observed with L-tryptophan in the presence of $DMPH_4$, 6-MPH_4, and BH_4 respectively. Even in the latter case, the affinity of tryptophan hydroxylase for L-phenylalanine ($K_m = 286$ μM) was still much lower than for L-tryptophan ($K_m = 32$ μM), so that competition of tryptophan hydroxylation by L-phenylalanine might occur only in pathological states such as in phenylketonuria (Tong and Kaufman, 1975). In other cases, the role of the cofactor has been ruled out. Thus, using the same pterin cofactor, 6-MPH_4, Gál and Christiansen (1975) still observed differences in the relative velocities of tryptophan and α-methyltryptophan hydroxylation by the soluble enzyme extracted from various brain areas. Whereas the septal enzyme was much more active with α-methyltryptophan as the substrate, the reverse was true for tryptophan hydroxylase prepared from the pons: in the latter case, tryptophan was hydroxylated 3.5 times more rapidly than α-methyltryptophan. The observation that the pineal enzyme was also much more active with α-methyltryptophan than with tryptophan as the amino acid substrate (Gál and Christiansen, 1975) in the presence of 6-MPH_4 as the cofactor is of interest.

In conclusion, although the substrate specificity of aromatic amino acids hydroxylases seems to depend largely on the nature of the pterin cofactor (Kaufman, 1974; Tong and Kaufman, 1975), other factors, including the possible occurrence of isoenzymes and allosteric effectors, may be of some importance in this respect, particularly for tryptophan hydroxylase.

8 Inhibitors

Extensive studies by McGeer and Peters (1969) have shown that compounds which inhibit tryptophan hydroxylase belong to three classes. They are: *cate-*

chols, *phenylalanine and ring-substituted phenylalanines*, and *6-substituted tryptophans*. Only a few compounds inhibit the enzyme specifically *in vitro* and *in vivo*.

(a) *p-Chlorophenylalanine*

The administration of *p*-chlorophenylalanine (generally 300 mg (kg body weight)$^{-1}$) induces a large (90%) and long-lasting (about 12 days) decrease of 5-HT levels in brain (Koe and Weissman, 1966). In contrast, only a slight and transient decrease of norepinephrine levels are observed after this treatment (Koe and Weissman, 1966). Mechanisms of action of *p*-chlorophenylalanine on the synthesis of serotonin were recently reviewed by Knapp and Mandell (1972a): the drug inhibits competitively the uptake of tryptophan by 5-HT neurons; it is a potent reversible competitive inhibitor of tryptophan hydroxylase *in vitro* with a K_i of 0.3 mM (Jequier *et al.*, 1967); *in vivo*, it induces a long-lasting inactivation of the enzyme. The latter effect occurs mainly in areas containing 5-HT terminals. The pineal enzyme (Deguchi and Barchas, 1972a, c) as well as tryptophan hydroxylase in 5-HT cell bodies of the raphe (Deguchi *et al.*, 1973; Aghajanian *et al.*, 1973) are not irreversibly inactivated since tryptophan loading can reverse the inhibition induced by *p*-chlorophenylalanine administration. According to Gál *et al.* (1970) the inactivation of tryptophan hydroxylase would result from the incorporation of *p*-chlorophenylalanine at the active site of the enzyme in the course of protein synthesis. With the extensive purification of tryptophan hydroxylase (Tong and Kaufman, 1975), the direct demonstration of incorporation of radioactive *p*-chlorophenylalanine into the enzyme molecule may now be possible. Indirect evidence to support Gál's hypothesis have appeared. Among them is the observation of Gál and Millard (1971) that inhibition of protein synthesis (i.e. the blockade of possible incorporation of *p*-chlorophenylalanine in tryptophan hydroxylase) by cycloheximide significantly delays the appearance of inactivated enzyme. However, other reports (Andén and Modigh, 1972; Bennett and Aghajanian, 1974) mention that the inactivation of tryptophan hydroxylase still occurs in lesioned axones, i.e. in absence of any synthesis of new molecules of the enzyme (only localized in cell bodies). In addition, since D-*p*-chlorophenylalanine, which is not incorporated into proteins, decreases the levels of 5-HT and 5-HIAA to the same extent as the L-amino acid (Koe and Weissman, 1966; Fratta *et al.*, 1973), the first hypothesis of Gál *et al.* (1970) is rather doubtful. Another possibility would be that the inhibitor binds to the existing enzyme at or near the catalytic site. Indeed, dialysis fails to remove ^{14}C-*p*-chlorophenylalanine from proteins (including tryptophan hydroxylase) previously incubated with the labelled inhibitor (Gál *et al.*, 1970). In any case, *p*-chlorophenylalanine is a very useful pharmacological tool for inhibiting 5-HT synthesis selectively in brain. Moreover, its inactivating effect on tryptophan hydroxylase can be of great interest in estimating the half-life of the enzyme: on the basis of the reappearance of newly synthesized enzyme after the administration of the drug, Meek and Neff (1972) reported a half-life of 2–3 days and Knapp and Mandell (1972a) calculated a value of about 5 days. Other procedures based on the decline of tryptophan hydroxylase activity in lesioned fibres in brain have also given values of slightly more than 2 days (Kuhar *et al.*, 1971, Meek and Neef, 1972).

(b) *Halogenoamino acids*

Other amino acids with a halogen moiety were looked at for specific inhibition of tryptophan hydroxylase by McGeer *et al.* (1968). Among the compounds tested were D,L-6-fluoro- and D,L-6-chlorotryptophan, which are as efficient as *p*-chlorophenylalanine in inhibiting tryptophan hydroxylase. Their specificity was higher: in particular they have lesser inhibitory effect on tyrosine hydroxylase activity. *In vivo*, the peak effect of 6-fluorotryptophan on tryptophan hydroxylation occurs 1–2 h. after its administration. In contrast to *p*-chlorophenylalanine, the action of the fluoro derivative does not last more than 16 h. According to Peters (1971), injection of 1 mmol of this compound per kilogram body weight induces a 77% decrease in brain tryptophan hydroxylase activity 2 h. after its administration. Therefore, 6-fluoro- and 6-chlorotryptophan are useful for the induction of large and transient specific inhibition of serotonin synthesis in pharmacological studies.

Recently, Gál and Christiansen (1975) studied the effect of α-methyltryptophan on rat brain tryptophan hydroxylase. Although this compound preferentially inhibits tryptophan hydroxylase when compared to tyrosine hydroxylase, its use in pharmacological studies is limited on account of its complex mechanism of action (Sourkes, 1974; Gál and Christiansen, 1975).

(c) *Catechols*

Catechol compounds such as norepinephrine, dopamine, and dihydroxyphenylalanine inhibit the soluble tryptophan hydroxylase activity. Using the pineal enzyme, Lovenberg *et al.* (1968) report that norepinephrine is a noncompetitive inhibitor when tested either with various substrate concentrations or with differing concentrations of the pterin cofactor ($DMPH_4$). In contrast, Mandell *et al.* (1972, 1974) found that norepinephrine and dopamine are competitive inhibitors of soluble brain tryptophan hydroxylase with regard to $DMPH_4$, whereas Karobath *et al.* (1972) observed that the inhibition of 5-HT synthesis by dihydroxyphenylalanine in synaptosomes is competitive to tryptophan. However, since addition of Fe^{2+} can reverse the inhibition induced by catechols it might be that these compounds interfere with tryptophan hydroxylase activity by chelating the cation possibly present in a prosthetic group of the enzyme (Lovenberg *et al.*, 1968). In synaptosomes (Karobath *et al.*, 1972), as well as in brain slices (Goldstein and Frenkel, 1971), the synthesis of serotonin from tryptophan is impaired by catechols; notably after *in vivo* administration of dihydroxyphenylalanine. A catechol derivative, α-propyl dopacetamide (H22/54 or α-propyl-3,4,dihydroxyphenylacetamide) inhibits tryptophan hydroxylase activity *in vivo*. This effect is not specific since tyrosine hydroxylase, phenylalanine hydroxylase, and catechol-*o*-methyl transferase are inhibited as well (Carlsson *et al*., 1963). However, α-propyldopacetamide is very often used to measure the turnover rate of 5-HT *in vivo* in brain (see Andén, 1974).

(d) p-Chloroamphetamine

In contrast to the compounds described previously, *p*-chloroamphetamine does not inhibit tryptophan hydroxylase activity *in vitro* (Sanders-Bush *et al*., 1972a, b). However, when the enzyme is assayed in brain extracts of rats previously treated with *p*-chloroamphetamine, a dose-related reduction of tryptophan hydroxylase activity is observed. This effect persists for at least 4 months after a single injection of 10 mg of *p*-chloroamphetamine per kilogram body weight (Sanders-Bush *et al*., 1972b). In fact, as confirmed by the decrease in the other specific marker of serotoninergic neurons, i.e. the synaptosomal uptake of the indoleamine, this irreversible effect of *p*-chloroamphetamine results from the chemical lesion of 5-HT neurons. Histological (Harvey *et al*., 1975) and biochemical (Bertilsson *et al*., 1975) evidence for a specific lesion of 5-HT cells restricted to the B9 group of Dahlström and Fuxe (1964) were recently published. However, direct electrical lesion of B9 nuclei failed to reproduce the large alteration in 5-HT metabolism observed after *p*-chloroamphetamine injection (Massari and Sanders-Bush, 1975; L. M. Neckers, personal communication). Therefore, the possible formation of dihydroxyindoles analogous to the highly toxic 5,6- and 5,7-dihydroxytryptamines (Baumgarten *et al*., 1972, 1973) from *p*-chloroamphetamine (as proposed by Gál *et al*., 1975a) does not occur only in 5-HT cells located in the B9 nuclei.

G REGULATORY PROPERTIES OF TRYPTOPHAN HYDROXYLASE

1 Effects of substrates and cofactors *in vitro*

(a) The role of tryptophan availability in the regulation of 5-HT synthesis in the central nervous system

The K_m of tryptophan hydroxylase for tryptophan (50 μM), even with the naturally occurring cofactor BH_4, is still above the physiological concentration of the amino acid in brain tissues (about 20 μM). As a consequence, the rate of tryptophan hydroxylation in brain would fluctuate with the concentration of the precursor amino acid according to first-order kinetics. Indeed, central (Consolo *et al.*, 1965; Grahame *et al.*, 1975) as well as peripheral (Moir and Eccleston, 1968; Fernstrom and Wurtman, 1971; Carlsson and Lindqvist, 1972; Bourgoin *et al.*, 1974) administrations of tryptophan induce large increases in the concentrations of 5-HT and 5-HIAA in the brain. As demonstrated by Carlsson and Lindqvist (1972) the rate of 5-hydroxytryptophan accumulation in the brain is increased by the intraperitoneal administration of large doses of tryptophan. According to these authors the saturation of tryptophan hydroxylase activity *in vivo* in brain is reached after intraperitoneal injection of 300 mg of tryptophan per kilogram body weight. Recently Bourgoin *et al.* (1975) mention that this effect is more pronounced in the forebrain when compared to the brainstem of adult rats, thus suggesting that tryptophan hydroxylase activity is far less than saturation rate in the latter region. The increased *in vivo* enzymic activity induced by peripheral administration of a large dose of tryptophan is much lower in newborn rats than in adults (Bourgoin *et al.*, 1974; Hoff and Goodrich, 1975). This is related to the fact that, in very young animals, tryptophan hydroxylase is almost saturated by tryptophan (Bourgoin *et al.*, 1974; Hamon and Glowinski, 1974).

Tagliamonte *et al.* (1971) and various authors thereafter have shown that when the synthesis of serotonin in the brain (i.e. the rate of tryptophan hydroxylation) was affected by various pharmacological or physical treatments, this was related to changes in the concentration of tryptophan in the brain. Since only free tryptophan in plasma can enter the brain, several authors (Curzon and Knott, 1974; Gessa and Tagliamonte, 1974; Hamon and Glowinski, 1974) have proposed that the rate of tryptophan hydroxylation in brain largely depends on the concentration of this particular form of the peripheral amino acid.

Using tetrahydrobiopterin as the cofactor, Friedman *et al.* (1972) and Tong and Kaufman (1975) observed a significant inhibition of tryptophan hydroxylase activity *in vitro* when the tryptophan concentration was higher than 0.2 mM. Similarly, Bensinger *et al.* (1974) mentioned a reduction in tryptophan hydroxylase activity of intact rat pineal gland (containing the endogenous cofactor) incubated with 0.75–1 mM when compared to that occurring with 0.5 mM tryptophan. On the other hand, no substrate inhibition was seen when $DMPH_4$ replaced BH_4 as the cofactor (Friedman *et al.*, 1972; Youdim *et al.*, 1974; 1975a). In agreement with Gál and Patterson (1973) and Kizer *et al.* (1975), M. Hamon and S. Bourgoin (unpublished observations) failed to detect any reduction of soluble tryptophan hydroxylase activity by high tryptophan concentrations (up to 0.5 mM) in presence of the pterin cofactor 6-MPH_4. The physiological importance of substrate inhibition *in vitro* of aromatic amino acids hydroxylases is still an open question. In the case of rat liver phenylalanine hydroxylase, the inhibition of the *in vitro* enzymic activity by phenylalanine at concentrations above 2 mM (Nielsen, 1969; Tourian *et al.*, 1969; Kaufman, 1970) was not observed by all authors (Stone and Townsley, 1973). Recent studies on the perfused isolated rat liver suggested that the inhibition found *in vitro* of phenylalanine hydroxylase by its own substrate might not be relevant to the regulation of enzyme activity *in vivo* (see Youdim and Woods, 1975; Youdim *et al.*, 1975b; Woods and Youdim, 1977). Similarly, by injecting tyrosine to rats, Dairman (1972) failed to induce any inhibition of *in vivo* tyrosine hydroxylase activity in the brain as well as in the heart.

Using brain slices (from hippocampus, brainstem, etc.) a decrease in the synthesis of ^{3}H-5-hydroxyindoles from ^{3}H-tryptophan when the concentration of this amino acid was raised to 1 mM (M. Hamon and S. Bourgoin, unpublished observations) was not detected. In synaptosomes (containing the actual endogenous pterin cofactor) conflicting results were reported. A tryptophan load of 50 mg (kg body weight)$^{-1}$ 3 h. before death, induced a significant increase in tryptophan hydroxylation in septal synaptosomes (Mandell *et al.*, 1972); whereas Azmitia and McEwen (1974) observed a significant decrease in the enzymic activity in forebrain synaptosomes 4 h. after the administration of 100 mg of L-tryptophan per kilogram body weight. Under the same time conditions, but during the course of a constant intravenous infusion of L-tryptophan (70 mg (kg body weight)$^{-1}$ h^{-1}), Ternaux *et al.* (1976) observed first an acceleration in the rate of tryptophan hydroxylation (on the 2nd hour after the onset of the infusion) then followed by a progressive return to the control level (Table 6.5). Since the concentration of tryptophan remained constant all along the perfusion period following the second hour, the relative decrease in the rate of 5-HT synthesis occurring 3 hours later (Table 6.5) could not be related to a possible substrate inhibition of tryptophan hydroxylase (J. P. Ternaux, S. Bourgoin, F. Héry and M. Hamon, unpublished observation). In agreement with these findings, Carlsson and Lindqvist (1972) observed no decrease in the rate of 5-

Table 6.5 Changes in 5-HT synthesis rate induced by an intravenous perfusion of L-tryptophan

	Tryptophan levels (μg g^{-1})	5-HT levels (μg g^{-1})	5-HT synthesis rate (μg g^{-1} h^{-1})
Control	6.349 ± 0.329	0.635 ± 0.065	0.637 ± 0.115
2 h.	42.691 ± 0.551*	0.869 ± 0.029*	1.673 ± 0.165*
5 h.	42.956 ± 3.200*	0.803 ± 0.032*	0.812 ± 0.197

The synthesis rate of 5-HT in the whole brain of rats anaesthetized with halothane was determined in the course of a constant intravenous perfusion of tryptophan (70 mg (kg body weight)$^{-1}$ h^{-1}). The '5-HT-MAOI' method proposed by Morot-Gaudry *et al.* (1974) was used for this calculation. Although tryptophan levels were not significantly different at the second or the fifth hour after the onset of perfusion, the synthesis of 5-HT was significantly increased only at the second hour (J. P. Ternaux *et al.*, unpublished observations).
* $P<0.05$.

hydroxytryptophan accumulation in brain when the concentration of tryptophan largely exceeded the level (0.5 mM) corresponding to *in vivo* saturation of tryptophan hydroxylase. Similarly, chronic treatment with tryptophan (75 mg (kg body weight)$^{-1}$ for 3 days) failed to alter 'particulate' tryptophan hydroxylase activity in the whole mouse brain (Ho *et al.*, 1975). Although discrepancies between *in vitro* and *in vivo* findings are difficult to explain, one possibility would be that tetrahydrobiopterin is not the endogenous cofactor of these hydroxylating enzymes (Youdim and Woods, 1975; Woods and Youdim, 1977).

Recently it has been observed that double reciprocal plots of tryptophan hydroxylase activity, with 6-MPH_4 as the cofactor, versus the amino acid concentration were not linear with a substrate concentration lower than 30 μM. The curve obtained by plotting $1/v$ versus $1/S$ was in fact parabolic. A possible activation of tryptophan hydroxylase by its substrate might be suggested (M. Hamon and S. Bourgoin, in preparation) by analogy with the effect of phenylalanine on rat liver phenylalanine hydroxylase (Tourian, 1971). Tryptophan cooperativity is totally supressed by previous activation of tryptophan hydroxylase with low concentration (0.01%) of sodium dodecyl sulphate (Figure 6.13)

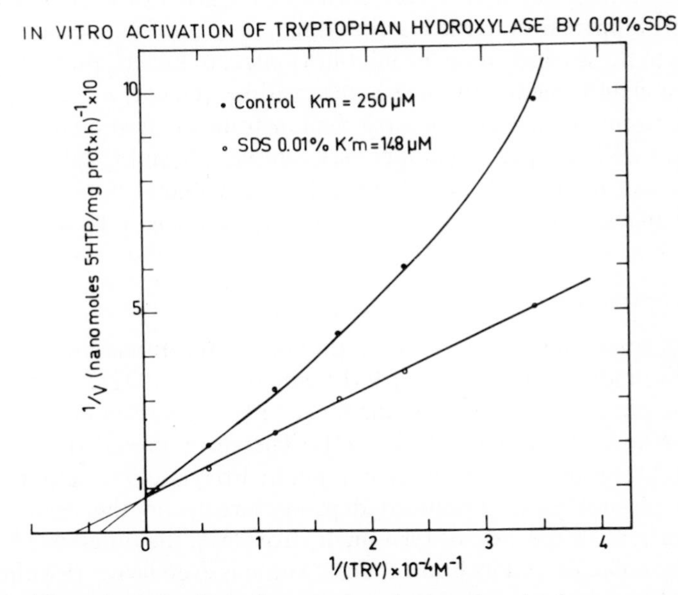

Figure 6.13 Allosteric activation of tryptophan hydroxylase by low concentration of sodium dodecyl sulphate (SDS) (0.01%).

Rat brainstem soluble tryptophan hydroxylase was estimated according to Gál and Patterson (1973) in the presence or absence of 0.01% of the detergent with various concentrations of tryptophan. In the presence of this low concentration of sodium dodecyl sulphate, tryptophan hydroxylase obeys classical Michaelis–Menten kinetics (M. Hamon and S. Bourgoin, unpublished observations)

(Hamon *et al.*, 1977). These results might explain the allosteric properties of tryptophan hydroxylase detected in some pharmacological conditions (see below).

(b) The role of oxygen

When tryptophan hydroxylase activity is measured with BH_4 as the cofactor, its K_m for oxygen is 2.5%, a value somewhat less than the oxygen concentration of 5% in the brain of animals breathing air (see Kaufman, 1974). Thus the enzyme may not be fully saturated with regards to oxygen in physiological conditions. In fact, when rats had to breathe pure oxygen, the rate of 5-HT synthesis in brain increased (Diaz *et al.*, 1968). Inversely, Davis and Carlsson (1973) have shown that the rate of central 5-HTP accumulation dropped significantly following exposure of rats in an atmosphere of 5.6% oxygen in nitrogen for 60–120 min. Therefore, one can conclude that the oxygen concentration may be of physiological importance to modulate the rate of tryptophan hydroxylase activity *in vivo*.

Unlike the tyrosine and phenylalanine hydroxylases, brain tryptophan hydroxylase is not inhibited by an excess of oxygen in the presence of either tetrahydrobiopterin or $DMPH_4$ (Friedman *et al.*, 1972). These *in vitro* data, contrary to those concerning the inhibitory effect of high tryptophan concentrations, are in agreement with *in vivo* observations (Diaz *et al.*, 1968). In spite of these clear-cut *in vivo* and *in vitro* demonstrations, it should be noted that hypoxia as well as hyperoxia induce very complex physiological changes. Therefore, fluctuations in tryptophan hydroxylase activity may not be directly related to the changes in the oxygen partial pressure occurring in tissues.

(c) Cofactor availability

The K_m for tetrahydrobiopterin of the soluble tryptophan hydroxylase from the rabbit hindbrain is about 31 μM (Friedman *et al.*, 1972) and the concentration of biopterin in rat brain is in the range of 1–10 μM (Rembold, 1964; Guroff *et al.*, 1967a, b; Musacchio *et al.*, 1971). Therefore, provided that BH_4 is the actual endogenous cofactor of tryptophan hydroxylase, the activity of this enzyme in physiological conditions depends largely on the reduced cofactor concentration. In the case of tyrosine hydroxylase, the difference between the K_m for the cofactor and its brain concentration is even larger (Kaufman, 1974). Hence, Musacchio *et al.* (1971) proposed that the enzyme which catalyses the synthesis of reduced pterin, quinonoid dihydropteridine reductase, might be the regulator of catecholamine synthesis in certain circumstances. However, since the regional distribution of quinonoid dihydropteridine reductase parallels neither the distribution of tyrosine hydroxylase nor that of tryptophan hydroxylase in brain, this possibility is very unlikely. Moreover, the activity of the reductase enzyme in brain is several thousand times greater than the reported maximal activities of tyrosine and tryptophan hydroxylases. This would suggest that the rate of production of reduced pterin may not be a limiting factor in the hydroxylation processes (Turner *et al.*, 1974). At first sight, the lack of stimulation of tryptophan hydroxylase activity in synaptosomes by exogenous pterin cofactor (Grahame-Smith, 1967; Ichiyama *et al.*, 1970; Youdim *et al.*, 1974, 1975a) could mean that the enzyme is not dependent on the availability of this cofactor in intact microcells; this should be in contrast to tyrosine hydroxylase, since *in vitro* addition of $DMPH_4$ to intact vas deferens (Cloutier and Weiner, 1973) or rat striatal slices (T. C. Westfall, personal communication) does stimulate its activity. Moreover, intraventricular injection of tetrahydrobiopterin (0.5 mg per rat) induces a significant increase in *in vivo* tyrosine hydroxylase activity in the striatum (Kettler *et al.*, 1974). Although much less is known concerning the effects of exogenous pterin cofactor on tryptophan hydroxylase activity *in vivo*, it would be hazardous to claim that the enzyme is not dependent on the cofactor availability in physiological conditions. Indeed, the V_{max} of tryptophan hydroxylase activity in crude extracts of the rat brainstem in the presence of 6-MPH_4 is 25–30 nmol of 5-hydroxytryptophan formed *per milligram of protein* and per hour (M. Hamon and S. Bourgoin, unpublished observations; Gál and Christiansen, 1975). Since this is greater than the rate of tryptophan hydroxylation *in vivo* in the same brain area after saturating tryptophan loading (Bourgoin *et al.*, 1975), the endogenous concentration of the cofactor may be one important limiting factor of the enzymic activity *in vivo*. If this assumption is correct, the measurement of tryptophan hydroxylase activity in the presence of large concentrations of exogenous cofactor would be unrelated to the actual *in vivo* enzyme activity (Youdim and Woods, 1975; Woods and Youdim, 1977). Moreover, the possible fluctuations of tryptophan hydroxylase activity induced by changes in the pterin cofactor concentration in brain tissues would be ignored as well.

(d) The question of feedback end-product inhibition

(i) The role of serotonin The role of intraneuronal 5-HT in the regulation of its synthesis was first suggested by estimations of the rate of *in vivo* 5-HT synthesis in animals previously treated with reserpine or *p*-chlorophenylalanine to deplete central 5-HT levels. Indeed, Tozer *et al.* (1966) observed an increased 5-HT synthesis after reserpine administration. Similarly, the stimulation of 5-HT synthesis induced by electric foot shocks stress was more pronounced in rats pretreated with *p*-chlorophenylalanine than in saline-treated animals (Thierry *et al.*, 1968). Conversely, raising the concentration of 5-HT in brain by inhibiting monoamine oxidase activity with pargyline or pheniprazine induced a significant reduction in the rate of conversion of intracisternally administered ^{3}H-tryptophan into ^{3}H-5-HT (Figure 6.14) (Macon *et al.*, 1971). In contrast the

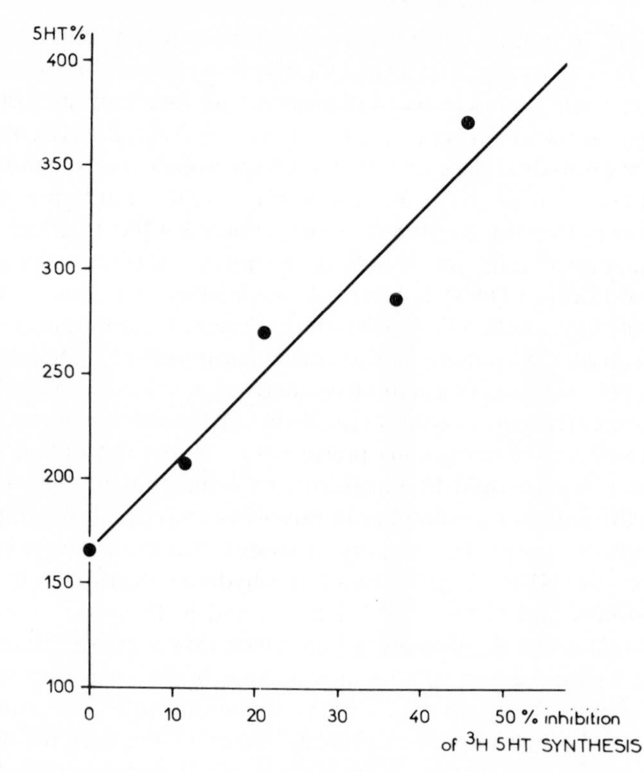

Figure 6.14 Progressive serotonin synthesis inhibition after monoamine oxidase blockade by pheniprazine. Correlation between serotonin concentrations and ^{3}H-5-HT synthesis inhibition in the rat brainstem. (The synthesis of ^{3}H-5-HT was estimated after an intracisternal injection of ^{3}H-tryptophan.) (J. B. Macon *et al*., unpublished data)

synthesis of ^{3}H-5-HT after the central administration of ^{3}H-5-hydroxytryptophan was not altered by the same treatment (Macon *et al*., 1971). Similarly, in mice, 3 h. after pargyline administration the synthesis of ^{3}H-5-HT from intravenously injected ^{3}H-tryptophan was significantly reduced (Glowinski *et al*., 1972). These results were then confirmed by Carlsson and Lindqvist (1973), who reported that the rate of 5-hydroxytryptophan accumulation in brain after the blockade of 5-hydroxytryptophan decarboxylase was lower in animals pretreated with pargyline provided that the latter drug was injected at least 60 min. before the 5-hydroxytryptophan decarboxylase inhibitor. In these conditions, brain 5-HT levels were doubled as compared to control mice. Similar findings were obtained *in vitro* with striatal slices of rats pretreated with pargyline or pheniprazine 3 h. before death (Figure 6.15) (Hamon *et al*., 1972, 1973). These tissues contained 2–2.5 times more endogenous 5-HT than the respective controls and converted ^{3}H-tryptophan into ^{3}H-5-HT at a reduced rate. Using

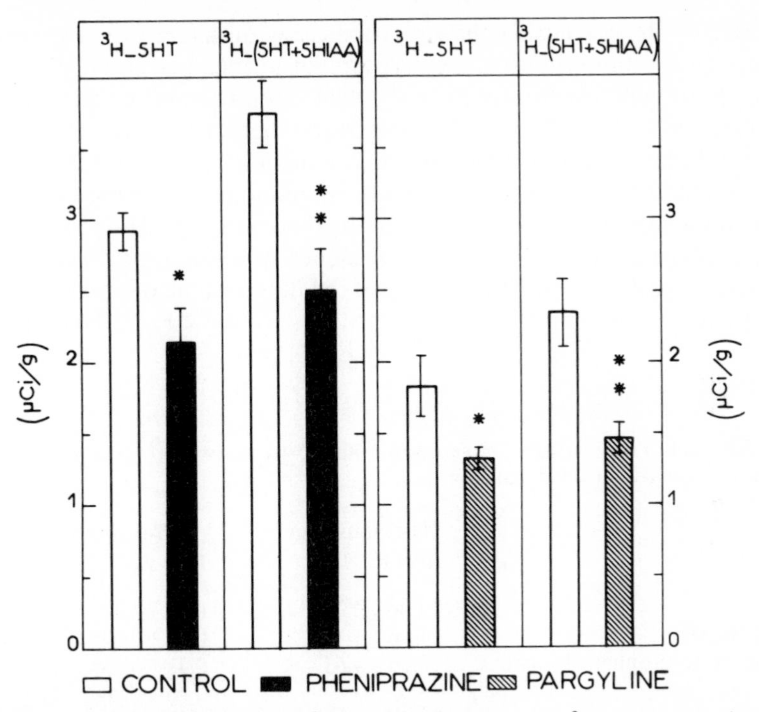

Figure 6.15 Feedback inhibition of ^{3}H-5-HT synthesis from ^{3}H-tryptophan in striatal slices of rats pretreated with a monoamine oxidase inhibitor.

Groups of rats received pargyline (75 mg (kg body weight)$^{-1}$, intraperitoneally), pheniprazine (10 mg (kg body weight)$^{-1}$, intraperitoneally), or saline, and were killed 3 h. later. Slices of striatum were incubated with ^{3}H-tryptophan, and ^{3}H-5-HT and ^{3}H-5-HIAA were estimated in slices and their incubating medium. In both experiments, the formation of ^{3}H-5-hydroxyindoles was reduced by the treatment which increased the endogenous 5-HT concentration (Hamon *et al.*, 1972). Reproduced by permission of Macmillan Journals Ltd

different MAO inhibitors, nialamide, and tranylcypromine, Renson (1973) failed to detect any significant reduction of tryptophan hydroxylation in brainstem slices. The difference between these contradictory results is difficult to ascertain. Whereas Hamon *et al.* (1972, 1973) injected MAO inhibitors 3 h. before death, Renson (1973) administered nialamide 24 h. before killing. In addition, the dose of tranylcypromine used by the latter author was so high that it induced not only MAO inhibition but also caused an enhanced release of 5-HT (Goodrich, 1969; M. Hamon *et al.*, unpublished observations). Therefore, experimental conditions used by Renson (1973) would trigger other regulatory mechanisms, thus masking the possible occurrence of 'feedback' inhibition. Another discrepancy between the data of Hamon *et al.* (1972, 1973) and those of Renson (1973) concerns the absolute rates of tryptophan hydroxylation. In the conditions used by Renson (1973), the V_{max} of the enzymic activity in brainstem slices was less than 1 nmol of 5-hydroxytryptophan formed per gram of fresh tissue and per hour, whereas Bourgoin *et al.* (1974) found a value equal to 12.6 nmol (g fresh tissue)$^{-1}$ h^{-1}. These differences might suggest that experimental conditions would be critical to demonstrate a reduction of 5-HT synthesis in brain slices several hours after the blockade of monoamine oxidase activity. To ascertain the inhibitory role of 5-HT on its own synthesis, Hamon *et al.* (1972, 1973) incubated rat striatal slices in presence of exogenous 5-HT. They still observed a significant inhibitory effect of the indoleamine on its synthesis from tryptophan (Table 6.6). Similar observations were made by Herr and Roth

Table 6.6 Inhibitory effects of serotonin and 5-hydroxytryptophan on tryptophan hydroxylase activity in striatal slices

Addition	^{3}H-tryptophan (μCi g^{-1})	^{3}H-5-hydroxytryptophan (μCi g^{-1})
None	22.46±2.37	0.77±0.11
5-HT (5.6 μM)	21.16±1.15	0.33±0.04*
L-5-hydroxytryptophan (10 μM)	18.90±1.61	0.41±0.05*

Slices were incubated in presence of Ro 4-4602 (20 μM) to inhibit the decarboxylation of ^{3}H-5-hydroxytryptophan synthesized from ^{3}H-tryptophan. In the presence of 5-HT (5.6 μM) or L-5-hydroxytryptophan (10 μM), the amount of ^{3}H-5-hydroxytryptophan finally recovered in slices was significantly reduced (M. Hamon and S. Bourgoin, unpublished observations).
* $P<0.05$.

(1972) and Renson (1973) with higher concentrations of 5-HT. Inhibition of 5-hydroxytryptamine formation by the indoleamine could also be demonstrated on the synaptosomal conversion of tryptophan into 5-HT (Karobath, 1972). At 10 μM 5-HT, a 25% inhibition occurred and at 0.1 mM it reached 40%. M. Hamon and S. Bourgoin (unpublished observations) observed only a 22% inhibition of ^{3}H-5-HT synthesis from ^{3}H-tryptophan in rat brain synaptosomes in the presence of 0.5 mM 5-HT. As previously discussed (Glowinski *et al.*, 1973), these results suggest that *in vivo* as well as in slices the normal environment of serotoninergic neurons, particularly the glial cells (Bauman *et al.*, 1974), might be necessary for the control of 5-HT synthesis. Since the effect of 5-HT is low on synaptosomes, it is thus not surprising that no significant inhibition of the activity of soluble tryptophan hydroxylase by the indoleamine was observed (Jequier *et al.*, 1969; Kaufman, 1974; Youdim *et al.*, 1974, 1975a; Table 6.7). However, one group (Mandell *et al.*, 1972) reported a strong non-

Table 6.7 Effect of various monoamines on the soluble and 'particulate' pig brain tryptophan hydroxylases

Addition	Tryptophan hydroxylase activity (%) G-25	P
None	100	100
0.1 mM serotonin	108	85
0.5 mM serotonin	112	78
0.5 mM tryptamine	95	—
0.5 mM tyramine	94	—
0.1 mM dopamine	51	45

Soluble tryptophan hydroxylase (partially purified up to the Sephadex G-25 step; Youdim *et al.*, 1975a) and the enzymic activity of the crude mitochondrial fraction (P) were assayed according to Youdim *et al.* (1975a). Monoamines were added to the assay mixture 5 min. before ^{3}H-tryptophan (3.5 μM). Results are expressed as a percentage of control values obtained in standard conditions (M. B. H. Youdim *et al.*, unpublished observations).

competitive inhibition of soluble tryptophan hydroxylase by 1 mM 5-HT with regard to $DMPH^4$. Using similar assay conditions, Kaufman (1974) could observe only a slight inhibition.

In conclusion, the inhibition by 5-HT of its own synthesis from tryptophan *in vivo* or in slices does occur but may not be related to a biochemical end-product inhibition acting at the level of tryptophan hydroxylase.

(ii) The role of 5-hydroxytryptophan Addition of exogenous 5-hydroxytryptophan (10 μM) to the incubating medium induced a 47% reduction in the rate of conversion of ^{3}H-tryptophan into ^{3}H-5-hydroxytryptophan in rat striatal slices (Table 6.6) (M. Hamon and S. Bourgoin, unpublished observations). Tryptophan hydroxylase activity in cultured pineal glands was also significantly decreased when 1.0 mM 5-hydroxytryptophan was added to the medium (Bensinger *et al.*, 1974). Jequier *et al.* (1969) observed a significant inhibition of partially purified soluble tryptophan hydroxylase from the rat brainstem by L-5-hydroxytryptophan at concentrations higher than 0.1 mM. In contrast, D-5-hydroxytryptophan, up to 2 mM had no effect. More recently, Kaufman (1974) has explored this phenomenon: he reported that the inhibition by 5-hydroxytryptophan was much more pronounced when the partially purified soluble tryptophan hydroxylase from the rabbit hindbrain was assayed with BH_4 instead of $DMPH_4$ as the cofactor. In agreement with Jequier *et al.* (1969), he found about a 50% inhibition by 1 mM of 5-hydroxytryptophan in the presence of $DMPH_4$ (0.5–1 mM). In contrast, the same concentration of the hydroxylated amino acid induced almost a complete inhibition of the enzyme when assayed with BH_4.

Another interesting observation made by Kaufman (1974) was that the inhibitory effect of 5-hydroxytryptophan increased in parallel with the concentration of the cofactor in the assay. With either $DMPH_4$ or BH_4 the decrease in tryptophan hydroxylase activity induced by 0.1–1 mM 5-HTP was more than doubled when the concentration of the pterin cofactor increased from 0.10 mM up to 0.50 mM. The formation of complexes between indoles and pterins might explain this phenomenom (Fujimori, 1959). Such atypical kinetics would suggest that regulatory properties of tryptophan hydroxylase may be related to allosteric modulations. In this respect, altered effects of phenylalanine upon tryptophan hydroxylation by the rat liver enzyme when the pterin concentration varied from 0.7 mM up to 3.5 mM were considered already as the result of allosteric alterations of 'phenylalanine, tryptophan hydroxylase' (Sullivan *et al.*, 1973). More recently, Numata and Nagatsu (1975) reported that the kinetics of inhibition of tyrosine hydroxylase by L-5-hydroxytryptophan were complex. Uncompetitive inhibition was observed with tyrosine in all cases, but, with BH_4, competitive inhibition occurred when the concentration of the pterin cofactor was higher than 90 μM, whereas it became nearly uncompetitive at lower concentrations. In agreement with the allosteric hypothesis, several authors have reported that various pharmacological treatments could induce changes in the activity of aromatic amino acid hydroxylating enzymes, which strongly suggests the occurrence of allosteric regulation of these enzymes.

In contrast to these *in vitro* data, the *in vivo* administration of a large dose of 5-hydroxytryptophan (300 mg (kg body weight)$^{-1}$, 90 min. before death) failed to alter tryptophan hydroxylase activity in brainstem slices (Renson, 1973). In addition, when 5-hydroxytryptophan accumulated in the brain after blockade of central 5-hydroxytryptophan decarboxylase, its level increased linearly for at least 30 min., suggesting that the rate of tryptophan hydroxylation remained constant for this period of time (Carlsson and Lindqvist, 1973; Morot-Gaudry *et al.*, 1974). Therefore, although the concentration of 5-hydroxytryptophan in serotoninergic neurons is still unknown (in the whole brain it ranges between 0.02 and 0.1 μM: Lindqvist *et al.*, 1975 and 0.3 μM: Aprison *et al.*, 1974), it is doubtful that feedback regulation of tryptophan hydroxylase by 5-hydroxytryptophan occurs in physiological conditions.

(e) *Other compounds*

Although the nature of the molecules is unknown, several authors have mentioned the presence of endogenous activators or inhibitors of tryptophan hydroxylase in various tissues. Thus, dialysis of a beef pineal homogenate (Jequier *et al.*, 1969) or a crude homogenate of the rat brainstem (Robinson *et al.*, 1968) resulted in a marked increase of tryptophan hydroxylase activity in these extracts. Similarly, Youdim *et al.* (1974, 1975a) reported a stimulating effect of dialysis upon the enzyme isolated from the pig brain. However, in the latter case, removal of excess ammonium sulphate rather than possible endogenous inhibitors could not be ruled out. In contrast to these data, Gál *et al.* (1970) and Sanders-Bush *et al.* (1972a) observed a significant loss of tryptophan hydroxylase activity with dialysis. However, as pointed out by the latter authors, since a similar decrease in activity occurred when the enzyme preparation was allowed to stand overnight at 5 °C, the effect observed was very likely related to the instability of the enzyme and not to the removal of an hypothetical stimulating molecule by dialysis. In the intestinal tract and carcinoid tumours, Lovenberg *et al.* (1971) mentioned the presence of a natural inhibitor of tryptophan hydroxylase. Preliminary studies made by these authors were in favour of a protein inhibitor. Finally the occurrence of a stimulating factor of tryptophan hydroxylase in protein-free, 30 000 **g** supernatant of brain homogenates has been reported by Gál and Roggeveen (1973). This molecule appeared to be thermostable, alkali-labile, dialysable, and light-sensitive, with properties of a pterin-like structure. Moreover, Gál (1974) observed a higher affinity of brain soluble tryptophan hydroxylase for tryptophan in the presence of tetrahydrobiopterin and this factor ($K_m = 25$ μM) than in presence of tetrahydrobiopterin alone ($K_m = 38$ μM). Hence, tryptophan hydroxylase resembles other aromatic amino acid hydroxylases in this respect. A protein-stimulating rat liver phenylalanine hydroxylase has been characterized by Kaufman and coworkers (Kaufman, 1970; Huang *et al.*, 1973; Huang and Kaufman, 1973). Non-protein-stimulating molecules (phospholipids) have also been isolated (Fisher and Kaufman, 1972, 1973). These lipids (mainly phosphatidylserine) also activate tryptophan hydroxylase from the rat (Hamon *et al.*, 1978a) but not from the rabbit hind brain (Tong and Kaufman, 1975).

2 *In vivo* changes in tryptophan hydroxylase activity

(a) *Tryptophan hydroxylase as an allosteric enzyme*

As in the case of tyrosine hydroxylase in noradrenergic (Morgenroth *et al.*, 1974, Roth *et al.*, 1974, 1975) and dopaminergic (Murrin *et al.*, 1974) neurones, several observations strongly suggest a close dependence of tryptophan hydroxylase activity on the firing rate along specific serotoninergic fibres. Thus, 1–2 h. after spinal cord transection, the rate of 5-hydroxytryptophan accumulation below the lesion after the blockade of 5-hydroxytryptophan decarboxylase was reduced by 50% (Carlsson *et al.*, 1973). This effect appeared well before any loss of the enzyme activity since the enzyme half-life is about 2–3 days (see above). In the brain, cerebral hemisection disconnecting the forebrain monoamine nerve terminals from their cell bodies in the lower brainstem also induced partial inhibition of tryptophan hydroxylation *in vivo* in the lesioned hemisphere when estimated 30–90 min. after the surgical manipulation. However, the effect was less pronounced than in the spinal cord (Carlsson *et al.*, 1972). Conversely, direct electrical stimulation of median raphe serotoninergic neurons induced a significant increase in the synthesis of 5-HT in the forebrain (Aghajanian *et al.*, 1967; Shields and Eccleston, 1972). Moreover, Eccleston *et al.* (1970) demonstrated that this stimulating effect was still detected after large peripheral tryptophan loading, thus eliminating the influence of possible alterations in the transport mechanism for tryptophan caused by electrical stimulation.

Although the same kind of experiments were repeated *in vitro* by Andén *et al.* (1964), who observed an increased synthesis of serotonin in isolated spinal cords of mice submitted to electrical field stimulation, no direct evidence of enhanced activity of tryptophan hydroxylase extracted from stimulated tissues has yet been obtained. The recent demonstration by Roth and his group (Roth *et al.*, 1974, 1975) of an allosteric activation of hippocampal tyrosine hydroxylase induced by the electrical stimulation of the locus coeruleus is a very interesting finding in this respect.

Extensive studies on the effects of various drugs on the firing rate of serotoninergic neurons in the median raphe and on that of dopaminergic cell bodies in the substantia nigra (Aghajanian and Bunney, 1974) confirmed that a direct relationship exists between the nerve impulse flow in monoaminergic fibres and the synthesis rate of their respective neurotransmitters in terminals. Thus administrations of monoamine oxidase inhibitors (Aghajanian *et al.*, 1970), chlorimipramine (Sheard *et al.*, 1972), and LSD (Aghajanian *et al.*, 1968) depressed the firing rate of 5-HT neurons as they decreased the rate of 5-HT synthesis (Table 6.8) (see Hamon and Glowinski, 1974). Inversely, the firing

Table 6.8 Inhibition of serotonin synthesis in the mouse brain by *in vivo* treatment with LSD (0.5 mg (kg body weight)$^{-1}$)

	Saline	LSD
^{3}H-Tryptophan (μCi g^{-1})	0.218±0.015	0.180±0.008
^{3}H-5-HT (nCi g^{-1})	8.91±0.63	6.20±0.38*
^{3}H-5-HIAA (nCi g^{-1})	1.94±0.08	1.62±0.08*
^{3}H-5-HT+^{3}H-5-HIAA (nCi g^{-1})	10.84±0.60	7.82±0.31*

Animals were killed 10 min. after the intravenous injection of ^{3}H-tryptophan, and each labelled molecule was determined in the whole brain. The formation of ^{3}H-5-hydroxyindoles was significantly reduced in animals pretreated with LSD 60 min. before injection with ^{3}H-tryptophan (Y. Morot-Gaudry, M. Hamon, and S. Bourgoin, unpublished observations).
* $P<0.05$.

rate of serotoninergic raphe cells was accelerated in rats maintained in a hot environment (Weiss and Aghajanian, 1971). The increased synthesis of 5-HT in the brain of these animals persisted even after large peripheral tryptophan loading. Thus, Squires (1974) demonstrated that tryptophan hydroxylase activity, and not possible alterations of brain tryptophan concentration, was responsible for the stimulating effect of heat exposition.

In the case of dopaminergic neurons, neuroleptics increased both the rate of dopamine synthesis in the striatum (Javoy *et al.*, 1970) and the nerve impulse flow of cells in the zona compacta of the substantia nigra (Bunney *et al.*, 1973). In the striatum, soluble tyrosine hydroxylase activity was increased after the administration of neuroleptics (methiothepin, haloperidol, pimozide and reserpine; Zivcović *et al.*, 1974a). This effect resulted from an increased affinity of the enzyme for the pterin cofactor (Zivcović *et al.*, 1974a). The story is much less clear regarding tryptophan hydroxylase. Chronic treatments with drugs which alter the activity of serotoninergic neurones such as reserpine and imipramine were reported to exert no significant effect on fore- and midbrain tryptophan hydroxylase activity (Hulme *et al.*, 1974). In contrast, Zivcović *et al.* (1974b) observed significant changes of brainstem soluble tryptophan hydroxylase activity after long-term treatments with reserpine and LSD. However, these latter effects were very likely related to changes in the concentrations of the enzyme molecule in brain tissues (see below). Indeed, acute treatment of rats with LSD failed to alter the activity of brainstem soluble tryptophan hydroxylase (M. Hamon and S. Bourgoin, unpublished observations). In contrast, administration of methiothepin, a potent antagonist of central 5-HT receptors (Monachon *et al.*, 1972; Enjalbert *et al.*, 1978) induced a large increase in 5-HT synthesis *in vivo* as well as *in vitro*, and a significant alteration of soluble brainstem tryptophan hydroxylase activity (Hamon *et al.*, 1975; Hamon *et al.*, 1977). This enzymic activity was increased by 30–40% for at least 5 h. following the drug injection (Figure 6.16). In fact methiothepin administration induced an enhanced affinity of tryptophan hydroxylase for tryptophan, the K_m value changed from 234 μM to 139 μM. Moreover, as shown by the straight line obtained by plotting $1/V$ versus $1/S$, tryptophan hydroxylase extracted from methiothepin-treated animals obeyed classical Michaelis–Menten kinetics. In contrast to the soluble enzyme from control rats, no substrate cooperativity occurred with tryptophan concentrations lower than 30 μM, suggesting that the administration of the 5-HT receptor blocking agent has induced a fully activated form of tryptophan hydroxylase. The direct addition of methiothepin (1 μM–0.1 mM) to the assay mixture did not reproduce the effect induced by *in vivo* injection of the drug. The change induced *in vivo* on tryptophan hydroxylase activity was very likely not related to the enhanced level of brain tryptophan after methiothepin administration since it was completely reversed by an injection of LSD, a treatment which induced a further increase in the amino acid concentration (Table 6.9). In fact, LSD counteracted the chance induced by the blockade of 5-HT receptors by methiothepin by the play of its well-known central 5-HT agonistic effect (Andén *et al.*, 1968).

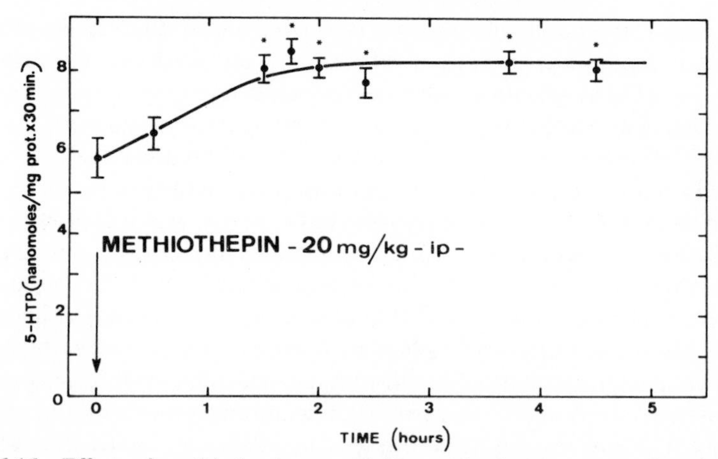

Figure 6.16 Effect of methiothepin on soluble rat brainstem tryptophan hydroxylase activity.

Rats were killed at various times after the treatment and the enzyme was assayed according to Gál and Patterson (1973) in the presence of 0.25 mM tryptophan. Points marked with an asterisk had $P<0.05$ when compared to the tryptophan hydroxylase activity of control animals (M. Hamon and S. Bourgoin, unpublished observations)

These data represent convincing evidence for the allosteric nature of tryptophan hydroxylase. In addition they reveal that changes in tryptophan hydroxylase activity may occur very rapidly after an alteration of central 5-HT synapses induced at least by pharmacological agents.

The administration of a 5-HT agonist like LSD induced a marked decrease both in the nerve impulse flow of serotoninergic neurones in the midbrain

Table 6.9 Activity of soluble tryptophan hydroxylase extracted from the rat brainstem after methiothepin (20 mg (kg body weight)$^{-1}$ at 120 min.), LSD (2 × 1 mg (kg body weight)$^{-1}$ at 140 and 100 min.), or combined treatments

Treatment	Tryptophan hydroxylase activity
Control	5.236 ± 0.427
Methiothepin	7.165 ± 0.199*
LSD	5.349 ± 0.238
LSD + methiothepin	4.875 ± 0.347

Tryptophan hydroxylase activity was measured according to Gál and Patterson (1973) in presence of 0.2 mM tryptophan and 0.15 mM 6-MPH$_4$. Each value is the mean ± S.E.M. of 5-hydroxytryptophan formed per milligram of protein and per h. in extracts of 6–7 animals (M. Hamon and S. Bourgoin, unpublished observations).
* $P<0.05$.

raphe (Aghajanian *et al*., 1968) and the synthesis rate of 5-HT in the brain (Lin *et al*., 1969; Hamon *et al*., 1974; Table 6.8). However, the activity of the rat brainstem soluble tryptophan hydroxylase was not altered simultaneously with these changes (Table 6.9). Zivcović *et al*. (1974b) observed a significant decrease of the enzymic activity, but only after the administration of 2 mg of LSD per kilogram body weight for 2 days. Although no detailed kinetic analyses were made to characterize the alteration induced by the treatment, it is very likely that a change in the concentration of tryptophan hydroxylase occurred in the conditions described by Zivcović *et al*. (1974b).

The present results demonstrate that the activity of tryptophan hydroxylase, like that of tyrosine hydroxylase (Dairman and Udenfriend 1970; Zivcović *et al*., 1974a) and histidine decarboxylase (Maudsley *et al*., 1973), can be modified with respect to the activity of specific receptors. At present the sequence of cellular events occurring between changes in receptor activity and tryptophan hydroxylase activity is still unknown. As for allosteric activation of tyrosine hydroxylase (Morgenroth *et al*., 1974; Roth *et al*., 1974, 1975; Harris *et al*., 1975), the flux of cations, in particular Ca^{2+} (Table 6.3), and/or nucleotides such as cyclic AMP might be involved in this process.

According to Harris *et al*. (1975) cyclic AMP may be responsible for the allosteric activation of tyrosine hydroxylase from depolarized tissues. Indeed, the cyclic nucleotide affects the enzymic activity directly by increasing its affinity for both the substrate tyrosine and the reduced cofactor tetrahydrobiopterin or dimethyltetrahydropterin. In agreement with Harris *et al*. (1975) we observed that the conversion of ^{3}H-tryptophan into ^{3}H-serotonin in striatal synaptosomes was also slightly increased in the presence of dibutyryl cyclic AMP (1 mM). Since the accumulation of ^{3}H-tryptophan by these particles was not altered by the cyclic nucleotide (Bauman *et al*., 1974) a direct role of this molecule as an allosteric activator might also be proposed in the case of striatal tryptophan hydroxylase. However, preliminary data do not support this hypothesis since 0.1 mM dibutyryl cAMP slightly inhibits the activity of the soluble enzyme (M. Hamon and S. Bourgoin, unpublished observations). The question is also far from solved for tyrosine hydroxylase since Lovenberg and Bruckwick (1975) reported an inhibitory effect of cAMP on tyrosine hydroxylase activity, a finding in contrast with Harris *et al*. (1975). In fact, Lovenberg *et al*. (1975) demonstrated that the effect of cAMP requires a protein kinase in the assay. Similarly, Hamon *et al*. (1978b) recently mentioned that tryptophan hydroxylase activity can be controlled by a phosphorylation-dephosphorylation process catalysed by a (Ca^{2+}-dependent) endogenous protein kinase.

Allosteric modification of tryptophan hydroxylase can be induced *in vitro* by more classical agents; thus treatments with low concentrations of detergent (sodium dodecyl sulphate, 0.01%) or trypsin stimulated tryptophan hydroxylase activity (Hamon *et al*., 1976). Moreover, since the effects induced by *in vivo* treatment with methiothepin and *in vitro* treatment with sodium dodecyl

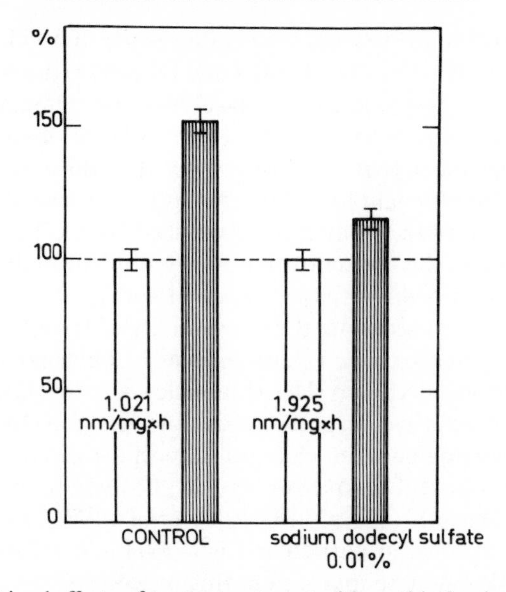

Figure 6.17 Combined effects of *in vivo* treatment with methiothepin and *in vitro* addition of 0.01% sodium dodecyl sulphate on the soluble tryptophan hydroxylase activity from the rat brainstem.

The enzyme was assayed according to Gál and Patterson (1973) in the presence of 30 μM tryptophan. The stimulatory effect of methiothepin treatment (hatched bars) was no longer detected on the enzyme activated by the detergent (M. Hamon and S. Bourgoin, unpublished observations)

sulphate were not entirely additive (Figure 6.17), it may be suggested that the allosteric changes induced by both treatments were related.

(b) 'Trans-synaptic' control of tryptophan hydroxylase

Peripheral administration of 6-hydroxydopamine at birth resulted in a marked increase in tryptophan hydroxylase activity in the brainstem and hypothalamus which could be detected during the adult life, 100–200 days after birth (Peters *et al.*, 1974). No change in enzymic activity occurred in the cerebral cortex. Although also studying the particulate (synaptosomal) tryptophan hydroxylase, Smith *et al.* (1973) failed to observe any change in the enzymic activity of the whole brain of 40–60-day-old rats treated intracisternally at birth with 6-hydroxydopamine. These data suggest that tryptophan hydroxylase activity might be modulated only in discrete areas by a trans-synaptic mechanism. Indeed, even in an anatomically well-defined area such as the raphe, various groups of serotoninergic cells may be differently affected by 6-hydroxydopamine treatment. Thus, Pujol *et al.* (1975) observed a significant increase of tryptophan hydroxylase activity in B4 and B7 (raphe dorsalis) nuclei 48 h. after an intracisternal injection of 6-hydroxydopamine. In contrast, the enzymic activity in B6 nucleus was reduced by 76% and remained unaltered in the B8 group. These results suggest that the control of tryptophan hydroxylase activity might be specific for each group of serotoninergic cells, depending on its neuronal inputs. Although no detailed kinetic analyses were made, it is very likely that long-term changes detected in these experimental conditions resulted from altered concentrations of tryptophan hydroxylase in brain tissues. In some particular cases, it was definitely established.

(c) Changes in the enzyme concentration

The best way to demonstrate that changes in tryptophan hydroxylase activity are related to altered concentrations of the enzyme in tissues would be to precipitate this protein selectively by a specific antibody. Thus, in the case of tyrosine hydroxylase, by immunoprecipitating the enzyme, Joh *et al.* (1973) have demonstrated that reserpine treatment induces an increased accumulation of the enzyme in sympathetic ganglia and adrenal medulla. Reis *et al.* (1974) showed that the same was true for tyrosine hydroxylase in the central nervous system. The same group (Joh *et al.*, 1975) has obtained a specific antibody against tryptophan hydroxylase. Therefore, direct demonstration of changes in the enzyme concentration in tissues in some pharmacological situations (see below) is now possible. However, in most cases, only indirect evidences were adduced.

(i) The role of corticosteroids Reserpine induces a significant increase in the activity of soluble tryptophan hydroxylase extracted from the rat brainstem 48 h. after its administration. Neither the affinity for tryptophan nor that for 6-MPH_4 is altered by the treatment (Zivcović *et al.*, 1973, 1974b, M. Hamon and S. Bourgoin, unpublished observations). However, the maximum velocity of the enzyme is considerably higher in brainstem 30 000 **g** supernatant from reserpine-treated than from control rats.

Repeated injections of the alkaloid produce a greater increase of brainstem tryptophan hydroxylase activity. Thus, when the first dose (5 mg (kg body weight)$^{-1}$) is followed 24 h. later by a second dose of reserpine (2.5 mg (kg body weight)$^{-1}$), the enzymic activity estimated 24 h. after the last injection is 2.4 times that of control rats. Neither single nor repeated administrations affect tryptophan hydroxylase activity in the pineal gland (Zivcović *et al.*, 1973). The induced increased activity is not abolished by dialysis of the brainstem enzyme solution for 6 h., but is greatly reduced when the animals receive an intraventricular injection of cychloheximide 12 h. after reserpine administration. Therefore, the enhanced tryptophan hydroxylase activity observed in reserpine-treated rats is very likely related to an increased synthesis of enzyme molecules.

Indeed the delay occurring between the onset of increased tryptophan hydroxylase activity in cell bodies (brainstem) as compared to that observed in nerve terminals (spinal cord) is quite compatible with the rate of axonal transport of enzyme molecules (Meek and Neff, 1972). At present, the mechanism involved in the effect of reserpine is not totally understood. The increased tryptophan hydroxylase activity might be mediated either by an interneuronal loop set out by the decrease in the functional activity of serotoninergic postsynaptic receptors, or by the increased level of corticosterone in rats receiving reserpine (Zivcović *et al.*, 1974b). The latter hypothesis finds some support in the literature. Thus, Azmitia and McEwen (1969) have shown that adrenalectomy decreased the tryptophan hydroxylase activity of brainstem isotonic homogenates (i.e. synaptosomes). In these animals, administration of corticosterone partly restored the activity of the enzyme up to the control level. Since the corticosterone effect was blocked by cycloheximide, Azmitia and McEwen (1969) suggested that the hormone might stimulate the synthesis of the enzyme protein. Another way these authors chose to demonstrate the role of corticosteroids in the control of tryptophan hydroxylase activity consisted of establishing the absolute requirement of intact adrenals to induce the enzyme by stress. In 1968, Gál *et al.* found a 40% increase in whole brain tryptophan hydroxylase activity after 8 h. of cold exposure. Azmitia and McEwen (1974) confirmed and extended this finding to other stressful situations. Thus, electric foot shock and ether anaesthesia also induced a significant increase of tryptophan hydroxylase activity in the midbrain of intact rats. In all cases, bilateral adrenalectomy prevented this change. The authors therefore suggested that the enhanced corticosterone levels in plasma of intact stressed rats were responsible for the increase in enzyme activity. However, no clear-cut evidence was presented that the hormonal effect was mediated by an increased synthesis of the protein enzyme. In particular, no attempt was made to study the soluble enzyme. Indeed, Lovenberg *et al.* (1973) have attempted to repeat the experiments of Azmitia and McEwen (1969), measuring the soluble tryptophan hydroxylase activity. In all cases, they failed to detect any significant alteration of the activity of the soluble enzyme extracted from the rat brainstem after adrenalectomy and/or corticosterone administration. Therefore, one wonders if the changes observed by Azmitia and McEwen (1969, 1974) are not related to the effect of corticosteroids on the tryptophan uptake mechanism in synaptosomes. Support for this contention comes from the work of Neckers and Sze (1975) who have shown that corticosterone stimulated the accumulation of tryptophan in synaptosomes. Similarly, fluctuations in the rate of synaptosomal conversion of ^{14}C-tryptophan into ^{14}C-serotonin occurring during the daytime (Azmitia and McEwen, 1974) might well be due to nycthemeral modulations of synaptosomal tryptophan transport rather than to actual changes of tryptophan hydroxylase activity. In this respect, Héry *et al.* (1972, 1974) have demonstrated that the transport mechanism for the amino acid in brain plays a major role in the diurnal rhythm of central serotonin synthesis. In conclusion, the exact mechanism of action of corticosteroids on serotonin synthesis in brain is largely unknown.

(ii) The role of thyroid hormone Neonatal hypothyroidism (Rastogi and Singhal, 1974a, b) largely impaired the normal developmental increase in brain tryptophan hydroxylase activity. This effect occurred only when hypothyroidism was induced before the 20th day following birth. Subsequent administration of L-triiodothyronine to these cretinous rats could restore tryptophan hydroxylase activity up to the control level (Rastogi and Singhal, 1974a, b). These changes were related to the well-known effect of thyroid hormone on protein synthesis. Thus, Gelber *et al.* (1964) have established that it acts at the level of translation of genetic message in maturing brain. Therefore, changes in tryptophan hydroxylase activity occurring in these particular endocrinological situations reflect simply those of protein synthesis in brain.

(iii) The role of axoplasmic flow By measuring the rate of tryptophan hydroxylation after various pharmacological treatments in areas containing the serotoninergic cell bodies (midbrain) and in structures enriched in 5-HT terminals (striatum, septum, etc.), Knapp and Mandell (1972a, b, 1973) have demonstrated that changes in the enzymic activity could be asynchronous in these brain regions. Thus, they observed (Knapp and Mandell, 1972a) that the 'defective' enzyme induced by *p*-chlorophenylalanine administration appeared first in the cell body area and a few days later in the septal area. Similarly, in the course of chronic treatment with lithium, the decrease of tryptophan hydroxylase activity observed in the midbrain (containing cell bodies) occurred a few days before a similar change in the rate of conversion of ^{14}C-tryptophan into ^{14}C-serotonin in striatal synaptosomes. According to Knapp and Mandell (1972a, b, 1973), the delayed appearance of fluctuations in the activity of 'particulate' enzyme corresponds to the axoplasmic flow of tryptophan hydroxylase, since, a few days later, the altered enzymic activity from the midbrain cell bodies reached the 5-HT terminals. The temporal relationship between changes in soluble and 'particulate' enzyme activity was consistent with an axonal flow rate of 1–2 mm d^{-1}, a value reported for soluble proteins (Barondes, 1969). The demonstration of tryptophan hydroxylase transport by axonal flow was made by Clineschmidt *et al.* (1971). These authors reported that chronic transection of the cat spinal cord induced a simultaneous reduction of tryptophan hydroxylase activity below the lesion and a higher enzymic activity in the brainstem. Although the occurrence of sprouting of 5-HT neurones cannot be excluded in these conditions, these opposing changes of tryptophan hydroxylase activity on both lesion sides indicate that the enzyme in cell bodies normally migrates to the serotoninergic terminals. Recent observations (J. P. Kan and J. F. Pujol, personal communication) also suggest that the temporal shift in the

diurnal rhythms of tryptophan hydroxylase activity in cell bodies (dorsal raphe nucleus) and terminals (striatum) would be related to the axonal flow of the enzyme along 5-HT neurones.

Knapp *et al.* (1974) reported that the delay between changes in tryptophan hydroxylase activity in cell bodies and in terminals is not always in agreement with the rate of axonal flow. For instance, the decrease in enzymic activity detected in the lateral midbrain 30 min. after amphetamine administration occurs in the striatum as rapidly as 1 h. after the drug treatment.

Therefore, the control of serotonin synthesis in serotoninergic neurons most likely involves some biochemical link(s) other than the axonal flow of soluble proteins between the cell bodies and their terminals.

H THE ROLES OF THE TRYPTOPHAN UPTAKE MECHANISM AND TRYPTOPHAN HYDROXYLASE ACTIVITY IN THE CONTROL OF 5-HT BIOSYNTHESIS IN CENTRAL SEROTONINERGIC NEURONS

The role of tryptophan uptake in the control of 5-HT synthesis in physiological (Héry *et al.*, 1970, 1972, 1974) and in pharmacological (Knapp and Mandell, 1974a, b, Hamon *et al.*, 1974a, b; Hamon and Glowinski, 1974, Mandell *et al.*, 1974; Grahame-Smith, 1974; Green and Grahame-Smith, 1975) situations has been already pointed out. In numerous situations, the rate of tryptophan uptake and that of tryptophan hydroxylase activity were changed in parallel. Thus, paradoxical sleep deprivation enhanced both tryptophan accumulation and tryptophan hydroxylase activity in slices of the spinal cord of rats (Héry *et al.*, 1970). Similarly, the administration of reserpine induced an increased rate of tryptophan uptake (Belin *et al.*, 1974), and a higher activity of soluble tryptophan hydroxylase in synaptosomes (Zivcović *et al.*, 1974b), both effects being observed in the same time conditions. Recent observations indicate that the blockade of central 5-HT receptors by methiothepin resulted not only in a stimulation of the tryptophan uptake mechanism in brainstem slices (Hamon *et al.*, 1975), but also in an allosteric activation of soluble tryptophan hydroxylase in the same area (M. Hamon and S. Bourgoin, see above). Conversely, the administration of LSD, a potent 5-HT agonist, induced a parallel decrease in the accumulation of tryptophan in rat striatal slices (Hamon *et al.*, 1974a) and in the activity of soluble tryptophan hydroxylase (Zivcović *et al.*, 1974b).

Another interesting observation concerns the effect of dibutyryl cyclic AMP which stimulates both tryptophan accumulation in brain slices (Hamon *et al.*, 1973, 1974b) and tryptophan hydroxylase activity in synaptosomes (Harris *et al.*, 1975; M. Hamon and S. Bourgoin, unpublished observations).

According to data reported by Mandell *et al.* (1974) and Gerson *et al.* (1974), tryptophan uptake mechanism might be partially specific of serotoninergic neurones. Thus, 8 days after lesion of the median and dorsal raphe nuclei, the V_{max} of tryptophan uptake in septal synaptosomes decreased by 25% (Mandell *et al.*, 1974). Similarly, the extensive chemical lesion of spinal 5-HT neurons by 5,6-dihydroxytryptamine induced a significant decrease of tryptophan uptake in spinal synaptosomes (Gerson *et al.*, 1974). These data indicate that, although the high affinity mechanism for tryptophan accumulation in synaptosomes is not exclusively localized in 5-HT terminals (Bauman *et al.*, 1974), there is, however, a preferential uptake of tryptophan in these particles. The particular role played by tryptophan in the control of serotonin synthesis (Fernstrom and Wurtman, 1971; Hamon and Glowinski, 1974; Parfitt and Grahame-Smith, 1974) on one hand, and the frequent parallel changes in both tryptophan uptake and tryptophan hydroxylase activity on the other hand, might suggest that both enzymic activities would be associated in serotoninergic neurones. A similar hypothesis has already been postulated by Mandell *et al.* (1972). Since Ichiyama *et al.* (1968) presented some experimental data in favour of the existence of a particular assembly between tryptophan hydroxylase and 5-hydroxytryptophan decarboxylase in synaptosomes, it is very likely that the synthesis of serotonin from tryptophan in serotoninergic neurones is facilitated by *a particular organization of the three enzymes involved in this biochemical process*. In this respect, the existence of serotoninergic lines among the C1300 neuroblastoma (Knapp and Mandell, 1974; Narotzky and Bondareff, 1974) and the recent isolation of one of them (Breakefield and Nirenberg, 1974) would be very helpful. The use of these clones should establish the particular organization of the enzymic complexes involved in serotonin synthesis, i.e. the tryptophan uptake process, tryptophan hydroxylase, and 5-hydroxytryptophan decarboxylase, in serotoninergic neurones. Such an hypothesis has already found strong support in the case of cholinergic neurones. The association between the choline pump and choline acetyl transferase activity has been demonstrated in various experimental conditions (Lefresne, 1974).

I CONCLUSION

It is over 10 years since the discovery of tryptophan hydroxylase activity in brain; however, the properties of this enzyme are still largely unknown. This situation can be explained by the failure of attempts to purify the enzyme. However, quite recently some efforts have been made by several groups (Tong and Kaufman, 1975; Youdim *et al.*, 1974, 1975a; Joh *et al.*, 1975). The regulatory mechanisms of the enzyme are particularly complex. As with most polymeric enzymes, tryptophan hydroxylase may be an allosteric enzyme, exhibiting substrate cooperativity with L-tryptophan. In addition, its properties depend largely on the nature of the reduced pterin cofactor involved in the hydroxylation process.

In physiological conditions, its activity is controlled by tryptophan concentration, availability of oxygen, and pterin cofactor. However, the nature of its naturally occurring cofactor is unknown. Pharmacological treatments which induce changes in tryptophan hydroxylase activity suggest that the enzyme is biochemically associated with 5-HT receptors and the tryptophan uptake process. In addition, these kinds of experiments have shown various types of regulatory mechanisms, for instance allosteric activation of the existing enzyme after blockade of central 5-HT receptors by methiothepin or synthesis of new enzyme molecules after reserpine treatment. In serotoninergic neurones, 5-HT synthesis is thus controlled by short- and long-term processes. However, the sequence of events occurring between the stimulus (for instance the electrical stimulation of 5-HT cells) and the observed increase in tryptophan hydroxylase activity remains to be established. It is believed that the preparation of a highly purified enzyme and the isolation of a serotoninergic neuroblastoma clone would facilitate our knowledge regarding tryptophan hydroxylation, both *in vitro* and *in vivo*.

This research was mainly supported by grants of l'Institut National de la Santé et de la Recherche Médicale (INSERM), la Direction de la Recherche et Moyens d'Essais (DRME), and la Société des Usines Chimiques Rhône-Poulenc.

First draft submitted 1975.

J REFERENCES

Aghajanian, G. K., and Asher, I. M. (1971). *Science*, **172**, 1159–1161.
Aghajanian, G. K., and Bunney, B. S. (1974). In *Frontiers of Neurology and Neuroscience Research* (Eds., P. Seeman and G. M. Brown), University of Toronto Press, Toronto, pp. 4–11.
Aghajanian, G. K., Foote, W. E., and Sheard, M. H. (1968). *Science*, **161**, 706–708.
Aghajanian, G. K., Graham, A. W., and Sheard, M. H. (1970). *Science*, **169**, 1100–1102.
Aghajanian, G. K., Kuhar, M. J., and Roth, R. H. (1973). *Brain Res.*, **54**, 85–101.
Aghajanian, G. K., Rosecrans, J. A., and Sheard, M. H. (1967). *Science*, **156**, 402–403.
Andén, N. E. (1974). *Adv. Biochem. Psychopharmacol.*, **10**, 35–43.
Andén, N. E., and Modigh, K.(1972). *J. Neural Trans.*, **33**, 211–222.
Andén, N. E., Carlsson, A., Hillarp, N. A., and Magnusson, T. (1964). *Life Sci.*, **3**, 473–478.
Andén, N. E., Corrodi, H., Fuxe, K., and Hökfelt, T. (1968). *Brit. J. Pharmacol.*, **34**, 1–7.
Aprison, M. H., Tachiki, K. H., Smith, J. E., Lane, J. D., and McBride, W. J. (1974). *Adv. Biochem. Psychopharmacol.*, **11**, 31–41.
Azmitia, E. C.., Jr., and McEwen, B. S. (1969). *Science*, **166**, 1274–1276.
Azmitia, E. C., Jr., and McEwen, B. S. (1974). *Brain Res.*, **78**, 291–302.
Barondes, S. H. (1969). In *Neuroscience Research Symposium Summaries III* (Eds., F. O. Schmitt, T. Melnechuck, G. C. Quarton, and G. Adelman), MIT Press, Cambridge, Mass., p. 191.
Bauman, A., Bourgoin, S., Benda, P., Glowinski, J., and Hamon, M. (1974). *Brain Res.*, **66**, 253–263.
Baumgarten, H. G., Lachenmayer, L., and Schlossberger, H. G. (1972). *Zeit. Zellforsch.*, **125**, 553–569.

Baumgarten, H. G., Victor, S. J., and Lovenberg, W. (1975). *Psychopharmacol. Commun.*, **1**, 75–88.
Baumgarten, H. G., Björklund, A., Lachenmayer, L., and Nobin, A. (1973). *Acta Physiol. Scand.* Suppl. 391, 1–22.
Belin, M. F., Chouvet, G., and Pujol, J. F. (1974). *Biochem. Pharmacol.*, **23**, 587–597.
Bennett, J. L., and Aghajanian, G. K. (1974). *Brain Res.*, **65**, 537–541.
Bensinger, R. E., Klein, D. C., Weller, J. L., and Lovenberg, W. (1974). *J. Neurochem.*, **23**, 111–117.
Bertilsson, L., Koslow, S. H., and Costa, E. (1975). *Brain Res.*, **91**, 348–350.
Boadle-Biber, M. C. (1975). *Biochem. Pharmacol.*, **24**, 1445–1460.
Bogdanski, D. F., Pletscher, A., Brodie, B. B., and Udenfriend, S. (1956). *J. Pharmacol. Exptl Ther.*, **117**, 82–88.
Boireau, A., Ternaux, J. P., Bourgoin, S., Héry, F., Glowinski, J., and Hamon, M. (1976). *J. Neurochem.*, **26**, 201–204.
Bourgoin, S., Faivre-Bauman, A., Benda, P., Glowinski, J., and Hamon, M. (1974). *J. Neurochem.*, **23**, 319–327.
Bourgoin, S., Ternaux, J. P., Boireau, A., Héry, F., and Hamon, M. (1975). *Naunyn-Schmiedeberg's Arch. Pharmacol.*, **288**, 109–121.
Breakefield, X. O., and Nirenberg, M. W. (1974). *Proc. Nat. Acad. Sci., USA*, **71**, 2530–2533.
Brodal, A., Walberg, F., and Taber, E. (1960). *J. Comp. Neurol.*, **11**, 239–259.
Brownstein, M. J., Saavedra, J. M., Axelrod, J., Zeman, G. H., and Carpenter, D. O. (1974). *Proc. Nat. Acad. Sci., USA*, **71**, 4662–4665.
Brownstein, M. J., Palkovits, M., Saavedra, J. M., and Kizer, J. S. (1975). *Brain Res.*, **97**, 163–166.
Bunney, B. S., Walters, J. R., Roth, R. H., and Aghajanian, G. K. (1973). *J. Pharmacol. Exptl Ther.*, **185**, 560–571.
Cardot, J. (1972). *C. R. Acad. Sci. Paris*, **274**, 1935–1937.
Carlsson, A., and Lindqvist, M. (1972). *J. Neural Trans.*, **33**, 23–43.
Carlsson, A., and Lindqvist, M. (1973). *J. Neural Trans.*, **34**, 79–91.
Carlsson, A., Corrodi, H., and Waldeck, B. (1963). *Helv. Chim. Acta*, **46**, 2271–2285.
Carlsson, A., Falck, B., and Hillarp, N. A. (1962). *Acta Physiol. Scand.*, **56**, Suppl., 196, 1–28.
Carlsson, A., Bedard, P., Lindqvist, M., and Magnusson, T. (1972). In *Neurotransmitters and Metabolic Regulation* (Ed., R. M. S. Smellie). The Biochemical Society, London, pp. 17–32.
Carlsson, A., Lindqvist, M., Magnusson, T., and Atack, C. (1973). *Naunyn-Schmiedeberg's Arch. Pharmacol.*, **277**, 1–12.
Clineschmidt, B. V., Pierce, J. E., and Lovenberg, W. (1971). *J. Neurochem.*, **18**, 1593–1596.
Cloutier, G., and Weiner, N. (1973). *J. Pharmacol. Exptl Ther.*, **186**, 75–85.
Consolo, S., Garattini, S., Ghielmetti, R., Morselli, P., and Valzelli, L. (1965). *Life Sci.*, **4**, 625–630.
Cooper, J. R., and Melcer, I. (1961). *J. Pharmacol. Exptl Ther.*, **132**, 265–268.
Coulson, W. F., Wardle, E., and Jepson, J. B. (1968). *Biochim. Biophys. Acta*, **167**, 99–109.
Coyle, J. T., and Axelrod, J. (1972). *J. Neurochem.*, **19**, 1117–1123.
Curzon, G., and Knott, P. J. (1974). In *Aromatic Amino Acids in the Brain. Ciba Symposium 22*, Elsevier Excerpta Medica North Holland, Amsterdam, pp. 217–229.
Dahlström, A., and Fuxe, K. (1964). *Acta Physiol. Scand.*, **62**, Suppl. 232, 1–55.
Dairman, W. (1972). *Brit. J. Pharmacol.*, **44**, 307–310.
Dairman, W., and Udenfriend, S. (1970). *Mol. Pharmacol.*, **6**, 350–356.
Davis, J. N., and Carlsson, A. (1973). *J. Neurochem.*, **21**, 783–790.

Deguchi, T., and Barchas, J. (1972a). *Nature New Biol.*, **235**, 92–93.
Deguchi, T., and Barchas, J. (1972b). *J. Neurochem.*, **19**, 927–929.
Deguchi, T., and Barchas, J. (1972c). *Mol. Pharmacol.*, **8**, 770–779.
Deguchi, T., Sinha, A. K., and Barchas, J. D. (1973). *J. Neurochem.*, **20**, 1329–1336.
Diaz, P. M., Ngai, S. H., and Costa, E. (1968). *Adv. Pharmacol.*, **6B**, 75–92.
Eccleston, D., Ritchie, I. M., and Roberts, M. H. T. (1970). Nature, **226**, 84–85.
Eisenstadt, M., Goldman, J. E., Kandel, E. R., Koike, H., Koester, J., and Schwartz, J. H. (1973). *Proc. Nat. Acad. Sci., USA*, **70**, 3371–3375.
Enjalbert, A., Hamon, M., Bourgoin, S., and Bockoert, J. (1978). *Mol. Pharmacol.*, **14**, 11–23.
Fernstrom, J. D., and Wurtman, R. J. (1971). *Science*, **173**, 149–152.
Fisher, D. B., and Kaufman, S. (1972). *J. Biol. Chem.*, **247**, 2250–2252.
Fisher, D. B., and Kaufman, S. (1973). *J. Biol. Chem.*, **248**, 4345–4353.
Fisher, D. B., Kirkwood, R., and Kaufman, S. (1972). *J. Biol. Chem.*, **247**, 5161–5167.
Fratta, W., Biggio, G., Mercuro, G., Di Vittorio, P., Tagliamonte, A., and Gessa, G. L. (1973). *J. Pharm. Pharmacol.*, **25**, 908–909.
Friedman, P. A., Kappelman, A. H., and Kaufman, S. (1972). *J. Biol. Chem.*, **247**, 4165–4173.
Fujimori, E. (1959). *Proc. Nat. Acad. Sci., USA*, **45**, 133–136.
Fuller, R. W. (1965). *Fed. Proc. Fed. Amer. Socs Exptl Biol.*, **24**, Abstract 2482.
Gál, E. M., Poczik, M., and Marshall, F. D., Jr. (1963). *Biochem. Biophys. Res. Commun.*, **12**, 39–43.
Gál, E. M. (1965). *Fed. Proc. Fed. Amer. Socs Exptl Biol.*, **24**, Abstract 2480.
Gál, E. M. (1974). In *Adv. Biochem. Psychopharmacol.*, **11**, 1–11.
Gál, E. M., and Christiansen, P. A. (1975). *J. Neurochem.*, **24**, 89–95.
Gál, E. M., and Millard, S. A. (1971). *Biochim. Biophys. Acta*, **227**, 32–41.
Gál, E. M., and Moses, F. C. (1974). *Fed. Proc. Fed. Amer. Socs Exptl Biol.*, **33**, 1587.
Gál, E. M., and Patterson, K. (1973). *Anal. Biochem.*, **52**, 625–629.
Gál, E. M., and Roggeveen, A. E. (1973). *Science*, **179**, 809–811.
Gál, E. M., Amstrong, J. C., and Ginsberg, B. (1966). *J. Neurochem.*, **13**, 643–654.
Gál, E. M. Christiansen, P. A., and Yunger, L. M. (1975a). *Neuropharmacology*, **14**, 31–39.
Gál, E. M., Heater, R. D., and Millard, S. A. (1968). *Proc. Soc. Exptl Biol.*, **128**, 412–415.
Gál, E. M., Roggeveen, A. E., and Millard, S. A. (1970). *J. Neurochem.*, **17**, 1221–1235.
Gál, E. M., Yang, S. L., and Moses, F. (1975b). *5th Int. Meet. Int. Soc. Neurochem., Barcelona*, Abstract 24.
Gelber, S., Campbell, P. L., Deibler, G. E., and Sokoloff, L. (1964). *J. Neurochem.*, **11**, 221–229.
Gerson, S., Baldessarini, R. J., and Wheeler, S. C. (1974). *Neuropharmacology*, **13**, 987–1004.
Gessa, G. L., and Tagliamonte, A. (1974). In *Aromatic Amino acids in the Brain. Ciba Symposium 22*, Elsevier Excerpta Medica North Holland, Amsterdam, pp. 207–216.
Glowinski, J., Hamon, M., and Héry, F. (1973). In *New Concepts in Neurotransmitter Regulation* (Ed., A. J. Mandell), Plenum, New York, pp. 239–257.
Glowinski, J., Hamon, M., Javoy, F., and Morot-Gaudry, Y. (1972). *Adv. Biochem. Psychopharmacol.*, **5**, 423–439.
Goldstein, M., and Frenkel, R. (1971). *Nature New Biol.*, **233**, 179–180.
Goodrich, C. A. (1969). *Brit. J. Pharmacol.*, **37**, 87–93.
Grahame, C., Green, R., Woods, H. F., and Youdim, M. B. H. (1975). *Brit. J. Pharmacol.*, **53**, 450.
Grahame-Smith, D. G. (1964). *Biochem. Biophys. Res. Commun.*, **16**, 586–592.
Grahame-Smith, D. G. (1967). *Biochem. J.*, **105**, 351–360.

Grahame-Smith, D. G. (1972). *The Carcinoid Syndrome*, William Heinemann Medical Books, London.
Grahame-Smith, D. G. (1974). *Adv. Biochem. Psychopharmacol.*, **10**, 82–92.
Green, A. R., and Grahame-Smith, D. G. (1975). In *Handbook of Psychopharmacology*, Vol. 3 (Eds., L. L. Iversen, S. Iversen, and S. H. Snyder), Plenum, New York, pp. 169–245.
Green, H., and Sawyer, J. L. (1966). *Anal. Biochem.*, **15**, 53–64.
Guroff, G., Rhoads, C. A., and Abramowitz, A. (1967b). *Anal. Biochem.*, **21**, 273–278.
Guroff, G., Daly, J. W., Jerina, D. M., Renson, J., Witkop, B., and Udenfriend, S. (1967a). *Science*, **158**, 1524–1530.
Håkanson, R., and Hoffman, G. J. (1967). *Biochem. Pharmacol.*, **16**, 1677–1680.
Hamon, M., and Glowinski, J. (1974). *Life Sci.*, **15**, 1533–1548.
Hamon, M., Bourgoin, S., and Glowinski, J. (1973). *J. Neurochem.*, **20**, 1727–1745.
Hamon, M., Bourgoin, S., Héry, F., and Glowinski, J. (1977). In *Structure and Function of Monoamine Enzymes* (Eds., E. Usdin, N. Weiner, and M. B. H. Youdim), Dekker, New York, pp. 59–90.
Hamon, M., Bourgoin, S., Héry, F., and Simonnet, G. (1978a). *Biochem. Pharmacol.*, **27**, 915–922.
Hamon, M., Bourgoin, S., Héry, F., and Simonnet, G. (1978b). *Mol. Pharmacol.*, **14**, 99–110.
Hamon, M., Bourgoin, S., Héry, F., Ternaux, J. P., and Hamon, M. (1976). *Nature*, **260**, 61–63.
Hamon, M., Bourgoin, S., Jagger, J., and Glowinski, J. (1974a). *Brain Res.*, **69**, 265–280.
Hamon, M., Bourgoin, S., Morot-Gaudry, Y., and Glowinski, J. (1972). *Nature New Biol.*, **237**, 184–187.
Hamon, M., Javoy, F., Kordon, C., and Glowinski, J. (1970). *Life Sci.*, **9**, 167–173.
Hamon, M., Bourgoin, S., Héry, F., Ternaux, J. P., and Glowinski, J. (1975). *5th Int. Meet. Int. Soc. Neurochem., Barcelona*, Abstract 423.
Hamon, M., Bourgoin, S., Morot-Gaudry, Y., Héry, F., and Glowinski, J. (1974b). *Adv. Biochem. Psychopharmacol.*, **11**, 153–162.
Hanley, M. R., Cottrell, G. A., Emson, P. C., and Fonnum, F. (1974). *Nature*, **251**, 631–633.
Harris, J. E., Baldessarini, R. J., Morgenroth, V. H., and Roth, R. H. (1975). *Proc. Nat. Acad. Sci., USA*, **72**, 789–793.
Harvey, J. A., and Gál, E. M. (1974). *Science*, **183**, 869–871.
Harvey, J. A., McMaster, S. E., and Yunger, L. M. (1975). *Science*, **187**, 841–843.
Herr, B. E., and Roth, R. H. (1972). *Vth Int. Congr. Pharmacol., San Francisco*, Abstract 600.
Héry, F., Pujol, J. F., Lopez, M., Macon, J., and Glowinski, J. (1970). *Brain Res.*, **21**, 391–403.
Héry, F., Rouer, E., and Glowinski, J. (1972). *Brain Res.*, **43**, 445–465.
Héry, F., Rouer, E., Kan, J. P., and Glowinski, J. (1974). *Adv. Biochem. Psychopharmacol.*, **11**, 163–167.
Hoff, K. M., and Goodrich, C. (1975). *Fedn. Proc. Fedn. Amer. Socs Exptl Biol.*, **34**, 565.
Ho, I. K., Brase, D. A., Loh, H. H., and Way, E. L. (1975). *J. Pharmacol. Exptl Ther.*, **193**, 35–43.
Hosada, S., and Glick, D. (1966). *J. Biol. Chem.*, **241**, 192–196.
Huang, C. Y., and Kaufman, S. (1973). *J. Biol. Chem.*, **248**, 4242–4251.
Huang, C. Y., Max, E. E., and Kaufman, S. (1973). *J. Biol. Chem.*, **248**, 4235–4241.
Hulme, E. C., Hill, R., North, M., and Kibby, M. R. (1974). *Biochem. Pharmacol.*, **23**, 1393–1404.
Ichiyama, A., Nakamura, S., Nishizuka, Y., and Hayaishi, O. (1968). *Adv. Pharmacol.*, **6A**, 5–17.

Ichiyama, A., Nakamura, S., Nishizuka, Y., and Hayaishi, O. (1970). *J. Biol. Chem.*, **245**, 1699–1709.

Ikeda, M., Levitt, M., and Udenfriend, S. (1965). *Biochem. Biophys. Res. Commun.*, **18**, 482–488.

Jalfre, M., Ruch-Monachon, M. A., and Haefely, W. (1974). *Adv. Biochem. Psychopharmacol.*, **10**, 121–134.

Jamieson, D., and Van den Brenk, H. A. J. (1965). *J. Appl. Physiol.*, **20**, 514–518.

Javoy, F., Hamon, M., and Glowinski, J. (1970). *Eur. J. Pharmacol.*, **10**, 178–188.

Jequier, E., Lovenberg, W., and Sjoerdsma, A. (1967). *Mol. Pharmacol.*, **3**, 274–278.

Jequier, E., Robinson, D. S., Lovenberg, W., and Sjoerdsma, A. (1969). *Biochem. Pharmacol.*, **18**, 1071–1081.

Jerina, D. M., Daly, J. W., Witkop, B. W., Zaltzman-Nirenberg, P., and Udenfriend, S. (1970). *Biochemistry*, **9**, 147–156.

Joh, T. H., Geghman, C., and Reis, D. J. (1973). *Proc. Nat. Acad. Sci., USA*, **70**, 2767–2771.

Joh, T., Shikimi, T., Pickel, V., and Reis, D. J. (1975). *Fed. Proc. Fed. Amer. Socs Exptl Biol.*, **34**, 3281.

Kan, J. P., Buda, M., and Pujol, J. F. (1975). *Brain Res.*, **93**, 353–357.

Karobath, M. (1972). *Biochem. Pharmacol.*, **21**, 1253–1263.

Karobath, M., and Baldessarini, R. J. (1972). *Nature New Biol.*, **236**, 206–208.

Karobath, M., Diaz, J. L., and Huttunen, M. (1972). *Biochem. Pharmacol.*, **21**, 1245–1251.

Kaufman, S. (1970). *J. Biol. Chem.*, **245**, 4751–4759.

Kaufman, S. (1974). In *Aromatic Amino Acids in the Brain. Ciba Symposium 22*, Elsevier Excerpta Medica North Holland, Amsterdam, pp. 85–108.

Kettler, R., Bartholini, G., and Pletscher, A. (1974). *Nature*, **249**, 476–478.

Kizer, J. S., Zivin, J. A., Saavedra, J. M., and Brownstein, M. J. (1975). *J. Neurochem.*, **24**, 779–785.

Knapp, S., and Mandell, A. J. (1972a). *Life Sci.*, **11**, 761–771.

Knapp, S., and Mandell, A. J. (1972b). *Science*, **177**, 1209–1211.

Knapp, S., and Mandell, A. J. (1973). *Science*, **180**, 645–647.

Knapp, S., and Mandell, A. J. (1974). *Brain Res.*, **66**, 547–551.

Knapp, S., Mandell, A. J., and Bullard, W. P. (1975). *Life Sci.*, **16**, 1583–1594.

Knapp, S., Mandell, A. J., and Geyer, M. A. (1974). *J. Pharmacol. Exptl Ther.*, **189**, 676–689.

Koe, B. K., and Weissman, A. (1966). *J. Pharmacol. Exptl Ther.*, **154**, 499–516.

Kuczenski, R. (1973a). *J. Biol. Chem.*, **248**, 2261–2265.

Kuczenski, R. (1973b). *J. Biol. Chem.*, **248**, 5074–5080.

Kuczenski, R. T., and Mandell, A. J. (1972). *J. Biol. Chem.*, **247**, 3114–3122.

Kuhar, M. J., Aghajanian, G. K., and Roth, R. H. (1972). *Brain Res.*, **44**, 165–176.

Kuhar, M. J., Roth, R. H., and Aghajanian, G. K. (1971). *Brain Res.*, **35**, 167–176.

Lefresne, P. (1974). PhD Thesis, Paris VI Univ.

Lin, R. C., Ngai, S. H., and Costa, E. (1969). *Science*, **166**, 237–239.

Lindqvist, M., Kehr, W., and Carlsson, A. (1975). *J. Neural Trans.*, **36**, 161–176.

Loizou, L. A. (1972). *Brain Res.*, **40**, 395–418.

Lovenberg, W. (1977). In *Structure and Function of Monoamine Enzymes* (Eds., E. Usdin, N. Weiner, and M. B. H. Youdim), Dekker, New York, pp. 43–58.

Lovenberg, W., and Bruckwick, E. A. (1975). In *Pre- and Post-synaptic Receptors* (Eds., E. Usdin, and W. E. Bunney, Jr.), Marcel Dekker, New York, pp. 149–168.

Lovenberg, W., Bruckwick, E. A., and Hanbauer, I. (1975). *Proc. Nat. Acad. Sci., USA*, **72**, 2955–2958.

Lovenberg, W., and Victor, S. J. (1974a). *Adv. Biochem. Psychopharmacol.*, **10**, 93–101.

Lovenberg, W., and Victor, S. J. (1974b). *Life Sci.*, **14**, 2337–2353.

Lovenberg, W., Bensinger, R. E., Jackson, R. L., and Daly, J. W. (1971). *Anal. Biochem.*, **43**, 269–274.
Lovenberg, W., Besselaar, G. H., Bensinger, R. E., and Jackson, R. L. (1973). In *Serotonin and Behavior* (Eds., J. Barchas, and E. Usdin), Academic Press, New York, pp. 49–54.
Lovenberg, W., Jequier, E., and Sjoerdsma, A. (1967). *Science*, **155**, 217–219.
Lovenberg, W., Jequier, E., and Sjoerdsma, A. (1968). *Adv. Pharmacol.*, **6A**, 21–36.
Lovenberg, W., Levine, R. J., and Sjoerdsma, A. (1965). *Biochem. Pharmacol.*, **14**, 887–889.
McGeer, E. G., and Peters, D. A. V. (1969). *Canad. J. Biochem.*, **47**, 501–506.
McGeer, E. G., Peters, D. A. V., and McGeer, P. L. (1968). *Life Sci.*, **7**, 605–615.
Macon, J. B., Sokoloff, L., and Glowinski, J. (1971). *J. Neurochem.*, **18**, 323–331.
Maickel, R. P., Cox, R. H., Saillant, J., and Miller, F. P. (1968). *Int. J. Neuropharmacol.*, **7**, 275–281.
Mandell, A. J., Knapp, S., and Hsu, L. L. (1974). *Life Sci.*, **14**, 1–17.
Mandell, A. J., Segal, D. S., Kuczenski, R. T., and Knapp, S. (1972). In *The Biology of Behavior* (Ed., J. A. Kiger, Jr.), Oregon State University Press, Portland, pp. 11–65.
Massari, V. J., and Sanders-Bush, E. (1975). *Eur. J. Pharmacol.*, **33**, 419–422.
Maudsley, D. V., Kobayashi, Y., Williamson, E., and Bovaird, L. (1973). *Nature*, **245**, 148–149.
Meek, J. L., and Neckers, L. M. (1975). *Brain Res.*, **91**, 336–340.
Meek, J. L., and Neff, N. H. (1972). *J. Neurochem.*, **19**, 1519–1525.
Millard, S. A., and Gál, E. M. (1972). *J. Neurochem.*, **19**, 2461–2464.
Moir, A. T. B., and Eccleston, D. J. (1968). *J. Neurochem.*, **15**, 1093–1108.
Monachon, M. A., Burkard, W. P., Jalfre, M., and Haefely, W. (1972). *Naunyn-Schmiedeberg's Arch. Pharmacol.*, **274**, 192–197.
Morgenroth, V. H. III, Boadle-Biber, M., and Roth, R. H. (1974). *Proc. Nat. Acad. Sci., USA*, **71**, 4283–4287.
Morot-Gaudry, Y., Hamon, M., Bourgoin, S., Ley, J. P., and Glowinski, J. (1974). *Naunyn-Schmiedeberg's Arch. Pharmacol.*, **282**, 223–238.
Morrissey, J. J., Walker, M. N., and Lovenberg, W. (1977). *Proc. Soc. Exptl. Biol. Med.*, **154**, 469–499.
Murrin, L. C., Morgenroth, V. H., III, and Roth, R. H. (1974). *Pharmacologist*, **16**, 213.
Musacchio, J. M., Craviso, G. L., and Wurzburger, R. J. (1972). *Life Sci.*, **11**, 267–276.
Musacchio, J. M., D'Angelo, G. L., and McQueen, C. A. (1971). *Proc. Nat. Acad. Sci., USA*, **68**, 2087–2091.
Narotzky, R., and Bondareff, W. (1974). *J. Cell. Biol.*, **63**, 64–70.
Neckers, L., and Sze, P. Y. (1975). *Brain Res.*, **93**, 123–132.
Nielsen, K. H. (1969). *Eur. J. Biochem.*, **7**, 360–369.
Noguchi, T., Nishino, M., and Kido, R. (1973). *Biochem. J.*, **131**, 375–380.
Numata (Sudo), Y., and Nagatsu, T. (1975). *J. Neurochem.*, **24**, 317–322.
Palkovits, M. (1973). *Brain Res.*, **59**, 449–450.
Parfitt, A., and Grahame-Smith, D. G. (1974). In *Aromatic Amino Acids in the Brain. Ciba Symposium 22*, Elsevier Excerpta Medica North Holland, Amsterdam, pp. 175–192.
Peters, D. A. V. (1971). *Biochem. Pharmacol.*, **20**, 1413–1420.
Peters, D. A. V. (1972). *Biochem. Pharmacol.*, **21**, 1051–1053.
Peters, D. A. V., McGeer, P. L., and McGeer, E. G. (1968). *J. Neurochem.*, **15**, 1431–1435.
Peters, D. A. V., Mazurkiewicz-Kwilecki, I. M., and Pappas, B. A. (1974). *Biochem. Pharmacol.*, **23**, 2395–2401.
Pujol, J. F., Buguet, A., Froment, J. L., Jones, B., and Jouvet, M. (1971). *Brain Res.*, **29**, 195–212.

Pujol, J. F., Kan, J. P., Buda, M., Petit-Jean, F., Mouret, J., and Jouvet, M. (1975). In *Chemical Tools in Catecholamine Research*, Vol. I (Eds., G. Jonsson, T. Malmfors, and C. Sachs), pp. 259–266.

Rastogi, R. B., and Singhal, R. L. (1974a). *J. Pharmacol. Exptl Ther.*, **191**, 72–81.

Rastogi, R. B., and Singhal, R. L. (1974b). *Endocr. Res. Commun.*, **1**, 261–270.

Reis, D. J., Joh, T. H., Ross, R. A., and Pickel, V. A. (1974). *Brain Res.*, **81**, 380–386.

Rembold, H. (1964). In *Pteridine Chemistry* (Eds., W. Pfleiderer, and E. C. Taylor), Pergamon Press, New York, pp. 465–484.

Renson, J. (1973). In *Serotonin and Behavior* (Eds., J. Barchas, and E. Usdin), Academic Press, New York, pp. 19–32.

Renson, J., Weissbach, H., and Udenfriend, S. (1962). *J. Biol. Chem.*, **237**, 2261–2264.

Renson, J., Daly, J., Weissbach, H., Witkop, B., and Udenfriend, S. (1966). *Biochem. Biophys. Res. Commun.*, **25**, 504–513.

Robinson, D., Lovenberg, W., and Sjoerdsma, A. (1968). *Arch. Biochem. Biophys.*, **123**, 419–421.

Rogawski, M. A., Knapp, S., and Mandell, A. J. (1974). *Biochem. Pharmacol.*, **23**, 1955–1962.

Roth, R. H., Morgenroth, V. H., III, and Salzman, P. (1975). *Naunyn-Schmiedeberg's Arch. Pharmacol.*, **289**, 327–343.

Roth, R. H., Salzman, P. M., and Morgenroth, V. H., III (1974). *Biochem. Pharmacol.*, **23**, 2779–2784.

Saavedra, J. M., Brownstein, M., and Axelrod, J. (1973). *J. Pharmacol. Exptl Ther.*, **186**, 508–515.

Saavedra, J. M., Palkovits, M., Brownstein, M., and Axelrod, J. (1974). *Brain Res.*, **77**, 157–165.

Sanders-Bush, E., Bushing, J. A., and Sulser, F. (1972a). *Biochem. Pharmacol.*, **21**, 1501–1510.

Sanders-Bush, E., Bushing, J. A., and Sulser, F. (1972b). *Eur. J. Pharmacol.*, **20**, 385–388.

Sanders-Bush, E., Bushing, J. A., and Sulser, F. (1975). *J. Pharmacol. Exptl Ther.*, **192**, 33–41.

Sato, T. L., Jequier, E., Lovenberg, W., and Sjoerdsma, A. (1967). *Eur. J. Pharmacol.*, **1**, 18–25.

Schmidt, M. J., and Sanders-Bush, E. (1971). *J. Neurochem.*, **18**, 2549–2551.

Sheard, M. H., Zolovick, A., and Aghajanian, G. K. (1972). *Brain Res.*, **43**, 690–694.

Shields, P. J., and Eccleston, D. (1972). *J. Neurochem.*, **19**, 265–272.

Shiman, R., Akino, M., and Kaufman, S. (1971). *J. Biol. Chem.*, **246**, 1330–1340.

Smith, R. D., Cooper, B. R., and Breese, G. R. (1973). *J. Pharmacol. Exptl Ther.*, **185**, 609–619.

Sourkes, T. L. (1974). In *Aromatic Amino Acids in the Brain. Ciba Symposium 22*, Elsevier Excerpta Medica North Holland, Amsterdam, pp. 361–378.

Squires, R. F. (1974). *Adv. Biochem. Psychopharmacol.*, **10**, 207–211.

Stone, K. J., and Townsley, B. H. (1973). *Biochem. J.*, **131**, 611–613.

Sullivan, P. T., Kester, M. V., and Norton, S. J. (1973). *Biochim. Biophys. Acta*, **293**, 343–350.

Tagliamonte, A., Biggio, G., Vargiu, L., and Gessa, G. L. (1973). *Life Sci.*, **12**, 277–287.

Tagliamonte, A., Tagliamonte, P., Perez-Cruet, J., Stern, S., and Gessa, G. L. (1971). *J. Pharmacol. Exptl Ther.*, **177**, 475–480.

Ternaux, J. P., Boireau, A., Bourgoin, S., Hamon, M., Héry, F., and Glowinski, J. (1976). *Brain Res.*, **101**, 533–548.

Thierry, A. M., Fekete, M., and Glowinski, J. (1968). *Eur. J. Pharmacol.*, **4**, 384–389.

Tong, J. H., and Kaufman, S. (1975). *J. Biol. Chem.*, **250**, 4152–4158.
Tourian, A. (1971). *Biochim. Biophys. Acta*, **242**, 345–354.
Tourian, A., Goddard, J., and Puck, T. T. (1969). *J. Cell Physiol.*, **73**, 159–170.
Tozer, T. N., Neff, N. H., and Brodie, B. B. (1966). *J. Pharmacol. Exptl Ther.*, **153**, 177–182.
Turner, A. J., Ponzio, F., and Algeri, S. (1974). *Brain Res.*, **70**, 553–558.
Twarog, B. N., and Page, I. H. (1953). *Amer. J. Physiol.*, **175**, 157–161.
Victor, S. J., Baumgarten, H. G., and Lovenberg, W. (1974). *J. Neurochem.*, **22**, 541–546.
Wapnir, R. A., Hawkins, R. L., and Stevenson, J. H. (1971). *Biol. Neonate*, **18**, 85–93.
Weiss, B. L., and Aghajanian, G. K. (1971). *Brain Res.*, **26**, 37–48.
Woods, H. F., and Youdim, M. B. N. (1977). In *Structure and Function of Monoamine Enzymes* (Eds., E. Usdin, N. Weiner, and M. B. H. Youdim), Dekker, New York, pp. 263–278.
Youdim, M. B. H., and Woods, H. F. (1975). *Biochem. Pharmacol.*, **24**, 317–323.
Youdim., M. B. H., Hamon, M., and Bourgoin, S. (1974). *Adv. Biochem. Psychopharmacol.*, **11**, 13–17.
Youdim, M. B. H., Hamon, M., and Bourgoin, S. (1975a). *J. Neurochem.*, **25**, 407–414.
Youdim, M. B. H., Mitchell, B., and Woods, H. F. (1976b). *Biochem. Soc. Trans.*, **3**, 683–684.
Zivcović, B., Guidotti, A., and Costa, E. (1973). *Brain Res.*, **57**, 522–526.
Zivcović, B., Guidotti, A., and Costa, E. (1974a). *Mol. Pharmacol.*, **10**, 727–735.
Zivcović, B., Guidotti, A., and Costa, E. (1974b). *Adv. Biochem. Psychopharmacol.*, **11**, 19–30.

Aromatic Amino Acid Hydroxylases and Mental Disease
Edited by M. B. H. Youdim

CHAPTER 7

Serotonin and the Pathogenesis of Affective Disorders

H. M. van Praag

A INTRODUCTION

In 1958 a new therapy for depression was introduced: medication with antidepressants. Almost simultaneously, but independently, two types of compounds with an antidepressant effect were discovered: the tricyclic antidepressants (prototype: imipramine) and the monoamine oxidase (MAO) inhibitors (prototype: iproniazid). Previously, only amphetamine derivatives and opiates had been available for antidepressant medication. The effect of amphetamines on affective life was variable and brief, and often followed by a hangover; opiates actually acted mainly on anxiety and agitation. The antidepressants, with an effect on mood regulation which in suitable cases was primary and lasting, represented a new pharmacotherapeutic principle.

Although tricyclic antidepressants and MAO inhibitors are chemically unrelated, they proved to have two characteristics in common. In psychopathological terms: they exert a beneficial influence on depressions, particularly on a given syndromal type: the vital depression.* In biochemical terms: in the brain they behave like monoamine (MA) agonists, albeit via different mechanisms. This led to the hypothesis that the therapeutic effect of antidepressants is related to their ability to potentiate MA. This hypothesis was supported by the discovery that reserpine is a depressogenic substance, capable of provoking typical vital depressions, while it reduced the amount of MA available in the brain. Psychopathologically and biochemically, reserpine proved to be a counterpart of the antidepressants.

The suspected relation between antidepressant effect and MA potentiation raised the question whether depressive patients who show a favourable response to antidepressants are suffering from a functional deficiency in catecholamines (CA) and/or 5-hydroxytryptamine (5-HT), or rather from a reduced susceptibility of postsynaptic CA and/or 5-HT receptors. In the past few years this question has been approached from several different angles. Initially, the overall metabolism of MA was studied; there was not much else that could be done. Later investigations focused in particular on the cerebrospinal fluid

* This syndrome is described in Anglo-American literature as endogenous depression.

(CSF), and on post-mortem findings in the brain in suicide victims. Yet another strategy was pharmacological manipulation of the central MA metabolism, followed by studies to establish whether this manipulation relieves, aggravates, or provokes depressive syndromes. An important product of this strategy was the development of drugs which can more or less selectively influence the metabolism of a given MA. Finally, efforts were made to establish whether therapeutic methods with an established antidepressant effect, such as electroconvulsive therapy (ECT), exert an influence on central MA.

Depression and mania, apparent antipodes, are often encountered in the same patients, and in psychodynamic views are closely related as to pathogenesis. It is therefore not surprising that biological depression research has also included manic patients, although their numbers have been small—probably because the manic patient does not readily yield to the exigencies of a strict research protocol.

The research discussed above led to two hypotheses: a CA hypothesis (Schildkraut, 1965; Bunney and Davis, 1965) and a 5-HT hypothesis (van Praag, 1962; Ashcroft *et al.*, 1965; Lapin and Oxenkrug, 1969; Coppen, 1967), which related the depressive syndrome to a central CA deficiency and 5-HT deficiency, respectively. Initially these hypotheses stood side by side. The monistic view has now been virtually abolished, and few doubt that it is more meaningful to think in terms of disturbed balances. In the context of this review, however, I will discuss only the 5-HT hypothesis, that is to say the arguments pro and con so far as they are derived from human studies. This isolation is somewhat artificial, the more so because several investigators now tend to determine the MA 'profile' in depressive patients rather than focusing on either CA or indoleamines (IA). To be sure, I shall permit myself an occasional 'excursion' to the CA.

The argumentation will be preceded by a brief outline of the diagnosis of depressions, including a brief discussion of the starting points of biological research in psychiatry. This is done in order to disprove the contention that biological psychiatric research has monopolist tendencies and claims to trace 'the' cause of behaviour disorders while ignoring the importance of social and psychological determinants.

The influence of tricyclic antidepressants, MAO inhibitors, and reserpine on the central 5-HT metabolism will be left undiscussed. The current views on neurotransmission in central serotonergic systems are assumed to be known to the reader.

B CLASSIFICATION OF DEPRESSIONS

1 Multidimensional diagnosis of depressions

For many years the pathological forms of depression (the distinction between physiological and pathological depressivity is left undiscussed here) were classified exclusively on the basis of aetiological criteria. It was customary, and in some textbooks it still is, to refer to neurotic (psychogenic) depression, involutional depression, arteriosclerotic depression, endogenous depression, etc. (The list of 'diagnoses' can be expanded *ad infinitum* because depressogenic factors are virtually innumerable.) This seemed to suggest that a depression can be adequately characterized with its aetiological adjective. This classification, however, is too superficial for adequate diagnosis, and adequate diagnosis is a prerequisite for the planning of both a therapy and a research protocol. In actual fact this classification has become a source of confusion. Particularly in the fields of biological psychiatry and psychopharmacology there have been investigators who have stressed the importance of adopting, in psychiatry also, some of the basic principles of medical nosology. They have insisted that: (i) disease pictures, in this case depressions, should be classified according to their symptoms, course, and causation, and (ii) the concept 'disease cause' should be divided into two components: aetiology and pathogenesis (van Praag and Leijnse, 1963a, b, 1965). Pathogenesis was defined as the constellation of cerebral dysfunctions which enable behavioural disorders to occur; aetiology was defined as all factors—hereditary, acquired somatic, psychological, and social—which have contributed to the occurrence of the cerebral dysfunctions.

The following sections briefly discuss a number of aspects of the diagnosis of depressions (Table 7.1), and for the unwary (non-psychiatric) reader three points should be made clear in advance: (i) the multidimensional diagnosis of depressions advocated here is not general usage and is not employed in all publications; (ii) the terminology has not yet been standardized, primarily I am discussing the concepts which I myself use, but I shall attempt to 'translate' them into the terms conventionally used in Anglo-American literature; (iii) the differentiation of depressive syndromes is still in its initial stage, the dichotomy discussed here is merely a preliminary attempt to chart the field.

2 Syndromal points of view

From a strictly syndromal point of view I distinguish two types of depression: vital and personal depression (van Praag, 1962; van Praag *et al.*, 1965). Mixed forms are quite common; in other words, between the two prototypes there is a broad area which has not yet been adequately reconnoitred in terms of classification. Efforts to this effect are in full progress (e.g. Paykel, 1972; Garside, 1971; Becker, 1974). The following five subsections describe what I regard as the cardinal symptoms of the vital depressive syndrome.

1. *Despondency* The mood is dejected in greater or lesser degree. The future seems obscured, and there are often marked suicidal tendencies. A typical feature is that the sense of comprehensibility is often absent: either no

Table 7.1 Classification of depressions

Syndrome	Aetiology		Course	Pathogenesis
Vital	Endogenous or Exogenous or Psycho(socio)genic or Idiopathic	Often combinations of these factors	Monophasic or Unipolar or Bipolar	Forms with and without disturbances in central MA metabolism
Personal	Psycho(socio)genic		Not well-known. Recurrence is common	Unknown
Mixed forms	Not well-known. Probably as in vital depression		Not well-known. Recurrence is common	Unknown

motive for the depression is experienced, and the origin of the symptoms is an enigma to the patient; or an objective motive is evident in the history (e.g. the death of a dear one), but the patient does not experience this (anymore) as an adequate explanation. The comprehensible correlation between motive and depression has been lost. This is the case in the so-called vitalized personal depression (cf. section B.3).

2. *Motor disorders* In principle, these carry a minus sign (retarded vital depression). In severe cases they are objectively observable (in facial expression and gestures) and the retardation is apparent also in speech and thinking (slow thought processes, impoverished dialogue, monosyllabic answers). In the (quite common) milder cases, the retardation is apparent only in subjective experiencing. The patient finds that his work is no longer going smoothly and that he cannot cope with it unless he makes a determined effort. Nothing comes easily anymore. The ideas are reluctant to come and can be retained only with difficulty. The patient himself speaks of disturbed concentration. In some cases there may be agitation at the same time (agitated vital depression): the patient feels tense and anxious, and may show manifest motor unrest. I have said 'at the same time': in these cases retardation and agitation are not polar factors, but the agitation is superimposed on the state of retardation.
3. *Hypoaesthesia* The ability to respond emotionally to exogenous stimuli is diminished. Things from which the patient used to derive pleasure—his work, his family, his hobbies—no longer mean much to him. He can no longer enjoy things. He is listless in the true sense of the word. In a later stage, his ability to feel grief also diminishes. Hypoaesthesia is not to be equated to depersonalization. In the former, emotional experiencing is quantitatively reduced; in the latter it is qualitatively changed.
4. *Somatic disorders* An obligatory triad, in my opinion, is that of disorders of sleep (early morning awakening), fatigue disproportionate to performance, and reduced appetite. The actual amount of food taken may still be normal, but it no longer tastes well. Sexual disorders (loss of libido, impotence, amenorrhoea) are common features but not, in my view, essential to the diagnosis. In some cases somatic complaints are hypochondriacally exaggerated and given the focus of attention. This entails the risk of failure to recognize their origin (the depression); the resulting condition is a so-called masked depression.
5. *Rhythmicity* The symptoms are most severe upon awakening and in the morning hours, and show some improvement in the course of the day. In more severe cases this fluctuation diminishes and finally disappears completely, the patient being despondent throughout the day. Seasonal influences (appearance in spring and/or autumn) are classical but by no means evident in all cases.

In severe vital depressions, delusions can develop; in particular delusions of guilt, sin, poverty, and hypochondriacal complaints. For these cases I use the designation melancholia.

In the following five subsections the cardinal symptoms of the personal depressive syndrome will be outlined as parallels of the features of the vital depressive syndrome.

1. In these cases, too, the patient is despondent, but the mood anomaly is comprehensible to him in that there is always a motive which is experienced as an adequate explanation of the gloomy mood.
2. Disorders of motor activity and intrinsic motivation are much less pronounced. Significant retardation is rare, but agitation and anxiety are common.
3. There is no hypoaesthesia, but depersonalization may occur.
4. Somatic disorders are common, but mainly involve expressions of autonomic dysregulation such as palpitations, atypical anginous pains, disagreeable sensations at various sites in the body, etc. Sleep is often disturbed, particularly onset of sleep and continued sleep.
5. There is no circadian rhythm of the type observed in vital depression.

The syndrome of personal depression is less clearly defined than that of vital depression, because there are transitions to the vast field of the psychosomatic diseases, functional somatic disorders, anxiety neuroses, etc. As long as this is not more accurately charted, the description of a personal depression cannot be more exact.

Roughly, the vital depressive syndrome corresponds with the syndrome described in Anglo-American literature under the heading endogenous depression, while the personal depressive syndrome corresponds with what is described under such headings as neurotic, reactive, or psychogenic depression. What I call melancholia would seem to me to be identical with the so-called 'psychotic depression'.

Some authors have maintained that the difference between vital (endogenous) and personal 'neurotic) depression is exclusively quantitative (Becker, 1974): personal depressions are alleged to be relatively mild, whereas vital depressions are always severe. I do not share this view. In my opinion the difference between the two types of depression is a qualitative one, and either type can occur in a mild or in a severe form.

3 Aetiological points of view

In the context of this approach, an estimate is made of the extent to which endogenous (hereditary), exogenous (acquired somatic), psychogenic (intrapsychic), or sociogenic (relational and environmental) factors have contributed to the development of a given syndrome.

In the individual case, the sole indication of the activity of an endogenous factor is a tainted family history; and the taint should not be exclusively apparent in the direct line because in that case pseudo-heredity (transmission via a learning process) cannot be eliminated. Exogenous factors are investigated on the basis of the history and physical examination.

A psychogenic 'load' becomes plausible when there are positive indications of an unresolved mental conflict and a personality structure with weak spots which make it understandable why the conflict should remained unresolved (Chodoff, 1972).

Sociogenic stress factors have their origin in the patient's life environment (Brown *et al.*, 1973). However, their pathogenic action is always a resultant of the interaction between psychotrauma and personality structure. Most people can cope independently with the majority of their frustrations. A minority shows decompensation. There should therefore be demonstrable features in the personality structure which explain its 'susceptibility'

The vital depression is an aetiologically non-specific syndrome which can be caused by endogenous as well as by exogenous or psychosocial factors. The vital depression of chiefly endogenous determination is identical to the classical endogenous depression in conventional psychiatric literature. I prefer the former designation because: (i) it is quite possible that an endogenous factor also plays a role in non-vital depressions, and (ii) vital depression can also be provoked by non-endogenous factors. Virus infections such as influenza, infectious mononucleosis, and infectious hepatitis are an example of an exogenous causation of vital depressions. The non-psychiatrist notes that these diseases have a prolonged period of convalescence. Psychiatrically speaking, these periods are often characterized by a mild vital depressive syndrome. Vital depressions can also be psychogenic and sociogenic (van Praag, 1962). The depression can have a vital character from the onset or initially be of a personal character and gradually change into a vital depression: the depression becomes 'vitalized'.

In many actual cases combinations of factors are involved. For example, an individual living under stress, with a depressively tainted family history, develops a vital depression after experiencing a physical illness. Finally, a vital depression can develop without any discernible aetiology; the term idiopathic applies to such cases.

In the origin of the personal depressive syndrome, psychogenic and sociogenic factors invariably play an important role. In actual fact this syndrome corresponds with that of the psychogenic and neurotic depression in the 'official' psychiatric literature. I prefer the designation psychogenic (or neurotic) personal depression, because psychogenic influences can also lead to a vital depression.

A vital depression is a self-limiting disease, which can last from a few days to many years but tends towards spontaneous recovery. It seems as if, once it develops, it takes an independent course following its own rules, quite regardless of the causative factors. The reserpine depression provides an excellent example. Discontinuation of reserpine usually does not ensure recovery. Antidepressant treatment is required to achieve this. For this reason one may pose the question: is it correct to distinguish exogenous and psychogenic vital depressions; is not rather every vital depression essentially endogenous, with the exogenous and psycho(socio)genic factors merely as subordinate, precipitating elements? This question is unanswerable but, in view of recent biochemical research, I regard this possibility as by no means excluded.

4 Classification according to course

In terms of their clinical course, two types of depression are distinguished: unipolar and bipolar depressions. A unipolar depression is defined as a recurrent vital depression without (hypo)manic phases. A bipolar depression is a recurrent vital depression in the course of which (hypo)manic phases also occur. Unipolar and bipolar depressions differ in other features as well. The mode of hereditary transmission is not the same (it is to be understood that genetic factors are by no means always demonstrable); but there are differences also in age of onset of the first phase, the duration of symptom-free intervals, and the premorbid personality structure (Perris, 1966; Angst, 1966).

The terms unipolar and bipolar hold no aetiological implications. In the classical cases the phases develop for no apparent reason and the family history is depressively tainted; they show the classical endogenous aetiology. However, cases in which the phases are provoked (or caused?) by exogenous and/or psychosocial factors are far from exceptional, particularly in the earlier stage. Not infrequently, later phases tend to develop more and more spontaneously.

The term manic-depressive psychosis can be regarded as a collective term for unipolar and bipolar depressions. I am disinclined to use this designation because: (i) a majority of patients never develop (hypo)manic phases, and (ii) a majority of patients never become psychotic either in their depressive or in their (hypo)manic phases.

If left untreated, the psychogenic (neurotic) personal depressive syndrome also tends towards recurrence. This is to be expected. As long as the patient's personality structure and life situation fail to change, there is a continued risk of minor or major conflicts which can result in a depression. Unlike the natural course of the vital depression, that of the personal depression has so far been given hardly any systematic investigation.

I personally confine the term unipolar depression to recurrent vital depressions, never using it to refer to recurrent personal depressions. In the literature it is by no means always clear whether or not this definition is accepted and used. In such cases the term unipolar (one-dimensional as it actually is) is confusing.

5 Pathogenetic points of view

The introduction of the pathogenesis concept in psychiatry (van Praag and Leijnse, 1963a, b, 1965) implies the assumption that: (i) the disease concept is of a material nature in psychiatry also, and (ii) every state of behaviour, disturbed or undisturbed, is dependent on a given functional state of the brain—a state which in principle can be analysed and influenced by biochemical (drugs) or physiological (depth electrodes) means. To put it briefly: there is no disturbed behaviour without corresponding cerebral substrate. This statement is not a revival of nineteenth-century materialism. It does not hold that behaviour disorders *are* diseases of the brain. What it does hold is that pathogenic influences of any kind—be they psychological, environmental, or somatic—do not affect mental life directly, via some sort of vacuum, but indirectly via changes in cerebral organization. In this vision the brain takes the role of an intermediary agent.

Anyone who accepts this vision thereby declares himself an opponent to the way of thinking in alternatives which is still common usage in psychiatry. An example: 'Is behaviour disorder A a biochemical or a psychosocial disease?' In my opinion this question poses a spurious problem. If a dichotomy is at all required, then the separation should be made as follows:

1. Behaviour disorders in which psychological and social factors contribute in an important degree to the development of the cerebral dysfunctions which generate these disorders.
2. Behaviour disorders based on cerebral dysfunctions which to an important extent must be ascribed to factors other than psychosocial factors: e.g. acquired somatic diseases which involve damage to the brain, and hereditary factors due to which, say, a given enzyme primordium is marginal.

Another example: 'Behaviour disorder A can only be treated by psychotherapy, and behaviour disorder B responds only to pharmacotherapy'. This statement also contains an inconsistency of logic. The purposes of pharmacotherapy and psychotherapy (here meant to include sociotherapy) are disparate. Drugs are resorted to in an effort to control cerebral dysfunctions. They are aimed at the substrate which generates the disease symptoms. Psychotherapy is aimed at the pathogenic input. Its purpose is to attenuate tensions generated within the individual or by the interaction between individual and environment. In other words: the objectives of pharmacotherapy and those of psychotherapy are quite different. The former aims at the pathogenesis, and the latter at the aetiology of the syndrome presented. As such they are complements, and they must be applied jointly, at least in so far as adequate methods are already available for a given diagnostic category.

A pathogenetic classification of depressions is based on data of cerebral function. Research on the basis of this criterion is still in its initial stage. Preliminary results, however, are encouraging in that they tend to show that this approach is not based on a fiction. I shall not discuss these findings here, for they are the terminal point and the quintessence of the arguments presented in subsequent sections.

6 Other diagnostic classifications

Chiefly on the basis of life history and the premorbid personality structure, Robins *et al.* (1972) made a distinction between primary and secondary affective disorders. The primary depression occurs in individuals with a more or less normal personality structure and with a psychiatric history which is 'clean', apart from possible depressions or manic phases. The depression is called secondary in patients with pre-existent psychiatric disorders other than depression and mania, and with a disturbed personality structure.

The primary depression concept in actual fact comprises the classical vital depression with a unipolar or bipolar course in non-neurotic individuals; the secondary depression concept covers the neurotic depression, but gives insufficient information on the symptomatology. I am not too happy with these two concepts. They are not multidimensional, ignore the possibility that vital depressions (with a unipolar or bipolar course) can also occur in neurotic individuals, and there are no indications that predominantly neurotic and predominantly non-neurotic vital depressions are fundamentally different in therapeutic or prognostic terms.

On the basis of an exhaustive family study, Winokur (1971) distinguished two subtypes within the unipolar group: depression spectrum disease and pure depressive disease. Prototype of the former group is a woman whose first depressive phase developed before the age of 40, in whose family depression is more common among women than among men, and in whose family the 'depression deficit' in the male line was 'repleted' by alcoholism and sociopathy. A prototype of the latter group is a man whose first depressive phase started after the age of 40 and in whose family depressions are as common in men as in women, while alcoholism and sociopathy are not over-represented.

No biochemical research has yet been done on the basis of the latter classification.

7 Conclusions

In this section a plea has been presented in favour of a multidimensional classification of depressions (and other psychiatric categories) according to symptomatology, aetiology, course, and pathogenesis. This approach is a necessity because: (i) the nature of the syndrome gives insufficient information on its aetiology; (ii) a given aetiology does not necessarily lead to a given syndrome; (iii) the course of a depression cannot be reliably predicted either on the basis of its aetiology or on that of the syndromal features. A multidimensional approach is of importance for clinical practice; it facilitates therapeutic planning. Schematically, and by way of example, the decision 'predominantly vital' or 'predominantly personal' depression determines whether antidepressants will or will not be given; the decision about the 'weight' of the psychosocial factors determines whether focused psychotherapy will be resorted to; and the decision 'unipolar' or 'bipolar' depression determines whether lithium prophylaxis will be used.

It seems to me that a multidimensional approach is indispensable for research purposes.

C PERIPHERAL 5-HT METABOLISM IN AFFECTIVE DISORDERS

1 Urinary excretion of indolamines

5-Hydroxyindoleacetic acid (5-HIAA) is the principal oxidatively deaminated metabolite of 5-HT. The amount excreted in urine is a gross index of the overall turnover of 5-HT. This is why the urinary 5-HIAA excretion has been determined in depressions, particularly in the early years of MA research in depressions. It was in fact the first strategy used to test the MA hypothesis. Several authors found the 24-h. excretion diminished (Pare and Sandler, 1959; van Praag and Leijnse, 1963a), but normal values have also been reported (Cazzullo *et al.*, 1966) and some authors found increased values (Tissot, 1962). Longitudinal studies showed that the level of excretion during manic phases exceeded that during depressive phases (Ström-Olsen and Weil-Malherbe, 1958). Moreover, it was believed that patients with a low 5-HIAA excretion showed a more favourable response to MAO inhibitors than those with higher excretion values (Pare and Sandler, 1959; van Praag and Leijnse, 1963b).

Little is known about the renal excretion of 5-HT. The tryptamine excretion can be either increased (McNamee *et al.*, 1972a) or decreased (Coppen *et al.*, 1965b) in depressions.

I do not think that too much value should be attached to observations of this type. Most of the 5-HIAA (and tryptamine) in the urine originates from the periphery, and it is unlikely that the peripheral 5-HT turnover could be indicative of the 5-HT turnover in the central nervous system (CNS). Inversely, it is very questionable whether changes in central 5-HT turnover (given an unchanged peripheral turnover) would lead to measurable changes in renal 5-HIAA excretion. In many of these studies, moreover, the diet was not standardized; and it was later found that 5-HT occurs in certain foods, particularly fruits (Crout and Sjoerdsma, 1959). This may have influenced the results.

2 5-HT synthesis

Coppen *et al.* (1965a) made the interesting observation that, in depressive patients, the production of (^{14}C) carbon dioxide (in expiratory air) from intravenously injected ^{14}C-labelled 5-HTP is decreased as compared with that in a control group. This might suggest a defect in the conversion of 5-hydroxytryptophan (5-HTP) to 5-HT. However, these investigators were unable to reproduce their results (Coppen, 1967).

There has since been no reduplication of this study, but a variant was carried out. Coppen *et al.* (1974a) administered ^{14}C-labelled 1-tryptophan to five severely depressed patients, and measured the urinary excretion of (^{14}C) 5-HIAA in the depressive phase and after recovery. They found no significant differences between pre-therapeutic and post-therapeutic values, nor between patients and controls. The results showed marked interindividual and intraindividual variations. These observations showed that an overall disorder in 5-HT synthesis in depressed patients is very unlikely. A similar indication is found in the fact that the urinary 5-HIAA excretion after oral administration of tryptophan to depressive patients is not subnormal (Cazzullo *et al.*, 1966). However, Coppen *et al.* (1974a) did demonstrate a slightly decreased tryptophan tolerance in depressions. They measured the total blood tryptophan level at different intervals after intravenous tryptophan loading and found that, during the first 3 h. after loading, the tryptophan concentrations in depressive patients were slightly higher than those in the controls. This might indicate a retardation of the chemical 'processing' of tryptophan, in whatever direction; but it might also be that in depressive patients tryptophan is distributed over a smaller volume.

All in all, there is no reason at this time to assume that in depressions there is a defect in the 5-hydroxylation and decarboxylation of tryptophan in the organism in its totality. This need not apply to the CNS. It is to be borne in mind, moreover, that the abovementioned studies involved only a small number of patients, and that on the other hand it has been demonstrated in recent years that disorders of MA metabolism in depressive patients are not a universal phenomenon, but occur in certain subcategories who are sometimes, but not always, identifiable on psychopathological grounds. Studies in small groups carry a risk of missing such biochemical disorders.

3 Availability of tryptophan

Only a small amount of tryptophan is used for the synthesis of the indole derivatives 5-HT and tryptamine. A much larger amount is converted via kinurenin and xanthurenic acid to the B-vitamin nicotinic acid, or utilized in protein synthesis. Two types of observation suggest that the amount of tryptophan available for 5-HT synthesis is possibly diminished in depressions.

(a) *Activation of the kynurenin route*

Depressive patients convert more tryptophan via the kynurenin route than normal subjects. This was concluded from the fact that in these patients the urinary xanthurenic acid excretion is increased, both basally and after tryptophan loading (Cazzullo *et al.*, 1966). Curzon (1969) reported the same in women with endogenous depressions. In two patients suffering from manic-depressive changes who were intravenously injected with radioactive tryptophan, moreover, the excretion of radioactive kynurenin was higher during the depressive than during the manic and the normal phases (Rubin, 1967). Why should an abnormally large amount of tryptophan be converted via the kynurenin route? A possible explanation is a high level of circulating hydrocortisone (cortisol), which is a common phenomenon in depression (Hullin *et al.*, 1967). Corticosteroids from the adrenal cortex stimulate the synthesis of tryptophan pyrrolase in the liver (Knox and Auerbach, 1955); tryptophan pyrrolase is the first enzyme in the conversion of tryptophan via the kynurenin route. In corroboration of this, Rubin (1967) found that an increased plasma cortisol level and increased kynurenin excretion go hand in hand.

A cardinal question in this context is, of course, whether activation of tryptophan pyrrolase does indeed withdraw a substantial amount of tryptophan from 5-HT synthesis. This is plausible. In the acute experiment, hydrocortisone induces in rats: (i) activation of liver pyrrolase; (ii) a decrease in plasma tryptophan; and (iii) a decrease by about 30% of the cerebral 5-HT and 5-HIAA concentrations (Curzon and Green, 1968). When the animals are given a pyrrolase inhibitor (e.g. allopurinol or yohimbine) in advance, hydrocortisone no longer causes a decrease in intracerebral 5-HT (Green and Curzon, 1968). Also, α-methyltryptophan—a compound which, like hydrocortisone, activates liver pyrrolase, but independent of the adrenal cortex—likewise causes a decrease in central 5-HT concentration (Curzon, 1969). It is therefore justifiable to conclude that an increased plasma concentration of adrenocortical hormones causes a decrease in intracerebral 5-HT via an increase of the pyrrolase activity in the liver. An observation of interest in this respect was reported by Mangoni (1974), who found that imipramine and the MAO inhibitor tranylcypromine (both antidepressants) inhibit the tryptophan pyrrolase activity in the rat liver.

(b) *Plasma tryptophan concentration*

According to Coppen *et al.* (1973), the plasma concentration of free trypto-

phan in women with 'depressive illness' (with no manic phases in the history) was decreased as compared with that in a matched control group. The total tryptophan concentration was not decreased. This implies that the tryptophan fraction bound to plasma proteins was increased at the expense of the unbound (free) tryptophan. After clinical recovery the free tryptophan concentration showed a significant increase without attaining total normalization. Rees *et al.* (1974) studied two patients with rapidly alternating manic and depressive phases and found a decreased plasma tryptophan concentration in the depressive, and an increased level in the manic phases. They determined only total tryptophan. In some patients with slower alternation of phases this phenomenon was less pronounced.

Coppen's findings seem important, but need verification and corroboration. The food intake was neither standardized nor controlled, and this may have influenced the results. The test subjects were not receiving drugs which change the degree of tryptophan binding (e.g. salicylates). However, there are other potential interfering factors. Free fatty acids in the plasma, for example, influence the protein binding of tryptophan (Curzon *et al.*, 1973; Lipsett *et al.*, 1973), and this fat fraction responds strongly to psychological stress; it is therefore not excluded with certainty that the decreased free tryptophan concentration in these cases was a non-specific phenomenon in relation to the depression.

4 Monoamine oxidase (MAO)

The data on this enzyme, which plays an important role in MA degradation, are controversial. In women with severe 'endogenous depressions' (all premenopausal), a plasma MAO activity was measured which was far higher (six times) than the normal value (Klaiber *et al.*, 1972). No information was given on the occurrence or non-occurrence of (hypo)manic phases. Now, it is a fact that peripheral as well as central MAO activity increases with increasing age (Robinson *et al.*, 1972); but age differences could not explain the difference between the depressive and the control group. Oral administration of conjugated oestrogens led to normalization of the MAO activity and abatement of the depression. In this study, however, there was no control group. The authors concluded that their findings are consistent with the hypothesis that a CA deficiency (possibly therefore as a result of MAO overactivity) plays a role in the pathogenesis of depressions. The findings of Sandler *et al.* (1975) point in the same direction. They found a deficit in tyramine conjugation in a group of depressed patients. Most of the amine was degraded via oxidative deamination. They postulated that the conjugation deficit was secondary to increased MAO activity.

Murphy and Weiss (1972), however, found in fact a 50% decrease in MAO activity in depressions of the bipolar type. A corresponding finding was an increased renal excretion of tryptamine—an amine whose degradation is entirely dependent on MAO. These authors measured the MAO activity, not in plasma but in blood platelets. In the four patients in whom this question was studied, the MAO activity continued to be decreased also after subsidence of the symptoms, and also in the (hypo)manic phase. In patients with depressions of the unipolar type the average MAO activity was slightly (10%) but not significantly increased. The MAO activity of blood platelets in women exceeds that in men. This applies to all age categories (Robinson *et al.*, 1971). However, since the three groups studied were matched as to sex and age distribution, this factor cannot have influenced the findings obtained.

Gottfries *et al.* (1974) recently found that the MAO activity in various areas of the brain in suicide victims ($n = 15$) was 20–40% lower than that in a comparable control group ($n = 20$) with a different cause of death. The decrease was demonstrable regardless of whether tryptamine or β-phenylethylamine was used as substrate for MAO. However, it seems questionable whether these observations indeed corroborate those reported by Murphy and Weiss. It is unknown whether the patients in the suicide group had all been suffering from depressions, let alone whether these depressions were of the bipolar type.

Is a decreased central MAO activity, if any, inconsistent with the 5-HT deficiency hypothesis? It need not necessarily be. In the first place, the phenomenon could be a secondary one—some sort of compensatory mechanism aimed at abolition of the assumed 5-HT deficiency. In that case, however, it would hardly have any physiological significance, because a 20–40% reduction of MAO activity it not likely to be sufficient to retard 5-HT degradation. Another possibility would be that, as a result of primarily decreased MAO activity, a non-transmitter amine such as octopamine would accumulate in serotonergic neurons. Such a compound could function as false transmitter, and reduce or arrest the activity in the corresponding circuit. An argument in favour of such a possibility is the fact that octopamine accumulates in blood platelets following administration of MAO inhibitors, and that the same has been observed in depressive patients with an (endogenously) decreased MAO activity in the blood platelets (Murphy 1972).

It is not clear why the data on MAO are so controversial. Perhaps the characteristics of MAO in plasma differ from those of MAO in blood platelets. It has been known for some years now that MAO comprises a complex of enzymes which have different properties, and that the enzyme's composition can vary from one organ to the next. Moreover, MAO activity can be temporarily influenced by hormonal influences (e.g. oestrogens) and—at least in animals—by certain states of behaviour (Klaiber *et al.*, 1971; Eleftherion and Boehlke, 1967). In human subjects these factors are not entirely controllable, and may have influenced results.

5 Conclusions

Studies of the periphery have so far failed to produce convincing indications of a specific disturbance in 5-HT metabolism in depressions. They have revealed a number of suspicious phenomena: a decreased concentration of free tryptophan in plasma, and indications that an abnormally large amount of tryptophan enters the kynurenin 'channel'. Perhaps the plasma tryptophan concentration is decreased because its removal via the kynurenin route is increased. It is by no means excluded that these phenomena correspond more with the degree of anxiety than with the depressive affect as such.

In view of the available data an overall defect in the conversion of tryptophan to 5-HT cannot be excluded with absolute certainty, and it would be worthwhile to continue investigations along these lines.

The data concerning the activity of MAO are contradictory and impossible to interpret.

D POST-MORTEM STUDY OF THE CENTRAL 5-HT METABOLISM IN DEPRESSIONS

1 Introduction

It is conceivable that the CNS can supply more information on the central MA metabolism than the periphery. The problem is how to obtain this information in human subjects. This has been tried in three different ways: post-mortem examination of the brains of suicide victims, determination of the concentration of MA metabolites in the CSF (the mother substances do not occur in the CSF in measurable amounts), and determination of the accumulation of acid MA metabolites in the CSF after probenecid loading. The results thus obtained will be discussed in this section and in sections E and F.

2 Post-mortem studies

The first study of this type was published in 1967, when Shaw *et al.* reported that the 5-HT concentration in the hindbrain was lower in a group of suicide victims than in a control group of individuals who had died from accidents or from acute somatic non-neurological diseases. The 5-HIAA concentration was not measured. Pare *et al.* (1969) likewise found a decreased 5-HT concentration, but reported normal values for 5-HIAA. In the study of Bourne *et al.* (1968), no differences in 5-HT concentration were found, but the 5-HIAA concentration was found decreased in the suicide group. The last study published on this subject was of great importance (Lloyd *et al.*, 1974): the brainstem was studied not in its entirety but in separate components. A significant decrease in 5-HT concentration was found exclusively in the raphe nuclei; not in all nuclei but only in the dorsal and the inferior central nucleus. In both nuclei the 5-HIAA concentration was lower than that in the control group, but the difference was not statistically significant. It may be mentioned in passing that the system of raphe nuclei is the site of predilection of 5-HT in the CNS.

Gottfries *et al.* (1975) compared MAO activity in the brains of 15 suicides of whom eight were alcoholics to a control material of 20 individuals without known mental disorders. MAO activity was determined in 13 different parts of the brain with β-phenylethylamine and tryptamine as substrates. The MAO activity in all parts of the brain investigated was found to be significantly lower in the alcoholic suicides as compared to controls, while there was no significant difference between the non-alcoholic suicides and controls.

Apparently there exists a connection between low MAO activity in the brain and suicidal behaviour among alcoholics.

3 Significance of post-mortem findings

Chemical data obtained after death are delicate material in terms of interpretation. They can have been influenced by a variety of uncontrolled or uncontrollable factors such as: use of drugs prior to death (several of the test subjects had committed suicide with the aid of drugs); death causes in the control group; interval between death and post-mortem examination; influences of age and sex; the duration of the final agony, etc. On the other hand: doubts about the method as such cannot obliterate the fact that the results of the four studies all point in the same direction: that of a deficiency of 5-HT and 5-HIAA in the suicide group. This situation might be based on a diminished 5-HT turnover.

That the results of these studies show a fair degree of agreement is surprising, particularly from a psychiatric point of view. Suicide is not a diagnosis but the fateful terminal point of all sorts of developments: overwhelming problems of life; a depression in the psychiatric sense; addiction; a psychotic state, etc. Information on the histories of the suicide victims involved was frequently inadequate. It seems unlikely that they had all been suffering from pathological depressions. The biochemical data, therefore, would seem to be more homogeneous than the psychopathological features. An explanation of this paradox might lie in the fact that disorders of 5-HT metabolism are related not so much to the clinical entity known as depression as to a decrease in the mood level, regardless of the syndrome within which this occurs, and its aetiology. This would be in agreement with my view (cf. section F.8) that the biochemical dysfunctions found in the brains of psychiatric patients in the past few years can be better and more sensibly correlated with disturbances in separate mental functions than with syndromes, i.e. complex patterns of disturbed behaviour, or nosological entities.

4 Conclusions

Post-mortem findings indicated that the 5-HT and 5-HIAA concentrations in the lower portion of the brainstem in suicide victims were decreased. MAO activity was found to be lowered in the brains of alcoholic suicides, a finding which is of course difficult to reconcile with decreased levels of 5-HT. The decrease in indolamines might indicate that the 5-HT turnover had been low prior to death. Since a group of suicide victims is probably not very homogeneous in psychopathological terms, the suspicion arises that the disorders of central 5-HT metabolism observed were related less to the syndrome depression, or a given nosological concept of depression, than to lowered mood level, regardless of the syndrome within which this phenomenon occurs, or its aetiology.

E CENTRAL 5-HT METABOLISM IN AFFECTIVE DISORDERS 1. CSF STUDIES WITHOUT PROBENECID

1 MA metabolites in CSF: do they reflect the central MA metabolism?

The concentration of MA metabolites is usually measured in lumbar CSF. Do these values give a more or less reliable impression of the metabolism of the mother amines in the CNS? Let me list the principal arguments indicating that indeed they do.

In the CNS itself, there is of course a relation between the concentration of MA metabolite in a given area, and the amount of amine locally metabolized. Animal experiments have demonstrated a correlation between the 5-HIAA and homovanillic acid (HVA) concentrations in the CSF, and their concentrations in adjacent parts of the CNS (Moir *et al.*, 1970; Papeschi *et al.*, 1971; Eccleston *et al.*, 1968). Under normal conditions, moreover, there is no transport of any significance of amine metabolites from the periphery to the CNS. This has been demonstrated for 5-HIAA and HVA in test animals (Bartholini *et al.*, 1966; Bulat and Zivković, 1971), and for (conjugated) MHPG in human individuals (Chase *et al.*, 1973).

The principal argument that the human lumbar CSF can supply relevant information on central MA metabolism is that, in response to factors which influence the central MA turnover, the CSF concentration of MA metabolites changes in the expected direction. The 5-HT precursors tryptophan and 5-HTP increase 5-HT synthesis. After administration to human subjects, the 5-HIAA concentration in the CSF increase in accordance with expectation (Dunner and Goodwin, 1972; van Praag *et al.*, 1973b; van Praag, 1975). Moreover, there is a significant correlation between tryptophan and 5-HIAA concentrations (Ashcroft *et al.*, 1973b), and this is an additional argument that the 5-HIAA concentration in the CSF and the 5-HT turnover in the CNS are linked.

After L-dihydroxyphenylalanine (L-DOPA) administration, the concentration of HVA in the CSF increases (Goodwin *et al.*, 1970). Exogenous L-DOPA is converted for the most part to dopamine (DA); only a small amount is converted to noradrenaline (NA). The only exogenous method to cause a selective increase in NA synthesis is administration of dihydroxyphenylserine (DOPS)—a compound which is directly converted to NA without intermediate DA production (Creveling *et al.*, 1968). To my knowledge there have been no efforts to establish whether DOPS does indeed increase the concentration of 3-methoxy-4-hydroxyphenyl glycol (MHPG) in human CSF.

Reduction of MA synthesis also has its reflection in the CSF. In response to *p*-chlorophenylalanine (PCPA)—an inhibitor of tryptophan hydroxylase and therefore of 5-HT synthesis—the 5-HIAA concentration in lumbar CSF decreases (Chase, 1972). The same happens with MHPG in response to α-methyl-*p*-tyrosine (α-MT), an inhibitor of tyrosine hydroxylase, and fusaric acid, an inhibitor of dopamine β-hydroxylase (Chase *et al.*, 1973). In Parkinson's disease, the DA concentration in the basal ganglia and the HVA concentration in the CSF both show a marked decrease (Bernheimer *et al.*, 1966).

2 MA metabolites in lumbar CSF: do they reflect the cerebral or the spinal MA metabolism?

As already pointed out, there are strong arguments in favour of a relation between the concentration of MA metabolites in lumbar CSF and the metabolism of the mother amines in the CNS in human individuals. Another question is whether it is indeed the cerebral MA metabolism that is reflected in the lumbar CSF. Could it be that mainly the spinal MA metabolism is measured in lumbar CSF? Even if this were true, such determinations would not be without value. If depressions involve a defect in the MA metabolism, then there is no reason to believe in advance that this disorder would suddenly cease at the level of the spinal cord. On the other hand, the value of determinations in CSF would increase if it could be shown that these probably do give an impression of the cerebral metabolism also.

(a) Homovanillic acid (HVA)

The most informative test in the latter sense is determination of HVA in lumbar CSF. For this HVA is really largely of cerebral origin. This is not surprising in view of the fact that, so far as we know, few or no dopaminergic neurons or nerve endings are contained in the spinal cord (Carlsson *et al.*, 1964). Accordingly, the gradient between ventricular and lumbar HVA concentrations is high (10:1) (Sourkes, 1973). There are other strong arguments in favour of the view that lumbar HVA is representative of the cerebral DA metabolism. Partial block (Curzon *et al.*, 1971; Garelis and Sourkes, 1973; Post *et al.*,

1973a) and total block of the spinal subarachnoidal space (Young *et al.*, 1973) leads to a decrease of HVA (respectively, its total disappearance from) in the CSF caudal to the block. After occlusion of both interventricular foramina, only very little HVA was demonstrable in the lumbar CSF (Garelis *et al.*, 1974). This demonstrates that HVA largely originates from structures in the vicinity of the lateral ventricles. These findings give substance to the view that HVA in the human and animal lumbar CSF (Papeschi *et al.*, 1971) originates from DA-rich areas close to the lateral ventricles, mainly from the caudate nucleus (Garelis *et al.*, 1974). According to Sourkes (1973), an important amount of the HVA there produced is drained into the lateral ventricles (not into the bloodstream). From the lateral ventricles it reaches the lumbar CSF partly by diffusion and partly in the flow of CSF. Under more or less basal conditions the time required for this is estimated to be 4 h. (Korf and van Praag, 1970; Pletscher *et al.*, 1967).

While under normal conditions the spinal cord hardly contributes to the HVA concentration in lumbar CSF, this changes when large doses of L-DOPA are administered. In the rat, at least, this leads to substantial DA synthesis in the spinal cord (Andén *et al.*, 1972b). It seems plausible that, in human subjects also, the spinal cord contributes substantially to the HVA in lumbar CSF in this case (Garelis *et al.*, 1974).

(b) 5-Hydroxyindoleacetic acid (5-HIAA)

5-HIAA in lumbar CSF is of mixed spinal/cerebral origin. It could hardly be otherwise, for serotonergic nerve endings occur in the spinal cord. Accordingly, there is a ventriculo-lumbar gradient of 5-HIAA concentration, but one which is less high than that of HVA concentrations (10:3) (Sourkes, 1973). It is none the less plausible that some of the lumbar 5-HIAA is of cerebral origin. The first argument is that, after oral administration of tryptophan to patients, the tryptophan concentrations in lumbar CSF begins to rise after 2 h., whereas that of 5-HIAA does not begin to rise until after 6 h. The 5-HIAA accumulation is believed to be delayed because some of the 5-HIAA must first be transported from the brain to the lumbar CSF (Eccleston *et al.*, 1970b). When transport of 5-HIAA from the CNS is inhibited with the aid of probenecid, the rise of the 5-HIAA concentration in lumbar CSF also occurs only after a few hours (Korf and van Praag, 1970). However, a different interpretation of this phenomenon is possible; namely that spinal 5-HT has lower rates of synthesis and degradation than cerebral 5-HT.

Another argument lies in the fact that in children with non-communicating hydrocephalus the ventriculo-lumbar 5-HIAA gradient is higher than that in hydrocephalic children in whom the CSF compartments do communicate (Andersson and Roos, 1969), so that ventricular 5-HIAA has free access to the spinal subarachnoidal space. The value of this argument should not be overrated. In non-communicating hydrocephalus, the CSF pressure above the obstruction is increased. In dogs, an increase in CSF pressure was found to delay the 5-HIAA transport from the CSF (Anderson and Roos, 1968). It is therefore possible that differences between children with communicating and with non-communicating hydrocephalus are explained by differences in hydrostatic pressure.

A final argument can be derived from studies of patients with transverse cord lesions or CSF block. In the former patients the 5-HIAA concentration in lumbar CSF is not decreased (Post *et al.*, 1973a). In test animals, severance of the spinal cord leads to about 80% reduction of the 5-HT concentration (Carlsson *et al.*, 1964; Shibuya and Andersson, 1968). If this applies to human subjects also, then the abovementioned observation is indicative of a substantial contribution of higher levels to the lumbar 5-HIAA level. Given a total block high in the spinal subarachnoidal space, one would expect a decrease in 5-HIAA concentration caudal to the block if a substantial amount of the lumbar 5-HIAA is of cerebral origin, and little change in this concentration if it largely originates from the spinal cord. The relevant data are controversial: Curzon *et al.* (1971) found decreased 5-HIAA values, but Garelis and Sourkes (1973) and Young *et al.* (1973) reported normal values. The discrepancy may be based on differences in the state of the cord below the block. It is quite conceivable that production and drainage of CSF at this level could be disturbed in individually different degrees (Garelis *et al.*, 1974).

One argument seemed to be an unqualified refutation of cerebral participation in the 5-HIAA concentration in lumbar CSF. It caused some misgivings in the ranks of investigators of the biological substrates of depressions, because part of their arguments that cerebral 5-HT plays a role in the pathogenesis of depressions derived from studies of lumbar CSF. Bulat and Zivković (1971) submitted cats to intracisternal injection of 5-HIAA, and found no increase of 5-HIAA in lumbar CSF. Their conclusion was: all lumbar 5-HIAA is of spinal origin. A rather hasty conclusion, because it is quite possible that the 5-HIAA injected is effectively eliminated from the CSF and disappears before it can reach the lumbar level. Weir *et al.* (1973) repeated the experiment in a slightly modified form, and obtained different results. A solution containing labelled 5-HIAA was infused into the basal cisterns of the cat at a constant rate. Calculations on the basis of the amount of 5-HIAA and the amount of label in the lumbar CSF demonstrated that a certain amount of the lumbar 5-HIAA (40–70% of its total amount) originated from structures rostral to the foramen magnum.

(c) 3-Methoxy-4-hydroxyphenyl glycol (MHPG)

Much less is known about the origin of MHPG in lumbar CSF. In man, there is no gradient between ventricular and lumbar CSF concentrations (Chase *et al.*, 1973), and it is therefore probable that a substantial amount of lumbar

MHPG originates from the spinal cord. An argument which points in the same direction is the fact that, in patients with a transverse lesion (whose spinal NA concentration can be assumed to have markedly decreased), the lumbar MHPG concentration is decreased, whereas this is not found in patients with a high-level spinal CSF block (in whom MHPG from rostral areas cannot reach the spinal subarachnoidal space) (Post *et al.*, 1973c).

3 Shortcomings of the CSF strategy

Determination of the concentration of MA metabolites in lumbar CSF (a procedure which I shall henceforth call CSF determination for the sake of brevity) affords information on the metabolism of the mother amines in the CNS, but the method shows some serious shortcomings.

To begin with, it gives an impression of the gross degradation of an amine at a given moment, but not of the amount of amine degraded within a given unit of time (i.e. not of the turnover). Another disadvantage is that the MA metabolites in lumbar CSF are not all of the same origin. MHPG is of spinal, 5-HIAA of mixed spinal/cerebral, and HVA of cerebral origin. In actual cases, the information obtained by CSF determination is rather heterogeneous. The last disadvantage I would mention is the most serious flaw: the method is not very reliable, and for several reasons. To begin with, MA metabolites are not altogether transported to the bloodstream via the CSF. A certain amount enters the bloodstream directly; for 5-HIAA this amount is estimated to be no less than 90% (Meek and Neff, 1973). For other metabolites this fraction is unknown. As such, a substantial transport of MA metabolites directly to the bloodstream does not devaluate CSF determination, provided the ratio between directly and indirectly transported material is constant; but it is precisely on this point that we have no information whatever.

Another factor detrimental to the reliability of CSF determination is the fact that the flow rate of the CSF is subject to the influence of CSF pressure (Davson, 1967). The amount of an MA metabolite encountered in the lumbar CSF at a given moment is in part determined by the CSF flow. This means that the concentration is affected by a variety of factors which influence CSF pressure: e.g. postural changes, straining, sneezing, coughing—all factors which it is difficult either to eliminate or to normalize.

The abovementioned factors probably explain the substantial day-to-day differences in the CSF concentration of MA metabolites in a given individual (H. M. van Praag *et al.*, unpublished data); they also explain why, as we shall see, the findings in depressive patients are not unequivocal.

The shortcomings of CSF determination can be largely avoided by premedication with probenecid. This procedure has consequently improved the study of the central MA metabolism in living human individuals to a considerable extent.

4 5-HT metabolites in the CSF in affective disorders

(a) Depression

In 1960 Ashcroft and Sharman reported that in depressive patients the CSF concentration of substances with a 5-hydroxyindole structure is decreased. In the course of subsequent years it was demonstrated that 5-HIAA is involved in these cases, and several investigators corroborated this observation (Ashcroft *et al.*, 1966; Denker *et al.*, 1966; van Praag *et al.*, 1970; Coppen *et al.*, 1972a; Mendels *et al.*, 1972). This phenomenon can indicate a diminished 5-HT metabolism in the CNS. However, not all results are unequivocal. Some authors described the decrease as slight and not significant (Bowers *et al.*, 1969; Papeschi and McClure, 1971); others found no decrease at all (Roos and Sjöström, 1969; Goodwin and Post, 1973; Wilk *et al.*, 1972; Sjöström and Roos, 1972).

(b) Mania

Manic phases have been much less fully studied than depressive phases, probably because manic patients cannot be deprived of medication for several days without difficulty, and do not readily adjust themselves to a strict research protocol. Several investigators reported a decreased lumbar 5-HIAA concentration (Denker *et al.*, 1966; Bowers *et al.*, 1969; Coppen *et al.*, 1972; Mendels *et al.*, 1972). This is the more remarkable because patients of this category often show motor unrest—a factor in response to which lumbar 5-HIAA concentration often tends to assume higher values, probably due to more effective 'mixing' of the CSF (Post *et al.*, 1973b). This finding could be a biological argument that mania and depression do not represent a polarity but have the same roots—a view which psychodynamics investigators have advocated on quite different grounds (Mendels and Frazer, 1973).

However, the findings in mania are not consistent either: normal and increased 5-HIAA values have also been reported (Bowers *et al.*, 1969; Goodwin and Post, 1973; Wilk *et al.*, 1972).

5 Causes of the variability of the results of CSF studies

It can be stated that more or less decreased 5-HIAA concentration in lumbar CSF have been fairly regularly found in depressive and manic patients; but these findings have by no means been consistent. There can be several reasons for this.

1. In many cases the syndromes studied were not systematically and con-

sistently differentiated by three criteria: symptomatology, aetiology, and course. Instead, the results in the depression group as a whole were averaged and compared with those in a control group. This method is apt to conceal possible differences between subgroups, and there are indications that this did happen. We ourselves, for example, regularly observed a decreased lumbar 5-HIAA concentration in the group of vital depressions, but only sporadically in the group of personal depressions (van Praag *et al.*, 1970, 1973)—findings which were corroborated by Mendels *et al.* (1972). The criterion 'course' is probably also important: in unipolar depressions the decrease is believed to be more pronounced than that in the bipolar group (Mendels *et al.*, 1972; Ashcroft *et al.*, 1973b).

So far, no correlations have been established between CSF concentration of 5-HIAA on the one hand, and severity or aetiology of the depressions on the other. All in all, there are indications that the concentrations of 5-HT metabolites in CSF differ in different categories of depression. This is why generalizing statements on the biochemical characteristics of 'depression' should be viewed with the necessary reservations.

2. Differences in age might play a role. According to Bowers and Gerbode (1968a) the 5-HIAA concentration in CSF increases with increasing age, and is significantly higher over rather than under the age of 60. Ashcroft *et al.* (1973b) found a more complicated relation between age and 5-HIAA, with minimal 5-HIAA values in the 40–50-year-old age group and higher values in both younger and older individuals. Other authors have found no age correlation (Goodwin and Post, 1973; Papeschi and McClure, 1971; Nordin *et al.*, 1971).
3. The control groups generally comprised no normal test subjects but consisted of hospitalized patients with neurological or psychiatric conditions. None showed marked abnormalities of mood, but as a group they were far from homogeneous.
4. The baseline 5-HIAA concentration proved to show substantial day-to-day variations in the same individual. Possible causes were discussed in a previous section. Perhaps this also explains why the subnormal values which we initially found in vital depressions (van Praag, 1969; van Praag *et al.*, 1970) were not reproduced in a reduplication study (van Praag *et al.*, 1973a, b).
5. It is questionable whether all studies were made under standardized conditions. An important point, for example, is that the patient be confined to bed during a few hours preceding the CSF determination. For 5-HIAA, after all, there is a ventriculo-lumbar concentration gradient, and motor activity promotes CSF mixing with, as a possible result, an increase in lumbar concentration (Post *et al.*, 1973a). To ensure comparable results, moreover, it is necessary always to examine the same CSF sample, e.g. the

first 5 ml obtained. For as a result of the gradient the concentration of MA metabolites increases as the CSF originates from a higher level. Finally, it is to be taken into account that food can contain MA, and that the diet must therefore be standardized in this respect (Erspamer, 1966).

6 Are the CSF changes syndrome-dependent or syndrome-independent

This question has so far received but scanty attention, and no unequivocal answer can therefore as yet be given. Mendels *et al.* (1972) reported that, after recovery, the 5-HIAA concentration increased only slightly. However, not all patients they examined for the second time were free from medication. The findings reported by Coppen *et al.* (1972b) and Ashcroft *et al.* (1973b) are less equivocal: no increase in CSF after recovery. At the time of the second examination the patient had been without medication for 5 days. It remains to be established whether this period is sufficiently long. Tricyclic antidepressants, which some of the patients had received, lower the 5-HIAA level, and a continuation of this effect may have concealed possible differences in concentration before and after treatment.

Another objection can be made to both studies. The second CSF determination was made during the hospital period, when the patient was considered to be free from symptoms and a few days after discontinuation of medication. In unipolar and bipolar depressions the depressive phase has an average duration of 5–6 months (van Praag, 1962; Angst *et al.*, 1973). Antidepressants bridge the depressive phase but do not arrest it: early discontinuation of medication is often followed by a relapse within one or a few weeks. In the light of this fact it is questionable whether the abovementioned patients could justifiably be described as having 'recovered'.

Even if further investigation would show that the disorders of the 5-HT metabolism are not syndrome-dependent but persist after recovery, this would not necessarily imply lack of correlation between the two phenomena (van Praag, 1974). The metabolic disorders might be factors predisposing to the behaviour disorders (cf. section F.10).

7 Specificity of CSF findings

A decrease in the 5-HIAA concentration in lumbar CSF is not specific for the group of affective disorders, but is observed also in Parkinson's disease (Olsson and Roos, 1968; Bernheimer *et al.*, 1966; Lakke *et al.*, 1972; Puite *et al.*, 1973; Godwin-Austen *et al.*, 1971), in seriously disabled (Claveria *et al.*, 1974) or bedridden multiple sclerosis patients (Sonninen *et al.*, 1973) (the phenomenon is not observed in less serious cases), in motor neuron disease, supranuclear palsy (Claveria *et al.*, 1974), and in pre-senile and senile dementia (Gottfries and Roos, 1969, 1970).

In Parkinson's disease the decreased levels are likely to be due to a reduced metabolism of 5-HT, for the phenomenon corresponds with the decreased 5-HT concentration found *post mortem* in the basal ganglia (Ehringer and Hornykiewicz, 1960). In depressions, a metabolic disorder is likewise plausible: first of all because post-mortem examination of suicide victims has revealed a decreased 5-HT concentration in the brainstem (cf. section D.2), and second because the 5-HIAA concentration in the CSF increases in response to 5-HT precursors—compounds which promote the synthesis of the mother substances (Goodwin *et al.*, 1970; Dunner and Goodwin, 1972; van Praag, 1975).

Low CSF 5-HIAA levels based on a deficient 5-HT metabolism in the CNS, therefore, are not specific of a given nosological entity of syndrome. But this does not mean that they are non-specific. They could be *symptomatologically specific*, that is to say related to particular disorders of motor or psychological function. For example, the decreased 5-HT turnover might be related to a pathological lowering of the mood level, regardless of the syndromal and nosological context in which this phenomenon occurs (van Praag *et al.*, 1975).

Of the other syndromes under discussion it is unknown whether the low 5-HIAA values are based on metabolic disorders. Post-mortem 5-HT concentration, and the effects of precursors, are not known. It is also possible that, in chronic degenerative diseases of the CNS such as multiple sclerosis, Pick's disease, and Alzheimer's disease, the transport of MA metabolites from the CNS is disrupted. Moreover, immobilization with inadequate CSF mixing can have played a role. An argument in favour of the latter possibility is that low values were only observed in seriously disabled patients with multiple sclerosis. Gottfries and Roos (1970), however, found no evident impoverization of motility in their (pre-)senile patients. Moreover, the probenecid-induced accumulation of 5-HIAA was reduced also, and this indicates the likelihood of a decreased metabolism of the mother amines (Gottfries and Roos, 1973).

In psychiatric syndromes other than the depressive and the manic, few CSF studies have been made. In acute schizophrenia there have been reports on decreased 5-HIAA values (Bowers *et al.*, 1969; Ashcroft *et al.*, 1966), but also on normal values (Rimon *et al.*, 1971; van Praag and Korf, 1975). However, the number of patients examined is too small to warrant definite conclusions.

8 The ability of the CNS to synthesize MA

In a group of 10 patients with unspecified depressive syndromes, Coppen *et al.* (1972b) found a decreased tryptophan concentration in the lumbar CSF. The enzyme tryptophan-5-hydroxylase is normally not saturated with tryptophan, and consequently the 5-HT synthesis in the brain varies with the amount of tryptophan available (Jéquier *et al.*, 1967). If the decreased tryptophan concentration in the CSF is representative of the situation in the CNS, then this could explain the suspected decrease in 5-HT synthesis. However, Ashcroft *et al.* (1973) found normal tryptophan values in unipolar as well as in bipolar depressions, and their findings are in agreement with ours.

The CSF 5-HIAA concentration following tryptophan loading gives an impression of the capacity of the CNS to convert tryptophan to 5-HT. Ashcroft *et al.* (1973a) found no differences in this respect between depressive patients in the unipolar and the bipolar group, and neurological controls. They determined the 5-HIAA concentration 10 h. after tryptophan administration. We carried out a similar experiment, but measured the CSF 5-HIAA and tryptophan concentrations at different times (and of course in different patients). A lumbar puncture was carried out after 1, 2, 5, and 8 h. per group of five patients. Eccleston *et al.* (1970b) made a similar study of neurological patients without psychiatric changes. In the depression group, the CSF tryptophan concentration was found to be tripled 1 h. after loading; in the non-depressive patients it had hardly changed by that time. Moreover, 8 h. after loading the CSF 5-HIAA concentration showed a much less marked increase in the depressive patients than in the control group (van Praag *et al.*, 1973a, b). More tryptophan in the CSF and less 5-HIAA. This suggests that the tryptophan administered so to speak 'congests' in the CNS in depressive patients: is less readily converted to 5-HT. This phenomenon could explain a decreased 5-HT turnover.

9 Conclusions

There are indications that the baseline concentration of 5-HIAA in the CSF can be decreased in depressions. The indications are most convincing for the group of the vital depressions with a unipolar course, and for the so-called primary depressions. Similar abnormalities have been found in manic patients (although less widely studied), and this suggests that mania and (vital) depression are pathogenetically related rather than representing the extremes of a range.

The sparse data available point in the direction of a persistence of the low 5-HIAA values, after the patient's clinical recovery. Provided it is corroborated, this suggestion permits of two interpretations: (i) the biochemical and the behaviour changes are unrelated, or (ii) the biochemical changes have not a directly causative but a predisposing significance: they render the patient susceptible, as it were, to the occurrence of depressions.

I must stress again that the CSF findings are far from consistent. This can have several causes, but undoubtedly an important one lies in the fact that the method is weak in the sense that it yields results which are not very reliable, and that it provides no more than a vague impression of the MA turnover in the brain and spinal cord. The probenecid technique is much less marred by these imperfections.

F CENTRAL 5-HT METABOLISM IN AFFECTIVE DISORDERS 2. CSF STUDY AFTER PROBENECID LOADING

1 Principles of the probenecid technique

It was explained in the preceding chapter that the concentration of an MA metabolite in the lumbar CSF reflects the metabolism of the mother amines in the CNS, but that the reflection is a rather blurred one. Its definition can be substantially enhanced by administering probenecid prior to CSF determination.

Probenecid intrarenally inhibits the transport of various organic acids, including 5-HIAA and HVA (Despopoulos and Weissbach, 1957; Werdinius, 1967). It was found that these acids are also eliminated from the CNS by an active transport process, which is inhibited by probenecid (Ashcroft *et al.*, 1968; Neff *et al.*, 1967). Consequently, probenecid administration is followed in all test animals studied by an increase in the concentration of 5-HIAA and HVA in the CNS, both in the brain (Werdinius, 1967) and in the ventricular and cisternal CSF (Guldberg *et al.*, 1966; Bowers and Gerbode, 1968b). The increase is a linear one, at least for several hours. This suggests that the degradation of the amines continues undisturbed for some time, and is not immediately reduced by feedback mechanisms.

Assuming that transport from the CNS is totally inhibited, the rate of accumulation of 5-HIAA and HVA should equal the rate of synthesis of these acids, that is to say: it should equal the rate of degradation of the mother amines. Under steady-state conditions, the rates of degradation and synthesis of the mother amines are equal. This means that the rate of accumulation of 5-HIAA and HVA in the CNS following probenecid administration can be accepted as a measure of the turnover of the mother amines. The turnover values measured with the probenecid technique, proved in fact to correspond well with those obtained by other methods such as the rate of accumulation of an amine after MAO inhibition or the rate of its disappearance after inhibition of synthesis (Neff *et al.*, 1967).

MHPG shows only a slight degree of accumulation following probenecid administration, and is therefore unsuitable for use as a method to study the central NA turnover (Korf *et al.*, 1971; Gordon *et al.*, 1973). The transport system of MHPG apparently has a low sensitivity to probenecid. This applies only to free MHPG, for the transport of MHPG sulphate certainly is probenecid-sensitive (Extein *et al.*, 1973). In the human CNS, however, MHPG is present for the most part in the unconjugated form.

2 Advantages of the probenecid technique over 'plain' CSF determination

The use of probenecid eliminates the principal limitations of 'plain' CSF determination (cf. section E.3). Its advantages can be summarized as follows.

1. The method is not a 'snapshot' of the amine degradation, but gives information on the total metabolite production in a given period of time, i.e. on the turnover of the amine in question.
2. The method gives an impression of the total metabolite production, whereas 'plain' CSF determination gives an impression only of the fraction which leaves the CNS via the CSF. For probenecid probably inhibits the direct transport of 5-HIAA and HVA to the bloodstream as well as their indirect transport, via the CSF to the bloodstream. This can be concluded from the fact that, in animals, 5-HT and DA turnover values measured with the aid of probenecid show good agreement with the values obtained by other methods (cf. section F.1). The probenecid technique, therefore, eliminates the uncertainty factor which lies in our ignorance of the amounts of 5-HIAA and HVA which are transported directly to the bloodstream and indirectly, via the CSF.
3. As a result of the transport block, 5-HIAA and HVA are caught in the CNS, so to speak, and the ventriculo-lumbar concentration gradient diminishes. In this way the lumbar CSF concentration of a metabolite gives a more faithful impression of the degradation of the mother amines in the CNS in its totality. For the same reason the influence of local changes in CSF pressure on the metabolite concentration in lumbar CSF diminishes. Consequently the intra-individual day-to-day variations in 5-HIAA and HVA concentrations after probenecid are much smaller than those in baseline concentrations.

The postprobenecid accumulation of 5-HIAA and HVA in lumbar CSF thus provides an indication of the turnover of 5-HT and DA in the CNS as a whole. I wrote *indication*, not *measure*. The lumbar punctures are generally confined to one before and one after administration of probenecid. This limitation is determined by ethical considerations. The accumulation 'curve' so obtained comprises too few points for calculation of the turnover. The probenecid technique would probably provide a true measure of central 5-HT and DA turnover if an acceptable method could be evolved to sample human CSF regularly over a period of, say, 12 h.

3 Procedure of the probenecid test

Unfortunately, the procedure of the probenecid test has not yet been standardized, and the various procedures have yet to be compared. The first studies in human individuals were made with small amounts of probenecid (a total of a few grams), distributed over several days (Roos and Sjöström, 1969; Tamarkin *et al.*, 1970; van Praag, 1969; van Praag *et al.*, 1970; Bowers, 1972a). With these amounts the concentrations of 5-HIAA and HVA in lumbar CSF were approximately doubled. In animal experiments, much larger amounts of probenecid were given, and the increase was many times as large. This indicated that transport inhibition in the human subjects was far from total, and that larger amounts of probenecid would have to be given.

Two dosage schedules are currently in use: that of Goodwin *et al.* (1973) and that of van Praag *et al.* (1973). The former procedure calls for administration of a large dose of probenecid in the course of 18 h. The first lumbar puncture (LP) is performed at 09.00, and the second at 15.00 on the following day. In the course of the 18 h. preceding the second LP, the patient receives 100 mg probenecid per kilogram body weight, divided into four fractional doses (given at 09.00, 14.00, 07.00 and 12.00). The patient is confined to bed during the 8 h. preceding the first LP and throughout the period of probenecid administration. Each LP yields 10 ml CSF, which is collected in 20 mg ascorbic acid and stored at −20 °C until determinations are made.

We give a large dose of probenecid (5 g) within a much shorter time (5 h.) and partly (1 g) by intravenous drip. This mode of administration was chosen in order to achieve optimal transport inhibition within the shortest possible time, thus minimizing the risk that accumulating metabolites start to influence the metabolism of the mother substances. The following is a detailed description of the procedure we have adopted.

Day 1 The patient is on a standard hospital menu, without fruits containing CA or IA (breakfast at 08.00, lunch at 12.30 and dinner at 17.30). He participates in all the usual ward activities, retires at the usual hour and remains in bed until 4 h. after the first LP on Day 2.

Day 2 The patient is on a standard hospital menu and remains in bed until 4 h. after the LP (at 08.30, still fasting). Breakfast immediately after the LP. Four hours after the LP the patient gets up and participates in the usual ward activities until he retires at the usual hour.

Day 3 The patient is on a standard hospital menu and participates in the usual ward activities until he retires at the usual hour. He remains in bed until 4 h. after the second LP on Day 4.

Day 4 The patient is on a standard hospital menu and remains in bed until 4 h. after the LP.

08.00	breakfast
08.30	1 g probenecid in 150 ml physiological saline, given by intravenous drip within 20 min.
09.30	1 g probenecid by mouth
10.30	1 g probenecid by mouth
11.30	1 g probenecid by mouth
12.30	1 g probenecid by mouth
16.30	lumbar puncture

The LP is carried out on the prone patient. The first 10 ml CSF is obtained and collected in one tube. The tube is carefully shaken manually to ensure optimal mixing, for there is a concentration gradient for 5-HIAA and HVA between ventricular, cisternal, and lumbar CSF, and the concentrations of these compounds increase in CSF from higher levels.

As soon as it is obtained the CSF is frozen, and duplicate determinations are made within a week of the following concentrations: 5-HIAA (Korf *et al.*, 1973), HVA (Westerink and Korf, 1975), and probenecid (Korf and van Praag, 1971). This is done because probenecid concentration and 5-HIAA/HVA accumulation are positively correlated (Korf and van Praag, 1971; Sjöström, 1972): the more probenecid in the CSF, the higher the 5-HIAA and HVA accumulation. This correlation of course ceases to exist when transport inhibition becomes total. It is uncertain whether this stage is actually reached in human subjects, but there are indications that it is (cf. section F.5). We therefore consider it necessary to have a check that a possible increased or decreased accumulation is in fact due to an increased or decreased production of 5-HIAA and HVA in the CNS, and not to increased or decreased penetration of probenecid into the CNS.

A disadvantage of our procedure is that test subjects not infrequently complain of a heavy sensation in the upper abdomen, nausea, and sometimes headaches in the course of the morning (in about one out of three cases). However, vomiting (which makes the test a failure) occurs only sporadically (in about one out of 25 cases). The abovementioned side effects gradually subside in the course of the 3–6 h. after the last probenecid administration.

We have never observed side effects from the intravenous probenecid drip. Intravenous injection of probenecid gives rise to local pain, possibly due to irritation of the vascular wall; this is why this practically simple mode of administration cannot be used (Korf and van Praag, 1970).

So far we have made the following observations in all psychiatric (psychosis, depressions) and neurological syndromes (Parkinson's disease) with disturbances in the central 5-HT and DA turnover (van Praag *et al.*, 1973; Lakke *et al.*, 1972; Korf *et al.*, 1974a). When after probenecid the accumulation of 5-HIAA (i.e. the difference between postprobenecid value and baseline concentration) is decreased or increased, then the postprobenecid value is likewise decreased or increased; and the levels of significance are virtually the same. The same was found to apply to HVA. This implies that it is not absolutely necessary to measure *accumulations*: it is sufficient to determine post-probenecid 5-HIAA and HVA concentrations. In the abovementioned syndromes we therefore consider it justifiable to perform only one lumbar puncture, particularly when diagnostic rather than research purposes are involved. This is of course a great practical advantage, and makes it easier to obtain permission to repeat the test.

4 Does the probenecid technique afford information on human DA and 5-HT turnover?

In test animals there is good agreement between MA turnover calculated with the aid of the probenecid technique, and values obtained by other methods such as determination of the rate of disappearance or accumulation of MA after inhibition of their synthesis or degradation, respectively. This means that the probenecid technique really yields turnover values. In human subjects this cannot be proven because there are no other methods to measure central MA turnover. The finding that 5-HIAA and HVA accumulation would increase after procedures which increase the turnover of the mother substances, and would decrease after procedures with the opposite effect, would be a strong argument that turnover is measured with the probenecid technique in human individuals also. The following subsections show that this argument is indeed valid.

(a) Decrease of 5-HT and DA turnover

Inhibition of the synthesis of 5-HT and DA leads by definition to a decrease of their turnover. If the probenecid technique really measures turnovers, then the inhibition of synthesis can be expected to be associated with a decrease in 5-HIAA and HVA accumulation. To begin with, 5-HT and DA synthesis can be inhibited *directly* with the aid of enzyme inhibitors such as PCPA and α-MT; they inhibit tryptophan and tyrosine hydroxylase, respectively, and these enzymes catalyse the first step in the synthesis of 5-HT and DA, respectively.

There are several indications that the synthesis of 5-HT and DA can also be inhibited *indirectly*, by stimulating postsynaptic 5-HT and DA receptors. Via a feedback mechanism, the firing rate in the presynaptic element is thus reduced, and the synthesis of 5-HT and DA diminishes (Sulser and Sanders-Bush, 1971; Schubert *et al.*, 1970; Sheard *et al.*, 1972; Thoenen, 1972). The 5-HT and DA receptors in turn can be stimulated either directly or indirectly. Directly: by means of substances which, like the transmitters themselves, stimulate these receptors, e.g. apomorphine, a DA receptor stimulator (Andén *et al.*, 1967), and quipazine (Grabowska *et al.*, 1974) and LSD (Andén *et al.*, 1968), which stimulate 5-HT receptors. Indirectly: by increasing the amount of 5-HT and

DA available at the corresponding receptors. This can be done in several ways:

1. By reducing the re-uptake of 5-HT and DA into the neuron; for 5-HT this can be effected with the aid of tricyclic antidepressants (Carlsson *et al.*, 1969a, b).
2. By reducing the intraneuronal degradation of 5-HT and DA by administration of MAO inhibitors (Pletscher *et al.*, 1960).
3. By facilitating the release of transmitters from the synaptic vesicles into the synaptic cleft; for 5-HT this effect can be obtained with certain 4-chloramphetamines (cf. section H.5).

In human individuals, the synthesis inhibitors PCPA and α-MT suppress the probenecid-induced accumulation of 5-HIAA and HVA almost entirely, and they do this selectively in that PCPA leaves the accumulation of HVA intact, and α-MT that of 5-HIAA (Goodwin *et al.*, 1973).

Of the substances which directly stimulate 5-HT and DA receptors, only LSD has been studied in human subjects. It was found to reduce the 5-HIAA accumulation (Bowers, 1972a). More information is available on the indirect stimulators. Two tricyclic compounds have been studied in human individuals: amitriptyline (Bowers, 1972; Post and Goodwin, 1974) and imipramine (Post and Goodwin, 1974). Both compounds reduce the accumulation of 5-HIAA but do not influence that of HVA. It has been found in animal experiments that tricyclic compounds reduce the intracerebral 5-HT turnover (Corrodi and Fuxe, 1969; Schildkraut *et al.*, 1969a). As pointed out, this effect is regarded as secondary to inhibition of the 5-HT uptake into the neuron. A direct influence on 5-HT synthesis is not excluded, however, for these substances also inhibit the uptake of the mother substance tryptophan into the neuron (Bruinvels, 1972). Be this as it may, diminution of 5-HT synthesis proves to be associated with decreased 5-HIAA accumulation after probenecid. The unchanged HVA accumulation is consistent with the fact that tricyclic compounds do not influence the (re-)uptake of DA.

Phenelzine, the only MAO inhibitor studied in human subjects, reduces the HVA accumulation but not that of 5-HIAA (Kupfer and Bowers, 1972). This is against expectation, for phenelzine blocks the degradation of DA and 5-HT and reduces the 5-HT turnover. A possible explanation of this discrepancy lies in the observation of Ashcroft *et al.* (1969) that phenelzine inhibits the transport of 5-HIAA from the CSF. Assuming that the inhibition of 5-HIAA transport by probenecid has been subtotal, it is conceivable that after phenelzine the 5-HT turnover is more markedly inhibited than the 5-HIAA accumulation in the CSF suggests.

4-Chlormethylamphetamine and 4-chloramphetamine reduce the accumulation of 5-HIAA, but not that of HVA (van Praag, 1975). In test animals they reduce the 5-HT turnover (Korf and van Praag, 1972; Sanders-Bush and

Sulser, 1970). In this case, too, decreased 5-HT turnover and decreased 5-HIAA accumulation therefore go hand in hand.

(b) Increase of 5-HT and DA turnover

The synthesis of 5-HT and DA can be directly stimulated by administering the corresponding precursors. There is also an indirect way, at least so far as CA are concerned: blocking the postsynaptic receptors. This leads to an increased firing rate in the presynaptic element, and an increase in transmitter synthesis (van Rossum, 1967; Thoenen, 1972; Bunney *et al.*, 1973). Via this mechanism, neuroleptics of the phenothiazine and the butyrophenone type are believed to increase the intracerebral turnover of DA and NA (Andén *et al.*, 1972a). They do not influence the 5-HT turnover. There are no known compounds which more or less selectively block 5-HT receptors.

The 5-HT precursors tryptophan (Goodwin *et al.*, 1973) and 5-HTP (van Praag, 1975) both cause a substantial increase in 5-HIAA accumulation after probenecid, without exerting a significant influence on that of HVA. The DA precursor L-DOPA causes a very marked increase in HVA accumulation (Goodwin *et al.*, 1973), but it is to be borne in mind that after administration of large doses of L-DOPA some of the peripherally formed HVA penetrates into the CNS; consequently the HVA concentration in the CSF is no longer a faithful reflection of the DA degradation in the CNS (Prockop *et al.*, 1974). In addition, L-DOPA induces a moderate but significant decrease of the 5-HIAA accumulation. A possible explanation could be that DOPA enters serotonergic neurons and is there converted by the aromatic amino acid decarboxylase to DA, which in its turn supersedes 5-HT from the stores (Ng *et al.* (1970).

In human individuals, neuroleptics of the phenothiazine and butyrophenone type cause a dose-dependent increase of HVA accumulation after probenecid; this is indicative of an increased DA turnover (Bowers, 1973; van Praag and Korf, 1975). They do not influence the 5-HIAA accumulation. No indications of an increased central NA turnover were found (van Praag and Korf, 1975). It is to be noted, however, that this statement is based exclusively on determination of the baseline MHPG concentration (MHPG transport being probenecid-insensitive); and this indicator is less reliable than determination of accumulation after transport block (cf. section E.3).

The 5-HT and DA turnover and the accumulation of 5-HIAA and HVA in the CSF after probenecid thus prove to be positively correlated. This is a strong argument that the accumulation is an indicator of the turnover of the mother substance.

5 Shortcomings of the probenecid technique

The probenecid technique enables us to make a gross estimate of the 5-HT and DA turnover in the human CNS as a whole, but the technique is by no means flawless. It is unsuitable for the study of the NA metabolism, for example, and for 5-HIAA and HVA it measures the total production in the CNS, which means that relatively small local changes can remain unnoticed.

A second imperfection is uncertainty about the question whether 5-HIAA and HVA transport block can be approximately total. There are indications that in fact it is. After administration of 5 g probenecid in 5 h. the accumulation of 5-HIAA and HVA does not significantly exceed that after 4 g in 5 h. Moreover, the rising curve shows a plateau at a time when the probenecid concentration in the CSF is still high (Tamarkin *et al.*, 1970; van Praag *et al.*, 1973). Exact data on the degree of inhibition achieved, however, are not available. In any case the doses given are the maximum amounts acceptable: more probenecid could not be given. For this reason the probenecid concentration in the CSF is routinely included in the determinations in our research unit, in order to exclude the possibility that an abnormal accumulation is due to an abnormal probenecid concentration rather than to an abnormal production of metabolites in the CNS.

Another vulnerable feature of this technique is that probenecid also influences the 5-HT metabolism in a manner other than by blocking 5-HIAA transport. In animals (Tagliamonte *et al.*, 1971b) and in human individuals (Korf *et al.*, 1972b) probenecid reduces the plasma total (free and protein-bound) tryptophan concentration. The plasma free tryptophan concentration increases (Lewander and Sjöström, 1973). Probenecid probably interferes with the binding of tryptophan to serum albumins. This explains the increase of the free fraction. Free tryptophan disappears to the tissues, and consequently the total plasma concentration diminishes. In agreement with this hypothesis, the intracerebral tryptophan concentration increases (Tagliamonte *et al.*, 1971a; Korf *et al.*, 1972). The 5-HT turnover rate, however, shows no or hardly any increase (Korf *et al.*, 1972; Barkai *et al.*, 1972). In view of the fact that the intracerebral 5-HT turnover can be reliably measured in animals with the aid of probenecid (Neff *et al.*, 1967), this was to be expected. It thus proves to be possible to vary the intracerebral tryptophan concentration without influencing the rate of 5-HT synthesis. This phenomenon is inconsistent with the theory of Fernström and Wurtman (1971), which maintains that intracerebral tryptophan is an important regulator of the rate of 5-HT synthesis. Probenecid does not influence the human CSF tryptophan concentration (Korf *et al.*, 1972; Lewander and Sjöström, 1973).

There are therefore no indications that changes in the tryptophan metabolism devaluate the probenecid test.

5-HIAA and HVA are the principal metabolites of 5-HT and DA in the human CNS. There is little doubt about this. However, some synthesis of alcohols such as 5-hydroxytryptophol or 3-methoxy-4-hydroxyphenylethanol cannot be excluded. This would mean that the turnover of 5-HT and DA exceeds the value indicated by the probenecid technique. Finally, it is to be noted in this context that, after probenecid premedication, a small fraction of large doses of intravenously given 5-HIAA enters the CNS (Bulat and Zivković, 1973). Since (surprisingly) there have been no studies of the influence of probenecid on the peripheral 5-HT metabolism, the possible practical significance of this factor in human individuals cannot be assessed.

The principle of the probenecid test is undoubtedly important: accumulation of a metabolite in the CNS as an index of its rate of synthesis. Probenecid itself, however, is not an ideal transport blocker; there is certainly a need for a better one.

6 Central 5-HT turnover in depressions

(a) Decrease of 5-HT turnover

The probenecid technique, evolved in animal experiments by Neff *et al.* (1967), was first applied to human individuals by Roos *et al.* in Sweden (1968, 1969), by van Praag *et al.* in The Netherlands (1969, 1970) and by Bowers (1969) and Goodwin *et al.* (1970) in the USA, in virtually simultaneous, but independent investigations.

Towards the end of the 1960s Roos and Sjöström (1969) and van Praag *et al.* (1969, 1970) reported that the 5-HIAA concentration in the human lumbar CSF increases in response to probenecid, and that the increase is less pronounced in depressive patients than in controls. As explained in section F.1, a low accumulation of 5-HIAA after probenecid administration is indicative of a low 5-HT turnover in the CNS. Van Praag *et al.* studied patients with recurrent vital ('endogenous') depressions of varying aetiology, and so did Roos and Sjöström. Tamarkin *et al.* (1970) and Bowers (1969) likewise observed 5-HIAA accumulation after probenecid, but they reported no results obtained in depressive patients.

The initial studies were done with relatively small amounts of probenecid. To maximize transport block, much larger doses were later used (cf. section F.3). The results thus obtained were in principle similar: an averagely decreased 5-HIAA accumulation in vital depressions of both the unipolar and the bipolar type. In this context van Praag *et al.* (1973) compared vital depressive with personal depressive syndromes, and depressive patients with non-depressive controls. Sjöström and Roos (1972) in addition studied psychotics and patients with neurotic personality disorders. Both groups of investigators reported that the decrease of accumulation was more pronounced in the bipolar than in the unipolar group. Goodwin *et al.* (1973) likewise reported lower accumulation values in vital depressive patients than in non-depressive controls. The difference was not significant, but this may have been due to the smallness of the depression group ($n = 6$).

(b) Negative findings

So far, only Bowers (1972a) has reported an averagely normal 5-HIAA accumulation in depressions. He studied 10 patients with depressions of the unipolar type. The literature currently shows an inclination to accept this designation as an adequate qualification of a depression. I do not share this view. The term unipolar denotes a clinical course, and gives no information on the nature of the syndrome. Vital depressions often tend to be recurrent; but so do personal depressions, which are often provoked by chronic neurotic personality factors. Moreover, in the confrontation with a first phase it is uncertain whether a depression will take recurrent course or whether (hypo)manic phases will or will not follow. The term unipolar is therefore not an adequate definition of a depression. A syndromal and aetiological qualification must be added. When dealing with a first depressive period, moreover, the term is useless. Several of Bowers' patients were described as 'anxious' and 'hostile'. It is therefore quite possible that personal depressions were involved—a category in which the phenomenon in question is hardly ever observed, if at all.

(c) Vital depressions with and without disturbed 5-HT metabolism

We ourselves found indications that the decreased 5-HIAA accumulation is not a universal phenomenon in the group of vital depressions either. We found subnormal values, i.e. values below the range of variation in the control group, in some 40% of cases (van Praag and Korf, 1971; van Praag *et al.*, 1973). Patients with and without demonstrable disorders of central 5-HT metabolism were indistinguishable in terms of psychopathological symptoms. From these facts we concluded that the group of vital depressions, although it tends to be homogeneous in psychopathological terms, is heterogeneous in biochemical terms, and comprises patients with as well as patients without disturbances in central 5-HT metabolism.

Our findings were confirmed by Goodwin (1976) and, in a CSF study without probenecid, by Bertilson and Åsberg (1975). In a group of patients with endogenous depression, both groups reported a bimodal distribution of CSF 5-HIAA values.

We have no indication that the presence or absence of an apparent central 5-HT deficit is related to the anxiety level. We were particularly interested in this question because in dogs (Pointers) of a nervous strain, the 5-HIAA accumulation in the CSF after probenecid loading was found to be lower than in normal congeners (Angel *et al.*, 1976).

That disparate processes can produce the same syndrome is not an unknown phenomenon in medicine. On the whole, all jaundiced patients show the same symptoms: yet the pathogenesis of the syndrome can be widely different. Another example is the anaemia syndrome. Differences in pathogenesis could also exist in the vital depressive syndrome. This would have important therapeutic implications.

(d) Cause or effect?

It is of course a cardinal question whether the disturbances observed are *primary* or *secondary*; whether they contribute to the development of a depression, or result from it. In human individuals, conclusive information on this question can be obtained in only two ways. By establishing whether the metabolic disturbances preceded manifestation of the depression or followed it; and by establishing whether abolition of the metabolic disorder causes total or partial subsidence of the depression. The former method is virtually impracticable with human patients. The second is feasible: for example, 5-HT synthesis can be increased with the aid of 5-HT precursors. In section G.2 the observation that the 5-HT precursor 5-HTP does indeed produce a therapeutic effect in 5-HT-deficient patients will be discussed (van Praag *et al.*, 1972, 1974). This is indicative of the primary character of the metabolic disorder.

(e) Causative or predisposing factor

In the majority of patients the signs of central 5-HT deficiency persist during the symptom-free/medication-free intervals (van Praag, 1977). This seemed to suggest that the suspected central 5-HT deficiency is not so much a causal as a predisposing factor, a factor increasing the susceptibility to depression. Corroboration of this hypothesis was found in two facts. First: 5-HTP was demonstrated to have a prophylactic effect against unipolar and bipolar vital depression, especially in patients with signs of a persistent central 5-HT deficiency (van Praag 1978). Second: in patients with persistent disturbances in central 5-HT, the depression frequency in the course of their lives exceeded that found in patients with the same depressive syndrome but no demonstrable disturbances in central 5-HT (van Praag en de Haan, 1979).

These findings support the hypothesis that a central 5-HT defect could be a manifestation, on a biological level, of the increased tendency of certain individuals to respond to threatening stimuli from the outer and inner world by a pathological depression of mood.

7 Central 5-HT turnover in mania

Manic patients, like those with vital depressions, have been found to show a decreased accumulation of 5-HIAA (Sjöström and Roos, 1972; Post and Goodwin, 1974). This finding is the more remarkable because the so-called mixing effect acts precisely in the opposite direction: increased concentration of 5-

HIAA in lumbar CSF due to better CSF mixing as a result of hyperactivity (Post *et al.*, 1973b). The similarity in metabolic findings indicates that the same pathogenetic mechanisms can be involved in depression and mania.

8 Selectivity of the probenecid findings

Diminished 5-HIAA accumulation has been demonstrated mainly within the group of the vital depressions, but not (or only sporadically) in patients with personal depressions, psychotic reactions, or non-depressive personality disorders. So far it seems, therefore, that the phenomenon is fairly selective within the range of psychiatric disorders.

Disorders of 5-HT metabolism do occur in diseases which entail severe anatomical degeneration of brain tissue. First of all in *Parkinson's disease*, in which 5-HIAA accumulation is markedly decreased (Olsson and Roos, 1968). This is an *in vivo* indication of the central 5-HT deficiency initially demonstrated in post-mortem studies (Bernheimer and Hornykiewicz, 1964). The significance of this metabolic disorder for motor pathology is obscure. We have been unable to establish a correlation with the mood level (Puite *et al.*, 1973).

Decreased 5-HIAA accumulation found also in *pre-senile dementia type Alzheimer* (Gottfries and Roos, 1973); the findings corresponded with the patient's degree of dementia. It is therefore plausible that the decreased 5-HT turnover is based on involvement of serotonergic systems in the degenerative process. Post-mortem studies could verify this hypothesis, but have not so far been carried out. The same applies to *multiple sclerosis*, in which low 5-HIAA accumulation values have also been found. In these cases, however, another explanation is to be considered in addition. The phenomenon occurred in only very severely disabled, largely bedridden multiple sclerosis patients (Sonninen *et al.*, 1973), but not in the less severe cases. It might therefore well be that immobility and poor CSF mixing rather than decreased 5-HT turnover were the cause of the low accumulation in these cases.

The question of the specificity of disorders of the 5-HT metabolism has not yet been exhaustively studied. With regard to future research into this question I should like to make one remark. The starting-point in studies aimed at the cerebral substrate of neurological and psychiatric abnormalities is often a given clinical entity (e.g. vital depression or Parkinson's disease), i.e. a group of disorders in heterogeneous functions brought under the same denominator because they are usually observed together in the clinical setting. The dysfunction or group of interrelated dysfunctions (e.g. hypokinesia or mood abnormalities) is much less commonly taken as a starting-point in these studies. It might well be, however, that metabolic (as well as physiological) disturbances correlate much more closely with certain motor, sensory of mental dysfunctions than with nosological entities or syndromes which represent a whole range of such dysfunctions. The concept of symptomatological specificity of a given metabolic disorder has been neglected in biological psychiatry and experimental neurology. Reorientation on the symptom, meant as a well-defined dysfunction —be it motor, sensory or mental—could well be a productive stimulant for these areas of research (van Praag, 1972; van Praag *et al.*, 1975).

9 Are the disorders of 5-HT metabolism syndrome-dependent or syndrome-independent?

The question whether the disorders of 5-HT metabolism are primary features, which contribute to the occurrence of the depression, or secondary features, which result from it, can in my view perhaps be answered in favour of the former possibility. This view is based on the fact that compounds which promote 5-HT synthesis exert a favourable influence on certain depressive symptoms. Another question is whether the occurrence of the biochemical symptoms is dependent or independent of the depressive syndrome. If they are syndrome-dependent, primarily or secondarily, then they disappear together with the depressive symptoms. In the opposite case there are two possible explanations: (i) the biochemical disorders are unrelated to the depression, and the therapeutic effect of the precursor is not based on abolition of the MA deficiency; (ii) the biochemical defect is related to the depressive syndrome, but in a predisposing rather than in a causative sense; it renders the individual more susceptible to the occurrence of depressions, but as such is not sufficient to cause manifestation of the depressive syndrome.

Decreased serotonergic transmission as a factor predisposing to depression and mania has been postulated as theoretical model by Kety (1971) and Prange *et al.* (1974). They believed that the depression would become manifest when the catecholaminergic transmission also diminishes, and that an abnormally increased activity in the CA system, combined with subnormal serotonergic activity, would underline the manic syndrome.

We reinvestigated a group of depressive patients after a symptom-free period of at least 6 months, with or without lithium prophylaxis. Such lithium prophylaxis as was given was discontinued 2 weeks before the investigation. None of these patients had received antidepressants during the preceding 3 months. The majority of low responders had shown an increase after recovery, but not always to a normal average level. In some patients the accumulation had remained rather low, and three patients had shown no increase at all so far. Should persistent subnormal 5-HIAA accumulation indeed be a factor predisposing to manifestation of depression, then chronic medication with 5-HT precursors might have a preventive effect. The literature comprises a report on one patient who experienced a depressive phase every year over a 17-year period. During treatment with tryptophan in large doses, the expected depressions did not occur. She was considered to be emotionally much more stable than previously (Hertz and Sulman, 1968). We studied the prophylactic value of 5-HTP in five patients suffering from recurrent vital depressions with (i) a high relapse rate and (ii) a low 5-HIAA accumulation after probenecid during the depressive phase and after clinical recovery. The course of the depression was bipolar in three and unipolar in two patients. They were successfully treated with tricyclic antidepressants, and 1 month after discontinuation of this therapy were started on 200 mg L-5-HTP and 150 mg MK 486 (a peripheral decarboxylase inhibitor) daily. After 1 month the 5-HIAA response to probenecid was found to be amply above normal—a sign of their ability to convert 5-HTP to 5-HT. No relapse occurred over a period of 8–12 months. In view of the history of these patients this was a striking result which, however, will have to be verified by interposing placebo periods; this will be done. No similar study has so far been made in manic patients.

10 Conclusions

The probenecid-induced accumulation of 5-HIAA and HVA in the lumbar CSF is a better yardstick of the central turnover of 5-HT and DA than the baseline level of these metabolites, even though the probenecid technique is far from flawless. Several investigators have established that the 5-HIAA accumulation in response to probenecid can be decreased in depressions. This is indicative of a decreased turnover of the mother amines in the CNS. The 5-HT defect was most pronounced in vital depressions with a bipolar course. According to van Praag (1973) the 5-HT defect is not a universal phenomenon within the category of vital depressions but occurs in only 40–50% of cases. They postulated the existence of two pathogenetically different types of vital depression: a type with and a type without a demonstrable defect in 5-HT metabolism.

Since abolition of the suspected 5-HT deficiency with the aid of precursors causes alleviation of the depressive syndrome or some of its features, the metabolic disorders probably play a role in the pathogenesis of the depressions. A fact which at first glance seems inconsistent with this is that, after disappearance of the psychopathological symptoms, the disorders of the 5-HT metabolism do not always, or not always entirely, disappear. It is quite possible, however, that the biochemical defects are not so much direct causative as rather predisposing factors.

The few studies so far carried out in manic patients indicate similarities to depression in that decreased 5-HIAA accumulation values have been found.

Defects in central 5-HT turnover have been established also in degenerative disease of the CNS, e.g. Parkinson's disease, Alzheimer's disease, and severe cases of multiple sclerosis. They are possibly a result of involvement of serotonergic systems in the degenerative process. We do not know to which psychological disorders the serotonergic dysfunctions lead in these cases.

G VERIFICATION OF THE 5-HT HYPOTHESIS WITH THE AID OF DRUGS AND OTHER METHODS OF ANTIDEPRESSANT THERAPY 1. 5-HT PRECURSORS AND SYNTHESIS INHIBITORS

1 Justification of the strategy

With the aid of drugs, transmission in the central serotonergic synapses can be influenced in two ways: *directly*, i.e. by influencing, say, the uptake or release of the transmitter or the susceptibility of the postsynaptic receptors, or *indirectly*, i.e. via interference with the synthesis or degradation of 5-HT. These two strategies, so far as they are acceptable for use on human individuals, have also been applied to patients with affective disorders in order to test the 5-HT hypothesis. The argumentation was as follows. If the 5-HT hypothesis contains a kernel of truth, then drugs which increase the amount of 5-HT available at the central postsynaptic receptors can be expected to produce an antidepressant effect, whereas drugs with the opposite action should promote depression. Reversely, it would reinforce the 5-HT hypothesis if therapeutic methods with an unmistakable antidepressant effect, such as electroconvulsive therapy (ECT), could be shown to promote transmission, even though this is not a *conditio sine qua non* for the correctness of the hypothesis. It is conceivable, after all, that mood regulation can be influenced also via non-serotoninergic systems.

It is more difficult to give a prediction with regard to manic syndromes. If one regards depression and mania as counterparts, then an excess of 5-HT should promote mania, while a decrease in this compound should act against mania. On the other hand, the few data available on the 5-HT metabolism in mania are more indicative of pathogenetic similarities to depression than of extreme opposites.

This chapter discusses the 'indirect strategy'. The next chapter deals with the 'direct strategy' and with the influence of certain therapeutic methods which influence mood (ECT; thyrotropin-releasing hormone; lithium) on central serotoninergic transmission.

2 Studies with drugs which increase 5-HT synthesis

(a) Tryptophan

The enzyme tryptophan hydroxylase is normally not saturated with substrate. Consequently administration of tryptophan to test animal results in increased 5-HT synthesis, e.g. in the brain (Fernström and Wurtman, 1972). There are sound reasons for the assumption that conversion of tryptophan to 5-HTP takes place exclusively in serotoninergic neurons (Lovenberg *et al.*, 1968). This is a great advantage of tryptophan over 5-HTP, which can be decarboxylated in non-serotoninergic neurons. In human individuals the probenecid-induced accumulation of 5-HIAA in the lumbar CSF shows a marked increase in response to tryptophan (Dunner and Goodwin, 1972). This indicates that 5-HT synthesis is stimulated in the human organism also. Tryptophan therefore seemed a suitable agent to be used in efforts to establish whether an increase of cerebral 5-HT can have a therapeutic effect in *depressions*—even though the method is inefficient because only a small fraction of each tryptophan dose is utilized for central 5-HT synthesis (cf. section G.4).

The problem statement in the first tryptophan experiment was whether tryptophan can potentiate the therapeutic effect of a MAO inhibitor (in this case iproniazid) (van Praag, 1962). A group of 20 vital depressive patients were treated for 10–14 days with 1500 mg L-tryptophan, combined with a MAO inhibitor. The two drugs were simultaneously started in 10 patients; in the remaining 10, tryptophan was started after a few weeks' unsuccessful or insufficiently successful treatment with a MAO inhibitor. No potentiation of the specific effects of the antidepressant was observed. Coppen *et al.* (1963), who gave larger doses of tryptophan (214 mg racemic mixture per kilogram body weight), did observe an effect: depressive patients showed a better response to a combination of a MAO inhibitor (tranylcypromine) with tryptophan than to a combination of a MAO inhibitor with a placebo. Their findings were subsequently corroborated by several independent investigators (Glassman and Platman, 1969; Pare, 1963; Lopez-Ibor *et al.*, 1973). Rats become hyperirritable and active in response to a combination of a MAO inhibitor with tryptophan—a phenomenon not observed after administration of tryptophan only (Grahame-Smith, 1971).

These data were not analysed with a view to a possible influence of the type of MAO inhibitor used on the potentiating effect of tryptophan. It was recently established that MAO is not a simple enzyme but a complex of related enzymes which differ in substrate specificity. The effects of various MAO inhibitors on this system are not identical (Youdim *et al.*, 1972). It is therefore conceivable that, in therapeutic terms, it makes a difference whether tryptophan is combined with one MAO inhibitor or with the other (Green and Youdim, 1975).

The findings obtained in depressions with tryptophan alone are certainly less perspicuous. Coppen *et al.* (1967, 1972b) considered tryptophan to be a potent antidepressant, comparable in therapeutic potential with ECT and imipramine. Other investigators found no evidence of an antidepressant effect of L-tryptophan (Dunner and Goodwin, 1972; Bunney *et al.*, 1971; Mendels *et al.*, 1975; Carroll *et al.*, 1970; Dunner and Fieve, 1975). All these studies were controlled studies. The findings reported by Herrington *et al.* (1974) lay between these extremes. They studied 43 patients with severe depressions, all candidates for ECT. Only one half of this group did receive ECT, the other half being treated with 8 g L-tryptophan daily. Both groups had improved significantly after 2 weeks. Subsequently, however, the ECT group showed further improvement, but the tryptophan group did not. Assuming that the effects of both therapies were not based on non-specific factors, and that no spontaneous improvement had occurred, these findings show that L-tryptophan is therapeutically active in severe depressions, but inferior to ECT. Finally, in a recent multinational study (Jensen *et al.*, 1975) tryptophan was found as effective as imipramine in patients with endogenous depressions.

The tryptophan question can be described as still moot. I can discern three possible causes of this situation. To begin with, no attempt has so far been made to classify the depressive syndrome according to biochemical criteria. *A priori* it seems improbable that the natural amino acid L-tryptophan could be a universal antidepressant. On the other hand, it cannot be excluded that a given category of depressive patients (those in whom a central 5-HT deficiency probably exists) might benefit from treatment with this agent. Mean values of a biochemically undifferentiated group of depressions would conceal such an effect (van Praag and Korf, 1970, 1971, 1974). Second, there is uncertainty about the question whether certain tryptophan effects or their absence should in fact be ascribed to changes in central 5-HT metabolism—a question to be discussed in section G.4. Finally, in each of the three experiments tryptophan was combined with large daily doses of pyridoxine, which is a coenzyme in several tryptophan conversions. The doses given were many times as large as the therapeutic dose required in pyridoxine deficiency. In Parkinson patients, pyridoxine antagonizes the therapeutic effect of L-DOPA (Duvoisin *et al.*, 1969) by a mechanism which has not yet been entirely explained. It is therefore possible that pyridoxine has prevented an antidepressant action of tryptophan in these cases, but it is not likely: in rats, pyridoxine exerts no influence on the amount of tryptophan converted to 5-HT in the brain (Carroll and Dodge, 1971).

Even if it were established with certainty that tryptophan has no therapeutic effect on any type of depression, this would not automatically devaluate the hypothesis of a 5-HT deficiency in depressions. In fact this deficiency might be based on a defect in synthesis, which makes the patient unable to utilize tryptophan for 5-HT synthesis.

Prange *et al.* (1974) studied the effect of L-tryptophan in *manic* patients: five received 6 g L-tryptophan daily, while the other five were given 400 mg chlorpromazine (Largactil). Both compounds produced a therapeutic effect, and L-tryptophan was the more effective of the two. This is a remarkable observation, which corresponds with the fact that indications of a reduced 5-HT turnover have been found in mania as well as in certain types of depression. This observation was confirmed by Murphy *et al.* (1974).

(b) 5-Hydroxytryptophan

Unlike tryptophan, 5-HTP is converted to 5-HT almost in its entirety. And it is consequently a much more effective means of enlarging the amount of 5-HT in the brain.

Few studies have been made with 5-HTP, and their results were initially negative. Both alone and in combination with a MAO inhibitor it failed to produce an antidepressant effect (Pare and Sandler, 1959; Kline *et al.*, 1964; Glassman, 1969). However, the doses of 5-HTP administered were small, and CSF studies of 5-HT metabolites were omitted. Persson and Roos (1968) described one patient who showed a favourable response to 5-HTP. The striking feature of this case was that the patient had been resistant to ECT and amitriptyline medication.

Van Praag *et al.* (1972) gave larger doses of the racemic mixture of 5-HTP (3 g daily), i.e. the maximum tolerated by human individuals, and they investigated two questions: (i) whether the agent was of use in the treatment of vital depressive patients and (ii) whether this was true in particular for patients with a subnormal 5-HIAA response to probenecid—a phenomenon considered to be indicative of a central 5-HT deficiency. The findings showed that 5-HTP probably does have a therapeutic effect in patients with a subnormal 5-HIAA response to probenecid (van Praag *et al.*, 1972), though the therapeutic results were not spectacular. In this respect van Praag's study differed from that described by Sano (1972), who in a large group ($n = 107$) of patients with unipolar and bipolar depressions observed striking results after relatively small doses of L-5-HTP (50–300 mg daily), without addition of a peripheral decarboxylase inhibitor. The effect was quite pronounced and often ensued before the fourth day of medication. Van Praag *et al.* did, but Sano did not, use placebo controls, nor did Sano make CSF determinations. The data of van Praag *et al.* and Sano were confirmed by Takahashi *et al.* (1975): 5-HTP proved to be efficacious in a subgroup of unipolar depressives. In a small group of six patients with psychotic depressions, Brodie *et al.* (1973) were unsuccessful with 5-HTP. The patients were not classified on biochemical criteria; moreover, a negative selection was involved: none of them had responded to ECT or antidepressant medication. Trimble *et al.* (1975) reported that L-5-HTP given by intravenous drip had a positive effect on mood in non-depressive test subjects.

The above data certainly warrant further research into 5-HTP in depressions; it seems essential to me that further studies should involve probenecid tests, the results of which should be included in the analysis of results.

(c) 5-Hydroxytryptophan in combination with clomipramine

If the therapeutic effect of 5-HTP in depressions reported by Sano (1972) and van Praag *et al.* (1972) is based on an increase of the amount of 5-HT available at the central postsynaptic 5-HT receptors, then it is to be expected that this therapeutic effect should be potentiated by drugs which produce a similar biochemical effect. One such drug is clomipramine (Anafranil). This is an inhibitor of the re-uptake of 5-HT and NA, just like other tricyclic antidepressants, but a relatively selective one: the re-uptake of 5-HT is strongly inhibited, but that of NA only slightly. The selectivity towards 5-HT also applies to man (Bertilsson and Åsberg, 1975). In test animals, clomipramine does potentiate the motor activation caused by 5-HTP in combination with a peripheral decarboxylase inhibitor (Modigh, 1973).* A similar effect has been observed in human individuals: the antidepressant effect of 5-HTP is potentiated by clomipramine (van Praag *et al.*, 1974). Clomipramine was given in daily doses of 50 mg, which as such are suboptimal. This was done to minimize this antidepressant's own effect. The study in question was open, and is now being verified in a controlled set-up. The antidepressant effect of clomipramine is likewise potentiated by tryptophan (Wålinder *et al.*, 1975).

3 Imperfections of the precursor strategy

(a) Tryptophan

Administration of tryptophan is an inefficient method of increasing the central 5-HT concentration. No more than 3% of the tryptophan in the diet is converted to 5-HT, and only a small fraction of this becomes involved in central 5-HT production. Most of this tryptophan is subjected in the liver to the influence of the enzyme tryptophan pyrrolase, which splits the indole ring; this is accompanied by formation of kynurenin which, via a series of intermediary products, is converted to the B-vitamin nicotinic acid. Some tryptophan is used for protein synthesis, and a small fraction is finally converted to tryptamine. It is to be borne in mind, moreover, that the pyrrolase activity increases in response to large amounts of tryptophan (Knox, 1951), and that in depressive patients this activity is in any case often increased already as a result of an increased concentration of circulating corticosteroids (Knox, 1955). In rats, administration of tryptophan with a pyrrolase inhibitor (allopurinol) does indeed lead to a more marked increase in intracerebral tryptophan than is produced by tryptophan alone. On the other hand it is nevertheless doubtful (Fernando *et al.*, 1975) whether this factor of increased pyrrolase activity is of any practical significance. Coppen's data suggest that it is not. He found that allopurinol did not potentiate the effect of tryptophan in depressions (taken from Carroll, 1971). Be this as it may, the useful effect of a tryptophan dose administered is certainly small.

The effect of tryptophan on the central MA metabolism, moreover, is not 'punctiform'. Administration of tryptophan to animals leads not only to an

* Peripheral decarboxylase inhibitors are compounds which inhibit the conversion of 5-HTP to 5-HT (and of DOPA to DA), but do not penetrate into the brain. Peripherally, the amount of circulating precursor increases, and this results in an increased supply to the CNS. The useful effect of the precursor given is thus greatly enhanced.

increased 5-HT level but also to a decrease in CA concentration (Green *et al.*, 1962; Sourkes *et al.*, 1961), possibly because tryptophan and tyrosine compete for the same transport mechanism to convey them into the CNS. There are indications that the same phenomenon occurs in the human organism (van Praag *et al.*, 1973).

In view of the above it is hardly surprising that there are several known tryptophan effects which almost certainly are not based on 5-HT synthesis: namely sedation, prolongation of sleep, and reduction of the period of REM sleep. These effects are not antagonized by PCPA, an inhibitor of tryptophan hydroxylase (Wyatt *et al.*, 1970; Modigh, 1973). Moreover, 5-HTP (combined with a peripheral decarboxylase inhibitor) induces activation rather than sedation, and an increased instead of a decrease in REM sleep time (Modigh, 1972; van Praag *et al.*, 1972, 1974; Carroll, 1971).

With regard to the toxicity of unphysiologically large doses of tryptophan I refer the reader to a review published by Harper *et al.* (1970). The possible risks should be given serious consideration before one decides to give tryptophan for longer periods, e.g. prophylactically (cf. section F.10). In this context Carroll (1971) made special mention of the risk of carcinoma of the bladder. This type of carcinoma can be induced in test animals with the aid of the tryptophan metabolites 3-hydroxykynurenin, 3-hydroxyanthranilic acid and 2-amino-3-hydroxyacetophenone (Boyland, 1958); and their excretion is increased after tryptophan administration.

Large doses of L-tryptophan can produce side effects such as nausea, a light-headed sensation and blurred vision, but these symptoms seldom take a serious form. After combination with a MAO inhibitor side effects can be more inconvenient: tremor, hyperreflexia, flushing, hyperhidrosis, paraesthesias, orthostatic hypotension, and nystagmus. A large single dose of L-tryptophan (5 g or more), given to patients being treated with a MAO inhibitor, can provoke disintegration reminiscent of LSD psychosis—a syndrome I have described in detail elsewhere (van Praag, 1962).

(b) 5-Hydroxytryptophan

A handicap of 5-HTP medication is that 5-HTP decarboxylase, the enzyme which converts 5-HTP to 5-HT, is not specific and also regulates, say, the conversion of L-DOPA to DA (Yuwiler *et al.*, 1959). 5-HTP which enters catecholaminergic neurons is thus converted to 5-HT; this supersedes CA from the stores and might function as a false transmitter.

Two arguments can be presented to justify the 5-HTP strategy. To begin with, the conversion of 5-HTP in non-serotoninergic neurons is small when moderate doses are given. This has been established by histochemical and biochemical methods (Fuxe *et al.*, 1971; Korf *et al.*, 1974). The biochemical test arrangement was as follows. Rats whose raphe nuclei had been destroyed under stereotactic control, and intact rats, were both treated with small doses of L-5-HTP (12 mg (kg body weight)$^{-1}$) combined with a peripheral decarboxylase inhibitor. Both groups showed increasing 5-HT and 5-HIAA concentrations, but the increase was much more marked in the intact than in those with a destroyed raphe system. This indicates that the conversion of 5-HTP to 5-HT takes place largely in serotoninergic neurons of the raphe system. The second argument is that the human 5-HIAA response to probenecid greatly increases in response to 5-HTP, whereas the HVA response hardly increases (van Praag, 1975). This indicates an increased 5-HT turnover and an unchanged DA turnover.

Side effects of 5-HTP are largely gastrointestinal: nausea, vomiting, and a heavy sensation in the upper abdomen. When we used 5-HTP without peripheral decarboxylase inhibitor we found for the majority of patients an unmistakable threshold value at 3 g D,L-5-HTP per day. Gastrointestinal side effects were quite regularly seen at higher dosages. Contrary to expectation, the gastro-intestinal side effects exacerbated rather than improved when we combined L-5-HTP (400 mg daily) with a peripheral decarboxylase inhibitor (MK 486, 150 mg daily). Since the side effects often occurred within 10–15 min. of ingestion of the 5-HTP capsules and proved to be quite unrelated to the blood 5-HTP concentrations (determined by Dr J. M. Gaillard, Psychiatric Clinic, Geneva University), a direct effect of 5-HTP on the stomach and proximal segment of the small intestine seemed likely. For this reason 5-HTP was then supplied in coated capsules, which pass the stomach unchanged and disintegrate only as they reach the small intestine (at pH = 8.6). The side effects were thus substantially reduced in frequency as well as in intensity.

4 Studies with drugs which inhibit 5-HT synthesis

(a) Inhibition of DOPA (=5-HTP) decarboxylase

The widely used hypotensive drug methyl-DOPA (α-methyl-1-3,4-dihydroxyphenylalanine) inhibits DOPA (and 5-HTP) decarboxylase and thus induces a decrease of cerebral CA and 5-HT concentrations (Sourkes, 1965). Since no CSF studies have been carried out, there are no indications whether this also applies to human individuals.

Several authors consider methyl-DOPA to be a depression-provoking substance, just like reserpine. The incidence is estimated to be 4–6% (Johnson *et al.*, 1966; Hamilton, 1968). The *British Medical Journal* (1966) described methyl-DOPA as contraindicated in the treatment of patients with a history of depression. This statement was not based on a comparative study of several antihypertensive drugs. The possibility therefore remained that what was involved was not pharmacogenic depression but a depression provoked by the awareness of suffering from a relatively serious illness.

Three subsequent studies failed to establish a relation between methyl-DOPA and depression. Each of the three, however, invites some criticism. Prichard *et al.* (1968) continued the medication for 3 months—a period which may have been too short for the depressive syndrome to develop. The remaining two studies (Bullpitt and Dollery, 1973; Snaith and McCoubrue, 1974) concerned patients who for the most part had already had methyl-DOPA treatment for years. Selection may therefore have been involved: patients who developed a depression in the initial phase can have discontinued the medication or reduced the dose to an acceptable level. Obviously, therefore, there is a need for prospective studies.

(b) Inhibition of tryptophan hydroxylase

PCPA is a strong and relatively selective inhibitor of the enzyme tryptophan-5-hydroxylase, which therefore blocks the synthesis of 5-HT (Koe and Weissman, 1966). As judged by the renal excretion of 5-HIAA, 3–4 g PCPA daily reduces the overall 5-HT synthesis by 50–90% in normal test subjects (Cremata and Koe, 1966) and by 72–88% in carcinoid patients (Engelman *et al.*, 1967). The CSF 5-HIAA concentration likewise decreases, indicating central 5-HT synthesis to be blocked also (Goodwin *et al.*, 1973). The crucial question of the behavioural effects of PCPA in these cases cannot be unequivocally answered. Because of its toxic side effects, this compound has been used therapeutically only in patients suffering from a carcinoid: a malignant growth which usually produces large amounts of 5-hydroxyindoles. In the first systematic study of PCPA used for this indication, psychological effects 'ranging from depression to hallucination' were observed; but it was added that it was difficult to assess these symptoms because of the poor physical condition of these patients (Engelman *et al.*, 1967). Subsequent studies failed to add any new point of view. This means that an unequivocal depressogenic effect of PCPA has not been established in carcinoid patients. In subhuman primates (*Maccaca speciosa*), the CA synthesis inhibitor α-MT provoked a series of behavioural changes reminiscent of depression, as already mentioned; PCPA did not (Redmond *et al.*, 1971).

The PCPA data so far collected are not consistent with the MA hypothesis (assuming that the PCPA dosage was sufficient for a 5-HT deficiency with functional implications): it is to be noted in this context that, owing to its toxicity, PCPA has never been given for longer periods to depressive or manic patients, nor to normal test subjects. The above conclusion is subject to yet another restriction: in an elegant study, Shopsin *et al.* (1974) demonstrated that PCPA arrested the therapeutic effect of imipramine within a few days, whereas several weeks of α-MT medication failed to influence this therapeutic effect. This indicates that the central 5-HT metabolism in any case plays a role in the antidepressant action of imipramine.

5 Conclusions

Precursor therapy is intended to abolish a suspected central 5-HT deficiency. Although these precursors are far from selective in their effects, it is fairly certain that they stimulate the 5-HT synthesis in human individuals also. The results of precursor medication in depressions are not unequivocal. This applies to tryptophan as well as to 5-HTP. On the other hand, indications that they are not therapeutically inert are too numerous to warrant rejection of this strategy as ineffective and interpretation of the results as evidence against the MA hypothesis. Most precursor studies, moreover, have been made in biochemically undifferentiated groups of depressive patients. This might explain the variability of the results. The few data now available do indicate the likelihood that 5-HT precursors, although far from being universal antidepressants, can be effectively used in certain biochemical subtypes of depression.

There are no convincing indications that inhibition of 5-HT synthesis has a depressogenic effect. On the other hand, the fact that inhibition of 5-HT synthesis arrests the therapeutic effect of imipramine supplies an unmistakable indication of a relation between 5-HT and mood regulation.

To summarize: in my view the results of studies with precursors and synthesis blockers do not devalue the 5-HT hypothesis but in fact support it on certain points, albeit not in a conclusive manner.

H VERIFICATION OF THE 5-HT HYPOTHESIS WITH THE AID OF DRUGS AND OTHER METHODS OF ANTIDEPRESSANT THERAPY 2. DIRECT INFLUENCING OF THE TRANSMISSION PROCESS

1 Disorders in central 5-HT metabolism and therapeutic efficacy of tricyclic antidepressants

It is fairly generally accepted that the therapeutic effect of tricyclic antidepressants relates to their ability to inhibit the central (re-)uptake of 5-HT and NA into the neuron, thus enlarging the amount of transmitter available at the post-synaptic receptors. Some of these compounds have a certain selectivity in biochemical terms. Clomipramine (Anafranil) exerts a pronounced influence on the 5-HT 'pump' but has little effect on that of NA. The reverse is true for such compounds as nortriptyline and desipramine. Tricyclic antidepressants are believed to exert but little influence on the (re-)uptake of DA.

If the disorders in central 5-HT metabolism discussed do indeed play a role in the pathogenesis of depressions, then it is to be expected that not all tricyclic compounds can be equally effective, but that a depressive patient with a presumed central 5-HT deficiency would benefit most from a (relatively) selective inhibiton of 5-HT uptake. So far only one group of investigators have tried to test a hypothesis of this type.

Åsberg *et al.* (1972) studied the possibility of a relation between baseline CSF 5-HIAA concentration and therapeutic response to nortriptyline. They found the therapeutic effect of this agent to be less pronounced in patients with a 5-HIAA concentration of less than 15 ng (ml CSF)$^{-1}$ than in patients with higher 5-HIAA levels, even though the plasma nortriptyline level was well within the therapeutic range in both groups.

These findings can be made to fit the 5-HT hypothesis. Nortriptyline is a relatively selective compound: its inhibitory effect on the 5-HT 'pump' is slight, whereas that on the NA 'pump' is marked (Carlsson *et al.*, 1969a, b). If a low CSF 5-HIAA level is indicative of a central 5-HT deficiency, then it is to be expected that a compound with a mostly NA-potentiating effect cannot be very effective in these patients. They would probably benefit more from a 5-HT-potentiating antidepressant such as clomipramine. The correctness of this hypothesis is yet to be studied.

The research strategy described above is waiting for compounds which really approximate a 100% selectivity of their influence on the (re-)uptake of 5-HT and NA. Selective inhibitors of the 5-HT 'pump' are being introduced (Fuller *et al.*, 1974; Wong *et al.*, 1974), but have yet to be clinically tested.

2 Tricyclic antidepressants combined with MAO inhibitors

Because it is believed to give rise to serious cardiovascular complications, nearly all textbooks seriously caution against this combination, however logical it may seem to be (for a review, see Schukitt *et al.*, 1971). On the other hand there have been a few reports describing the (clinical) application of this combination without serious side effects, with good results in otherwise refractory depressions (Winston, 1971; Schukitt *et al.*, 1971; Sethna, 1974). The combination so far most thoroughly tested is that of amitriptyline (25–100 mg daily) with isocarboxazid (Marplan, 10–20 mg daily) or phenelzine (Nardil, 45 mg daily) as MAO inhibitor.

It would of course be unjustifiable simply to credit the MA hypothesis with this synergism. The MAO inhibitor could interfere with the microsomal liver enzymes which degrade tricyclic compounds, permitting the plasma level to rise higher. In view of the preliminary results reported by Snowdon and Braithwaite (1974), however, this is unlikely.

3 MAO inhibitors combined with reserpine

MAO inhibitors block the intraneuronal degradation of MA, and possibly also the (re-)uptake of MA through the neuronal membrane. Both processes are assumed to lead to an increase in the amount of MA available at the postsynaptic receptors. Reserpine blocks the uptake of MA into the synaptic vesicles, causing them to be metabolized by MAO, so that their concentration diminishes. If MAO is inhibited, however, then reserpine increases the MA concentration in the axoplasm and consequently the MA 'leakage' to the synaptic cleft is believed to increase markedly. This mechanism has been suggested as explanation of the fact that reserpine does not sedate MAO inhibitor-treated animals, but in fact stimulates them (Shore and Brodie, 1957).

If a deficiency of one or several MA plays a role in the pathogenesis of depressions, then it may be expected that the therapeutic effect of MAO inhibitors is potentiated by addition of reserpine. We tested this hypothesis in 32 patients with vital depressions; all were under 45 and had normal baseline blood pressure, which was carefully kept under control (van Praag, 1962). The daily reserpine dosage was 1 mg orally; the daily dosages of the MAO inhibitors tested (iproniazid and isocaboxazid) were 100-150 and 30–45 mg, respectively. In 15 patients the two drugs were started simultaneously; the remaining patients were given reserpine after 4–6 weeks' unsuccessful or insufficiently successful medication with a MAO inhibitor. The combined therapy lasted 4 weeks. Eight patients of the second group, aged 25–40, were again given reserpine (5 mg, by slow continuous drip) 1–2 weeks after discontinuation of oral reserpine, while the MAO inhibitor was still being given; a careful check was kept on the blood pressure.

The result of this experiment can be summarized as follows. No overall therapeutic effect of reserpine was observed in the three groups as a whole. The combined treatment did not average a stronger or more rapid effect than the MAO inhibitor alone. In the refractory patients, addition of reserpine caused no *average* improvement. I emphasize the word average because a few patients (four in the first and four in the second group) showed a strikingly favourable response. They were indistinguishable from the other patients in psychopathological terms. They may have been recognizable on the basis of a biochemical criterion, but CSF studies were not yet being made at the time of this experiment. We have not repeated the experiment in view of the hypotensive risk.

4 Tricyclic antidepressants combined with reserpine

Reserpine has also been combined with tricyclic antidepressants. Uncontrolled studies seemed to indicate success (Pöldinger, 1963; Hascovec and Rysanek, 1967), but controlled studies did not (Carney *et al.*, 1969). In any case, this combination is hardly tenable theoretically. Tricyclic antidepressants are assumed to increase the extraneuronal MA concentration. Reserpine releases MA *within* the neuron, but these are degraded by MAO before they can 'leak out' and produce an effect. There is consequently no theoretical ground for the hypothesis that reserpine could potentiate the therapeutic effect of the tricyclic compounds.

5 Enhancement of central 5-HT activity with the aid of chloramphetamines

(a) 4-Chloramphetamines and central 5-HT metabolism

In 1964 Pletscher *et al.* reported on a compound called 4-chloro-*N*-methylamphetamine (CMA), whose effect on the 5-HT metabolism tended towards selectivity. It causes a marked decrease in central 5-HT concentration without exerting any significant influence on the NA concentration. The 5-HIAA concentration likewise diminishes, giving rise to the suspicion that CMA inhibits 5-HT synthesis. However, tryptophan-5-hydroxylase in the liver was not inhibited by CMA. This suggested to Pletscher *et al.* (1966) that CMA releases 5-HT from the stores, but fails to cause an increase in 5-HIAA concentration because at the same time it is a (feeble) MAO inhibitor. No positive evidence of an alternative degradation of 5-HT was found.

It had meanwhile been established that tryptophan-5-hydroxylase in the brain was inhibited by CMA (Sanders-Bush *et al.*, 1972) and that CMA reduces the 5-HT turnover (Sanders-Bush and Sulser, 1970; Korf and van Praag, 1972). The two effects, probably interrelated, might be primary effects but also could be secondary, resulting from a feedback mechanism activated by excitation of postsynaptic 5-HT receptors. There were two phenomena indicative of greater likelihood of the latter mechanism because they suggested that CMA is more likely to increase than to decrease the activity in serotonergic systems.

To begin with, CMA was found to stimulate motor activity in test animals—a phenomenon observed also when central 5-HT activity is increased by means of 5-HTP in combination with a peripheral decarboxylase inhibitor (Modigh, 1972). Second, the motor-stimulating effect of CMA was inhibited by substances which reduce the amount of 5-HT available in the brain, i.e. cyproheptadine, a 5-HT antagonist, and PCPA, an inhibitor of 5-HT synthesis (Frey and Magnussen, 1968). Should CMA *primarily* block 5-HT synthesis, then cyproheptadine and PCPA could have been expected to potentiate the behavioural effects of CMA. The findings mentioned therefore suggest that the decrease in central 5-HT concentration in response to CMA is associated with increased serotonergic activity. A possible explanation of this paradox is that, in response to CMA, 5-HT is for the most part released not *within* the neuron (where it would be inactivated before it could develop any activity) but in the synaptic cleft. In that case the serotonergic activity would increase even though the absolute amount of 5-HT in the brain would diminish.

In the context of the MA hypothesis a compound with a 5-HT-potentiating effect can be expected to exert an antidepressant influence. This is why between 1967 and 1970 we investigated the effects of CMA and a substance with a related action (4-chloramphetamine) (Fuller *et al.*, 1965) in depressive patients (van Praag, 1967). An exhaustive review of these studies is presented in two recent publications (van Praag and Korf, 1973, 1975).

It has meanwhile been established that the effect of chloramphetamines on the central 5-HT system is a very complex one, which comprises at least four components: (i) inhibition of 5-HT synthesis; (ii) release of 5-HT from the synaptic vesicles; (iii) inhibition of 5-HT transport through the neuronal membrane; and (iv) MAO inhibition (Fuller and Molloy, 1974). However, these more recent findings have not upset the hypothesis that 4-chloramphetamine enhances the central 5-HT activity in the acute experiment.

(b) Chemical and psychopathological effects of 4-chloramphetamine in human individuals

Of course the foremost question was whether CMA and 4-chloramphetamine also influence the human 5-HT metabolism and, more specifically, whether there are indications of a depletion of 5-HT stores. In response to both compounds the renal excretion of 5-HT as well as 5-HIAA proved to show a gradual increase, and we interpreted this phenomenon as indicating that they do indeed interfere with 5-HT storage. The percentage of ^{14}C-labelled 5-HT converted to ^{14}C-labelled 5-HIAA did not diminish; overall MAO inhibition was therefore improbable. Finally, the overall 5-HT synthesis was assessed by successively loading test subjects with the 5-HT precursors (unlabelled) tryptophan and (^{14}C-labelled) 5-HTP and then measuring the fractions excreted in the urine as 5-HT and 5-HIAA. Tryptophan was not labelled because it is partly utilized in protein synthesis so that the radioisotope would long remain in the organism. Neither CMA nor 4-chloramphetamine reduced the 5-HT and 5-HIAA yield, and consequently (overall) inhibition of 5-HT synthesis was unlikely. As judged by the response of CSF 5-HIAA to probenecid, CMA as well as 4-chloramphetamine reduce the 5-HT turnover in the CNS.

All in all, therefore, it is quite likely that the chloramphetamines studied interfere with 5-HT storage without noticeably altering the overall ability to synthesize 5-HT. The decreased 5-HT turnover in the CNS could be secondary to 5-HT receptor stimulation. The conclusions from studies in human individuals are therefore in agreement with the findings of animal experiments.

When used in the treatment of depressive patients, CMA and 4-chloramphetamine proved to be significantly superior in therapeutic effect to a placebo. They did not mutually differ in efficacy. Vital depression proved not to be an indication of preference; improvements were also seen in the group of personal depressions. This would seem to be a striking difference from the tricyclic compounds, with a definite indication of preference in the vital depressive group. Another striking finding was the absence of typical side effects as produced by non-chlorated amphetamine derivatives, e.g. agitation, insomnia, anorexia, and hypertension. The side effects were found to be insignificant in general. In normal test subjects, too, the psychological effects of CMA were quite distinguishable from that of non-chlorated methylamphetamine.

We tend to interpret the results of this study as supporting the hypothesis that a central 5-HT deficiency plays a role in the pathogenesis of depressive symptoms.

6 Thyrotropin-releasing hormone (TRH)

In 1972 Prange *et al.* and Kastin *et al.* reported on the basis of independent studies that TRH very rapidly produces a therapeutic effect in depressions (within a few hours). The conventional antidepressants need 10–20 days to produce a therapeutic effect. In terms of chemical structure, TRH differs widely from the conventional agents with an antidepressant effect. It is a tripeptide produced in the hypothalamus, and already known to stimulate the hypophysis to release TSH (thyroid-stimulating hormone; thyrotropin), a hormone which enhances the hormone production in the thyroid. In animals, too, TRH induces behavioural effects. These probably result from a direct influence of TRH on the brain, for they also occur in hypophysectomized animals (Potnikoff *et al.*, 1972).

The results of reduplication studies were generally negative. Various investigators reported no (Coppen *et al.*, 1974b; Mountjoy *et al.*, 1974; Benkert *et al.*, 1974; Hollister *et al.*, 1974), or only a marginal, therapeutic effect (van den Burg *et al.*, 1975a, b; Takahashi *et al.*, 1973). On the other hand, the various series studied included a few individual patients who showed a favourable if not readily-defined response to TRH, but not to a placebo. Coincidence has not been established or ruled out with certainty. It would therefore be worthwhile to establish whether such patients can be psychopathologically or biochemically identified. Moreover, Wilson *et al.* (1973) observed an activating and euphorizing effect of TRH in normal women.

Assuming that TRH is not entirely devoid of an influence on mood and motor activity, the question arises whether (and if so in what direction) TRH influences the MA metabolism. TRH has no effect on the intracerebral MA concentration (Reigle *et al.*, 1974), nor does it change the intracerebral CA uptake (Horst and Spirt, 1974). However, it increases the CA turnover (Horst and Spirt, 1974; Keller *et al.*, 1974; Constantinidis *et al.*, 1974). Whether this is based on increased neuronal activity in catecholaminergic systems or represents a compensatory mechanism triggered by inhibition or block of transmission, remains to be established. Since TRH potentiates the motor-activating effect of L-DOPA combined with a MAO inhibitor in mice (Plotnikoff *et al.*, 1972, 1974), the former mechanism is more plausible than the latter. In chronic experiments, however, the effect of TRH on the NA turnover was found to be no more than marginal (Reigle *et al.*, 1974).

The effect of TRH on the 5-HT turnover is unknown, but it seems likely on indirect grounds that the 5-HT system is influenced. In rats treated with a

MAO inhibitor, tryptophan induces hyperactivity—probably as a result of an abnormally high 5-HT concentration at the postsynaptic 5-HT receptors (Grahame-Smith, 1971). This effect is potentiated by TRH (Green and Grahame-Smith, 1974), and this is an indirect indication that TRH enhances central serotonergic activity.

Consequently, neither the clinical nor the biochemical books on TRH can as yet be closed. It would be regrettable if the largely negatively-coloured publications of the past few years would frustrate continued research.

7 Methysergide

Methysergide (1-methyl-D-lysergic acid butanolamide), a derivative of LSD, is a 5-HT (and tryptamine) antagonist (Doepfner and Cerletti, 1958). This has been established in the periphery. It antagonizes the vasoconstrictive effect of 5-HT as well as a series of 5-HT effects on extravascular smooth muscle tissue. The mechanism of this antagonism is reversible block of 5-HT receptors. We do not know whether methysergide also behaves as antagonist of 5-HT in the brain. The compound has been used for years in the prophylaxis of migraine, but it is uncertain whether this effect is based on 5-HT antagonism.

Reports published by the end of the 1960s indicated that methysergide caused, within 48 h., dramatic subsidence of manic symptoms (Dewhurst, 1968; Hascovec and Soucek, 1968; Hascovec, 1969). In some cases these symptoms were in fact replaced by depressive symptoms (van Scheyen, 1971). The same has been reported of another 5-HT antagonist: cinanserin (Kane, 1970). These were uncontrolled observations, and the findings are therefore questionable because the manic syndrome can markedly differ in intensity from day to day. The findings have not been corroborated in reduplication studies (Fieve *et al.*, 1969; Grof and Foley, 1971; McNamee *et al.*, 1972; McCabe *et al.*, 1970). According to Coppen *et al.* (1969), methysergide can in fact cause exacerbation of manic symptoms.

The causes of these negative results can be widely varied. For example, manic symptoms may be unrelated to the functioning of central serotonergic systems; or the influence of methysergide on central 5-HT receptors may be too weak; or the doses given may have been too small (in any case the gradient in methysergide concentration between blood and brain is great: Doepfner, 1962); etc. The pre-therapeutic CSF 5-HIAA concentration was not determined, nor was the possible influence of methysergide on this parameter. Finally, it is possible also that the manic syndrome is heterogenous in pathogenetic (biochemical) terms—much as it is now considered to be likely for the depressive syndrome—and that only a certain subgroup is responsive to methysergide. In that case we must assume that this effect was concealed because group averages were used.

8 Lithium

(a) *Range of action*

Lithium is one of the most striking new assets in pharmacopsychiatry. (i) It is therapeutically effective in manic patients, but much less in agitations of a different nature (Rimon and Räköläinen, 1968; Lackroy and van Praag, 1971). (ii) It has a prophylactic effect in unipolar and bipolar depressions in that it reduces the number and severity of the phases (Schou, 1973). (iii) It is alleged to be therapeutically effective in certain depressions. This component of lithium action is the least certain. Its existence was initially doubted, but there are now indications that in particular the group of bipolar depressions could be lithium-susceptible (Goodwin *et al.*, 1972). Moreover, there have been reports on two chemical parameters believed to be predictive in this respect: the capacity of the erythrocytes to take up lithium—increased in lithium-susceptible depressions (Mendels and Fraser, 1973); and the increase or decrease of the plasma Mg and Ca levels during the first week of lithium medication, the lithium responders showing an increase (Carman *et al.*, 1974). Our findings indicate the improbability of an antidepressant effect of lithium. Should lithium be therapeutically active in bipolar depressions, then recurrence during lithium prophylaxis could be expected to be more frequent in the unipolar than in the bipolar group. We have no indications that this is true.

Studies of the influence of lithium on the central MA metabolism have yet to lead to unequivocal results and conclusions. It is impossible to summarize these studies. Suffice it to make mention of a few findings in order to demonstrate that the influence of lithium on central MA is a very complex one, and that it may be very difficult to identify the net effect of this complex influence. For a more detailed review, I refer the reader to publications by Schildkraut (1973) and Shaw (1975).

(b) *Noradrenaline*

Lithium influences noradrenalinergic transmission in several different ways. There are some indications that it reduces the sensitivity of the postsynaptic NA receptor (Douša and Hechter, 1970; Forn and Valdecasas, 1971). Moreover, in brain slices (Katz *et al.*, 1968) and in isolated perfused cat spleen (Bindler *et al.*, 1971) lithium reduces the release of NA provoked by direct electrical stimulation of the slices and stimulation of the splenic nerve, respectively. *In vitro*, finally, lithium increases the uptake of NA into synaptosomes (Baldessarini and York, 1970; Kuriyama and Speken, 1970). Each of the three effects mentioned is believed to result in a relative or absolute decrease of the amount of NA available at the corresponding receptors. Perhaps the increase in NA turnover observed after lithium administration both with the aid of tracer kinetics (Schildkraut *et al.*, 1969b) and by inhibition of degradation (Corrodi *et al.*, 1967) may be regarded as an attempt to compensate the reduced receptor stimulation. However, the plausibility of this interpretation diminishes in view of the fact that, *in vivo*, lithium does not increase the NA uptake in the brain of test animals (Schildkraut *et al.*, 1967, 1969a) and that the influence on the NA turnover in chronic experiments soon diminishes (Corrodi *et al.*, 1969), so that this effect cannot be used to explain the long-term effects of lithium.

(c) *Dopamine*

Studies of the influence of lithium on the central DA system have been less exhaustive. The results, moreover, are contradictory. Some authors maintain that DA synthesis is decreased (Friedman and Gershon, 1973; Corrodi *et al.*, 1969), whereas others hold that it remains uninfluenced (Ho *et al.*, 1970). Interest attaches to the observation of Friedman and Gershon (1973) that the rate of DA synthesis decreases only after chronic lithium administration; this might correspond with the clinical experience that the therapeutic effect of lithium in manic patients does not become manifest until after several days to a week.

(d) *Serotonin*

Lithium administered *in vivo* has no effect on the 5-HT uptake into brain synaptosomes (Kuriyama and Speken, 1970), but in brain slices it reduces the release of (radioactive) 5-HT following electrical stimulation (Katz *et al.*, 1968). In view of these findings one may be inclined to regard reduced availability of 5-HT at the central receptors as probable, just as for NA. However, this can hardly be reconciled with the data presented by Grahame-Smith and Green (1974): lithium combined with a MAO inhibitor made rats hyperactive, exactly as the combination of tryptophan and a MAO inhibitor does. The latter combination is believed to act via an increase of the 5-HT concentration at the postsynaptic receptors, and the author regards it as likely that the same applies to the former combination.

The intracerebral 5-HIAA concentration increases in response to lithium (Schildkraut *et al.*, 1969a; Sheard and Aghajanian, 1970; Perez-Cruet *et al.*, 1971), but it is uncertain whether this is based on an increased 5-HT turnover or on decreased transport of 5-HIAA from the brain. The available data on the influence of lithium on the 5-HT turnover are rather controversial. Mention has been made (but different methods of determination were used) of an increase (Sheard and Aghajanian, 1970; Perez-Cruet *et al.*, 1971; Schubert, 1973; Poitou *et al.*, 1974), of a decrease (Corrodi *et al.*, 1969; Ho *et al.*, 1970), and of an absence of any change (Genefke, 1972).

(e) Human studies

The findings obtained in human individuals are not consistent either. The renal excretion of MHPG increases in the initial phase of lithium medication, and then decreases (Schildkraut, 1973). The renal DA excretion likewise decreases (Messiha *et al.*, 1970). The MHPG concentration in the CSF decreases (Wilk *et al.*, 1972). The HVA concentration has been reported to increase (Wilk *et al.*, 1972) or to decrease (Bowers *et al.*, 1969). However, these studies encompass only a few patients, and probenecid responses were not studied. The same applies to the 5-HIAA concentration in the CSF, data on which are likewise confusing: some authors reported an increase during lithium medication (Mendels, 1971; Wilk *et al.*, 1972), but Bowers *et al.* (1969) described it as unchanged.

An important study was published by Beckmann *et al.* (1975), in which they demonstrated biochemical differences between lithium responders and non-responders. Comparing the pre-drug period with the third and fourth week of lithium treatment, all of the responders showed an increase in MHPG, while the non-responders showed no change or a decrease. In addition, there was a tendency for the pretreatment MHPG excretion to be low in the patients who went on to show a clear-cut antidepressant response to lithium compared to those who were unequivocal non-responders. So, biochemical heterogeneity of the depressive disorders could explain the variability of the biochemical effects of lithium reported in the literature.

Some peripheral observations, too, are sufficiently interesting to be mentioned here. In human individuals, lithium reduces the hypertension provoked by NA infusions. The pressor response caused by tyramine is not affected (Fann *et al.*, 1972). A possible explanation is diminished sensitivity of NA receptors. Another possibility is that lithium promotes inactivation of NA, either through increased degradation or through increased neuronal uptake. An argument in favour of increased degradation can be found: lithium increases MAO activity in human blood platelets (Bockar *et al.*, 1974). But increased re-uptake can also be argued: Murphy *et al.* (1969) demonstrated that lithium promotes the uptake of 5-HT into human blood platelets, and this cell type is considered to be a valid model of cerebral synaptosomes. These observations thus suggest that lithium reduces the amount of MA available at the receptors.

All in all, it seems justifiable to conclude that the influence of lithium on the MA metabolism is highly complex, and as yet insufficiently analysed to be used in testing the MA hypothesis.

9 Electroconvulsive therapy

(a) Range of action

Electroconvulsive therapy (ECT) is an effective therapeutic method in depressions, and particularly in vital depressions (Carney and Sheffield, 1974). Owing to its rather drastic character, however, it has been largely replaced by antidepressant medication. The vital depression which shows no or no adequate response to antidepressants, and melancholia (deep vital depression with delusions—a syndrome which often shows no or no adequate response to antidepressants) have remained indications for ECT.

In my view ECT is only of limited value in testing the MA hypothesis. It should be understood that the rush of current causes massive discharges in large areas of the brain and in the peripheral autonomic nervous system. During the convulsion, moreover, several endocrine glands are stimulated to secretion, and tonic and clonic contractions occur throughout large parts of the smooth and striated muscle systems (the last-mentioned factor can be reduced by establishing a neuromuscular block). The chemical homoeostasis of the organism is profoundly disturbed, and only an incorrigible optimist could hope to select from the numerous changes precisely those which determine the therapeutic effect.

(b) Animal experiments

Nevertheless, the question whether ECT influences the central MA metabolism seems relevant in this context. It does indeed. Immediately after a single electroshock the turnover rate of NA (Schildkraut *et al.*, 1971; Schildkraut and Draskoczy, 1974) as well as that of 5-HT (Engel *et al.*, 1971; Ebert *et al.*, 1973) in the rat brain is increased; however, the same phenomena are observed after all sorts of disagreeable interventions (Bliss *et al.*, 1972; Thierry *et al.*, 1968a, b). They can probably be regarded as non-specific stress effects. Moreover, in a human individual a single shock is rarely effective: a series of shocks are required for a therapeutic effect. The question to be raised in this context is therefore: does the central MA metabolism change after a *series* of shocks and, if so, how long does the change persist? The NA concentration is reported to remain unchanged under these conditions (Hinsley *et al.*, 1968); that of 5-HT has been described as normal (Bertaccini, 1959) or as increased (Garattini *et al.*, 1960; Hinsley *et al.*, 1968).

According to several authors, the rate of disappearance of intracisternally introduced ^{3}H-labelled NA increases after a series of shocks; moreover, a larger than normal fraction of it is converted to normetanephrine (Kety *et al.*, 1967; Ladisisch *et al.*, 1969; Ebert *et al.*, 1973), and tyrosine hydroxylase activity increases slightly (Musacchio *et al.*, 1969). This could mean that the NA turnover is increased and that more NA is released in the synaptic cleft (for NA methylation takes place extraneuronally). In addition there are indications that the uptake of NA into synaptosomes is decreased after a single shock as well as after a series of shocks (Hendley and Welch, 1975). Extrapolated to an *in vivo* situation, this would mean an increased NA concentration in the synaptic cleft.

These effects are demonstrable up to 3 days after completion of ECT, and they are more pronounced than those after other stress-producing interventions. Probably, therefore, they are shock-specific. Using the method of synthesis inhibition, however, Papeschi *et al.* (1974) found no influence of chronic ECT on the NA turnover.

The concentrations of 5-HT, 5-HIAA, and the mother substance tryptophan are likewise increased for a few days after a series of electroshocks (Ebert *et al.*, 1973). This might indicate an increased 5-HT turnover. However, the effect is not more marked than that after other stressors, and can therefore not be described as specific.

The effects of repeated electroshocks in test animals, therefore, are neither at variance with the MA hypothesis nor strongly in support of it.

(c) *Human studies*

Few human studies have so far been carried out. The serum DAβ-hydroxylase activity increased during ECT (Lamprecht *et al.*, 1974). Since this enzyme is released into the synaptic cleft together with the transmitter, this phenomenon might indicate that synaptic activity is in any case increased in the periphery. Stressors of a different nature, however, provoke the same phenomenon (Wooten and Cardon, 1973; Pflanz and Palm, 1973), and its specificity is therefore questionable. The CSF 5-HIAA and HVA concentration do not change in response to ECT (Nordin *et al.*, 1971). We ourselves found that the accumulation of these metabolites in the CSF after probenecid is increased on the day of ECT, regardless of its therapeutic effect. In patients whose 5-HIAA and HVA responses were diminished prior to treatment, normal responses were found 1 week after a successful ECT course. We have no data on patients who were refractory to ECT. In patients with normal pretherapeutic responses, these values were found slightly but not significantly increased 1 week after ECT.

In view of these data it is likely that the 5-HT and DA turnover can increase on response to ECT, particularly in patients in whom the turnover of these amines was decreased to begin with. We have no data on the further course of 5-HIAA and HVA responses in these patients.

10 Conclusions

Transmission in serotonergic synapses can be directly influenced with the aid of drugs, either in a facilitating or in an inhibitory sense. It has been shown that several procedures assumed to facilitate serotonergic transmission probably relieve depression, although the results are not always unequivocal. This is in accordance with expectation. There is much more uncertainty about the psychological consequences of inhibition of transmission. The only 5-HT antagonist investigated in this respect has no demonstrable consistent effect on mood regulation.

Therapeutic methods with an established effect on mood regulation, e.g. ECT and lithium medication, could on the basis of the 5-HT hypothesis be expected to exert an influence on the serotoninergic system. Such an influence is in fact demonstrable, but it is so complex and in part controversial that no net effect can be deduced from it and that for the time being it is impossible to use it in testing the MA hypothesis.

Direct influencing of monoaminergic transmission remains an important strategy in testing the MA hypothesis, but there is an urgent need for compounds which are more selective than the available substances. In this field, too, it seems to me to be of essential importance that in depression research classifications are made not only according to syndrome, aetiology, and course, but also according to pathogenesis, i.e. on the basis of biochemical points of view.

The favourable effect of liver extract in pernicious anaemia would probably not have been discovered if it had simply been given to all anaemic patients, without distinction, and if the therapeutic results had simply been averaged.

I THEORIES TO EXPLAIN THE 5-HT DEFICIENCY IN DEPRESSIONS

1 Possible causes of the 5-HT deficiency

The 5-HT hypothesis—the pros and cons of which have been discussed in previous sections—postulates a functional cerebral 5-HT deficiency as a (contributing) factor in the pathogenesis of certain types of depression. What can be the cause of this postulated deficiency? Generally speaking there are three possibilities: precursor deficiency, decreased synthetic capacity, and decreased firing rate in the serotoniergic system. The firing rate in the presynaptic element is a factor which is among the determinants of the rate of MA hypothesis. A decrease in firing rate is associated with a decrease in the activity of tryptophan hydroxylase. Less enzyme is produced, or its structure changes. In either case the functional capacity diminishes, and 5-HT synthesis decreases as a result (Thoenen, 1972; Bunney *et al.*, 1973; Sheard and Aghajanian, 1968; Shields and Eccleston, 1972).

With these factors as a starting-point, a number of theories to explain the 5-HT deficiency have been advanced.

2 Precursor deficiency

The amount of tryptophan in the brain is an important factor in the regulation of 5-HT synthesis. Central tryptophan deficiency leads to a decrease in 5-HT synthesis (Fernström and Wurtman, 1971). Coppen *et al.* (1973) found a decreased plasma-free tryptophan concentration in depressive patients. Moreover, the CSF tryptophan concentration was decreased (Coppen *et al.*, 1972b). The former finding was confirmed by Rees *et al.* (1974), and the latter was refuted by Ashcroft *et al.* (1973b). In our patients, too, we found no decreased CSF tryptophan concentration. As pointed out in section C.3, moreover, there are indications that depressive patients convert an abnormally large amount of tryptophan via the kynurenin route. This is believed to result from activation of tryptophan pyrrolase in the liver by circulating hydrocortisone, the amount of which is often increased in depressive patients. In test animals, tryptophan pyrrolase is so activated by hydrocortisone that the plasma tryptophan concentration decreases and the cerebral 5-HT concentration diminishes by about 30% (cf. section C.3).

In view of these findings it has been hypothesized (Curzon, 1969; Lapin and Oxenkrug, 1969) that depressions (or certain types of depression) involve primarily an increased secretion of adrenocortical hormone (possibly in response to increased ACTH production). This causes activation of tryptophan pyrrolase in the liver: more tryptophan is directed to the kynurenin route, and the plasma tryptophan concentration decreases. Plasma tryptophan (Tagliamonte *et al.*, 1971), or rather the ratio of plasma tryptophan to other neutral amino acids using the same transport mechanism to reach the brain (e.g. tyrosine and phenylalanine) (Fernström and Wurtman, 1972), is an important determinant of the intracerebral tryptophan concentration. It is therefore plausible that the intracerebral tryptophan concentration decreases and, with it, the 5-HT synthesis.

The theory seems to be fairly tenable at first glance. More careful consideration, however, reveals its weak spots. (i) As pointed out, administration of a single dose of hydrocortisone leads to a decrease in central 5-HT concentration, but this concentration is normalized as hydrocortisone administration is continued (Curzon and Green, 1968). (ii) Several authors have been unable to confirm the correlation between plasma tryptophan and intracerebral tryptophan concentrations (Morgan *et al.*, 1975). (iii) A high concentration of adrenocortical hormones in the plasma is not so much specific of depression as related to the factor anxiety (Sachar. 1967). (iv) Increased pyrrolase activity can be observed in a variety of conditions not associated with depression (Altman and Greengard, 1966), and nothing is known about pyrrolase activity in the liver in depressions. (v) There are investigators who have not measured increased kynurenin excretion in depressions (Birkmayer and Linauer, 1970). Consequently it would be premature to regard the abovementioned data as an outline of a biochemical *depression* theory. Perhaps they may later be found to have been more important for our understanding of the somatic substrate of anxiety.

Unlike tryptophan hydroxylase, tyrosine hydroxylase is normally saturated with substrate (Udenfriend, 1966), and it is therefore impossible to increase CA synthesis by supplying extra tyrosine. Yet it is quite conceivable that a tyrosine deficiency reduces CA synthesis, although there are hardly any indications that it really does. The peripheral data—plasma tyrosine concentration, baseline and after oral tyrosine loading—are inconsistent (Takahashi *et al.*, 1968; Birkmayer and Linauer, 1970; Benkert *et al.*, 1971). Moreover, the CSF tyrosine concentration is normal in depressive patients (van Praag *et al.*, 1973a, b). There is no theory which relates the supposed central CA deficiency to tyrosine deficiency.

3 Diminished synthetic capacity

There are indications that the 'tryptophan tolerance' is decreased in depressive patients. After oral loading with L-tryptophan, the blood tryptophan level was found to be above normal during the first few hours (Coppen *et al.*, 1974). This may be an indication that the chemical 'processing' of tryptophan, in whatever direction, is delayed; of course other explanations are conceivable also (cf. section G.2). A comparable phenomenon has been observed in the CNS. During the first few hours following administration of a large oral dose of tryptophan, the CSF tryptophan concentration showed a more marked, and the 5-HIAA concentration a less marked increase than that in the control group (van Praag *et al.*, 1973). This suggests that the tryptophan administered 'congests', so to speak, in the CNS in depressive patients; that it is less readily converted to 5-HT.

Since the conversion of 5-HTP to 5-HT is undisturbed (Coppen, 1967; van Praag, 1975), these findings are indicative of a decreased tryptophan hydroxylase activity, or of a diminished accessibility of the neuron to tryptophan.

There are no analogous data on the CA metabolism. In depressive patients the ability to convert L-DOPA to HVA is undisturbed (Goodwin *et al.*, 1970; van Praag, 1974).

4 Postsynaptic defects

In the accepted view, tricyclic antidepressants reduce the central re-uptake of 5-HT and NA into the nerve endings of serotoninergic and noradrenalinergic neurons. The postsynaptic receptors are activated as a result. This leads to a reduced firing rate in the presynaptic element, and subsequently to a gradual decrease of tryptophan hydroxylase and tyrosine hydroxylase activity. The last-mentioned phenomenon is regarded as a side effect of the receptor stimulation. It is the receptor stimulation that is held responsible for the therapeutic effect.

A fact which it is difficult to reconcile with this view is that MA uptake is inhibited within a few minutes by tricyclic antidepressants (Schildkraut and Kety, 1967), whereas the therapeutic effect as a rule does not become manifest until after a latent period of 10–20 days. The theory formulated by Mandell *et al.* (1975) offers an explanation. They confined their reasoning to the noradrenalinergic system, and postulated that the postsynaptic NA receptors are hyperactive rather than hypoactive in depression, either due to hypersensitivity or as a result of a chronic NA surplus, e.g. due to reduced extraneuronal NA degradation. By way of compensation the presynaptic firing rate decreases, and tyrosine hydroxylase activity gradually diminishes as a result. The last-mentioned phenomenon is reflected in low concentrations of CA metabolites in the CSF. Tricyclic antidepressants expose the postsynaptic receptors to acute activation, and this results in further reduction of tyrosine hydroxylase activity. The latter effect is considered not to be an epiphenomenon but the key to their therapeutic effect: it gives the hyperactive (hypersensitive) receptor extra protection.

This is an interesting theory, but one that is not readily reconciled with various other phenomena. For example, although little is known about the psychic action of the CA synthesis inhibitor α-MT in human subjects (cf. section G.5), there is no indication at all that it could have an antidepressant effect. On the other hand L-DOPA (a CA precursor which enhances CA synthesis in depressive patients) is of therapeutic value in the treatment of these patients. Moreover, according to Mandell's theory tricyclic compounds should initially cause exacerbation of the depression (receptor stimulation), but there is no evidence that they do.

Ashcroft *et al.* (1972) also encompassed the postsynaptic receptor in their arguments. On the basis of the observation that there are depressions with and without decreased concentrations of MA metabolites in the CSF, they distinguished two types of depression. In the one type (the so-called low-output depression) they believed the supply of stimuli to monoaminergic systems to be abnormally low, with consequently a low transmitter output and a decreased concentration of amine metabolites in the CSF. In the other type (the so-called low-sensitivity depression), the primary factor was believed to be reduced sensitivity of the postsynaptic receptors with, at least in the initial phase, normal transmitter output and normal CSF concentrations of metabolites.

The receptor theories are purely speculative for the time being, because we have no yardstick of human receptor activity. Determination of cyclic AMP in the CSF is not likely to open important perspectives in this context (Sebens and Korf, 1975).

5 Conclusions

Efforts to explain the 5-HT deficiency postulated in depression have so far remained futile. There are no convincing arguments in favour of a precursor deficiency. There are some indications of reduced synthetic capacity, but they are insufficiently strong to carry any theory. Finally, there is the view that the defect is postsynaptically localized rather than presynaptically, and relates to a changed sensitivity of 5-HT receptors. Interesting as they may be, these theories are still entirely speculative.

J GENERAL CONCLUSIONS

1. With regard to the hypothesis that a relation exists between behavioural disturbances in affective disorders on the one hand and disorders of the central 5-HT metabolism on the other, arguments in favour outweigh arguments against, I believe. The principal relevant data were obtained by CSF studies, post-mortem examinations of the brain, and studies of the psychological effects produced by drugs with an influence more or less focused on the central MA metabolism. The material available indicates the probability that depression as well as mania can be associated with a decreased turnover of 5-HT. According to the most widely accepted theory, the decreased 5-HT turnover reflects a decreased transmission in serotoninergic systems, and this phenomenon is directly linked to the occurrence of depressive symptoms. It is yet to be established whether the suspected decrease in MA turnover is based on a deficiency of the mother substance, an enzyme defect or a primary reduction of the firing rate in the monoaminergic system.
2. Disorders of the 5-HT metabolism have not been found in all types of depression, but in particular in depressions which are symptomatologically characterized by the so-called vital ('endogenous') syndrome, take a unipolar or bipolar course, and involve relatively little disturbance of the premorbid personality structure.
3. It is plausible that disorders of the 5-HT metabolism play a role in the pathogenesis of depressions, instead of resulting from them. This can be deduced from the fact that control of these disorders can abolish or alleviate the depressive syndrome or some of its elements.
4. The disorders of the 5-HT metabolism do not always, or not always entirely, disappear after clinical recovery from the depression. A possible interpretation of this fact is that the metabolic disorders represent a predisposing rather than a direct causative factor: they increase the individual's susceptibility to depression, so to speak. This would imply that abolition of a chronic 5-HT deficiency, e.g. with the aid of precursors, should have a

prophylactic effect. Our preliminary findings indicate that this is probably true.

5. There is a marked inclination to correlate biological factors found in behavioural disorders, e.g. in depressions, with nosological entities (e.g. manic-depressive psychosis) or syndromes (e.g. vital depression). The efficacy of this approach is questionable, and it may well be advisable to combine this procedure with one in which the syndrome's components (the disorders of psychological functions) are analysed and then correlated with biological functional disorders. In our studies of the DA metabolism this proved to be an effective procedure. An increase or decrease in DA turnover was found not to be correlatable with a given nosological or syndromal entity (e.g. Parkinson's disease, depression, psychosis) but rather with the patient's motor status, i.e. the degree of motor retardation or agitation. In my view it would certainly be unjustified to describe the metabolic disorders as non-specific on these grounds. They are specific on a symptomatological, but non-specific on a nosological and syndromal level, simply because the symptoms in question happen to be features of several different syndromes. In this context I define the term symptom as a well-defined disturbance of a psychological function.
6. There is a great need for drugs with an optimally selective and circumscribed influence on serotonergic systems. Not only for further verification of the 5-HT hypothesis but also, and to an equal extent, because I suspect that in this way the treatment of depression could be refined and better adjusted to the individual symptom pattern.
7. The data available would seem to suggest that depressions are not a homogeneous group in pathogenetic (biochemical) terms either; that in fact biochemical differences can exist even within a group which tends towards homogeneity in psychopathological terms. Within the group of vital depressions, for example, patients with and without demonstrable disorders of the central 5-HT metabolism can be found. In the study of depressive patients, particularly if aimed at testing methods of medication, data on the MA metabolism could advantageously be included in the analysis of results. Averages of results in biochemically undifferentiated groups can have the effect of a veil. This implies that analysis of MA metabolites in lumbar CSF will have to become a routine part of biologically oriented depression research (the CSF affords data of greater relevance to the functioning of the CNS than the periphery can yield). CSF studies after probenecid loading are far preferable to determination of baseline concentrations in CSF, because the probenecid test greatly enhances the instructive value of CSF studies.
8. There is no reason to assume that the pathogenesis of depressions could be understood exclusively in terms of disturbed 5-HT or even MA metabolism. It is, therefore, by all means advisable to include other transmitter systems

in research. The central cholinergic system is an obvious candidate in this respect, even though it will be far from simple to obtain information on this system in living human individuals.

9. Quite apart from the fact that it probably contains a kernel of truth, the heuristic value of the 5-HT hypothesis for psychiatry cannot be overestimated. It has catalysed an impressive amount of research, not only in the biochemical field but also in pharmacotherapy, in psychopathology, and in experimental behavioural research. The 5-HT hypothesis demonstrates that multidimensional but target-oriented experimental research is quite feasible in psychiatry as long as a plausible working hypothesis is available. It is in these terms that the 5-HT hypothesis possibly has its greatest importance.

REFERENCES

Altman, K., and Greengard, O. (1966). *J. Clin. Invest.*, **45**, 1527–1534.

Andén, N. E., Corrodi, H., and Fuxe, K. (1972a). *J. Pharm. Pharmacol.*, **24**, 177–182.

Andén, N. E., Engel, J., and Rubensson, A. (1972b). *Naunyn-Schmiedeberg's Arch. Pharmacol.*, **273**, 11–26.

Andén, N. E., Corrodi, H., Fuxe, K., and Hökfelt, T. (1968). *Brit. J. Pharmacol.*, **34**, 1–7.

Andén, N. E., Rubensson, A., Fuxe, K., and Hökfelt, T. (1967). *J. Pharm. Pharmacol.*, **19**, 627–629.

Andersson, H., and Roos, B-E. (1968). *Acta Pharmacol. (Kbh)*, **26**, 531–538.

Andersson, H., and Roos, B-E. (1969). *Acta Paediat. (Uppsala)*, **58**, 601–608.

Angel, Ch., Murphree, O. D., DeLuca, D. C., and Newton, J. E. O. (1977). *Biol. Psychiat.*, **12**, 573–576.

Angst, J. (1966). *Psychosen, eine genetische soziologische und klinische Studie*, Springer Verlag, Berlin.

Angst, J., Baastrup, P., Grof, P., Hippius, H., Pöldinger, W., and Weis, P. (1973). *Psychiat. Neurol. Neurochir. (Amsterdam)*, **76**, 489–500.

Åsberg, M., Bertilsson, L., Tuck, D., Cronholm, B., and Sjöqvist, F. (1972). *Clin. Pharm. Ther.*, **14**, 277–286.

Ashcroft, G. W., and Sharman, D. F. (1960). *Nature*, **186**, 1050–1051.

Ashcroft, G. W., Dow, R. C., and Moir, A. T. B. (1968). *J. Physiol. (London)*, **199**, 397–424.

Ashcroft, G. W., Crawford, T. B. B., Dow, R. C., and Moir, A. T. B. (1969). In *Metabolism of Amines in the Brain* (Ed., H. Cooper), Macmillan, London, pp. 65–69.

Ashcroft, G. W., Eccleston, D., Knight, F., McDougall, E. J., and Waddell, J. L. (1965). *J. Psychosom. Res.*, **9**, 129–136.

Ashcroft, G. W., Crawford, T. B. B., Eccleston, D., Sharman, D. F., McDougall, E. J., Stanton, J. B., and Binns, J. K. (1966). *Lancet ii*, 1049–1052.

Ashcroft, G. W., Blackburn, I. M., Eccleston, D., Glen, A. I. M., Hartley, W., Kinloch, N. E., Lonergan, M., Murray, L. G., and Pullar, I. A. (1973b). *Psychol. Med.*, **3**, 319–325.

Ashcroft, G. W., Crawford, T. B. B., Cundall, R. L., Davidson, D. L., Dobson, J., Dow, R. C., Eccleston, D., Loose, R. W., and Pullar, I. A. (1973a). *Psychol. Med.*, **3**, 326–332.

Ashcroft, G. W., Eccleston, D., Murray, L. G., Glen, A. I. M., Crawford, T. B. B., Pullar, I. A., Shields, P. J., Walter, D. S., Blackburn, I. M., Connechan, J., and Lonergan, M. (1972). *Lancet ii*, 573–577.

Baldessarini, R. J., and Yorke, C. (1970). *Nature*, **228**, 1301–1303.

Barkai, A., Glusman, M. and Rapport, M. M. (1972). *J. Pharmacol. Exptl Ther.*, **181**, 28–35.

Bartholini, G., Pletscher, A., and Tissot, R. (1966). *Experientia (Basel)*, **22**, 609–610.

Becker, J. (1974). *Depression: Theory and Research*, Winston, Washington.
Beckmann, H., St-Laurent, J., and Goodwin, F. K. (1975). *Psychopharmacologia (Berlin)*, **42**, 277–282.
Benkert, O., Renz, A., Marano, C., and Matussek, N. (1971). *Arch. Gen. Psychiat.*, **25**, 359–363.
Benkert, O., Martschke, D., and Gordon, A. (1974). *Lancet ii*, 1146.
Bernheimer, H., and Hornykiewicz, O. (1964). *Arch. Exptl Path. Pharmacol.*, **247**, 305–306.
Bernheimer, H., Birkmayer, W., and Hornyliewicz, O. (1966). *Wiener Klin. Wochenschr.*, **78**, 417–419.
Bertaccini, G. (1959). *J. Neurochem.*, **4**, 217–222.
Bertilsson, L., and Åsberg, M. (1975). Paper read at the Sixth International Congress of Pharmacology, Helsinki.
Bindler, E. H., Wallach, M. B., and Gershon, S. (1971). *Arch. Int. Pharmacodyn. Ther.*, **190**, 150–154.
Birkmayer, W., and Linauer, W. (1970). *Arch. Psychiat. Nervenkrankh*, **213**, 377–387.
Bliss, E. L., Thatcher, W., and Ailion, J. (1972). *J. Psychiat. Res.*, **9**, 71–80.
Bockar, J., Roth, R., and Heninger, G. (1974). *Life Sci.*, **15**, 2109–2118.
Bourne, H. R., Bunney, W. E., Jr., Colburn, R. W., Davis, J. M., Davis, J. N., Shaw, D. M., and Coppen, A. J. (1968). *Lancet ii*, 805–808.
Bowers, M. B., Jr. (1969). *Brain Res.*, **15**, 522–524.
Bowers, M. B., Jr. (1972a). *Psychopharmacologia (Berlin)*, **23**, 26–33.
Bowers, M. B., Jr. (1972b). *Arch. Gen. Psychiat.*, **27**, 440–442.
Bowers, M. B., Jr. (1972c). *Neuropharmacol.*, **11**, 101–111.
Bowers, M. B., Jr. (1973). *Psychopharmacologie (Berlin)*, **28**, 309.
Bowers, M. B., Jr., and Gerbode, F. (1968a). *Nature*, **219**, 1256–1257.
Bowers, M. B., Jr., and Gerbode, F. (1968b). *Life Sci.*, **7**, 773–776.
Bowers, M. B., Jr., Heninger, G. R., and Gerbode, F. (1969). *Int. J. Neuropharmacol.*, **8**, 255–262.
Boyland, E. (1958). *Brit. Med. Bull.*, **14**, 153–158.
British Medical Journal i (1966). 119–120.
Brodie, H. K. H., Sack, R., and Siever, L. (1973). In *Serotonin and Behavior* (Eds., J. Barchas and E. Usdin), Academic Press, New York, pp. 549–559.
Brown, G. W., Sklair, F., Harris, T. O., and Birley, J. L. T. (1973). *Psychol. Med.*, **3**, 74–87.
Bruinvels, J. (1972). *J. Pharmacol.*, **20**, 231–237.
Bulat, M., and Zivković, B. (1971). *Science*, **173**, 738–740.
Bulat, M., and Zivković, B. (1973). *J. Pharm. Pharmacol.*, **25**, 178–179.
Bullpitt, C. J., and Dollery, C. T. (1973). *Brit. Med. J.*, **3**, 485–490.
Bunney, W. E., Jr., and Davis, J. M. (1965). *Arch. Gen. Psychiat.*, **13**, 483–494.
Bunney, W. E., Jr., Brodie, H. K. H., Murphy, D. L., and Goodwin, F. K. (1971). *Amer. J. Psychiat.*, **127**, 872–881.
Bunney, B. S., Walters, J. R., Roth, R. H., and Aghajanian, G. K. (1973). *J. Pharmacol. Exptl. Ther.*, **185**, 560–571.
Burg, W. van den, Praag, H. M. van, Bos, E. R. H., Zanten, A. K. van, Piers, D. A., and Doorenbos, H. (1975). *Psychol. Med.*, **5**, 404–412.
Burg, W. van den, Praag, H. M. van, Bos, E. R. H., Zanten, A. K. van, Piers, D. A., and Doorenbos, H. (1976). *Psychol. Med.*, **6**, 393–397.
Carlsson, A., Falck, B., Fuxe, K., and Hillarp, N. A. (1964). *Acta Physiol. Scand.*, **60**, 112–119.
Carlsson, A., Corrodi, H., Fuxe, K., and Hökfelt, T. (1969a). *Eur. J. Pharmacol.*, **5**, 357–366.
Carlsson, A., Corrodi, H., Fuxe, K., and Hökfelt, T. (1969b). *Eur. J. Pharmacol.*, **5**, 367–373.

Carman, J. S., Post, R. M., Teplitz, T. A., and Goodwin, F. K. (1974). *Lancet ii*, 1454.
Carney, M. W. P., and Sheffield, B. F. (1974). *Brit. J. Psychiat.*, **125**, 91–94.
Carney, M. W. P., Thakurdas, H., and Sebastian, J. (1969). *Psychopharmacologia*, **14**, 349–350.
Carroll, B. J. (1971). *Clin. Pharmacol. Exptl Ther.*, **12**, 743–761.
Carroll, B. J., and Dodge, J. (1971). *Lancet i*, 915.
Carroll, B. J., Mowbray, R. M., and Davies, B. M. (1970). *Lancet i*, 967–969.
Cazzullo, C. L., Mangoni, A., and Mascherpa, G. (1966). *Brit. J. Psychiat.*, **112**, 157–162.
Chase, T. N. (1972). *Arch. Neurol. (Chicago)*, **27**, 354–356.
Chase, T. N., Gordon, E. K., and Ng, L. K. Y. (1973). *J. Neurochem.*, **21**, 581–587.
Chodoff, P. (1972). *Arch. Gen. Psychiat.*, **27**, 666–673.
Claveria, L. E., Curzon, G., Harrison, M. J. G., and Kantamaneni, B. D. (1974). *J. Neurol. Neurochir. Psychiat.*, **37**, 715–718.
Constantinides, J., Geissbühler, F., Gaillard, J. M., Hovaguimian, Th., and Tissot, R. (1974). *Experientia*, **30**, 1182.
Coppen, A. (1967). *Brit. J. Psychiat.*, **113**, 1237–1264.
Coppen, A., Brooksbank, B. W. G., and Eccleston, E. (1974a). *Psychol. Med.*, **4**, 164–173.
Coppen, A., Brooksbank, B. W. G., and Peet, M. (1972b). *Lancet i*, 1393.
Coppen, A., Eccleston, E. G., and Peet, M. (1973). *Lancet ii*, 60–63.
Coppen, A., Prange, A. J., Whybrow, P. C., and Noguera, R. (1972a). *Arch. Gen. Psychiat.*, **26**, 474–478.
Coppen, A., Prange, A. J., Whybrow, P. C., Noguera, R., and Paez, R. M. (1967). *Lancet ii*, 338–340.
Coppen, A., Shaw, D. M., and Farrell, J. P. (1963). *Lancet i*, 79–81.
Coppen, A., Shaw, D. M., and Malleson, A. (1965a). *Brit. J. Psychiat.*, **11**, 105–107.
Coppen, A., Shaw, D. M., Herzberg, B., and Maggs, R. (1967). *Lancet ii*, 1178–1180.
Coppen, A., Shaw, D. M., Malleson, A., Eccleston, E., and Gundy, G. (1965b). *Brit. J. Psychiat.*, **111**, 996–998.
Coppen, A., Montgomery, S., Peet, M., Bailey, J., Marks, V., and Woods, P. (1974b). *Lancet ii*, 433–435.
Corrodi, H., and Fuxe, K. (1969). *Eur. J. Pharmacol.*, **7**, 56–59.
Corrodi, H., Fuxe, K., and Schou, M. (1969). *Life Sci.*, **8**, 643–651.
Corrodi, H., Fuxe, K., Hökfelt, T., and Schou, M. (1967). *Psychopharmacologia*, **11**, 345–353.
Cremata, V. Y., and Koe, B. K. (1966). *Clin. Pharmacol. Exptl Ther.*, **7**, 768–776.
Creveling, C. R., Daly, J., Toluyama, T., and Witkop, B. (1968). *Biochem. Pharmacol.*, **17**, 65–70.
Crout, J. R., and Sjoerdsma, A. (1959). *New Engl. J. Med.*, **261**, 23–26.
Curzon, G., and Green, A. R. (1968). *Life Sci.*, **7**, 657–663.
Curzon, G. (1969). *Brit. J. Psychiat.*, **115**, 1367–1374.
Curzon, G., Friedel, J., and Knott, P. J. (1973). *Nature*, **242**, 198–200.
Curzon, G., Gumpert, E. J. W., and Sharpe, D. M. (1971). *Nature New Biol.*, **231**, 189–191.
Davson, H. (1967). *Physiology of the Cerebrospinal Fluid*, Churchill, London.
Denker, S. J., Malm, U., Roos, B.-E., and Werdenius, B. (1966). *J. Neurochem.*, **13**, 1545–1548.
Despopoulos, A., and Weissbach, H. (1957). *Amer. J. Physiol.*, **189**, 548–550.
Dewhurst, W. G. (1968). *Nature*, **219**, 506–507.
Doepfner, W. (1962). *Experientia (Basel)*, **18**, 256–257.
Doepfner, W., and Cerletti, A. M. S. (1958). *Int. Arch. Allergy*, **12**, 89–97.
Douš, T., and Hechter, O. (1970). *Lancet*, *i*, 834–835.
Dunner, D. L., and Goodwin, F. K. (1972). *Arch. Gen. Psychiat.*, **26**, 364–366.
Dunner, D. L., and Fieve, R. R. (1975). *Amer. J. Psychiat.*, **132**, 180–183.

Duvoisin, R. C., Yahr, M. D., and Cote, L. D. (1969). *Trans. Amer. Neurol. Assoc.*, **94**, 81–84.

Ebert, M. H., Baldessarini, R. J., Lipinski, J. F., and Berv, K. (1973). *Arch. Gen. Psychiat.*, **29**, 397–401.

Eccleston, D., Ashcroft, G. W., Moir, A. T. B., Parker-Rhodes, A., Lutz, W., and O'Mahoney, D. P. (1968). *J. Neurochem.*, **15**, 947–957.

Eccleston, D., Loose, R., Pullar, J. A., and Sugden, R. F. (1970a). *Lancet ii*, 612–613.

Eccleston, D., Ashcroft, G. W., Crawford, T. B. B., Stanton, J. B., Wood, D., and McTurk, P. H. (1970b). *J. Neurol. Neurosurg. Psychiat.*, **33**, 269–272.

Ehringer, H., and Hornykiewicz, O. (1960). *Klin. Wochenschr.*, **38**, 1236–1239.

Eleftherion, E. B., and Boehlke, K. W. (1967). *Science*, **155**, 1693–1694.

Engel, J., Hanson, L. C. F., and Roos, B.-E. (1971). *Psychopharmacologia*, **20**, 197–200.

Engelman, K., Lovenberg, W., and Sjoerdsma, A. (1967). *New Engl. J. Med.*, **277**, 1103–1108.

Erspamer, V. (1966). In *5-Hydroxytryptamine and Related Indolalkylamines*, Springer Verlag, Berlin, Heidelberg, New York, pp. 132–181.

Extein, I., Korf, J., Roth, R. H., and Bowers, M. B. Jr. (1973). *Brain Res.*, **54**, 403–407.

Fann, W. E., Davis, J. M., Janowski, D. S., Cavanaugh, J. H., Kaufman, J. S., Griffith, J. D., and Oates, J. A. (1972). *Clin. Pharmacol. Ther.*, **13**, 71–77.

Fernando, J. C., Joseph, M. H., and Curzon, G. (1975). *Lancet i*, 171.

Fernström, J. D., and Wurtman, R. J. (1971). *Science*, **173**, 149–152.

Fernström, J. D., and Wurtman, R. J. (1972). *Science*, **178**, 414–416.

Fieve, R. R., Platman, S. R., and Fliess, J. L. (1969). *Psychopharmacologia*, **15**, 425–429.

Forn, J., and Valdecasas, F. G. (1971). *Biochem. Pharmacol.*, **20**, 2773–2779.

Friedman, E., and Gershon, S. (1973). *Nature*, **243**, 520–521.

Frey, H. H., and Magnussen, M. P. (1968). *Biochem. Pharmacol.*, **17**, 1299–1308.

Fuller, R. W., and Molloy, B. B. (1974). *Adv. Biochem. Psychopharmacol.*, **10**, 195–205.

Fuller, R. W., Hines, C. W., and Mills, J. (1965). *Biochem. Pharmacol.*, **14**, 483–488.

Fuller, R. W., Perry, K. W., and Molloy, B. B. (1974). *Life Sci.*, **15**, 1161–1171.

Fuxe, K., Butcher, L. L., and Engel, J. (1971). *J. Pharm. Pharmacol.*, **23**, 420–424.

Garattini, S., Kato, R., Lamesta, L., and Valzelli, L. (1960). *Experientia*, **16**, 156–157.

Garelis, E., and Sourkes, T. L. (1973). *J. Neurol. Neurosurg. Psychiat.*, **4**, 625–629.

Garelis, E., Young, S. N., Lal, S., and Sourkes, T. L. (1974). *Brain. Res.*, **79**, 1–8.

Garside, R. F., Kay, D. W. K., Wilson, I. C., Deaton, I. D., and Roth, M. (1971). *Psychol. Med.*, **1**, 333–338.

Genefke, I. K. (1972). *Acta Psychiat. Scand.*, **48**, 400–404.

Glassman, A. (1969). *Psychosom. Med.*, **31**, 107–114.

Glassman, A., and Platman, S. R. (1969). *J. Psychiat. Res.*, **7**, 83–88.

Godwin-Austen, R. B., Kantamaneni, B. D., and Curzon, G. (1971). *J. Neurol. Neurosurg. Psychiat.*, **34**, 219–223.

Goodwin, F. K. (1977). Discussion remarks. In *Neuroregulators and Hypotheses of Psychiatric Disorders*. (Eds., J. Barchas, D. A. Hamburg, and E. Usdin), Oxford University Press, London.

Goodwin, F. K., and Post, R. M. (1973). In *Serotonin and Behavior* (Eds., J. Barchas and E. Usdin), Academic Press, New York, pp. 469–480.

Goodwin, F. K., Brodie, H. K. H., Murphy, D. L., and Bunney, W. E., Jr. (1970). *Biol. Psychiat.*, **2**, 341–366.

Goodwin, F. K., Murphy, D. L., Dunner, D. L., and Bunney, W. E., Jr. (1972). *Amer. J. Psychiat.*, **129**, 44–47.

Goodwin, F. K., Post, R. M., Dunner, D. L., and Gordon, E. K. (1973). *Amer. J. Psychiat.*, **130**, 73–79.

Gordon, E. K., Olivier, J., Goodwin, F. K., Chase, T. N., and Post, R. M. (1973). *Neuropharmacology*, **12**, 391–396.
Gottfries, C. G., and Roos, B.-E. (1969). *J. Neurochem.*, **16**, 1341–1345.
Gottfries, C. G., and Roos, B.-E. (1970). *Acta Psychiat. Scand.*, **46**, 99–105.
Gottfries, C. G., and Roos, B.-E. (1973). *Acta Psychiat. Scand.*, **49**, 257–263.
Gottfries, C. G., Oreland, L., and Wiberg, Å. (1974). *Lancet ii*, 360.
Gottfries, C. G., Oreland, L., Wiberg, Å., and Winblad, B. (1975). *J. Neurochem.*, **25**, 667–673.
Grabowska, M., Antkiewicz, L., and Michaluk, J. (1974). *Biochem. Pharmacol.*, **23**, 3211–3212.
Grahame-Smith, D. G. (1971). *J. Neurochem.*, **18**, 1053–1066.
Grahame-Smith, D. G., and Green, A. R. (1974). *Brit. J. Pharmacol.*, **52**, 19–26.
Green, A. R., and Curzon, G. (1968). *Nature*, **220**, 1095–1097.
Green, A. R., and Grahame-Smith, D. G. (1974). *Nature*, **251**, 524–526.
Green, A. R., and Youdim, M. B. H. (1975). In *Ciba Foundation Symposium, 22*. Monoamine Oxidase and its Inhibition. (Ed., J. Knight), North Holland, Amsterdam, pp. 231–246.
Green, H., Greenberg, S. M., and Erickson, R. W. (1962). *J. Pharmacol. Exptl Ther.*, **136**, 174–178.
Grof, P., and Foley, P. (1971). *Amer. J. Psychiat.*, **127**, 1573–1574.
Guldberg, H. C., Ashcroft, G. W., and Crawford, T. B. B. (1966). *Life Sci.*, **5**, 1571–1575.
Hamilton, M. (1968). *Postgrad. Med. J.*, **44**, 66–69.
Harper, A. E., Benevenga, N. J., and Wolhueter, R. M. (1970). *Physiol. Rev.*, **50**, 428–558.
Haskovec, L., and Rysanek, K. (1967). *Psychopharmacologia*, **11**, 18–30.
Haskovec, L., and Soucek, K. (1968). *Nature*, **219**, 507–508.
Haskovec, L. (1969). *Lancet ii*, 902.
Hendley, E. D., and Welch, B. L. (1975). *Life Sci.*, **16**, 45–54.
Herrington, R. N., Bruce, A., Johnstone, E. C., and Lader, M. H. (1974). *Lancet ii*, 731–734.
Hertz, D., and Sulman, F. G. (1968). *Lancet i*, 531–532.
Hinsley, R. K., Norton, J. A., and Aprison, M. H. (1968). *J. Psychiat. Res.*, **6**, 143–152.
Ho, H. K. S., Loh, H. H., Craves, F., Hitzeman, R. J., and Gershon, S. (1970). *Eur. J. Pharmacol.*, **10**, 72–78.
Hollister, L. E., Berger, P., Ogie, F. L., Arnold, R. C., and Johnson, A. (1974). *Arch. Gen. Psychiat.*, **31**, 468–470.
Horst, W. D., and Spirt, N. (1974). *Life Sci.*, 1073–1082.
Hullin, R. P., Bailey, A. D., McDonald, R., Dransfield, G. A., and Milne, H. B. (1967). *Brit. J. Psychiat.*, **113**, 593–600.
Jensen, K., Fruensgaard, K., Ahlfors, U. G., Pihkanen, T. A., Tuomikoski, S., Ose, E., Dencker, S. J., Lindberg, D., and Nagy, A. (1975). *Lancet ii*, 920.
Jéquier, E., Lovenberg, W., and Sjoerdsma, A. (1967). *Mol. Pharmacol.*, **3**, 274–278.
Johnson, P., Kitchin, A. H., Lowther, C. P., and Turner, R. W. D. (1966). *Brit. Med. J. i*, 133–137.
Kane, F. J. (1970). *Amer. J. Psychiat.*, **126**, 1020–1023.
Kastin, A. J., Ehrensing, R. H., Schalch, D. S., and Anderson, M. S. (1972). *Lancet ii*, 740–742.
Katz, E. I., Chase, T. N., and Kopin, I. J. (1968). *Science*, **162**, 466–467.
Keller, H. H., Bartholini, G., and Pletscher, A. (1974). *J. Pharm. Pharmacol.*, **26**, 649–651.
Kety, S. S., Javoy, F., Thierry, A. M., Julou, L., and Glowinski, J. (1967). *Proc. Nat. Acad. Sci., USA*, **58**, 1249–1254.
Kety, S. S. (1971). In *Brain Biochemistry and Mental Disease* (Eds., B. T. Ho and W. M. McIsaac), Plenum Press, New York, pp. 237–263.

Klaiber, E. L., Kobayashi, Y., Broverman, D. M., and Hall, F. (1971). *J. Clin. Endocrinol.*, **33**, 630–638.
Klaiber, E. L., Broverman, D. M., Vogel, W., Kobayashi, Y., and Moriarty, D. (1972). *Amer. J. Psychiat.*, **128**, 42–48.
Kline, N. S., Sacks, W., and Simpson, G. M. (1964). *Amer. J. Psychiat.*, **121**, 379–381.
Knox, W. E. (1951). *Brit. J. Exptl Pathol.*, **32**, 462–469.
Knox, W. E., and Auerbach, V. H. (1955). *J. Biol. Chem.*, **214**, 307–313.
Koe, B. K., and Weissman, A. (1966). *Fed. Proc. Fed. Amer. Socs Exptl Biol.*, **25**, 452.
Korf, J., and Praag, H. M. van. (1970). *Psychopharmacologia*, **18**, 129–132.
Korf, J., and Praag, H. M. van (1971b). *Brain Res.*, **35**, 221–230.
Korf, J., and Praag, H. M. van (1972a). *Neuropharmacologia*, **11**, 141–144.
Korf, J., Praag, H. M. van, and Sebens, J. B. (1971a). *Biochem. Pharmacol.*, **20**, 659–668.
Korf, J., Praag, H. M. van, and Sebens, J. B. (1972b). *Brain Res.*, **42**, 239–242.
Korf, J., Schutte, H. H., and Venema, K. (1973). *Anal. Biochem.*, **53**, 146–153.
Korf, J., Venema, K., and Postema, F. (1974b). *J. Neurochem.*, **23**, 249–252.
Korf, J., Praag, H. M. van, Schut, T., Nienhuis, R. J., and Lakke, J. P. W. F. (1974a). *Eur. Neurol.*, **12**, 340–350.
Kupfer, D. J., and Bowers, M. B., Jr. (1972). *Psychopharmacologia*, **27**, 183–190.
Kuriyama, K., and Speken, R. (1970). *Life Sci.*, **9**, 1213–1220.
Lackroy, G. H., and Praag, H. M. van (1971). *Acta Psychiat. Scand.*, **47**, 163–173.
Ladisisch, W., Steinhauff, N., and Matussek, N. (1969). *Psychopharmacologia*, **15**, 296–304.
Lakke, J. P. W. F., Korf, J., Praag, H. M. van, and Schut, T. (1972). *Nature New Biol.*, **236**, 208–209.
Lamprecht, F., Ebert, M. H., Turek, I., and Kopin, I. J. (1974). *Psychopharmacologia*, **40**, 241–248.
Lapin, I. P., and Oxenkrug, G. F. (1969). *Lancet i*, 132–136.
Lewander, T., and Sjöström, R. (1973). *Psychopharmacologia*, **33**, 81–86.
Lipsett, D., Madras, B. K., Wurtman, R. J., and Munro, H. N. (1973). *Life Sci.*, **12**, 57–64.
Lloyd, K. J., Farley, I. J., Deck, J. H. N., and Hornykiewicz, O. (1974). *Adv. Biochem. Psychopharm.*, **11**, 387–397.
Lopez-Ibor, A. J. J., Gutierrez, J. L. A., and Iglesias, M. L. M. (1973). *Int. Pharmacopsychiat.*, **8**, 145–151.
Lovenberg, W., Jéquier, E., and Sjoerdsma, A. (1968). *Adv. Pharmacol.*, **6A**, 21–36.
McCabe, M. S., Reich, T., and Winokur, G. (1970). *Amer. J. Psychiat.*, **127**, 354–356.
McNamee, H. M., Le Poidevin, D., and Naylor, G. J. (1972b). *Psychol. Med.*, **2**, 66–69.
McNamee, H. M., Moody, J. P., and Naylor, G. J. (1972a). *J. Psychosom. Res.*, **16**, 63–70.
Mangoni, A. (1974). *Adv. Biochem. Psychopharm.*, **11**, 293–298.
Meek, J. L., and Neff, N. H. (1973). *Neuropharmacology*, **12**, 497–499.
Mendels, J., and Frazer, A. J. (1973). *J. Psychiat. Res.*, **10**, 9–18.
Mendels, J., Stinnet, J. L., Burns, D., and Frazer, A. J. (1975). *Arch. Gen. Psychiat.*, **32**, 22–30.
Mendels, J., Frazer, A. J., Fitzgerald, R. G., Ramsey, T. A., and Stokes, J. W. (1972). *Science*, **175**, 1380–1382.
Messiha, F. S., Agallianos, D., and Clower, C. (1970). *Nature*, **225**, 868–869.
Modigh, K. (1972). *Psychopharmacologia*, **23**, 48–54.
Modigh, K. (1973). *Psychopharmacologia*, **30**, 123–134.
Moir, A. T. B., Ashcroft, T. B. B., Eccleston, D., and Guildberg, H. C. (1970). *Brain*, **93**, 357–368.
Morgan, W. W., Saldana, J. J., Yndo, C. A., and Morgan, J. F. (1975). *Brain Res.*, **84**, 75–86.
Mountjoy, C. Q., Price, J. S., Weller, M., Hunter, P., Hall, R., and Dewar, J. H. (1974). *Lancet i*, 958–960.

Murphy, D. L. (1972). *Amer. J. Psychiat.*, **129**, 141–148.
Murphy, D. L., and Weiss, R. (1972). *Amer. J. Psychiat.*, **128**, 35–41.
Murphy, D. L., Colburn, R. W., Davis, J. M., and Bunney, W. E., Jr. (1969). *Life Sci.*, **8**, 1187–1193.
Murphy, D. L., Baker, M., Goodwin, F. K., Miller, H., Kotin, J., and Bunney, W. E., Jr. (1974). *Psychopharmacologia*, **34**, 11–20.
Mussacchio, J. M., Julou, L., Kety, S. S., and Glowinski, J. (1969). *Proc. Nat. Acad. Sci., USA*, **63**, 1117–1119.
Neff, N. H., Tozer, T. N., and Brodie, B. B. (1967). *J. Pharmacol. Exptl Ther.*, **158**, 214–218.
Ng, K. Y., Chase, T. N., Colburn, R. W., and Kopin, I. J. (1970). *Science*, **170**, 76–77.
Nordin, G., Ottosson, J. O., and Roos, B.-E. (1971). *Psychopharmacologia*, **20**, 315–320.
Olsson, R., and Roos, B.-E. (1968). *Nature*, **219**, 502–503.
Papeschi, R., and McClure, D. J. (1971). *Arch. Gen. Psychiat.*, **25**, 354–358.
Papeschi, R., Randrup, A., and Munkvad, I. (1974). *Psychopharmacologia*, **35**, 159–168.
Papeschi, R., Sourkes, T. L., Poirier, J. L., and Boucher, R. (1971). *Brain Res.*, **28**, 527–533.
Pare, C. M. B. (1963). *Lancet ii*, 527–528.
Pare, C. M. B., and Sandler, M. (1959). *J. Neurol. Neurosurg. Psychiat.*, **22**, 247–251.
Pare, C. M. B., Yeung, D. P. H., Price, K., and Stacey, R. S. (1969). *Lancet ii*, 133–135.
Paykel, E. S. (1972). *Arch. Gen. Psychiat.*, **27**, 203–210.
Perez-Cruet, J., Tagliamonte, P., Tagliamonte, A., and Gessa, G. L. (1971). *J. Pharmacol. Exptl Ther.*, **178**, 325–330.
Perris, C. (1966). *Acta Psychiat. Scand.*, **42**, Suppl. 194, 1–189.
Persson, T., and Roos, B.-E. (1968). *Lancet ii*, 987–988.
Pflanz, G., and Palm, D. (1973). *Eur. J. Clin. Pharmacol.*, **5**, 555–558.
Pletscher, A., Bartholini, G., and Tissot, R. (1967). *Brain Res.*, **4**, 106–109.
Pletscher, A., Gey, K. F., and Burkard, W. P. (1966). *Handb. Exptl Pharmacol.*, **19**, 593–735.
Pletscher, A., Gey, K. F., and Zeller, P. (1960). *Progr. Drug Res.*, **2**, 417–590.
Pletscher, A., Bartholini, G., Bruderer, H., Buckard, W. P., and Gey, K. F. (1964). *J. Pharmacol. Exptl Ther.*, **145**, 344–350.
Plotnikoff, N. P., Prange, A. J., Jr., Breese, G. R., Anderson, M. S., and Wilson, I. C. (1972). *Science*, **178**, 417–418.
Plotnikoff, N. P., Prange, A. J., Jr., Breese, G. R., and Wilson, I. C. (1974). *Life Sci.*, **14**, 1271–1278.
Poitou, P., Guerinot, F., and Bohuon, C. (1974). *Psychopharmacologia*, **38**, 75–80.
Pöldinger, W. (1963). *Psychopharmacologia*, **4**, 308.
Post, R. M., Goodwin, F. K., Gordon, E. K., and Watkin, D. M. (1973a). *Science*, **179**, 897–899.
Post, R. M., Kotin, J., Goodwin, F. K., and Gordon, E. K. (1973b). *Amer. J. Psychiat.*, **130**, 67–72.
Post, R. M., Gordon, E. K., Goodwin, F. K., and Bunney, W. E., Jr. (1973c). *Science*, **179**, 1002–1003.
Post, R. M., and Goodwin, F. K. (1974). *Arch. Gen. Psychiat.*, **30**, 234–239.
Praag, H. M. van (1962). Thesis, Utrecht.
Praag, H. M. van (1967). *Psychiat. Neurol. Neurochir.*, **70**, 219–233.
Praag, H. M. van (1969). *Ned. Tijdschr. Geneesk.*, **113**, 2245–2247.
Praag, H. M. van (1972). *Compr. Psychiat.*, **13**, 401–410.
Praag, H. M. van. (1974). *Compr. Psychiat.*, **15**, 389–401.
Praag, H. M. van (1975). *Compr. Psychiat.*, **16**, 7–22.
Praag, H. M. van. (1977). *Biol. Psychiat.*, **12**, 101–131.

Praag, H. M. van (1978). Central serotonin: its relation to depression vulnerability and depression prophylaxis. Paper read at the Second World Congress of Biological Psychiatry, Barcelona. In press.

Praag, H. M. van, and Korf, J. (1975). *Pharmakopsychiat*, **8**, 321–326.

Praag, H. M. van, and Korf, J. (1970). *Lancet ii*, **612**.

Praag, H. M. van, and Korf, J. (1971a). *Psychopharmacologia*, **19**, 148–152.

Praag, H. M. van, and Korf, J. (1971b). *Psychopharmacologia*, **19**, 199–203.

Praag, H. M. van, and Korf, J. (1973). *J. Clin. Pharmacol.*, **13**, 3–14.

Praag, H. M. van, and Korf, J. (1974). *J. Nerv. Ment. Dis.*, **158**, 331–337.

Praag, H. M. van, and Korf, J. (1977). In *Psychotherapeutic Drugs* (Eds., E. Usdin and I. S. Forrest), Marcel Dekker, New York, pp. 1217–1251.

Praag, H. M. van, and Leijnse, B. (1963a). *Psychopharmacologia*, **4**, 1–14.

Praag, H. M. van, and Leijnse, B. (1963b). *Nervenarzt*, **34**, 530–537.

Praag, H. M. van, and Leijnse, B. (1965). *Psychiat. Neurol. Neurochir.*, **68**, 50–66.

Praag, H. M. van, Korf, J., and Puite, J. (1970). *Nature*, **225**, 1259–1260.

Praag, H. M. van, Korf, J., and Schut, T. (1973a). *Arch. gen. Psychiat.*, **28**, 827–831.

Praag, H. M. van, Uleman, A. M., and Spitz, J. C. (1965). *Psychiat. Neurol. Neurochir.*, **68**, 329–346.

Praag, H. M. van, Burg, W. van den, Bos, E. R. H., and Dols, L. C. W. (1974). *Psychopharmacologia*, **38**, 267–269.

Praag, H. M. van, Korf, J., Dols, L. C. W., and Schut, T. (1972). *Psychopharmacologia*, **25**, 14–21.

Praag, H. M. van, Korf, J., Lakke, J. P. W. F., and Schut, T. (1975). *Psychol. Med.*, **5**, 138–146.

Praag, H. M. van, Flentge, F., Korf, J., Dols, L. C. W., and Schut, T. (1973b). *Psychopharmacologia*, **33**, 141–151.

Praag, H. M. van, and Haan, S. de (1979). *Acta Psychiat. Scand.* In press.

Prange, A. J., Jr., Wilson, I. C., Lara, P. P., Alltopp, L. B., and Breese, G. R. (1972). *Lancet ii*, 999–1002.

Prange, A. J., Jr., Wilson, I. C., Lynn, C. W., Alltopp, L. B., Stikeleather, R. A., and Raleigh, N. C. (1974). *Arch. Gen. Psychiat.*, **30**, 52–62.

Prockop, L., Fahn, S., and Barbour, P. (1974). *Brain Res.*, **80**, 435–442.

Prichard, B. N. C., Johnston, A. W., Hill, I. D., and Rosenheim, M. L. (1968). *Brit. Med. J. i*, 135–144.

Puite, J. K., Schut, T., Praag, H. M. van, and Lakke, J. P. W. F. (1973). *Psychiat. Neurol. Neurochir.*, **76**, 61–70.

Redmond, E. E., Jr., Maas, J. W., Kling, A., Graham, C. W., and Dekirmenjian, H. (1971). *Science*, **174**, 428–430.

Rees, J. R., Alltopp, M. N. E., and Hullin, R. P. (1974). *Psychol. Med.*, **4**, 334–337.

Reigle, T. G., Avni, J., Platz, P. A., Schildkraut, J. J., and Plotnikoff, N. P. (1974). *Psychopharmacologia*, **37**, 1–6.

Rimón, R., and Räkköläinen, V. (1968). *Brit. J. Psychiat.*, **114**, 109–110.

Rimón, R., Roos, B.-E., Räkköläinen, V., and Alanen, Y. (1971). *J. Psychosom. Res.*, **15**, 375–378.

Robins, E., Munoz, R. A., Martin, S., and Gentry, K. A. (1972). In *Disorders of Mood* (Eds., J. Zubin, and F. A. Freyhan), Johns Hopkins University Press, Baltimore, pp. 33–45.

Robinson, D. S., Davis, J. M., Nies, A., Ravaris, C. G., and Sylwester, D. (1971). *Arch. Gen. Psychiat.*, **24**, 536–539.

Robinson, D. S., Davis, J. M., Nies, A., Colburn, R. W., Davis, J. N., Bourne, H. R., Bunney, W. E., Jr., Shaw, D. M., and Coppen, A. J. (1972). *Lancet i*, 290–291.

Roos, B.-E., and Sjöström, R. (1969). *J. Clin. Pharmacol.*, **1**, 153–155.

Rossum, J. M. van (1967). In *Neuropsychopharmacology* (Ed., H. Brill), Excerpta Medica Foundation, Den Haag, pp. 321–329.
Rubin, R. T. (1967). *Arch. Gen. Psychiat.*, **17**, 671–679.
Sachar, E. J. (1967). *Psychosom. Med.*, **30**, 162–171.
Sanders-Bush, E., and Sulser, F. (1970). *J. Pharmacol. Exptl Ther.*, **175**, 419–426.
Sanders-Bush, E., Bushing, J. A., and Sulser, F. (1972). *Biochem. Pharmacol.*, **21**, 1501–1510.
Sandler, M., Carter, S. B., Cuthbert, M. F., and Pare, C. M. B. (1975). *Lancet* i, 1045–1048.
Sano, I. (1972). *Münch. Med. Wochenschr.*, **144**, 1713–1716.
Scheyen, J. D. van (1971). *Ned. Tijdschr. Geneesk.*, **115**, 1634–1637.
Schildkraut, J. J. (1965). *Amer. J. Psychiat.*, **122**, 509–522.
Schildkraut, J. J. (1973a). In *Lithium. Its Role in Psychiatric Research and Treatment* (Eds., S. Gershon, and B. Shopsin), Plenum Press, New York, pp. 51–73.
Schildkraut, J. J. (1973b). *Amer. J. Psychiat.*, **130**, 695–698.
Schildkraut, J. J., and Draskoczy, P. R. (1974). In *Psychobiology of Convulsive Therapy* (Eds., M. Fink, S. Kety, J. McGaugh, and T. A. Williams), Wiley, New York, pp. 143–170.
Schildkraut, J. J., and Kety, S. S. (1967). *Science*, **156**, 21–30.
Schildkraut, J. J., Dodge, G. A., and Logue, M. A. (1969a). *J. Psychiat. Res.*, **7**, 29–34.
Schildkraut, J. J., Draskoczy, P. R., and Sun Lo, P. (1969). *Science*, **171**, 587–589.
Schildkraut, J. J., Schanberg, S. M., Breese, G. R., and Kopin, I. J. (1967). *Amer. J. Psychiat.*, **124**, 600–608.
Schildkraut, J. J., Schanberg, S. M., Breese, G. R., and Kopin, I. J. (1969b). *Biochem. Pharmacol.*, **18**, 1971–1978.
Schou, M. (1973). In *Lithium. Its Role in Psychiatric Research and Treatment* (Eds., S. Gershon, and B. Shopsin), Plenum Press, New York, pp. 269–294.
Schubert, J. (1973). Thesis, Stockholm.
Schubert, J., Nyback, H., and Sedvall, G. (1970). *J. Pharm. Pharmacol.*, **22**, 136–139.
Schuckit, M., Robins, E., and Feighner, J. (1971). *Arch. Gen. Psychiat.*, **24**, 509–514.
Sebens, J. B., and Korf, J. (1975). *Exptl Neurol.*, **46**, 333–344.
Sethna, E. R. (1974). *Brit. J. Psychiat.*, **124**, 265–272.
Shaw, D. M., Camps, F. E., and Eccleston, E. G. (1967). *Brit. J. Psychiat.*, **113**, 1407–1411.
Shaw, D. M. (1975). In *Lithium, Research and Therapy* (Ed., F. N. Johnson), Academic Press, New York, pp. 411–423.
Sheard, M. H., and Aghajanian, G. K. (1968). *J. Pharmacol. Exptl Ther.*, **163**, 425–430.
Sheard, M., and Aghajanian, G. K. (1970). *Life Sci.*, **9**, 285–290.
Sheard, M. H., Zolovick, A., and Aghajanian, G. K. (1972). *Brain Res.*, **43**, 690–694.
Shibuya, T., and Anderson, E. G. (1968). *J. Pharmacol. Exptl Ther.*, **164**, 185–190.
Shields, P. J., and Eccleston, D. (1972). *J. Neurochem.*, **19**, 265–272.
Shopsin, B., Gershon, S., Goldstein, M., Friedman, E., and Wilk, S. (1974). *Psychopharmacol. Bull.*, **10**, 52.
Shore, P. A., and Brodie, B. B. (1957). *Proc. Soc. Exptl Biol. Med.*, **94**, 433–435.
Sjöström, R. (1972). *Psychopharmacologia*, **25**, 96–100.
Sjöström, R., and Roos, B.-E. (1972). *Eur. J. Clin. Pharmacol.*, **4**, 170–176.
Snaith, R. P., and McCoubrie, M. (1974). *Psychol. Med.*, **4**, 393–398.
Snowdon, J., and Braithwaite, R. (1974). *Brit. J. Psychiat.*, **125**, 610–611.
Sonninen, V., Riekkinen, P., and Rinne, U. K. (1973). *Neurology*, **23**, 760–763.
Sourkes, T. L., Murphy, G. F., and Chavez, B. (1961). *J. Neurochem.*, **8**, 109–115.
Sourkes, T. L. (1965). *Brit. Med. Bull.*, **21**, 66–69.
Sourkes, T. L. (1973). *Adv. Neurol.*, **2**, 13–35.
Strom-Olsen, R., and Weil-Malherbe, H. (1958). *J. Ment. Dis.*, **104**, 696–704.
Sulser, F., and Sanders-Bush, E. (1971). *Ann. Rev. Pharmacol.*, **11**, 209–230.

Tagliamonte, A., Tagliamonte, P., Perez-Cruet, J., and Gessa, G. L. (1971a). *Nature New Biol.*, **299**, 125–126.
Tagliamonte, A., Tagliamonte, P., Perez-Cruet, J., Stern, S., and Gessa, G. L. (1971b). *J. Pharmacol Exptl Ther.*, **177**, 475–480.
Takahashi, S., Kondo, H., and Kato, N. (1975). *J. Psychiat. Res.*, **12**, 177–187.
Takahashi, S., Kondo, H., Yoshimura, M., and Ochi, Y. (1973). *Fol. Psychiat. Neurol. Japonica*, **27**, 305–314.
Takahashi, R., Utema, H., Machiyama, Y., Kurihama, M., Otsuka, T., Nakamura, T., and Konamura, H. (1968). *Life Sci., Part II*, **7**, 1219–1231.
Tamarkin, N. R., Goodwin, F. K., and Axelrod, J. (1970). *Life Sci.*, **9**, 1397–1408.
Thierry, A.-M., Fekete, M., and Glowinski, J. (1968a). *J. Clin. Pharmacol.*, **4**, 384–389.
Thierry, A.-M., Javoy, F., Glowinski, J., and Kety, S. S. (1968b). *J. Pharmacol. Exptl Ther.*, **163**, 163–171.
Thoenen, H. (1972). *Pharmacol. Rev.*, **24**, 255–267.
Tissot, R. (1962). In *Monoamines et système nerveux central* (Ed., J. de Ajuriaguerra), Georg & Cie, S.A., Geneva, pp. 169–207.
Trimble, M., Chadwick, D., Reynolds, E. H., and Marsden, C. D. (1975). *Lancet i*, 583.
Udenfriend, S. (1966). *Pharmacol. Rev.*, **18**, 43–51.
Wålinder, J., Skott, A., Nagy, A., Carlsson, A., and Roos, B.-E. (1975). *Lancet i*, 984.
Weir, R. L., Chase, T. N., Ng, L. K. Y., and Kopin, I. J. (1973). *Brain Res.*, **52**, 409–412.
Werdenius, B. (1967). *Acta Pharmacol. Toxicol.*, **25**, 18–23.
Westerink, B. H. C., and Korf, J. (1975). *Biochem. Med.*, **12**, 106–114.
Wilk, S., Shopsin, B., Gershon, S., and Suhl, M. (1972). *Nature*, **235**, 440–441.
Wilson, I. C., Prange, A. J., Jr., Lara, P. P., Alltopp, L. B., Stikeleather, R. A., Lipton, M. A., and Hill, C. (1973). *Arch. Gen. Psychiat.*, **29**, 15–21.
Winokur, G., Cadoret, R., Dorzab, J., and Baker, M. (1971). *Arch. Gen. Psychiat.*, **24**, 135–144.
Winston, F. (1971). *Brit. J. Psychiat.*, **118**, 301–304.
Wong, D. T., Horng, J. S., Bymaster, F. P., Hauser, K. L., and Molloy, B. B. (1974). *Life Sci.*, **15**, 471–479.
Wooten, G. F., and Cardon, V. P. (1973). *Arch. Neurol. (Chicago)*, **28**, 103–106.
Wyatt, R. J., Engelman, K., Kupfer, D. J., Fram, D. H., Sjoerdsma, A., and Snyder, F. (1970). *Lancet ii*, 842–846.
Youdim, M. B., Collins, G. G., Sandler, M. J., Bevan Jones, A. B., Pare, C. M. B., and Nicholson, W. J. (1972). *Nature*, **236**, 225–228.
Young, S. N., Lal, S., Martin, J. B., Ford, R. M., and Sourkes, T. L. (1973). *Psychiat. Neurol. Neurochir.*, **76**, 439–444.
Yuwiler, A., Geller, E., and Eiduson, S. (1959). *Arch. Biochem.*, **80**, 162–173.

Index

References to figures and tables are in **bold** type. The following abbreviations are used:

CNS1	central nervous system
CSF	cerebrospinal fluid
DA	dopamine
DOPA	dihydroxyphenylalanine
ECT	electroconvulsive therapy
5-HIAA	5-hydroxyindoleacetic acid
5-HT	5-hydroxytryptamine, serotonin
MAO	monoamine oxidase
NE	norepinephrine
PA	phenylalanine
PAH	phenylalanine hydroxylase
PK	phenylketonuria
TAD	tricyclic antidepressants
TYH	tyrosine hydroxylase, tyrosine-3-monooxygenase